HOMAGE TO THE DISCOVERY OF COSMIC RAYS, THE MESON-MUON AND SOLAR COSMIC RAYS

SPACE SCIENCE, EXPLORATION AND POLICIES

Additional books in this series can be found on Nova's website
under the Series tab.

Additional e-books in this series can be found on Nova's website
under the e-book tab.

SPACE SCIENCE, EXPLORATION AND POLICIES

HOMAGE TO THE DISCOVERY OF COSMIC RAYS, THE MESON-MUON AND SOLAR COSMIC RAYS

JORGE A. PEREZ-PERAZA

EDITOR

Novinka

New York

NOTICE TO THE READER

Library of Congress Cataloging-in-Publication Data

Homage to the discovery of cosmic rays, the meson-muon and solar cosmic rays / editor, Jorge A. Perez-Peraza (Instituto de Geofisica, UNAM, C.U., Coyoacan, Mexico, D.F. Mexico, 04510).
    pages cm
  Includes bibliographical references and index.
  ISBN 978-1-62618-998-0 (hardcover)
  ISBN 978-1-63483-084-3 (softcover)
  1. Cosmic rays. 2. Solar cosmic rays. 3. Cosmic rays--Research--History. I. Perez-Peraza, Jorge A., editor of compilation.
  QC485.H58 2013
  539.7'223--dc23
                            2013018507

Published by Nova Science Publishers, Inc. † New York

CONTENTS

PROLOGUE

This year, 2012, scientists in cosmic ray research celebrate the first centenary of Cosmic Rays discovery by the Austrian young scientist from the University of Graz, **Victor F. Hess**, who looking for sources of air ionization by means of Balloon flights, at different heights, he discovered another radiation that later was denominated as Cosmic Rays. As it is wonderful described in the first chapter by **Dorman & Dorman**, Victor Hess faced many problems to convince the scientific community that what he had discovered in 1912 were particles of extraterrestrial origin. In 1926 **Robert Milikan** call this radiation as ***Cosmic Rays***, but he associated it to gamma rays. There was even a famous debate at the beginning of the 30's between Milikan and **Arthur Compton** about the nature of Cosmic Rays (CR). Such a debate appeared in several media. Later in the 30's, after a number of experiments on the latitude effect it was made clear that CR are charged particles in their vast majority. Once the work of Compton and Clay had shown that cosmic rays were **charged particles**, the next enigma to be confronted concerned the nature of their charge, **was it positive or negative?** The answer to this question would have important implications for cosmological theories of the origin of the universe. Besides, in a series of papers (1933- 1937) Manuel Sandoval Vallarta and the Abbot George Lemaitre, based on the Störmer theory about the behavior of charged particles in the geomagnetic field (1910-1911), they interpreted the latitude and azimuth effects discovered by Clay and Compton: they showed that cosmic rays, since they are affected by the geomagnetic field, can only be charged particles, excluding the possibility that they were mainly gamma-rays, as had been suggested by Milikan and others. Based on the results of Fredrik Carl Stormer published years before and on the Liouville's theorem, Lemaitre and Vallarta also determined the minimal threshold value of particle magnetic rigidity for particles to be able to penetrate up to a given geomagnetic latitude, namely **the geomagnetic cutoffs**, and also determined the allowed **acceptance cones** for the arrival of cosmic rays on earth. One of the most important predictions of Lemaitre and Vallarta calculations was the **west–east asymmetry effect:** they showed that the effect of the geomagnetic field was not only the selective access to different latitudes according to their magnetic rigidity (momentum per charge unit), but it deviated particles in different directions according to the sign of the charge. Compton sent his student, **Luis W. Alvarez** (a future Nobel Prize Laureate) to conduct the experiments in the mountains around Mexico City and on the roof of the Geneve Hotel, near the center of the city: in the latter location he put the counters in a bricklayer's wheelbarrow and measured the cosmic ray intensity by varying the orientation of the wheelbarrow. Through these experiments, Compton and Alvarez determined that there was an excess of about **10% in the intensity of deviations to the west, implying that cosmic**

radiation consisted principally of protons. The associated anecdotal history is recounted in Adv. Space Res. 44, 1215-1220, 2009).

Measurements of cosmic ray has been done with a number of devices, among which, electroscopes, Wilson chambers, Geiger-Muller counters, riometers, Neutron Monitors, Super-Neutron Monitors underground muon detectors and so on, some of which are described in Chapters 1 and 2.

At the occasion of the 100 years of Cosmic Ray discovery, on May 1-to May 5, 2012, The Victor F. Hess Society and the European Physical Society/History of Physics group (EPS/HoP) of Autria organized a Conference in homage to this discovery by Victor G. HESS. The Conference was held under the auspices of the first European Centre for the History of Physics, within the frame of the series: *The Roots of Physics in Europe (RoPE)*. The Conference was held in two places, the University of Insbruck and in the Poellau Castle, in the city of Poellau. Scientists of many countries participated in this conference splendidly organized by Prof. Peter Maria Schuster, Director of ECHOPHYSICS, and collaborators. There were speakers from Austria, Brazil, Bulgaria, England, Finland, Germany, Greenland Ireland, Mexico, Poland, Russia, Spain, Sweden, Switzerland and USA among others. A nice visit to the Victor F. Hess Heritage and Research Centre in the local Mountains (where the Austrian scientist use to work), was organized as a part of the homage. The proceedings of this transcendental Conference will be published in the next future.

This year, it is also commemorated the 75 anniversary of anniversary of muon discovery in cosmic rays (1937), what is nicely described in Chapter 2 by Dorma & Dorman, and known by physicists as the hard component of cosmic rays.

In 2012 it is also celebrated the 70 years of Solar Cosmic Ray discovery - detections of cosmic ray in a continuous form began in 1936 on basis of to μ-meson counters, which use to respond to particles $\sim$ 10 GeV. However the three first anomalous enhancements, have been also detected by ionization chambers: an abnormal enhancement of cosmic ray intensity (8 % over the background of cosmic rays)) was registered on 28 February 1942, by Lange and Forbush (Terr. Mag. 47, 185 & 47, 331, 1942), and Berry and Hess (Terr. Mag. 47, 251,1942). For those days the hard component of cosmic rays, the μ-meson, was already continuously recorded at several stations of the world. Subsequent solar cosmic ray events were recorded with μ-meson detectors on 7 March 1942 (6 %) and 25 July 1946 (16 %) (Forbush Phys Rev. 70, 771, 1946). The 1946 event was also reported by Ehmert (Zeit Naturforsch 3a, 264, 1948). The next event, on 19 November 1949 present an unusual enhancement (563 %), was the first event measured not only with μ-meson counters but also with a Neutron Monitor (which response varies from station to station as 1-4 GeV) placed in Leeds England (Forbush et al., Phys Rev. 79, 79, 501, 1949), Adams (Phil. Mag. 41, 503, 1950) & Adams and Braddick (Zeit Naturforsch 6a, 592, 1951). Up to now the event of higher intensity took place on 23 February 1956 (4554 %) (Simpson, Nuovo Cim. Supp.8, seriesX, 1958). This event has produced a vast literature, even nowadays. Since cosmic ray observations initiated in 1936, there has been 71 of these events of these MultiGeV events, known in the literature as GLE's (Ground Level Enhancements) because they penetrate the earth magnetosphere and can be detected at ground level, normally at high latitudes where the geomagnetic cutoff is low, but occasionally, as the event of 23.02.1956, particles have been recorded at latitudes as low as Mexico City and even in India. With some few exceptions all these GLE's are associated to huge solar flares (3 and 3+), Radio IV noise and shock waves.

The average occurrence rate, if we consider the date of first registered event in 1942, or the debut of cosmic ray particles detection in 1936 is about 1 per year. However, the real occurrence rate is quite different, since there are years where several GLE occur, while there are periods, up to six years, where no event occurs. That was the case with GLE70 (2006) and GLE71 (201). Prediction of a GLE occurrence was done for the first time in the ICRC of Beijing in 2011 dor the occurrence of GL71 the last month of 2011 up to the first months of 2012. It effectively took place on 17 May 2012 (Proc. 32rd ICRC. Beijing, Vol. 10, p.149-152, 2011).

At present it is well known that solar particles are produced quasi continuously at energies around 0.1- 1 MeV, up to some hundreds of MeV. To differentiate with relativistic protons (GLE's) people often refer to those particles as ESP (Energetic Solar Particles) covering energies $< 10^3$ MeV. The study of the energy spectrum of relativistic solar protons gives information about their sources, the acceleration mechanisms, their time profile and anisotropy gives information about the temporal state of the Interplanetary magnetic field, as particles act as` tracers of the field lines. Their high energy cutoff teaches us about the capacity limits of the Sun as a producer engine of high energy charged particles. In a general context it must be emphasized, that the study of solar Cosmic rays has opened many clues for the understanding of cosmic ray physics within a wide range of astrophysical fields (see for instance "Astrophysical aspects in the studies of solar cosmic rays", *International Journal of Modern Physics 23-1, 1- 141, 2008.*)

But the impact of cosmic particles is not only in the Astrophysical context, but also at the level of the earth environment and the different layers of our planet. These aspects are ripping described here in chapters 3, 5 and 12. A summary of the present status in these disciplines is given in the book (*Highlights in Helioclimatology*, Ed. by Elsevier, July 2012).

Jorge A. Pérez-Peraza
Editor
Email: perperaz@geofisica.unam.mx

In: Homage to the Discovery of Cosmic Rays ...
Editor: Jorge A. Perez-Peraza

ISBN: 978-1-63483-084-3
© 2015 Nova Science Publishers, Inc.

Chapter 1

HISTORY OF COSMIC RAY DISCOVERY AND EXPERIMENTS SHOWING THEIR NATURE AND COSMIC ORIGIN (DEDICATED TO 100 YEARS OF DISCOVERY)

Lev I. Dorman[1,2] *and Irina V. Dorman*[3]

[1]Israel Cosmic Ray and Space Weather Centre with Emilio Segre
Israel-Italian Observatory, affiliated to Tel Aviv University,
Golan Research Institute, and Israel Space Agency, Israel
[2]IZMIRAN of Russian Academy of Sciences (RAN), Moscow, Russia
[3]Institute of History of Science and Technology of RAN, Moscow, Russia

ABSTRACT

As many great discoveries, the phenomenon of cosmic rays was discovered mainly accidentally, during investigations that sought to answer another question: what are sources of air ionization? This problem became interesting for science about 225 years ago at the end of the 18th century, when physics met with the problem of leakage of electrical charge from very good isolated bodies.

In the beginning of the 20th century, in connection with the discovery of natural radioactivity, it became apparent that this problem was solved: it was assumed that the main source of the air ionization were α, β, and γ – radiations from radioactive substances in the ground (γ-radiation was considered the most important cause because α- and β-radiations are rapidly absorbed in the air). Victor Hess, a young scientist from the Graz University, started to investigate how γ-radiations change their intensity with the distance from their sources, i.e. from the ground. When he performed his historical experiment in 1912 on balloons, Victor Hess found that at the beginning (up to approximately one km) ionization was unchanged, but with an increase of the altitude for up to 4-5 km, the ionization rate escalates several times. Victor Hess drew the conclusion that some new unknown source of ionization of extraterrestrial origin exists. He named it 'high altitude radiation'. Many scientists did not agree with his conclusion and tried to prove that the discovered new radiation has terrestrial origin (e.g., radium and other emanations from radioactive substances in the ground, particle acceleration up to high energies during thunderstorms, and so on). However, a lot of experiments showed that Victor Hess's findings were right: the discovered new radiation has extraterrestrial origin. In 1926 the great American scientist Robert Millikan named them 'cosmic rays': cosmic as coming from space, and rays because it was generally wrongly accepted at those time that new radiation mostly consisted of γ-rays). Robert Millikan believed that God exists and continues to work: in space God has created He atoms from four atoms of H with the generation high energy gamma rays (in contradiction with physical laws, as this reaction can occur only at very high temperature and great density, e.g., as inside stars). On this problem, interesting to many people, there was a famous public discussion between the two Nobel Prize laureates Arthur Compton and Robert Millikan, widely reported in many Newspapers. Only after a lot of latitude surveys in the 1930s, organized mostly by Compton and Millikan, it became clear that 'cosmic rays' are mostly not γ-rays, but rather charged particles (based on Störmer's theory about behavior of charged energetic particles in the geomagnetic field, developed in 1910-1911, before CR were discovered). Moreover, in the 1930s it was shown by investigations of West-East CR asymmetry that the majority of primary CR must be positive energetic particles. Later, in the 1940s – 1950s, it was established by direct measurements at high altitudes on balloons and rockets that for the most part cosmic rays are energetic protons, about 10% He nuclei, 1% more heavy nuclei, 1% energetic electrons, and only about 1% energetic gamma rays. Nevertheless, the name 'cosmic rays' (for short, CR) continues to be used up to now. The importance of CR to different branches of science was understood in the 1930s, when positrons ('positive electron') were discovered in CR as the first antiparticle. In 1936 Nobel Prize laureates in Physics were Victor Hess for CR discovery and Carl Anderson for discovery positron in CR. Later, new types of elementary particles such as mesons and hyperons, new types of nuclear reactions at high and very high energies, formation of nuclear-meson and electromagnetic cascades in the atmosphere were all discovered in CR. CR became considered a very important natural source of high and very high energies.

1. ELECTRICITY DISPERSION THROUGH AIR (18TH-19TH CENTURIES)

Like many other phenomena, the cosmic rays were fortuitously discovered when studying quite different phenomenon. The history of cosmic ray discovery may be treated as an example of significant consequences that may ensue from a thorough examination of an unknown and seemingly inconsiderable effect. The starting point was when the physicists got interested in the source of the constant, though weak, ionization of the ambient air. The X-rays and radioactive radiation causing the ionization of the ambient air had been discovered by the beginning of 20th Century. Somewhere near 1900 the study of those ionizing radiations stimulated the study of the atmospheric electricity. There were no doubts yet at that time that the natural conductivity of the air was induced by some unknown radiation and was not caused by spontaneous disintegration of the air molecules due to thermal excitation.

From the first steps on air electricity studying it was known that electroscopes with gold leaflets and other charged bodies gradually lose a charge, despite the precautions undertaken for maintenance of good isolation. Therefore, still Charles Coulomb (1785) came to the conclusion that charge loss by the charged body ("electricity dispersion") occurs not only because of a low insulation of supports, but mainly through air. Charles Coulomb considered that at contact with the charged bodies of a particle of air, the dust and water steams get a charge and carry away it, i.e. the charge as though flows down from the charged body in the air. Such explanation existed until 1880s because at the condition of knowledge of that time about electro-conductivity gases, another satisfactory interpretation of the phenomenon of dispersion of electricity in the air and could not be given.

2. EXPERIMENTS OF J.J. THOMSON IN CAVENDISH LABORATORY

Further research of atmospheric electricity was stimulated by the discovery of an ionizing radiation made in the end of the 19th century, and is closely connected with a series of experiments on studying of conductivity of gases which put J.J. Thomson in Cavendish Laboratory in Cambridge. He experimentally showed that under the influence of X-rays and radiation from radioactive elements, electric conductivity of gases strongly increases at the expense of a great number occurrence positively and negatively charged ions, and the theory of ionic conductivity of gases was created.

3. EXPERIMENTS OF J. ELSTER AND H. GEITEL (1900)

Application of this theory to an explanation of dispersion of electricity phenomenon in the air is a merit of two German scientists working in Berlin: J. Elster and H. Geitel (1900). They came to a conclusion about the existence of the radiation constantly ionizing air surrounding us. It became clear that the gradual discharge of electroscopes could be explained by the presence of insignificant number of ions in the air. There was a question: what nature of an unknown source of ions?

Discovery of new radiation was not in those days an extreme event and did not cause big surprise. This results from the fact that near 1900, soon after discovery of X-rays and radio-activity one another was followed by attempts to find still any new rays, there was present "a rays fever". Usually (the truth, sometime came later) it appeared that messages on again discovery new rays is a fruit of misunderstanding or supervision errors. For this reason the proof of existence of the radiation opened by J. Elster and H. Geitel, needed careful research. To have the possibility to judge the degree of dispersion of electricity (so, and conductivity of air) depending on a geographical position, height and atmospheric conditions and by that to define a site of an unknown source of ions, Elster and Geitel (1900) designed special "disseminating device" which is shown on Figure 1.

The device, shown in Figure 1, consisted of an electroscope on whose core was planted a "disseminating body" - needled copper cylinder in length of 10 cm and in diameter 5 cm. The disseminating body was charged by a column of Zamboni which is visible in Figure 1 to the left of electroscope. Then the potential difference right after charging of the disseminating body was measured through time t.

By means of this device, from April until August 1900, measurements of electro-conductivity of air in various points (Biskra, Algeria, island Capri, island Spitsbergen, Luchano), at various heights above sea level were made. However, errors of measurements were so big that it was not possible to find out any laws. At the same year Geitel (1900) established that weak ionization is observed also in the closed vessels, containing air cleared of dust and dried, even in the dark.

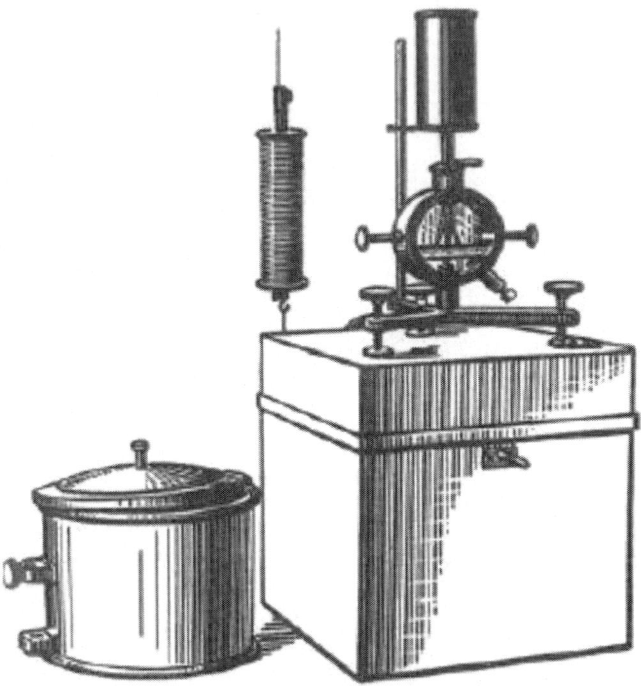

Figure 1. Disseminating device of J. Elster and H. Geitel (1900).

Further investigations showed that air in cellars and caves is especially strongly ionized; therefore, electricity dispersion there considerably increases. So, in Bauman's cave in Harz conductivity of air appeared 20 times more than in the open air. It was explained by the presence in earth's crust of very insignificant quantities of the radioactive substances which γ-radiation led to ionization increase. Studying of tests of soil and water confirmed the put forward assumption. Besides, by J. Elster and H. Geitel had been offered a way of definition of presence of an emanation (i.e. the gases allocated at disintegration radium, thorium, and actinium) in air. Over the earth on two isolated hooks the copper wire of length approximately 10 m was tense and about two hours on it negative potential of 2000 V. Then the wire was supported removed, reeled up on the mesh cylinder, and placed under a glass bell. Conductivity of air under the bell strongly increased because of an ionizing radiation of the radioactive deposits, which settled on the wire because of disintegration of an emanation. All these experiments, naturally, led to a conclusion that the basic source of ionization of air is the radioactive emanations accumulating in the atmosphere, and the radioactive substances contained in the earth crust. The numerous measurements made in those years by J. Elster and H. Geitel, showed that conductivity of air strongly fluctuates depending on atmospheric conditions, it much more over a land, than over the sea, and changes with change of height of a place of observations. To make any more certain conclusions was impossible because of the imperfection of the device which researchers used.

4. EXPERIMENTS OF C.T.R. WILSON (1900-1901) AND THOUGHTS ON POSSIBLE EXTRATERRESTRIAL ORIGIN OF AIR IONIZATION SOURCES

Also in 1900, independent of J. Elster and H. Geitel, the unknown source of ions in air was found by Wilson (1900) conducting research in Cavendish laboratory in Cambridge.

C.T.R. Wilson was a magnificent experimenter, worked very carefully and himself did majority of devices necessary for research. Having become interested in natural conductivity of the air, he first of all offered a method of measurement of the number of ions formed in 1 cm^3 of air in 1 c.

The electroscope, which Wilson (1900) used in his research, is shown in Figure 2. It differed from usual electroscopes only by the addition of the second insulator - a ball from sulphur S and very thin wire 1 which connects an insulator A, connected to one of battery poles, with platen P.

After charging the electroscope, the thin wire was moved aside by means of a thread 2, and contact between A and P interrupted. After some time it was possible to notice that the leaf G gradually loses the charge that was not a consequence of leak through an insulator as the insulator A was connected to a voltage source.

Charge leak could occur only through the air, and Wilson (1900) concluded that the speed of electroscope discharge measured the number of ions present in the vessel.

The method of definition of speed of the ionization offered by C.T.R. Wilson became standard in research of that time (many experimental methods, developed by C.T.R. Wilson, were used by physicists for the different purposes; so, in 1906 to measure a charge of electron, R. Millikan has altered and has improved the method offered by C.T.R. Wilson based on movement of charged droplets of water under the influence of electric field).

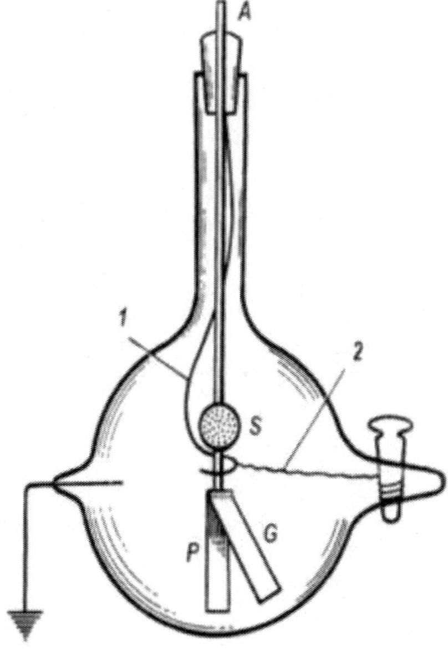

Figure 2. Electroscope of Wilson (1900) for measuring ionization rate in air.

Wilson (1900) measured conductivity of air cleared of dust in a closed vessel and found out that it always ionized. According to Wilson's measurements, every second in 1 sm^3 of air about 20 ions are formed. If one recollects that in a cubic centimeter of air at atmospheric pressure and room temperature there are 2.7×10^{19} molecules, the possibility to find out and quantitatively to study process at which in one second from such quantity of molecules collapses only 20, it shows unusual sensitivity of the offered way of registration of ions.

The Wilson's article "On the leakage of electricity through dust-free air" appeared in 1900, at once after the paper of Geitel (1900). C.T.R. Wilson noticed that the speed of leak for a positive and negative charge is identical, proportional to pressure in a vessel and does not depend on that, on light or in the dark there is in observation. It is necessary to notice that though J. Elster and H. Geitel in Germany, and C.T.R. Wilson in England worked independently, they well knew about the works of each other that are visible from numerous references in their articles.

The received results resulted in Wilson (1901) thinking that the ionizing radiation comes from extraterrestrial sources: "Experiments which will be made in the future, may be, will show that formation of ions in air deprived of any pollution, is caused by radiation which arises out of our atmosphere to similarly X-ray or cathodic rays, but possesses considerably bigger penetrating ability". However, having measured speed of formation of ions in the railway tunnel (experiment was spent at night; when movement of trains stopped) and without having found out reduction of speed of ionization in comparison with usual conditions though the screen was multimeter rock, C.T.R. Wilson changed the opinion: "It is improbable therefore that ionization is caused by radiation passing through our atmosphere. Most likely, as has concluded H. Geitel, this is a property of air". According to modern representations, the result received by C.T.R. Wilson it is possible to explain by some increase in radioactive pollution in the railway tunnel that almost compensated reduction of external radiation under the screen. But at that time erroneous interpretation of the received data induced C.T.R. Wilson mostly to refuse the prophetical assumption of an extraterrestrial origin of an ionizing radiation. However, the thought on the possibility of existence of the radiation coming from extraterrestrial sources was expressed before C.T.R. Wilson. Therefore, Marie Curie (1898), studying uranium and thorium radiation, wrote in the end of the 19th century "for interpretation of spontaneous radiation of these elements it is necessary to imagine that all space is crossed by the beams similar to beams of the X-ray, but considerably more penetrating. These beams can be absorbed only by certain elements with big nuclear weight, such, as uranium and thorium".

In other work, Curie (1899) continues the thought: "Uranium and thorium radiation is the secondary issue caused by beams, similar to γ-rays. If these beams exist, their source may be the Sun, and in that case will be various at midnight and at midday. However I could not find it". Subsequently the hypothesis about an extraterrestrial origin of radiation was absolutely forgotten as recollected R. Millikan (M1935, M1939): "Until 1910 there were no data about penetrating radiation coming to the Earth from the outside. I cannot find any information on the existence though any idea, even distantly concerning the phenomenon which we connect now with the term cosmic rays".

5. AIR IONIZATION AND RADIOACTIVE SUBSTANCES (1902-1910)

In 1902-1903 it was found that radioactive substances spontaneously break up with emission α-, β-, and γ-beams. Because γ-beams are considerably more penetrating than α- and β-beams, there was naturally a representation what exactly γ-radiation of radioactive substances creates observable ionization of air.

Discussed experiments concerned by then when measurements of speed of formation of ions in the air were often made in laboratories, before used for experiments with radioactive substances and consequently strongly polluted by radioactive deposits. The considerable share of measured ionization of air was caused by this pollution that strongly complicated searches of other sources of ionization of air. E. Rutherford and H. Cooke (1903) working at that time in Montreal, decided to surround an electroscope with a lead layer to get rid of radioactive radiation from laboratory walls. They with surprise found out that even in five-ton weight of lead the speed of ionization though considerably decreases, but remains to equal 6 pairs of ions in sm^3 per second. This residual ionization was attributed to radioactive pollution of the material of which the device (own radiation of the device), instead of to the radiation getting outside is made. In other words, all observable ionization, as Rutherford and Cooke (1903) believed, had a radioactive origin. Not casually in the detailed review of K. Kurz (1909) devoted to the analysis of results received up to 1909, three possible sources of observable ionization of air were specified: earth crust γ-radiation, the radiation coming from the atmosphere, and radiation from world space, and the two last possibilities were resolutely rejected as improbable.

In 1909 German physicist T. Wulf (1909) improved old-fashioned electroscopes, having replaced gold leaflets with two thin metal wires tense by means of an easy quartz thread (Figure 3). By distance change between the wires, measured by a microscope with the ocular micrometer of F it was possible to judge speed of discharge of an electrometer. Next two years, using this sensitive portable device, which gave the chance to fix changes of speed of ionization very precisely (with errors not more than only several ions $cm^{-3}s^{-1}$). By this device a series of worth-while experiments on observation of ionization of air over the seas, lakes and glaciers were made. The thought that led to these observations was simple: if the radiation causing the discharge of the electroscope is let out by a terrestrial surface, it should be much less over water containing less of the radioactive impurity.

Measurements of many scientists showed, however, that though the speed of ionization in this case considerably decreases, it nevertheless remains equal to several ions $cm^{-3}s^{-1}$. It seemed not clear, and there were foggy assumptions that, probably, part of the observable ionization is caused by radiation not having radioactive origin. However any explanations about a prospective additional source of ionization were not given.

At the same time, the first attempts to define how speed of ionization with height will change started. T. Wulf (1910) lifted the device onto the Eiffel tower (on height 330 m) and found that speed of ionization decreases with height much more slowly than was expected: at the bottom of a tower it equalled to 6 ions $cm^{-3}s^{-1}$, and at the top - 3.5 ions $cm^{-3}s^{-1}$. For an explanation of the received results, T. Wulf saw two possibilities: either γ-radiation absorption in the atmosphere is much less than was estimated earlier or iron parts of the tower radiate, being an additional source of ionization. Unfortunately, the decrease of speed of ionization appeared so insignificant that it was impossible what to draw any exact conclusions.

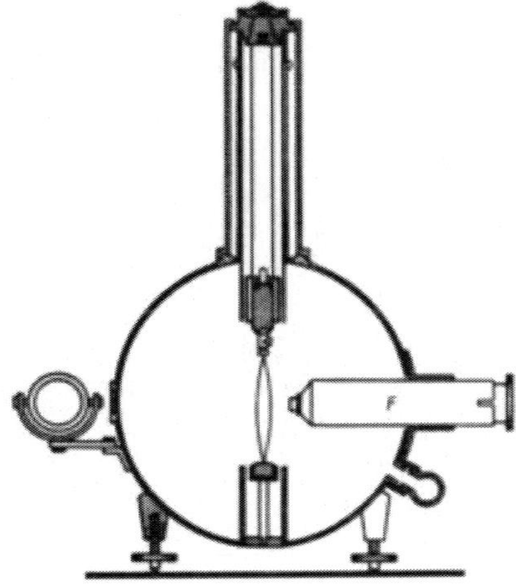

Figure 3. Electroscope of T. Wulf (1909).

6. The First Measurements of Air Ionization on Balloons by A. Gockel up to Altitude 4500 M (1909-1911)

In 1909-1911, the Swedish meteorologist A. Gockel (1911) made three flights on balloons, having reached heights of 4500 m. "It is revealed, - he wrote, — the ionization reduction, however is far not in that measure as it would be possible to expect in the assumption of the radioactivity proceeding from the earth ground". It is necessary to notice that pressure in the device that A. Gockel used fell in the process of the sphere lifting and consequently, the speed of ionization could decrease simply at the expense of the reduction in the number of atoms in each volume unit, instead of at the expense of ionizing radiation easing. A. Gockel understood that "recalculation of observable ionization on initial pressure can give even increase in ionization with height", but any definitive conclusions could not make. The results received by A. Gockel did not bring any clearness to the question on sources of ionization of air. At best, they drew the attention of researchers to the study of absorption of γ-radiation in air.

7. Discovery by Victor Hess of Radiation from Space (1911-1912)

"In general, before Victor Hess's flights, all experimental results, - as wrote R. Millikan (1930) - could be interpreted as proof of that all ionization of atmospheric air is caused by sources of a radioactive origin. Imperfection of used devices and impossibility at that time precisely to reproduce experimental conditions have appeared for physics of cosmic rays happy coincidence of circumstances as stimulated the further research under various conditions, especially at the big heights." So was the situation when in 1911 Victor Hess, employee of the Institute of Radium in Vienna, engaged in studying of absorption of γ-rays in air.

Victor Franz Hess was born on June 24, 1883 in Austria to the family of a forester. In 1905, he completed University in Graz and in the same place in 1910 received a PhD. From 1910, V. Hess under the direction of Stefan Meyer was engaged in radioactivity studying. Therefore, in 1911 he undertook two flights on balloons to learn at what height propagates γ-radiation from the radioactive substances which are in the earth crust. Based on these data, V. Hess planned to determine absorption factor of γ-radiation in air. Knowing the works of A. Gockel (see previous Section 5), V. Hess placed the device in a hermetic vessel in which pressure of air remains constant at all heights. He did it because he was surprised by the results of the flight of A. Gockel, and explains the absence of falling of air ionization rate with height exclusively by his device not being airtight. However, having reached heights of 1100 m, V. Hess in both cases, like A. Gockel, did not observe an appreciable fall in ionization rate in comparison with measurements near the terrestrial surface. Therefore V. Hess (1911) made the assumption that "there should be other source of a penetrating radiation in addition to γ-radiation from radioactive substances in the earth crust".

The problem of finding the absorption factor for γ-radiation in the air was left, and V. Hess, carefully thinking over all trifles, prepared for follow-up flights. In 1912, with assistance of the Viennese Academy of Sciences, he made seven more flights in which he used two-wired Wulf's electrometer (see above Figure 3) with walls in the thickness of 3 mm through which only γ-rays could pass, and both electrometers had hermetic cases. The third Wulf's electrometer with walls of 0.188 mm was in the thickness non-hermetic and intended for simultaneous studying of behavior of β-beams. Threads of electrometers were charged to voltage U ≈ 200 V to provide a saturation current in the chamber; then was observed continuously the speed of the system discharge. The ionization rate I in the volume chamber W was defined under the formula

$$I = \frac{CdU}{300eWdt},\qquad(1)$$

where C is the system electro-capacity, and e - charge of transferred one ion. For Hess's first device the voltage loss in 1 Volt/hour corresponded to $I = 1.56$ ion.cm^{-3}s^{-1}, and for the second $I = 0.74$ ion.cm^{-3}s^{-1}. In some hours before flight control measurements by all three devices, which were for this purpose fixed in a balloon's basket as during flight, were made. The average height to which the balloon rose was deduced by a graphic method from the barograph indication.

It is necessary to tell that before Hess's experiments nobody approached with such gravity to measurements during flight, and flights were carried out with more of an adventurous than scientific character. Besides, Victor Hess himself took part in all flights, which demanded in those conditions big courage, and wrote down indications of devices. To define the position of a mysterious source of radiation, he purposefully made flights at various times of the day, under different atmospheric conditions and in detail marked the strength of wind, overcast, and the temperature.

The first flight from this series was made on April 17, 1912 during a partial solar eclipse. Any reduction of ionization rate during the eclipse time was not observed, and since the height of 2000 m ionization rate increased, from which Victor Hess drew the conclusion that since the fixed measurements even during eclipse were not affected an ionizing radiation, the Sun cannot be its source. In other flights, Victor Hess did not find a difference in measurements of ionization rate between day and night that has confirmed its point of view.

The seventh and Victor Hess's most well-known, famous flight began on August 7, 1912 at 6 o'clock 12 minutes in the morning about the city of Aussiga in Austria. In the balloon's gondola, there was a pilot, the meteorologist and Victor Hess. At this time the balloon was filled by hydrogen (earlier Victor Hess filled balloons with the warmed-up air) and the record at that time for height was 5350 m. At midday the balloon therefore reached and landed near to the German city Piskov, 50 km to the east of Berlin, having flown 200 km. In a unique photo (on the front page) we see Victor Franz Hess in a balloon gondola right after landing. This flight was in detail described in an article which appeared in a November 1912 issue of the magazine "Physikalische Zeitschrift" (Hess, 1912). At the lifting of a balloon to 1000 m insignificant reduction of the ionization rate, caused by absorption of γ-radiation of radioactive substances which are in the earth crust was observed. After that air ionization rate started to an increase gradually with height as though the balloon came nearer to the radiation source, instead of away from it. In the range of heights from 4000 to 5200 m the ionization rate became much higher than at sea level. After landing the balloon cover was carefully investigated to define whether it is covered by the radioactive substances which settled during flight that could cause increase of ionization rate, and V. Hess came to the conclusion that the cover of the balloon does not radiate.

Great attention was cared to weather during flight because as it was already told, that V. Hess tried to find dependence of properties of unknown radiation on atmospheric conditions. The condition of weather during flight was fixed in great detail: reduction of barometric pressure; overcast weak to 4000 m and stronger above through which the sunlight, the strong wind quickly carrying a balloon aside poorly makes the way. To show how the ionizing radiation changes with height, V. Hess united 88 values of measurements of the ionization rate, made at various heights during all seven flights (for each height from several values received under various conditions, the average was taken).

The data are represented in Table 1.

Table 1. The dependence of ionization rate from the altitude. According to Hess (1912)

Average height from the earth, m	Observable ionization rate, ion.cm^{-3}s^{-1}		Average height from the earth, m	Observable ionization rate, ion.cm^{-3}s^{-1}	
	The first device	The second device		The first device	The second device
0	16.3 (18)*	11.8 (20)	1000-2000	15.9 (7)	12.1 (8)
Up to 200	15.4 (13)	11.1 (12)	2000-3000	17.3 (1)	13.3 (1)
200-500	15.5 (6)	10.4 (6)	3000-4000	19.8 (1)	16.5 (1)
500-1000	15.6 (3)	10.3 (4)	4000-5200	34.4 (2)	27.2 (2)

*The figures in brackets means the number of observations from which were obtained average.

From Table 1 it is visible that to height of 1000 m there was a reduction in ionization rate on average on 0.7-1.5 ion.cm^{-3}s^{-1} (in some flights it reached 3 ion.cm^{-3}s^{-1}) that is caused by absorption of γ-radiation of the earth crust. "From here, - V. Hess wrote, - we conclude that at earth crust radiation gives ionization rate only nearby 3 ion.cm^{-3}s^{-1} in zinc electrometer". It is interesting that V. Hess, as it was accepted at that time, specified from what the electrometer is made because he knew that each device has its own specific radiation.

Further, at the increase in height from 1000m to 2000 m, the ionization rate slowly increased, and in the range of heights between 4000 m and 5200 m it appeared already on 16-18 ion.cm^{-3}s^{-1} more than the ionization rate on the surface of the Earth.

What is the reason for so substantial a growth of ionization rate with height, which was observed by V. Hess many times, and simultaneously by all devices? V. Hess well understood that "if to adhere to the point of view that only known radioactive substances in the earth crust and in the atmosphere let out the γ-radiation making ionization in the closed vessel, there are serious difficulties at an explanation of the received results". Really, V. Hess experimentally defined that at the height of 500 m the earth crust γ-radiation decreases more than in 5 times and, certainly, cannot make considerable ionization at big heights. The congestion of radioactive emanations in the atmosphere, by V. Hess estimations, could cause only 1/20 from all ionization observed at height from 1 to 2 km, and with the increase in height an emanation role, naturally, should become even less. As a result V. Hess came to the conclusion that it is possible to explain all experiments only by the existence of the radiation coming from the outside, of extraterrestrial origin. Reporting in September 1912 at the Session in Munster results of the flights, V. Hess made following sensational conclusion: "Results of the presented observations are better explained by the assumption that radiation of big penetrating ability is coming into our atmosphere from above and even in its bottom layers makes a part of the ionization observed in closed vessels" (Hess, 1912). V. Hess named the discovered ionizing radiation ultra-gamma radiation to underline its big penetrating ability. For those times, the assumption of existence of radiation coming from world space on the border of the atmosphere was extraordinary courageous, and many years passed before it became commonly accepted. Many physicists, if not the majority, doubted that "Hess's" radiation (so named this radiation at the beginning) had an extraterrestrial origin, and attributed it's more habitually then to the radio-activity phenomena.

Now, nobody has any doubt that the discovery of cosmic rays (so this new radiation began to be named after 1925) belongs to V. Hess, from the first flights purposefully searching for it and believed in its existence. If before V. Hess's experiments various assumptions of a mysterious ionizing radiation also came out (see previous Sections), all of them had the characteristic of the hypotheses having few bases. Whether that fact speaks, what about them made them quickly forgotten and recollected only after Hess's discovery?

Hess's information on the discovery of the coming from the outside of radiation caused a great interest to German physicists. If one looks at the magazine "Physikalische Zeitschrift" for those years we will not find any issue in which there would be no V. Hess article or articles devoted to Hess's radiation. It is curious that in the scientific literature in England, USA, Russia, France and other countries at those times no any paper has the new penetrating radiation. In 1919 Victor Hess received the Liben award for discovery of "ultra-radiation" and soon after that became professor of experimental physics at University in Graz. In 1921 - 1923 Victor Hess worked in USA. In 1923 he came back to University in Graz, and in 1931 was appointed to be a director of just based Institute of Radiology in Innsbruck. Near Innsbruck Victor Hess based on a mountain Hafelekar station for continued observation and studying of cosmic rays (this station works until now). In 1932 Charl Zeis's Institute in Yen awarded Victor Hess with the memorial award and medal of Abbe; he became also a member-correspondent of the Viennese Academy of Sciences.

8. EXPERIMENTS OF W. KOLHÖRSTER AND PROOF OF EXISTING HIGH-ALTITUDE RADIATION (1913-1914)

It is not correct to think that scientists immediately agree with V. Hess that "Hess radiation" really exist and has an extraterrestrial origin. Just the opposite, even the problem of the existence of this radiation was discussed many years after the experiments of V. Hess in 1912. In the first, Hess's results should be repeated and checked by other scientists. In the second, before agreeing with the extraterrestrial origin of "Hess radiation", it was necessary to try to find other, not so radical origins. For example, a hypothesis was discussed that this radiation arises in the upper atmosphere during thunderstorms when charged particles can be accelerated by big electric fields. Other scientists suggested that radioactive emanations and radioactive particles might be concentrated in the upper atmosphere, which can explain an increasing of air ionization with increasing of altitude.

In the frame of both of these hypotheses it should be expected that on the intensity of "Hess radiation" will be a strong influence of weather, and big variations with time of the day and season should exist. The absence of those big time-variations was underlined in the first publications of V. Hess.

The biggest criticism of results of V. Hess was inflicted by German physicist W. Kolhörster, who was sure that the observed by V. Hess increasing of the air ionization rate with altitude is absolutely false and caused by the influence of changed air temperature during the balloon flight on data obtained by Hess's device. With the aim to 'shut up' Hess's discovery, W. Kolhörster achieved in 1913-1914 five flights on balloons and attained the maximal altitude 9,300 m. In Kolhörster (1913a) was described in detail the device used during the flights. The device scheme and appearance are shown in Figs. 4 and 5, correspondingly. The device represented a cylindrical chamber from steel of volume 4.5 l. In the centre of the device on a special framework, the most thin quartz silvered threads have been placed. For the elimination of any influence of temperature on instrument readings, Kolhörster used as holders of threads quartz handles serving by good insulators. After charging the threads, the size of their divergence was defined with the help of microscope at the top cover of the device. The device was very well sealed and that change of pressure upon the device case was not transferred to the framework with threads, it was attached to the chamber only in one place near microscope. Air which was in the chamber was preliminary carefully dried, and besides, in the bottom of the device there was a dehumidifier.

The device design satisfied to variety, apparently, enough inconsistent requirements, connecting high sensitivity of the device and independence of its indications of external conditions, portability and full reliability. At those years the device created by Kolhörster was the best among the numerous electrometers of various types, therefore it was not incidental that almost all European physicists used devices made on the sample of Kolhörster's device by the firm "Gunter and Tegetmajer" in Braunschweig (Germany).

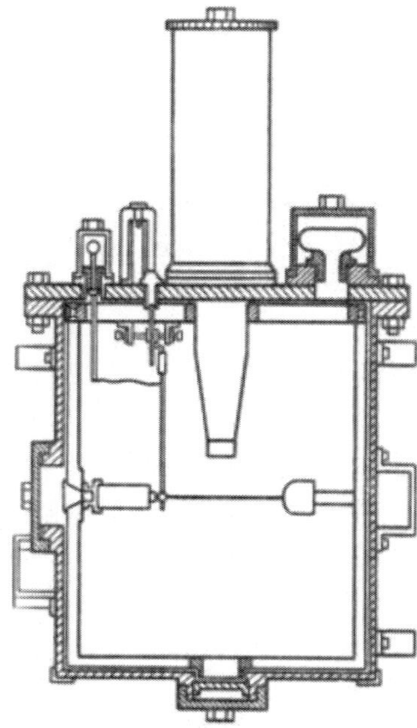

Figure 4. Scheme of W. Kolhörster's apparatuses for measuring air ionization rate.

Figure 5. Appearance of W. Kolhörster's apparatuses.

The results received by Kolhörster (1913b) during flights completely confirmed Hess's observation data, against Kolhörster's desire. At the lifting of the equipment up, the speed of ionization at first decreased, then at the height of 1700 m it accepted the same value as on the terrestrial surface, and then it began to increase considerably. At a height of 9,000 m the difference between ionization speed observed at the given height, and speed of ionization at sea level $\Delta \approx 80$ ион.см$^{-3}$.сек$^{-1}$. The main results obtained by Kolhörster (1913b) are shown in Table 2.

Table 2. Differences in ionization rates Δ in the air at altitudes H from 1 to 9 km in comparison with ionization rate at the ground near sea level. According to Kolhörster (1913b)

altitude, km	1	2	3	4	5	6	7	8	9
Δ, ion.cm^{-3}.s^{-1}	−1.5	1.2	4.2	8.8	16.9	28.7	44.2	61.3	80.4

So no doubts remained that with removal from the terrestrial surface the speed of ionization increases, Kolhörster assumed that the radiation causing the ionization of atmospheric air is absorbed by the atmosphere under the simple exponential law:

$$I = I_0 \exp(-\mu h), \tag{2}$$

where I_0 is the intensity of radiation on atmosphere border, I - intensity of radiation after passage of a layer of atmosphere of thickness h (in g/cm^2), and μ - absorption coefficient (in cm^2/g). Using this assumption, Kolhörster calculated that the coefficient of absorption of an ionizing radiation coming from above is equal to 5.7×10^{-3} cm^2/g. The received value was in 10 times less than the value of the absorption coefficient for the most rigid γ-radiation from radioactive elements (for example, RaC). "Thus, - wrote Kolhörster, existence of the very rigid radiation having absorption coefficient about 0.1 of absorption coefficient of known before γ-radiation is established". Extrapolating the obtained data, Kolhörster defined that at sea level the speed of ionizing radiation coming from above, makes about 2 ion.cm^{-3}.s^{-1}. Let us note that in 1973-1974 absolute measurements of ionization of air by cosmic rays at sea level were executed by Kyker and Lyboff (1978); it was found that the speed of ionization at sea level by cosmic rays is equal to 2.15 ± 0.05 ion.cm^{-3}.s^{-1}. Measurements were made by means of a 900-litre chamber filled with pure air. This chamber was calibrated in a hydrochloric mine on the depth of 600 m underground; apparently, the received value is very close to the value of the speed of the ionization defined by Kolhörster (1913b) for sea level about 60 years earlier.

This meant that at an altitude of 9000 m speed of ionization increases by 40 times! The first period of investigations, on which the extent of the problem of reality of existence of high-altitude ionizing radiation was the basic problem, came to an end with the experiments of Kolhörster. After that, to doubt that unknown radiation of big getting ability, apparently, comes from above - was not necessary any more. Kolhörster (1913b) named the radiation coming from above "Höhenstrahlung", i.e. high-altitude radiation.

9. EXPERIMENTS OF R.A. MILLIKAN WITH COLLABORATORS ON BALLOONS AND CONTRADICTION WITH W. KOLHÖRSTER'S RESULTS (1922-1923)

If the rough measurements of ionization by high-altitude radiation give only an order of values, the knowledge of its nature and an origin needed much more careful research than was necessary for an establishment of the existence of high-altitude radiation only.

It is impossible to forget that those methods which were successfully applied in laboratories to studying of X-rays and radio-activity were often unsuitable to high-altitude radiation which comes from above in different directions and are uncontrollable, as it is impossible to switch them on or off like , for example, X-rays.

Besides, many parameters of high-altitude radiation were unknown, and the source of this radiation unlike the sample of radioactive substance is inaccessible.

Therefore the experiments made were not quite unequivocal, and that created additional difficulties.

The First World War for a long time delayed the further study of high- altitude radiation. Interest in this radiation renewed again only in 1921 in the USA, and in Germany - in 1923. First of all it is necessary to mention the experiments of professor of Californian Institute of Technology R.A. Millikan and his employees.

It is necessary to tell that Prof. Millikan was skeptical about the conclusion of V. Hess and W. Kolhörster about an extraterrestrial origin of high-altitude radiation, and decided to check up the experimental results of V. Hess and W. Kolhörster, - how many are correct. Let us underline that V. Hess and W. Kolhörster were compelled to accompany devices on balloons during flight to write down their indications, but Millikan and Bowen (1923) in spring 1922 adapted meteorological balloons for the raising of devices. Especially for flight were designed four complete sets of devices, each of which included an electroscope, thermometer and barometer.

The dimensions of the collected device were about 15 cm, and its weight was only 190 g. Record of results was made automatically by fixation of a shade from electroscope's delays on a moving film. The device raised by two small balloons 47 cm in diameter while one of the balloons was broken off under the influence of external pressure of gas extending with reduction. After that, the remaining not-broken balloon played a parachute role for the descent, reducing speed of descent to a safe one for device integrity. Besides, the escaped balloon played a landmark role in that it was easy to find for equipment after its landing.

During the most successful flight lasting 3 hours 11 minutes, a maximum altitude of 15,500 m was obtained. Equipment was reached having landed at a distance of 100 km from the start; it was found, and the film was manifested. Use of sounding balloons appeared a valuable innovation, reduced the price of carrying out of experiments, and eliminated danger to the experimenter. Results of flights in general confirmed the existence of high-altitude radiation, but did not completely resolve doubts of Millikan as speed of ionization at altitudes of more than 10 km, obtained by means of sounding balloons, appeared 4 times smaller than what was expected on the basis of extrapolation of Eq. 2.

10. MEASUREMENTS OF ABSORPTION FACTOR IN ICE AND LEAD IN THE MOUNTAINS: IRREPARABLE BLOW STRUCK TO THE HYPOTHESIS ABOUT EXISTING RADIATION FROM SPACE (1923)

The reason for the big divergence between Millikan's and Kolhörster's results became clear after, in 1923, when Kolhörster (1923) defined the value of the absorption factor of high-altitude radiation more precisely. Measurements by W. Kolhörster were made on the mountain Jungfrau-Joh glacier in Switzerland at altitudes of 2300 m and 3500 m.

Placing the device in ice at a depth of 15 m, Kolhörster defined that the absorption factor μ at the altitude of 2300 m is equal to 1.61×10^{-3} cm^2.g^{-1}, and at the altitude of 3500 m $\mu = 2.7 \times 10^{-3}$ cm^2.g^{-1}. Without giving attention to that circumstance that at the height of 2300 m the absorption factor appeared smaller than at the top of a glacier, Kolhörster decided to take as the true value simply the arithmetic average of both received values.

The value of th absorption factor found thus appeared almost three times less than defined earlier, and the difference between values of the ionization rates, calculated by extrapolation given by Kolhörster's Eq. 2 and taking into account a more exact factor of absorption, and received experimentally by Millikan and Bowen (1923) above 10 km, decreased much more. Discussing the experimental data received in 1923 and the nature of high-altitude radiation, Kolhörster wrote, "Recently I more and more abandon the idea that high-altitude radiation represents the phenomenon the origin of which it is necessary to search in world space".

However, R. Millikan and his colleagues still were not agreeing with V. Hess and W. Kolhörster's conclusions.

They considered that nevertheless it is possible to explain the received results by radioactive pollutions. Doubts of R. Millikan and his scientific group got stronger after measurements of the absorption factor of high-altitude radiation in lead. R. Millikan (1924) with his student R. Otis made these measurements during a summer expedition in 1923 on the mountain Peak Pike (4300 m). They placed over the device measuring the speed of ionization a plate of lead of thickness 48 mm and found out that the factor of absorption of high-altitude radiation in lead is close to the factor of absorption of γ-rays that are let out by thorium.

On this basis the conclusion was drawn that "the radiation for the most part nevertheless has a local origin".

The scientific authority of Robert Millikan (Nobel prize winner in physics in 1923) was so huge that his skeptical relation to the existence of the radiation coming from space was reflected in the opinion of many physicists at that time. "Really, as talked Millikan in the speech at the Congress of the British association of assistance to science development in Leeds at September, 2nd 1927, up to 1925, among physicists, judging by the literature, was not felt confidence that existence of radiation of space origin is proved" (Millikan and Cameron, 1928c). So, 1923 hardly became a turning point in the history of studying of radiation from space. "It was possible to think, - wrote L.V. Myssowsky (M1929, page 27) - that to a hypothesis about existing radiation from space has been struck irreparable blow".

11. THE DEPENDENCE OF ABSORPTION FACTOR ON ATOMIC CHARGE NUMBER Z: REHABILITATION OF THE HYPOTHESES ON SPACE ORIGIN OF HIGH-ALTITUDE RADIATION (1925-1926)

As became clear later, the erroneous conclusion of R. Millikan was caused by special features of absorption of high-altitude radiation in substances with the various atomic charge number Z. Absorbing ability of substance sharply increases with growth of Z, and consequently lead is a much more effective absorber of high- altitude radiation than air at identical thickness in g/cm^2.

It is necessary to underline thus that the truth was found out by R. Millikan and his colleagues as a result of measurements of absorption factor of high-altitude radiation in water on high-mountainous lakes. In August 1925, Millikan and Cameron (1926) with the purpose of definitively denying the idea about the existence of the radiation having not radioactive origin, began a series of experiments in Southern California on lakes Muir and Arrowhead. The appearance of the device developed by Millikan and Cameron (1926) for underwater measurements of ionization speed is shown in Figure 6.

Figure 6. Designed by Millikan and Cameron (1926), a device for measurements of ionization rate in the air by high-altitude radiation under the water.

The device had the form of a sphere with volume about 1.5 l, three times less than the device of W. Kolhörster (see Section 8). Two hemispheres incorporated among themselves by 34 bolts, which pulled together flanges, condensed with rubber rings. Two quartz-silvered threads were suspended on a quartz crossbeam. Supervision was conducted by means of a microscope supplied with a periscope device. As well as in Kolhörster's device, the charging of threads was made by the special device from the auxiliary battery, not breaking tightness. The device worked at pressure in 8 atm that together with some improvement of the internal device allowed to raise its sensitivity by 8 times in comparison with Kolhörster's device, and self study of the device was rather insignificant, only 7 $ion.cm^{-3}.sec^{-1}$. The device of Millikan and Cameron (1926) for a long time was used by many American researchers, as it was convenient and reliable during use in different conditions.

R. Millikan chose mountain lakes Muir (altitude 3600 m above sea level) and Arrowhead (1530 m) because of the extraordinary cleanliness of their water, formed of thawed snow that excluded the presence in the water of radioactive impurity. The device was plunged into Lake Muir to a depth of 18 m, and the speed of ionization registered by the device fell from 13.3 $ion.cm^{-3}.sec^{-1}$ on the surface of the lake to 3.6 $ion.cm^{-3}.sec^{-1}$ on the depth of 18 m. Below this point the sensitivity of the devices used did not give the possibility to track the further reduction of ionization speed. Then the same measurements of speed of ionization at various depths were made in Lake Arrowhead, located 450 km to the south and 2060 m lower than Lake Muir. It appeared that devices in different lakes gave identical evidence in the case when the device in Lake Muir fell 1.85 m (6 feet) deeper than in lake Arrowhead. "We, - wrote Millikan, - have received the same curve, as well as in Lake Muir, with that only a difference that each readout is respective shaft to device moving just in 6 foot upwards".

To explain the received results, R. Millikan assumed that equal weights on the unit of the area of water and air absorb equally. Therefore, as the air layer between levels of both lakes weighs as much as 1.85 m of water, for the radiation coming from above through the atmosphere, the total weight of the absorber (air + water) is identical to when the device is shipped in lake Muir, 1.85 deeper than in lake Arrowhead. The measurements made on lakes Arrowhead and Muir convincingly proved that the observable radiation comes from above, and is not formed in the air layer between lakes (in that case the device on lake Arrowhead would fix an ionization rate bigger than the device on lake Muir). Let us note, by the way, that the basic argument of R. Millikan was not correct. Equal weights of air and water on an area unit absorb high-altitude radiation not absolutely equally. If the experimental techniques of that time had allowed the obtaining of more exact data, Millikan would have found a difference between corresponding instrument readings in both lakes. Moreover, it could have prevented the drawing of the correct conclusion! Results of measurements of Millikan and Cameron (1926) are presented in Figure 7.

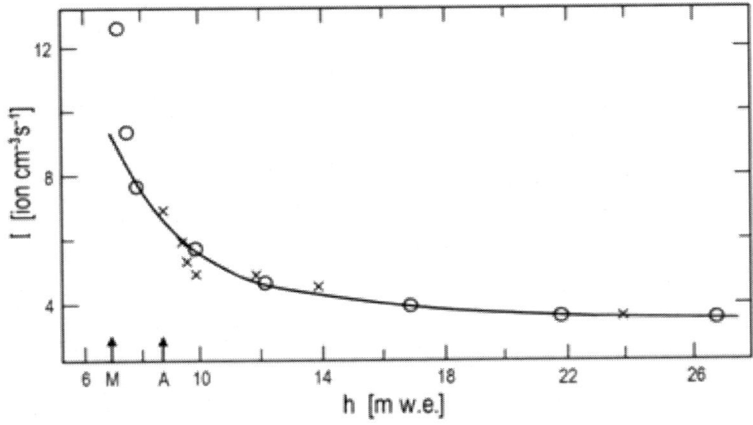

Figure 7. Results of measurements of ionization rate on various depths in lakes Muir (are designated by circles) and Arrowhead (are designated by daggers), received by Millikan and Cameron (1926). M and A - levels of a surface of lakes Muir and Arrowhead taking into account thickness of atmosphere in m w.e. (water equivalent).

If for the axes in Figure 7 of abscissas one presents depth of level of observations concerning the atmosphere border, expressed in m w.e., and on an axis of ordinates − the corresponding ionization rate in ion.cm^{-3}.sec^{-1}, then the data received from observations in various lakes lays down on one curve. For interpretation of the received experimental data, Millikan and Cameron (1926) for the first time took into consideration that the investigated radiation falls not only on a vertical, but comes also under various zenith angles θ. Therefore the observable intensity $I(x)$ at depth x of the atmosphere counted from the border in m w.e. will be described by the following expression:

$$I(x) = I_0 \int_0^{\pi/2} \exp(-\mu x/\cos\theta)\sin\theta d\theta, \tag{3}$$

where I_0 - intensity of high-altitude radiation on the atmospheric border. Entering a new variable $u = \mu x/\cos\theta$ and having changed accordingly the integration limits, Millikan and Cameron (1926) received that

$$I(x) = I_0\left[\exp(-\mu x) - \mu x \int_{\mu x}^{\infty} \frac{\exp(-u)}{u} du\right] = I_0\Phi(\mu x), \tag{4}$$

The function $\Phi(\mu x)$ in the American scientific literature received the name Gold function; it replaces the simple exponential law, fair only for vertically falling radiation. Using tables of values of $\Phi(\mu x)$, Millikan and Cameron (1926) defined that the factor of absorption of high-altitude radiation is equal in water to 3×10^{-3} cm^{2}g^{-1} that is once again evident that high-altitude radiation possesses delectability much bigger than γ-rays from radioactive elements. Let us note that the proof of the big penetrating ability of high-altitude radiation, given by R. Millikan and his employees in 1925, gave the occasion to such American magazines as "Science" and "Scientific Monthly" to name this radiation "Millikan's rays". But because R. Millikan and his employees, in essence, only confirmed expanded results of the discovery made by European researchers still in 1912-1913, subsequently physicists rejected this name as inexact and misleading.

12. Final Proof of Extra-Terrestrial Origin of High-Altitude Radiation and Millikan's Supposition to Call It Cosmic Rays (1926)

Experiments on lakes definitively convinced Millikan and Cameron (1926) that high-altitude radiation has an extraterrestrial origin, "These rays, - it was summarized in Millikan and Cameron (1926), - do not occur from our atmosphere and consequently can be rightfully named by 'cosmic rays', this most descriptive and most suitable name". Thus, Robert Millikan, long denying the existence of high-altitude radiation as coming from space, in 1926 fixed the name which is to ths day used all over the world.

Analyzing the curve of absorption of cosmic rays in water, Millikan and Cameron (1926) made the first attempt to distinguish absorption factors of cosmic rays at various depths. If for the top part of a curve of absorption (see Figure 7) they received the value 3×10^{-3} cm^{2}g^{-1}, in the bottom part of a curve the value of the absorption factor turned out slightly less: 1.8×10^{-3} cm^{2}g^{-1}. This meant that during propagation with absorption cosmic rays become more and more rigid, and Kolhörster's suggestion that high-altitude radiation (cosmic rays) is absorbed under the simple exponential law was erased and, hence, representation of cosmic rays as a stream of homogeneous photons was simply incorrect.

It is necessary to notice that in the earlier investigations of cosmic radiation (called Hess's rays – after V. Hess' experiments in 1912, high-altitude radiation – after W. Kolhörster's experiments in 1913, high frequency rays and penetrating radiation – after R. Millikan experiments in 1923), it was considered only as a geophysical phenomenon together with the magnetic field of the Earth and with atmospheric electricity (let us note that when one of the authors, Lev Dorman, started to work after graduation in 1950 at the Nuclear Division of Moscow State University, - in the Cosmic Ray Laboratory – this Laboratory was inside the Department of Atmospheric Electricity of the Research Institute of Terrestrial Magnetism, now IZMIRAN, near Moscow). It was investigated mainly by experts of radioactivity and by meteorologists, and the first who distinctly understood what important information for all physics can be taken from cosmic ray regular and general study was R. Millikan. The reason research of the nature of cosmic rays became the basic problem of the Physical Laboratory (California Institute of Technology), formed and lead by R. Millikan.

13. THE EARLIER COSMIC RAY INVESTIGATIONS IN THE FORMER USSR: ABSORPTION IN WATER AND ZENITH ANGLE DISTRIBUTION (1925-1926)

In 1924, the group of the Soviet physicists led by Lev Myssowsky, Head of the Physical Department of the State Radium Institute of Academy of Sciences USSR (RIAN, Leningrad) became interested in high-altitude radiation. In July 1925 Myssowsky and Tuwim (1926a) defined the absorption factor of high-altitude radiation in water, and irrespective of the group of R. Millikan (even two weeks earlier) showed that penetrating radiation different from γ-radiation of radioactive substances really exists. Early studies of absorption of high-altitude radiation were not casually made in water. This results from the fact that to detect the radiation, large thicknesses of material (if one excludes screens of lead and other heavy elements) are necessary for research of its absorption factor, and water appeared a most suitable substance. During study of absorption of high-altitude radiation in water, it was definitively proven that it comes from space.

Experiments of Myssowsky and Tuwim (1926a) were made on Lake Onega, near the Petrozavodsk bay. The place of measurements was chosen where the bay coast only slightly towered over the lake level. It was made so that influence of γ-rays from coastal breeds could be neglected. Experiments were made at a distance of 0.5 km from the coast where the depth of the lake equaled 19.5 m, and the device was plunged only to a depth of 10 m to avoid γ-radiation from the lake bottom. "All these precautions had at that time great value as then yet there was no confidence of a space origin of high-altitude radiation and it was possible to expect occurrence of same radiation from outside a bottom of the lake or surrounding heights", - recollected later Myssowsky (M1929, page 38). In experiments the device (№ 5050), designed by W. Kolhörster and produced by firm "Gunter and Tegetmajer" in Braunschweig (Germany), was used. Before starting measurements, L. Myssowsky and L. Tuwim conducted with this device special experiments for finding-out of influence of temperature on its indications. It appeared that change of temperature from −18°C to +35°C does not influence instrument readings. Results of the measurements of Myssowsky and Tuwim (1925) are presented on Figure 8 (curve I).

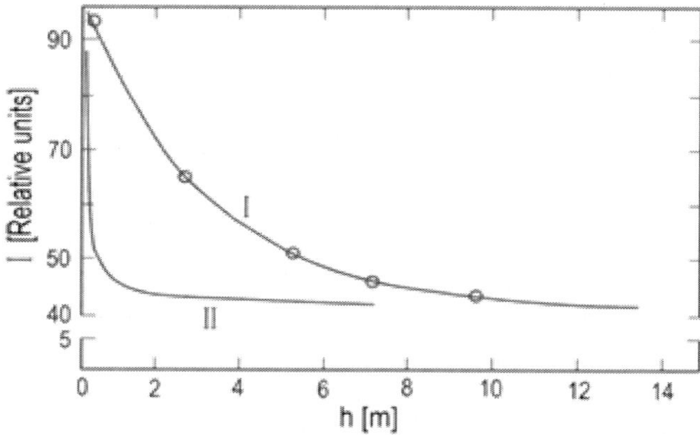

Figure 8. Results of measurements of ionization rate at various depths in Onega (curve I), obtained by Myssowsky and Tuwim (1926a). Curve II shows expected absorption of γ-rays from radioactive substances.

In Figure 8 on an axis of abscissas is the depth of the level of observation in meters, and on an axis of ordinates – the speed of ionization in any units. Curve II was calculated by Myssowsky and Tuwim (1925) in the assumption that high-altitude radiation consists of γ-rays having a radioactive origin: $\mu = 3.6\times10^{-2}$ cm^{2}g^{-1}. The difference between curves I and II was so great that L. Myssovsky and L. Tuwim did not doubt: "About the confusion of cosmic rays with usual γ-rays cannot be any speeches". Calculation of the absorption factor from the data received in the experiments on Lake Onega was made in the assumption that absorption of high-altitude radiation occurs under a simple exponential law. Its value appeared equal to 3.6×10^{-3} cm^{2}g^{-1}. Unfortunately, Myssovsky and Tuwim (1925) could not define the radiation of the device itself experimentally, and calculated it from a curve that strongly deformed the received result. At a comparison of values of cosmic ray absorption factors received by various experimenters, it was evident that they strongly differ from each other. First of all there was a question of whether it is necessary to take into consideration in calculations of the factor of absorption that cosmic rays come from all directions, as was made by Millikan and Cameron (1926), or if it does not follow. To answer this question, Myssowsky and Tuwim (1926b) made in December, 1925 experiments on a water tower of the Polytechnic Institute in Leningrad. In these experiments, witty use was made of the shielding action of big weights of water in the water pressure head tank at the height of 35 m over the surface of the earth. The water tower device allowed the making of research of distribution of intensity of space radiation in dependence of zenith angle (see Figure 9).

The device was suspended from the tower at various distances from the tank with water. Measurements were made at the empty tank and after filling of the tank with water. Experiments showed that "slanting rays" play a much bigger role than it was accepted to think earlier. "The received results, - Myssowsky and Tuwim (1926b) drew the conclusion, - it is already enough to be convinced of the necessity to make calculation of the absorption factor in the air and in water on $\Phi(\mu x)$, instead of on $\exp(-\mu x)$". In reality, if according to Myssowsky and Tuwim (1926b) the data received on Lake Onega one calculates the absorption factor using $\Phi(\mu x)$, determined by Eq. 4 (see Section 9), the value of the absorption factor considerably will decrease and will be equal to 2.8×10^{-3} cm^{2}g^{-1}, close to the value of the absorption factor obtained by Millikan and Cameron (1926).

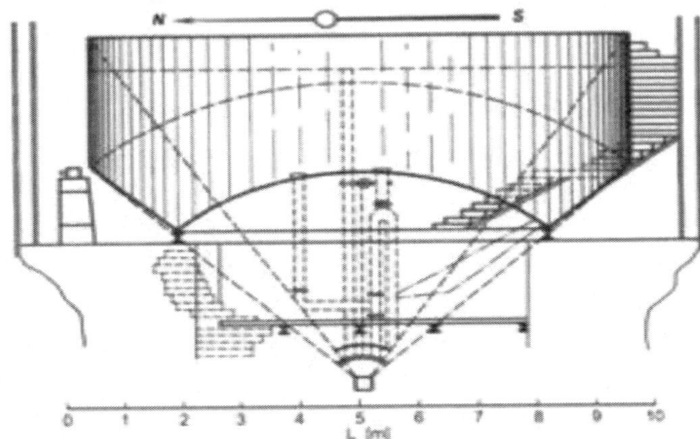

Figure 9. The scheme of experiments of Myssowsky and Tuwim (1926b) made on a water pressure tower of Polytechnic Institute in Leningrad.

14. FURTHER INVESTIGATIONS OF CR ABSORPTION IN THE AIR AND WATER BY MILLIKAN AND CAMERON: DISCOVERY OF CR HETEROGENEITY (1926-1928)

The further experiments of Millikan and Cameron (1928a) in mountain lakes focused and added to the results received earlier. In the autumn of 1926, they made measurements of cosmic ray absorption in Lake Miguilla (Bolivia), which is located in mountains at a height of 4570 m above sea level and has some tens of meters in depth. Millikan and Cameron (1928a) united results of measurements on Lakes Muir, Arrowhead and Miguilla and as all lakes are located at various heights, replaced absorbing layers of atmosphere above each of the lakes with their equivalent weight in sheets of water. Obtained results are shown in Figure 10.

In Figure 10 on the axis of abscissas depth concerning atmosphere border in m w.e. is presented, and on the axis of ordinates – measured values of ionization rate. It appeared that all obtained data lays down well on one curve, despite the fact that lakes are in different hemispheres. Knowing the value of the radiation of the device itself (7.4 ion.cm^{-3}s^{-1}), Millikan and Cameron (1928a) defined the ionization rate caused by cosmic rays atsea level on this curve (the thickness of the entire atmosphere is equivalent to a sheet of water of depth 10.33 m). It appeared equal to 1.4 ion.cm^{-3}s^{-1} that well coordinated with the value of the ionization rate at sea level, received by Kolhörster (1913b). The careful analysis of the curve of absorption allowed the defining of four values of cosmic ray absorption factors at various depths (Table 3).

Using the value of the cosmic ray absorption factor calculated for the top part of the curve, and the value of the ionization rate found from the same curve at sea level, Millikan and Cameron (1928a) calculated what will be the values of ionization rate in the terrestrial atmosphere at various heights.

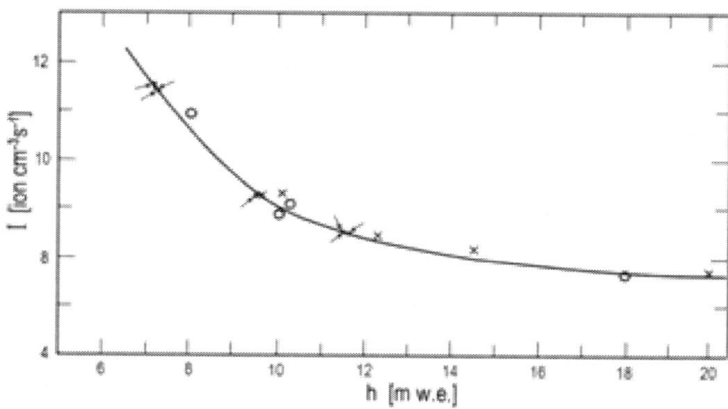

Figure 10. Results of measurements of ionization rate on various depths h (in m w.e.) in lakes Muir (circles), Arrowhead (daggers) and Miguilla (arrows) received by Millikan and Cameron (1928a).

Table 3. Cosmic ray absorption factor for different depths according to Millikan and Cameron (1928a)

Depth, m w.e.	Cosmic ray absorption factor, 10^{-3} cm^{2}g^{-1}
6.2-9.3	2.5
9.3-10.3	2.3
10.3-12.3	2.0
12.3-18.0	1.5

It is interesting to notice that the calculated values of the ionization rate at the height from 5 to 15 km coincided with the average values of the ionization rate, obtained by Millikan and Bowen (1923) on sounding balloons. The change of cosmic ray absorption factor with increasing of thickness of an absorbing layer confirmed preliminary assumptions of the authors about the heterogeneity of cosmic rays, whose structure, in their opinion, varied in the process of absorption owing to filtering of softer components. "Except us, - marked R. Millikan, - nobody had been found out heterogeneity of cosmic rays, though the last results of G. Hoffmann and E. Steinke (see below, Section 17) lead to their acknowledgement of our opening and to the assumption that in a mix of cosmic rays there can be even more rigid γ-rays, than found out by us".

15. THE CHECKING OF WILSON'S HYPOTHESES ON ELECTRON ACCELERATION IN ELECTRIC FIELDS IN THUNDERSTORM CLOUDS (1927-1928)

Millikan and Cameron (1928a) took advantage of the fact that the Lake Miguilla is surrounded from different directions by high mountains, and decided to check the hypothesis of Wilson (1925) according to which ionizing particles of high energy can arise in the upper atmosphere at the acceleration of β-particles (electrons) in an electric field of thunderstorm clouds. Though the lake was reliably protected from energetic ionizing particles, which could be formed from thunderstorms, the measured ionization rate appeared the same as on a Californian beach during a thunderstorm. "Thus, - Millikan and Cameron (1928a) have solved, - the Wilson's hypothesis is absolutely definitely excluded".

16. INCREASING OF MEASUREMENT DEPTH UP TO 50 M AND HYPOTHESES ON COSMIC RAYS CONSISTING OF THREE GROUPS OF PHOTONS WITH DIFFERENT ABSORPTION FACTORS

Having improved the devices, Millikan and Cameron (1928b) could make on the Californian mountain lakes in the summer of 1927 measurements of cosmic ray absorption up to a depth of 50 m. Even at big water depths it was possible to notice a gradual reduction of ionization rate. Cosmic rays, it appeared, possess still bigger getting ability, than it was possible to assume on the basis of all previous experiments. The absorption factor calculated from the curve of absorption for depths from 50 to 60 m w.e., was only 1.0×10^{-3} $cm^2 g^{-1}$. In those years, R. Millikan and G. Cameron adhered to the common standard point of view that cosmic rays are fluxes of photons of very high energy. They tried to present purely formally all curves of absorption in the form of the sum of three separate exponential components, each of which would describe absorption of the group of photons of certain energy. It turned out that cosmic rays consist of three various groups of photons with absorption factors of 3.5×10^{-3}, 0.8×10^{-3}, and 0.4×10^{-3} $cm^2 g^{-1}$ and in the upper atmosphere the first prevailed. It is the softest and most intensive component by which 90% of energy of cosmic rays was attributed.

The experiments of R. Millikan and G. Cameron on lakes for many years were considered classical because they were executed by means of the most perfect at that time equipment specially designed for underwater measurements, and they definitively proved the existence the radiation with complex structure coming from space and possessing specific properties.

17. UNDERWATER COSMIC RAY EXPERIMENTS AT THE DEPTH INTERVAL 30-230 M (1928-1930)

Further underwater measurements, considerably having improved their technique, continued by German physicist E. Regener (1929, 1932, 1933). He began the measurements in 1928 in the lake Boden in Switzerland, and continued them in 1929-1930. The measurements were carried out deeper than 30 m, as above this level the influence of waves affected still use in Boden lake, and were made up to a depth of 230.8 m. For underwater measurements by E. Regener sensitive, automatically working a data-acquisition equipment had been created (see Figure 11). The ionization chamber J, executed in the form of an iron bomb, had volume of 35 l and was filled with carbonic acid under a pressure of 30 atm. The chamber was connected with one-film string electrometers, and registration of its indications was carried out by photographing the position of the thread through certain time intervals on a motionless photographic plate of size 4.5×6.0 cm^2. Position of the thread of the electrometer was initially charged to some potential, and gradually varied in the process of its discharge, and the distance between the next positions of the thread defined the speed of discharge of the electrometer, or equivalently, intensity of ionization in the chamber.

The ionization chamber was located in a tank filled with water, which was take at the lake surface to exclude the possible influence of the water radioactivity which was unequal, as believed E. Regener, at various depths (the first measurements E. Regener made without this tank with water). Six floats provided floatability of all equipment which plunged when to it were suspended cargoes and anchors in weight of about 250 kg. The general idea of the form and sizes of the equipment applied to under-water measurements is given in a photo (Figure 12) in which E. Regener examines the device before its immersing in water is removed.

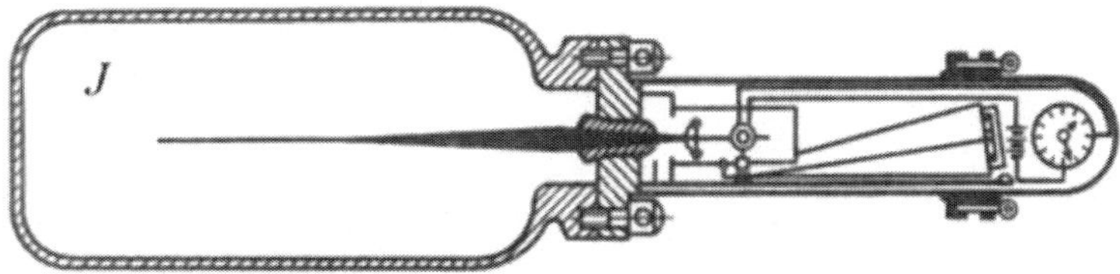

Figure 11. Regener's device for underwater measurements of cosmic ray intensity up to 230 m.

Figure 12. E. Regener with the device before its immersing in water.

Results of two series of measurements of absorption of cosmic rays in Lake Boden, received by E. Regener are presented in Figure 13. In this Figure on an axis of abscissas depth concerning atmosphere border in m w.e. is presented, and on an axis of ordinates - ionization current registered by the device. Results of separate measurements lay down well on one smooth curve. From this absorption curve it is visible that ionization current registered by the device continues to decrease up to a depth of 230 m. Such huge receiving ability of cosmic rays seemed at that time so surprising that Regener in the beginning "has not believed to his eyes" and made the assumption that reduction of ionization speaks to the reduction of the maintenance of the radioactive admixtures in water of the lake with depth increase. However, ionization measurements by the device surrounded with a sheet of water showed that it is not so. The site of the curve of absorption to a depth of up to 30 m was investigated later in one of the Alpine lakes by W. Kramer (1933). As Kramer used the same device as E. Regener, his data without any recalculations supplemented the curve of absorption received by E. Regener.

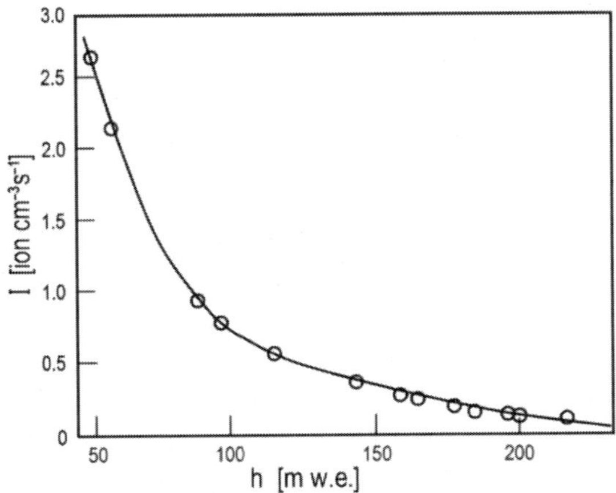

Figure 13. A curve of absorption of cosmic rays in Boden lake, obtained by E. Regener.

Regener's results were in good agreement with the results of R. Millikan and G. Cameron investigating absorption of cosmic rays to a depth of 50 m, as well as with results obtained in 1929 by Steinke (1929). For Steinke (1929) it became possible by means of complicated, but very sensitive equipment of Hoffmann's type (which in detail will be considered later, in Section 16) to make underwater measurements to a depth of 50 m. According to E. Regener's opinion, the character of the change of intensity of space radiation in water testified that it consists of photons characterized by various energies.

18. UNDERWATER COSMIC RAY EXPERIMENTS AND DISCUSSION: WHAT ARE COSMIC RAYS, ENERGETIC PHOTONS OR PARTICLES?

As we mentioned many times above, up to the end of 1920^s the common opinion was that cosmic rays are high-energy photons. However, to the beginning of 1930s, after D.V. Skobeltsyn and V. Bote and W. Kolhörster's experiments (see below, Chapter 3), a hypothesis became popular according to which cosmic rays consisted of charged high-energy particles. This hypothesis was widely discussed among the physicists who were engaged in cosmic rays, and had many supporters trying all new experiments to see the proof in its advantage. So, J. Clay (1932), discussing in 1932 on the nature of cosmic rays, concluded that "the curve absorption of E. Regener does not exclude the possibility of corpuscular radiation as corpuscular beams with Maxwell distribution on energies can give similar reduction of intensity by the big depths". Thus, the curve of absorption of cosmic rays in water, obtained by E. Regener, did not give an unambiguous answer on the question exciting physicists on the nature cosmic rays, and they could be interpreted as miscellaneous depending on the propensity of the author.

Absolutely distinct from all of the described above methods, A.B. Verigo (1934) used some new research methods. In 1930, he lowered the ionization chamber on a submarine to the depth of 40 m. Having united values of intensity of cosmic rays obtained earlier on mountain Elbrus at various heights up to 5.4 km and having obtained data on various depths, A.B. Verigo drew the conclusion that cosmic radiation is divided into two components: soft and hard. For the soft component he found an absorption factor $\mu = 3.8 \times 10^{-3}$ cm^{2}g^{-1}, and for the rigid component $\mu = 2.3 \times 10^{-4}$ cm^{2}g^{-1}.

After the numerous measurements made at big depths by E. Regener and other experimenters with the help of ionization chambers of various kinds, it appeared that to expect any unexpectedness was impossible. However, the results received in 1932 J. Clay (1933) caused great interest and again drew attention to studying of the curve of absorption of cosmic rays at big depths. During travel from Holland to the island Java, undertaken with a research objective of the intensity of cosmic rays at various latitudes, J. Clay with the help of the ionization chamber filled with argon under a pressure of 45 atm made measurements at depths up to 270 m. His measurements yielded an absolutely unexpected result: the ionization rate continuously decreased to a depth of 200 m and then began to increase, reaching a maximum at the depth of 250 m, and then sharply reduced to zero at the depth of 270 m (see Figure 14).

J. Clay repeated the measurements twice, but the initial result remained. The impression turned out that the curve has a ledge which shows the existing of a border of penetration of a considerable part of CR. J. Clay was a supporter of the corpuscular nature of CR and interpreted the received results as proof that the most incident part of cosmic radiation consists of charged particles of approximately one energy. However, repeatedly receiving similar results at measurements in the North Sea by J. Clay became impossible.

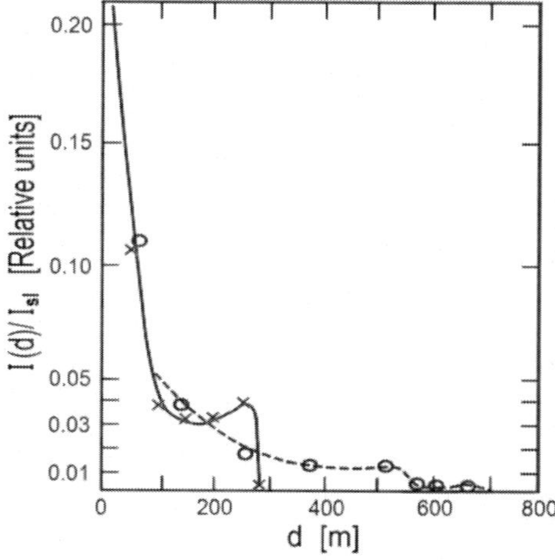

Figure 14. A curve of absorption of cosmic rays in the Red sea (continuous line), obtained by J. Clay (1933), and a curve of cosmic ray absorption in mines in Sweden (dashed line), obtained by A. Corlin (1934). $I(d)/I_{sl}$ - is the ratio of measured cosmic ray intensity at depth d (in m w.e.) to I_{sl} is the cosmic ray intensity at sea level.

Even more surprising results were received by A. Corlin (1934), investigating absorption of CR by layers of iron ore in mines in the north of Sweden. Having overcome the considerable difficulties connected with radioactive radiation of iron ore, A. Corlin made a number of measurements at various depths up to 157 m. Such a layer of ore corresponded to about 785 m of water. The measurements were not constrained in the size and weight of their installations, and could use the bulky, but much more sensitive equipment of E. Steinke. The gross weight of this equipment, including the ionization chamber, the lead for protection against soft radiation of radioactive impurity of ore, the iron box and batteries, reached 4.5 tons. Installation took place on two trolleys which could be transported from one gallery to another on the underground railway lines. Data of A. Corlin are shown on Figure 14 (dotted line). Unfortunately, there was no possibility to obtain points often enough at the depth at which measurements were made, as it was set by an arrangement of mines and horizontal galleries. Consequently on a curve in an area between 200 and 300 m w.e. the obtained points lay too far apart to answer with confidence the question of whether the maximum opened by J. Clay was observed or not. A. Corlin has drawn the conclusion that he has found one more maximum on the depth of 550 m w.e.

At that time the discussion of the absorption curves obtained by J. Clay and A. Corlin (see Figure 14) caused hot discussions. Supporters of the corpuscular nature of CR, everywhere searching for proofs of this point of view, considered that the results of J. Clay and A. Corlin testify to the presence in CR of two groups of charged particles with energies of 10^{10} and 10^{11} eV. However later, in the late 1930s, after numerous unsuccessful attempts to receive similar absorption curves, it became clear that anomalies in the curves of J. Clay and A. Corlin were simply errors of experiment.

19. THE PROBLEMS WITH INTERPRETATION ABSORPTION CURVES IN MATERIALS FROM HEAVY ELEMENTS (1925-1928)

In general, it is necessary to notice that already the first measurements of absorption of CR in substances with big atomic charge Z resulted in the researchers being in deadlock. As it was already told (see Section 9), the first attempt of R. Millikan and R. Otis to investigate absorption of high-altitude radiation in lead led to the negation of existence of radiation of extraterrestrial origin. To strange conclusions came G. Hoffmann (1925), based on experiments of cosmic ray absorption in lead. He worked in Königsberg and constructed a device that strongly differed from the design of the portable devices which W. Kolhörster and R. Millikan used. At construction of the electrometer which was offered to them in 1913, G. Hoffmann refused the requirement that the device must be portable and convenient for forwarding works. He achieved extremely high sensitivity, constructing a stationary bulky device. The section of Hoffmann's installation is shown in Figure 15.

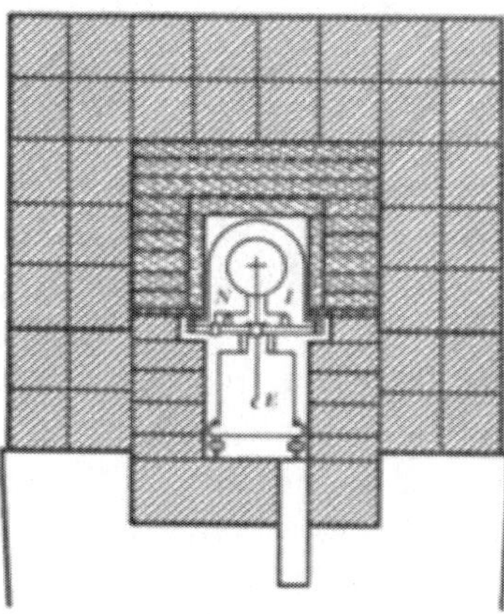

Figure 15. A schematic vertical section of installation of G. Hoffmann (1925).

Unlike the devices of W. Kolhörster and R. Millikan, the ionization chamber *I* in the device of G. Hoffmann was separated from the electrometer E. The volume of the ionization chamber was various in different experiments, sometimes reaching 40 l and considerably exceeding the volume of the chambers in W. Kolhörster's and R. Millikan's experiments. The chamber was filled with carbonic gas, and was covered by the copper bell put for tightness on a rubber ring that allowed to support for a long time in it a pressure nearly 2.6 atm. The useful volume of the chamber was limited by a spherical wire electrode *N*, in which there was a core collecting ions attached to the electrometer. Electrometer indications automatically registered on a moving photosensitive film. The shielding of the ionization chamber by lead of thickness 12 cm was made with the help of ring-shaped plates of thickness of 1 cm (the weight of the lead armour reached 356 kg). Besides this the ionization chamber and electrometer were shielded by lead blocks of the size 10 cm × 10 cm × 20 cm with everyone weight 22 kg (see Figure 15).

G. Hoffmann achieved that his device became the most sensitive of all existing at that time. Sensitivity of his device exceeded the sensitivity of Kolhörster's device by about 100 times! However, Hoffmann's device possessed a number of essential drawbacks: this difficult and expensive device was very cumbersome and suitable for work only in stationary conditions. The device could not be used for measurements at the big heights where cosmic ray intensity was several times greater than on the Earth. By this device it was impossible to measure the ionization rate deep in water for definition of its own radiation, and therefore always there was some doubt as to what part of ionization is caused by the radiation which has come from the outside through the lead screen, and what part is caused by the radiation of the device itself. This led to Hoffmann's interesting results on absorption of CR in the air and lead being difficult to compare to the results of other researchers and Hoffmann's works stood apart as such.

The results received by G. Hoffmann appeared absolutely unexpectedly: the measured ionization current did not decrease monotonically with growth of absorber thickness. Rather, it was sharply decreased at the lead screens in the thickess of up to 12 cm, and then remained almost constant. Analysing the absorption curve in lead, G. Hoffmann concluded that the initial sharp recession speaks to the absorption of γ-radiation of radioactive substances from surrounding subjects, and then, within the accuracy of the experiment, the reduction of ionization current was not observed. He summarized the conclusions by following words: "Thus, the assumption on the existence of radiation more penetrating than radiation from radioactive substances, does not prove to be true". F. Behounek (1926) joined Hoffmann's opinion. His experiments F. Behounek were made with the device of Kolhörster's type at the altitude of 729 m above sea level. The thickness of the lead screen over the device reached 20 cm. All observations made by F. Behounek led to the conclusion that during the change of thickness of the lead screen from 10 to 20 cm the ionization current remains constant. Besides, curve the absorption factor calculated from absorption in lead appeared equal to the absorption factor of γ-rays from radioactive substances.

Only a year after the first unsuccessful results, G. Hoffmann (1926), considerably having improved accuracy of the experiment, found insignificant reduction of ionization current under screens in the thickness more than 12 cm that testified specifically to the character of absorption of cosmic radiation in lead. After that, Hoffmann was compelled to recognize the existence of the radiation incidence much more strongly than known γ-rays from radioactive sources, i.e., having not radioactive, but extraterrestrial origin.

His student E. Steinke (1927) continued Hoffmann's experiments. The curve of absorption of CR in lead screens in the thickness of up to 60 cm, obtained by E. Steinke, was similar to the curve of absorption of G. Hoffmann. Analyzing the character of absorption of CR in the lead, E. Steinke concluded that abnormal absorption speaks to heterogeneity of CR. Unfortunately, E. Steinke could not calculate precisely the absorption factor in the lead as then it was not yet known the radiation of the device itself. Therefore, despite the big sensitivity of the device, conclusions carried more likely qualitative, rather than quantitative character. This position changed a little when, in 1927, E. Steinke (1928) placed the installation in a tunnel in Switzerland and defined the device's own radiation. Then, the factor of absorption of CR in lead was defined precisely.

It is necessary to notice that the absorption factors of CR in lead received by various experimenters so strongly differed from each other that they were difficult even to compare. Not without reason, L.V. Myssowsky (M1929, page 59) wrote: "Now the business with CR, apparently, is the same as before works of Ellisa and Meitner was with γ-rays of radioactive elements. Observing absorption of γ-rays in various thicknesses of lead and other elements, various researchers gave different values of absorption factor a little different among themselves. However even by absorption it was possible to establish that in the process of increase in absorbing thickness the increase of rigidity of γ-rays is observed".

20. THE CR TRANSITIVE EFFECT: BEHAVIOUR NEAR THE BOUNDARY BETWEEN MATERIALS WITH DIFFERENT ABSORPTION FACTORS (1928-1931)

L. Myssowsky and L. Tuwim (1928) started studying the effects arising at transition of CR from one environment to another (the so-called transitive effect). Created by them for this purpose was an installation that consisted of a measuring device (as in former experiments, the device was developed by W. Kolhörster), surrounded from different directions by a lead screen. The installation weight at the maximum thickness of the screen was about 18,000 kg, which in those days was grandiose.

The obtained experimental data basically confirmed G. Hoffmann's and E. Steinke's conclusions. The calculated absorption factor of CR in lead at screen thickness from 0 to 7 cm appeared in 5 times bigger than the absorption factor in lead screens from 7 to 23 cm. To find out the reason for the sharp initial falling of the curve of cosmic ray absorption in lead, Myssowsky and Tuwim, (1928) in the Park of Polytechnic Institute in Leningrad, made a special experiment. The installation used consisted of the following: in a wooden box of size 200 cm × 202 cm ×203 cm were laid blocks of ice from the nearby lake. Intervals between ice floes were filled with water, which at freezing fastened them together. In the ice monolith formed thus a cylindrical course in diameter of 58 cm and length 130 cm in which the device was located was cut lengthwise. Directly over the device through a crack (specially cut out in ice) lead plates of a various thickness were located. Based on the experiences with installation shown in Figure 16, Myssowsky and Tuwim (1928) made the following conclusions: the absorption factor of CR in thick lead screens was within errors of measurements the same as in water (at equal values of g/cm^2).

Figure 16. Appearance of installation of Myssowsky and Tuwim (1928) for studying cosmic ray absorption in ice and lead, and the transitive effect.

The business with absorption of CR by thin layers of lead is otherwise. In this case, first it appeared abnormally big, second, varied depending on whether there was lead under the ice layer or over this layer. The authors refused at once the most simple explanation that this abnormal absorption is possible to explain by an impurity to CR of softer γ-rays from radioactive pollution, because in this case at the passage of 65 cm thickness of ice, being over the device, the intensity of γ-rays should decrease several times. Meanwhile, results of the experiment showed that the sharp fall of the curve remains in the case when lead is in the ice. Myssowsky and Tuwim refused the assumption that the soft component of CR is more strongly absorbed by lead, because they considered that in this case the order of the arrangement of absorbing layers (under lead or under ice) the ice should not play any role. A unique possible explanation, according to Myssowsky and Tuwim (1928), was that the process of absorption of CR is connected with the occurrence in the lead of "secondary rays". Though the Myssowsky and Tuwim (1928) experiments are impossible to consider as proof in lead of secondary particles, the thought that CR can create secondary particles in a substance was very interesting. In general, in the late 1920s - the beginning of 1930s years there was a large quantity of works on studying the "anomaly of transitive zones". The numerous results received those years, and attempts at their interpretation, resulted in Schindler (1931) and in a review paper by Hoffmann (1932). As various researchers marked, experimental data specified that at the passage of CR from light substance to heavy its intensity did not decrease monotonically with the growth of the thickness of the absorber, passing instead through a maximum which took place in the beginning of the transitive curve. Presence in the transitive curve of a maximum was considered proof of the occurrence of secondary radiation treatment. Here Skobelzyn wrote about it (1934): "about the existence of rather intensive secondary radiation of high energy (irrespective of, ionized or not primary particles) testify, however, with full obviousness of anomaly of transitive zones".

As to a definitive explanation of the transitive curve, only in the mid-thirties, after the occurrence of the theory of Bete and Haitler, did it become clear that character of absorption of cosmic radiation, especially in substances with big Z, is distinct from what physicists used in the past.

So, to the beginning of 1930s, no doubts were raised as to the existence of cosmic radiation having huge incidence and complex structure. Any conclusions about the nature of CR and the mechanism of its absorption in a substance, being based on the analysis of curves of the absorption received exclusively by an ionization method, it was impossible to make.

21. Primary Cosmic Rays as High-Energy Gamma Rays with Absorption due Mostly to Compton Scattering (1912-1930)

If on the existence of CR of extraterrestrial origin there were long arguments, it now it seems strange that nobody doubted their nature. Within almost 18 years of Hess's discovery, the problem on the nature of CR (or, as it was called in the past, Hess's rays, high-altitude rays, penetrating radiation, ultra-gamma radiation) was not discussed seriously. This results from the fact that up to 1929-1930 it was considered that the nature of CR is already known: it was a commonly accepted representation that CR are none other than a stream of photons of huge energy. Not casually in the German literature until 1929-1930 CR usually were called as 'ultra-gamma radiation' which were entered still by V. Hess. The reason for this is clear. Of all types of radiations known before V. Hess's discovery in 1912, only γ-rays from radioactive substances were the most penetrating with the smallest absorption. The mechanism of γ-ray absorption in matter was not well studied, however the theoretical calculations of that time based on the assumption that γ-rays are absorbed mainly by interaction with electrons in atoms, mostly by the mechanism of Compton scattering, which predicts reduction of absorption factor with growth of energy E of γ-ray photons (the first relativity quantum theory of Compton effect, discovered in 1923 by A. Compton, has been offered in 1926 by P.A.M. Dirac). Let us note that if the energy of a γ-ray photon greatly exceeds the coupling energy of electrons in atoms, these electrons in relation to such photons it is possible to consider free particles. Collision of a photon with an electron in this case is likened to the impact of two billiard balls - moving and based. After impact the photon jumps aside at some angle to the initial direction with smaller energy, and the electron starts to move in some other direction taking part of the photon's energy. At the passage of a layer of substance by thickness L (in g/cm^2), intensity of an ionizing radiation weakens in e times. The factor of absorption μ is equal to $1/L$ with dimensions of cm^2/g.

Experiments on studying of absorption of CR showed that it is tens of times more penetrating, than γ-radiation from radioactive substances. From this, by analogy to radioactivity, the conclusion was made seeming naturally that CR are also γ-radiation but with energy much bigger than the γ - radiation of radioactive substances. "The question on the absorption mechanism, - testified Skobelzyn (M1936, page 18), - was not put almost, and the most this mechanism, appear, did not represent a particular interest as it was supposed that the cosmic ray phenomenon is identical with the phenomena, already studied on an example of γ-rays of radium and thorium".

22. PRIMARY CR AS THREE GROUPS OF GAMMA RAYS, GENERATED IN THE INTERSTELLAR SPACE AS THE "FIRST CRY" OF ATOMS (1928)

Millikan and Cameron (1928a) had a tempting idea to calculate energy of photons of CR experimentally by certain factor of absorption. Analysing the curve of absorption of CR in the air and water, R. Millikan and G. Cameron concluded that it can be presented as the sum of three exponential curves with different factors of absorption. From here the conclusion was drawn that CR consist of three groups of photons with factors of absorption 3.5×10^{-3} cm^2/g, 0.8×10^{-3} cm^2/g, and 0.4×10^{-3} cm^2/g. Having taken advantage of the formula of Dirac-Gordon which as it then seemed, unequivocally, gave dependence between energy of a photon and factor of absorption, R. Millikan and G. Cameron calculated values of energy for each group of photons. These values appeared accordingly 26, 110 and 220 MeV.

The received estimations laid down a hypothesis basis about the origin of CR, put forward by Millikan and Cameron (1928b), and caused in due time numerous rumours and discussions. R. Millikan and G. Cameron assumed that CR arise at the expense of the energy released at the synthesis of atoms of helium, nitrogen-oxygen, and silicon from atoms of the hydrogen filling interstellar space (neutrons were not yet discovered, and as there is a synthesis, of course, it was not specified). Therefore, for example, four atoms of hydrogen can incorporate and form an atom of helium. As the atom of helium weighs a bit less than four atoms of hydrogen, according to Einstein's principle about interrelation of weight and energy, at the formation of the atom of helium from four atoms of hydrogen some energy $\Delta E = \Delta m c^2$ (where $\Delta m = 4 m_H - m_{He}$ is a defect of weight) is released. This energy appeared equal to 27 MeV. The value of 27 MeV within experimental errors coincided with energy of the first of three groups of photons from which as R. Millikan and G. Cameron represented, CR consist. R. Millikan and G. Cameron tried to explain similarly the origin of the next group of photons. Nitrogen and oxygen concern the most widespread elements in the Universe. One atom of nitrogen is lighter than 14 atoms of hydrogen at value near 1.8×10^{-25} g. Knowing this defect of weight, it is possible to calculate that at the merging of 14 atoms of hydrogen into one atom of nitrogen, energy of 100 MeV is released. Similarly, at synthesis of an atom of oxygen from 16 atoms of hydrogen, energy of about 120 MeV is released. These values are close to each other, and will well enough be coordinated with the value of energy of the second group of photons. R. Millikan and G. Cameron drew the conclusion that photons of this group are the result of synthesis of nitrogen and oxygen.

The third group of photons R. Millikan and G. Cameron explained by energy allocation at the formation of an atom of silicon from 28 atoms of hydrogen, as the allocated energy was close to the energy of the third group photons. These coincidences, which as then was found out, appeared absolutely casual, led R. Millikan and G. Cameron to the conclusion that CR are the "first cry" of atoms which are born continuously in interstellar space.

The first strike to the hypothesis of R. Millikan and G. Cameron was obtained by the paper of O. Klein and Y. Nishina (1929), in which on the basis of relativistic equations of Dirac's theory the modern model of the Compton effect was developed (the relativistic theory of the Compton effect was developed also by soviet physicist I.E. Tamm, 1930, 1975).

The Klein-Nishina theory defining the formula of the connection between energy of a photon and absorption factor essentially differed from the earlier used Dirac-Gordon theory. So, in the beginning of 1929, Skobelzyn (1929) and then many other researchers showed that for experimentally obtained energies of the γ-spectrum, the data testifies to the suitability of the formula of Klein-Nishina for the quantitative description of the Compton effect and against the use of the formula of Dirac-Gordon. If the energy of photons were defined not under the formula of Dirac-Gordon (as was done by R. Millikan and G. Cameron) but according to the Klein-Nishina formula, the coincidence upon which the theory of the origin of CR was based will cease to be. It became clear that the hypothesis of R. Millikan and G. Cameron, in the well-aimed expression of one physicist, - "most improbable of all possible".

23. USING WILSON'S CHAMBER IN THE MAGNETIC FIELD FOR RESEARCH: DISCOVERY OF "ULTRA β-PARTICLES" AND "SHOWERS" IN CR

Only application of new experimental methods of research of CR essentially changed available representations about its nature. The method of research by means of Wilson's chamber appeared one of the most fruitful methods. This witty device invented by C.T.R. Wilson still in 1911, allowed seeing traces of the energetic charged ionizing particles. With a sharp movement of the piston, the volume of the chamber filled with a mix of steam of spirit with air, and owing to adiabatic gas expansions its temperature fell also, and liquid steams suddenly increased and were condensed on ions, forming the path of an ionizing particle. For a short time the particle trace appeared in the form of a strip of a fog and could be photographed. In 1923 Skobelzyn (1924) started studying Compton electrons from the γ-rays of radium formed as a result of absorption in the gas filling the Wilson chamber, and received interesting results. *However wide application of the Wilson's chamber for various researches became possible only after 1925-1927 when D.V. Skobelzyn in detail developed an essentially new technique: a chamber premise in a strong magnetic field (let us note that in the first time the Wilson's chamber was placed in a magnetic field in 1923 by P.L. Kapitza to measure energy of α-particles).

In Figure 17 is shown D.V. Skobelzyn in 1924 near the installation for research of Compton effect of γ-rays.

In Skobelzyn's installation, (1927) a source of γ-rays prepared from radium was located in a crack drilled in a piece of lead, and in the path of the γ-rays between Wilson's chamber and the crack a lead diaphragm was put to allocate the parallel efficient bunch of beams.

Figure 17. D.V. Skobelzyn in 1924 near the installation for research of γ-ray Compton-effect.

Besides, directly ahead of the preparation there was a lead filter in the thickness of 3 mm, serving for absorption of soft γ-rays. Wilson's chamber was surrounded by a solenoid of two coils, which were reeled up on the general cylinder. In the middle of the cylinder between coils there was a crack passing light for illumination of the chamber at the moment of the lowering of the piston. For a short time (from 2 to 3 sec) the current in the solenoid created in it a homogeneous joined magnetic field intensity of about 1000 Gs.

Simultaneously, there was a piston lowering in Wilson's chamber, and photographing by means of a three-dimensional device was made. When the bunch of fast photons passed through gas, the number of the molecules ionized as a result of the effect of Compton was not high enough. However, it appeared that the Compton electrons possessed sufficient energy that on the way to make appreciable ionization thanks to which their traces can be seen in photos. The magnetic field bends an electron's trajectory, therefore, knowing its weight and having measured the radius of curvature of its trajectory, it is possible to define its energy. As one would expect, in the obtained pictures a great number of electron trajectories with curvature radius of some cm can be seen. It was revealed there were traces of Compton electrons formed by interaction of γ-rays from radium with the gas filling the chamber. However, at more careful study of the obtained pictures, D. Skobelzyn noticed that on some of them there are rectilinear traces of particles of unknown origin, absolutely not bent by the magnetic field, but of the form not appreciably different from the traces of usual electrons. D. Skobelzyn named these particles "ultra β-particles" and defined that their energy exceeds 20 MeV. It was very surprising, as it was known that the energy of the γ-rays which are let out by radioactive substances does not exceed several MeV.

In Figure 18 is one of the photos received by D. Skobelzyn in 1927, in which the rectilinear trace of 'ultra β-particles' among curvilinear traces of Compton electrons from γ-beams is clearly visible. D. Skobelzyn drew the conclusion that "ultra β-particles" are Compton electrons, formed at absorption of CR by the gas filling the chamber.

Further observations, results of which were published after two years by Skobelzyn (1929), showed that the number of rectilinear traces was identical irrespective of, whether the pictures were taken in the presence of a preparation of radium about Wilson's chamber or in its absence. Some photos with rectilinear traces of the "ultra β-particles" obtained by D. Skobelzyn (1929a,b) are shown in Figure 19.

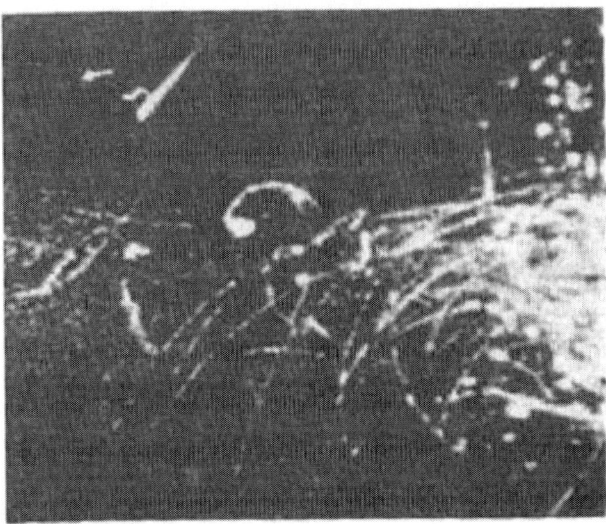

Figure 18. A photo of traces of the "ultra β-particles" obtained by D.V. Skobelzyn in 1927.

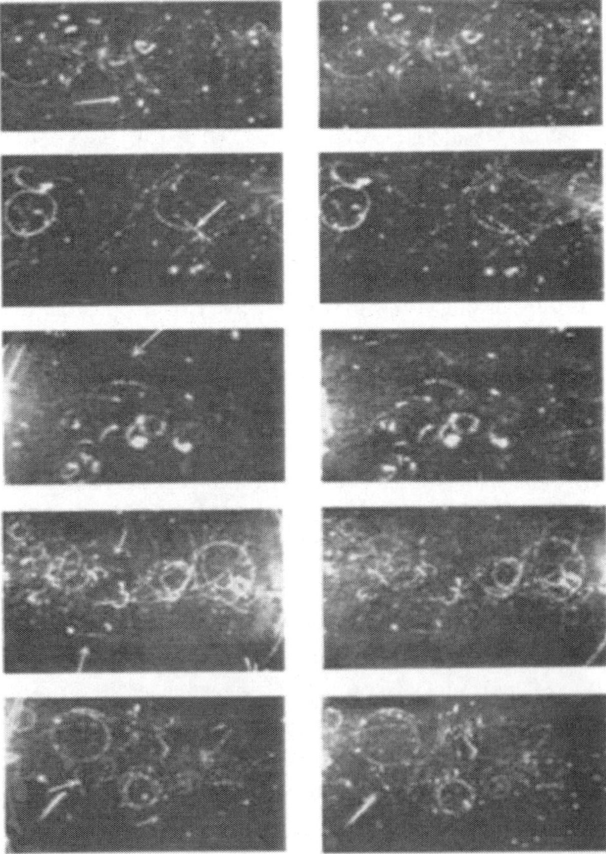

Figure 19. Traces of the "ultra β-particles" (are shown by errors in left panels) obtained by D.V. Skobelzyn in 1929. In rows a-e are shown photos in two projections.

D.V. Skobelzyn (1929a,b) counted up the number of the traces created by "ultra β-particles" in Wilson's chamber. The data was put in a basis of calculations about chamber working hours (an order 0.02 s) and its volume, frequency of occurrence of rectilinear traces in pictures (613 pictures by then were received, in 32 of which rectilinear traces were found) and the number of the "ultra β-particles" falling in 1 s on 1 cm^2 of a surface of the chamber is counted up.

Having estimated the ionization created by fast electrons (on average 40 ions on a way of 1 cm), D.V. Skobelzyn received that by observed "ultra β-particles" should be creating 1.1 ion cm^{-3}s^{-1}. If one takes into account that Skobelzyn's experiments were made in the ground floor of the laboratory of the Leningrad Polytechnic Institute, i.e. part of CR was absorbed by building, that within experimental errors the average ionization which was made by "ultra β-particles" in Wilson's chamber appeared to equal ionization caused by CR (defined by ionization chambers in experiments of L. Myssowsky and R. Millikan).

In his work, D.V. Skobelzyn did not reject at all the hypothesis about the photon nature of CR, and held the opinion that the reason for almost all ionization created by CR are the "ultra β-particles" formed at absorption of cosmic ray photons in the atmosphere. It is important to underline that in some of Skobelzyn's pictures were found two or three traces of "ultra β-particles" simultaneously (from all 32 pictures with rectilinear traces on three are traces of two "ultra β-particles", and in one picture - three "ultra β-particles"). In Skobelzyn (1929b,c) it was shown that the frequency of occurrence simultaneously of a "shower" of particles simultaneously surpasses all estimations, which it was possible to expect on the basis of statistical calculations (it testified to genetic relation of existence between these two or three simultaneously observed particles).

These classical works of D.V. Skobelzyn not only gave a new method of research of CR which became standard, but also drew the attention of researchers to the question of the nature of CR. It became clear that former representations about the mechanism of absorption of CR were incorrect, and for the question research studying "ultra β-particles" was necessary.

The value of these works of D.V. Skobelzyn is difficult to exaggerate. At the 50th anniversary of the publication of the first work of D.V. Skobelzyn (1927), there was an article of Vernov and Dobrotin (1977a,b) in which influence of research of D.V. Skobelzyn on the subsequent development of physics of CR and physics of high energies was analysed. "Rereading now, after 50 years of works of D.V. Skobelzyn (1927, 1929), - Vernov and Dobrotin (1977a,b) write, - are surprised boldness of their plan, art and intuition of their author, depth of the analysis made by him".

24. THE JUMP IN CR EXPERIMENT: GEIGER'S SHARPER AND GEIGER-MULLER THIN WIRE COUNTERS AS NEW DEVICES FOR COUNTING OF INDIVIDUAL ENERGETIC CHARGED PARTICLES

Farther progress in CR research became possible thanks to the invention of Geiger's pungent and Geiger-Muller thin wire counters, which allowed the registering of separate energetic charged particles. Ionization chambers with electroscopes, which were the oldest method of cosmic ray research, allow measuring the full ionization made in a closed vessel by all kinds of ionizing particles. The method of measurement of total ionization does not give any representation of radiation structure, as the number of ions in the chamber depends on the number, kind and energy of ionizing particles. Wilson's chamber, though it allowed registering separate ionizing particles, was in working order only for fractions of a second, not sensitive enough to track the flight of ionizing particles. Therefore, the invention of the device allowing the continuous registering of separate ionizing particles gave the chance to receive more detailed information about the structure of CR. Such a device with thin wire in a cylindrical tube was invented in the autumn of 1928 by German scientists H. Geiger and W. Muller (1928), and it became called the Geiger-Muller counter.

Let us note that some prototype of the Geiger-Muller counter was invented 15 years earlier, in 1913, by H. Geiger working at that time in the Laboratory of E. Rutherford at the Manchester University. Let us note that Hans Geiger was nearest to E. Rutherford, being his pupil and assistant. E. Rutherford named him "a demon of the account of α-particles" and, being surprised by his talent and indefatigability, liked to speak: "Geiger worked as the slave". During the First World War, H. Geiger, working that time in Germany, corresponded with his Teacher through friends in the neutral countries and helped very much captured English soldiers-physicists, pupils of his Teacher, and among them was young John Chadwik, - the future Nobel winner.

Hans Geiger offered a design of the device which with identical success registered α- and β- particles, as it was described in Geiger (1924). This device became in history named the "sharper counter". The counter consisted of a metal case in which on an insulator a pointed core was strengthened. Ionizing particles got in the case through a window closed by a thin foil before an edge. The counter case was connected to the negative pole of a battery of high voltage, and the core was connected with an electroscope, was grounded through big resistance. While the potential difference was less than a certain critical value, the counter worked as a usual ionization chamber, but at the increase in the potential difference above the critical value near to the edge (there where intensity of electric field was the greatest) the formed in counter gas electrons gained kinetic energy, sufficient for ionization of molecules of gas.

The new electrons thus arising also were accelerated by an electric field of the counter and in turn formed new ions. The avalanche of ions therefore got big electric conductivity, and a discharge happened. Thus, even for the small amount of the ions formed by the ionizing particle, the discharge in the counter can be easily registered with the help of the electroscope. By the time of flight of the following particle, the electroscope has time to be discharged completely through resistance on the earth. Ability to register separate particles also made the basic value of the counter offered by H. Geiger. Its weakness was that it registered only the particles lying near an edge, and therefore the device did not receive wide application in researches of radiation of such small intensity, as CR.

The new type of the counter offered in 1928 by H. Geiger and his pupil W. Muller (1928) was sensitive practically in its entire volume, and was more reliable in work than the "sharper counter". The Geiger-Muller counter (see Figure 20) represented a metal tube T along the axis of which a thin metal wire W was tensed. The tube was pumped out and filled with gas under pressure of about one tenth of atmosphere. During operating time on the wire, the positive potential concerning walls of the tube was supported and working electrical potential of the counter usually got out in the range from 1000 to 1500 V.

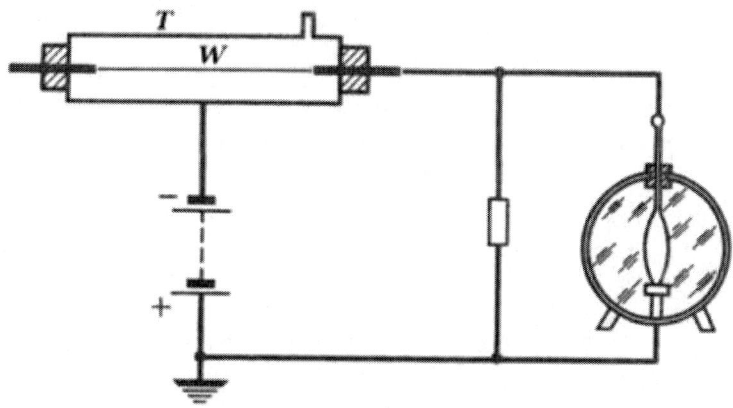

Figure 20. The scheme of Geiger-Muller counter. According to Geiger and Muller (1928).

The working principle of this counter was the same as that of "sharper counter", i.e. passage through the counter of even one ionizing particle causes in it a discharge. The Geiger-Muller counters had a number of advantages: they were simple to produce, they were sensitive practically in their entire volume, and they could have varied sizes, and therefore they could create a new era in the physics of CR. It is interesting that H. Geiger and W. Muller (1928) specified for the first time on the important possibility of application of their counters for research in CR.

25. IMPORTANCE OF USING GEIGER-MULLER COUNTERS IN CR RESEARCH: INVESTIGATION OF ENERGETIC CHARGED PARTICLES IN CR

Really the significance of Geiger-Muller counters was realized when W. Bothe and W. Kolhörster (1929), working in Germany at Physic-Technical Institute (in the same place where H. Geiger and W. Muller worked), noticed that in the counters placed at small distances one over another, discharges often occur simultaneously. Discharges of counters were registered by means of two electrometers on the same moving photosensitive film. Therefore it appeared possible to define the number of simultaneous deflections of threads of electrometers. As the distance between counters increased, the frequency of simultaneous discharges (or, as we talk now, coincidences),- decreased. Bothe and Kolhörster (1929) came to the conclusion that the coincidences could not be casual. Moreover, the probability that coincidences in counters could be caused by photons, realized two consecutive Compton collisions, is insignificantly small. Bothe and Kolhörster (1929) came to prepared from radium the final correct conclusion that "observable coincidences are caused by the passage of the same ionizing particle through both counters". The assumption that there are usual α- and β- rays from radioactive substances it was necessary to reject because they would be absorbed in the counter walls (1 mm of zinc).

The received result in and of itself did not contradict the point of view that CR consist of high-energy photons because the ionizing particles causing coincidences could appear as

Compton electrons, arising at absorption of high-energy photons in the atmosphere. According to theoretical calculations of those years, it was considered that the energy of hypothetical primary photons lies in the limits from 20 to 150 MeV. Compton electrons created by such photons Compton could have energy more than sufficient to penetrate through walls of the counters, however obviously insufficient to cause coincidence at the presence between counters Z_1 and Z_2 of absorbers of a considerable thickness. To check up such a possibility, Bothe and Kolhörster (1929) placed between counters in the beginning lead plates of various thicknesses, and then a gold plate A with a thickness of 4.1 cm (see Figure 21).

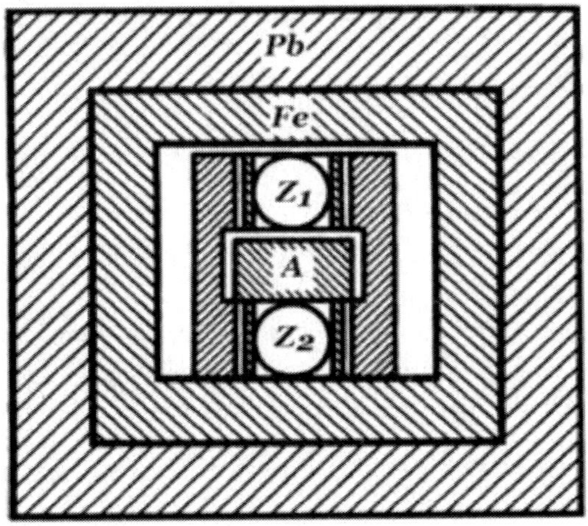

Figure 21. The scheme of experiments of Bothe and Kolhörster (1929).

The result was simply amazing. It appeared that the number of coincidences decreases very slightly: At the presence between counters of a gold plate, the number of coincidences decreased only by 24 %. In other words, 76 % of the charged particles which were registered by the device passed through the plate of gold at a thickness of 4.1 cm. In the opinion of Bothe and Kolhörster (1929), not only Compton electrons have such big energy. They came up with the idea that coincidences can be caused by charged high-energy particles that are some part of CR. "Now the basic problem consists in that, - wrote W. Bothe and W. Kolhörster in their work, - whether corpuscular radiation is secondary from γ-radiation as it is considered to be, or it is high-rise radiation". The factor of absorption of the charged particles defined by the method of coincidences appeared equal to $(3.5\pm0.5)\ 10^{-3}$ cm²/g and practically did not differ from the factor of absorption of the CR defined from curves of absorption, received by ionization chambers. To W. Bothe and W. Kolhörster did not remain any more doubts that "high-altitude radiation has most likely a corpuscular nature". By estimations of Bothe and Kolhörster (1929), the energy of the charged particles necessary for passage through the atmosphere reaches from 10^9 to 10^{10} eV.

Besides, Bothe and Kolhörster (1929) for the first time came up with the idea that if CR at least partially consist of charged particles, as a result of the rejecting action of the magnetic field of the Earth, intensity of CR at the poles should be much greater than at the equator, i.e. there must exist a latitude effect of CR. Detection of the latitude effect would be the experimental proof of the hypothesis about the corpuscular nature of CR. At those times

Bothe and Kolhörster (1929) knew that J. Clay, during the voyage from island Java to Holland, found a reduction of cosmic ray intensity around the equator approximately of 10 % (we will consider this experiment in detail a little later), and they hoped that "the last measurements of J. Clay, probably, specify the existence of this effect".

Thus, even though the common point of view at that time was that CR consists only of high-energy photons, there became a hypothesis of Bothe and Kolhörster (1929) about the corpuscular nature of CR. Let us underline that the experiments of Bothe and Kolhörster (1929) which opened a new page in the physics of CR became possible thanks to the invention of Geiger-Muller counters, allowing registering separate particles of CR.

Against conclusions of Bothe and Kolhörster (1929) there was one serious objection: interpretation of their experiments was based on absolutely any assumption that well-known properties of photons and electrons of low energies are fair and in the field of big energies.

Actually, the energy of photons of CR could appear much more than that which was calculated based on their factor of absorption on the basis of the Klein - Nishina formula, and correctness of this formula was established only in the area of energies of order of several MeV. In this case, secondary electrons could have energy sufficient for passage through a gold plate between counters. However, this objection arose later - after discovery of CR "showers". Prior to the beginning of the 1930s, it seemed doubtless that CR can be treated by analogy to usual radioactive radiation and that distinction came only in the scale. The question was reduced, as it seemed, to a choice between two possibilities: primary CR are either "ultra-γ rays" or "ultra - β-radiation".

Despite the fact that it was impossible to consider the results of Bothe and Kolhörster (1929) as experimental proof of the corpuscular nature of CR, the paper became a major landmark in the history of studying CR because they were the first attempt to experimentally study the nature of CR. The question of the nature of CR became the basic problem of physics of CR for many of the next years.

Now it is clear that both hypotheses about the nature of CR were far from true: though CR also consist of a stream of charged particles, the behavior of these particles during propagation in the atmosphere is far from that considered at that time.

26. Improving of Coincidence Method and Discovery of Soft and Hard Components in CR

The method of registration of coincidences was improved by B. Rossi (1932a), who worked at that time in Germany in the Laboratory of W. Bothe at Physic-Technical Institute. The scheme of coincidences on the electronic lamps, offered by B. Rossi (see Figure 22), strongly reduced the number of casual coincidences registered by three Geiger-Muller counters: G1, G2 and G3.

As can be seen from Figure 22, the simultaneous discharges in all three Geiger-Muller counters led to all three lamps: I1, I2 and I3 - ceasing to spend and a current through resistance R at once stopped. The potential of a point A changed jumpily to value of positive potential of the battery, and this jump registered by the voltmeter or caused the mechanical counter to work, signaling that there was a threefold coincidence. One of essential advantages of the scheme offered by B. Rossi, consisted that it could be applied to registration of coincidence in any number of counters.

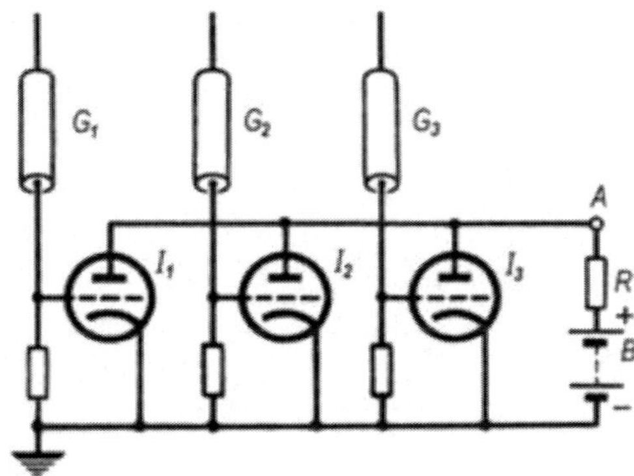

Figure 22. The scheme of coincidences on the electronic lamps, offered by B. Rossi (1932a).

B. Rossi (1932a) set for himself the problem to prove the big incidence of the charged particles which are a part of CR. For this purpose he arranged three counters, the coincidence included in the scheme, vertically one over another (the so-called telescope of CR) and investigated absorption of the particles causing coincidence in the lead plates of various thickness placed between counters (see Figure 23).

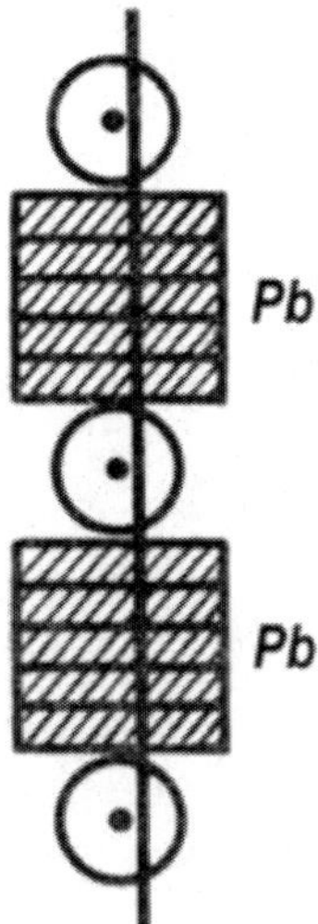

Figure 23. The scheme of installation of B. Rossi (1932b) for demonstration of penetrating ability of CR.

At first the total thickness of Pb between counters was 25 cm, and then it was increased to 1 m. Results of measurements are presented in Figure 24. In Figure 24 on the axis of abscissas, - the thickness of lead between counters is presented, on the axis of ordinates - the account rate of coincidences N in time units is presented. From the curve, it is visible that 25 cm of lead absorbs less than half of cosmic ray particles, and a considerable part of them at sea level is capable of getting through 1 m of lead. This result caused big surprise, as "energy of CR in this case should make billions eV (as it was known, the relativistic single charged particle lose about 1 MeV at crossing 1 g/cm^2 of lead; that gives for 1 m of lead minimum energy of about 10^9 eV). "These, my early experiments, - Rossi (M1966, page 45) recollected in his book, - only have strengthened my feeling of surprise, almost awe of the new phenomena which only started to appear before us. I have been convinced that CR can appear essentially new kind of radiation if only properties of photons and the charged particles do not change radically at the increase in their energy"

Such huge energy, certainly, could not be allocated at synthesis of heavy elements, and it was necessary to refuse definitively the beautiful theory of cosmic ray origin offered by R. Millikan. The received curve of absorption (see Figure 24) led B. Rossi (1933) to the conclusion that CR consists of two various components: a soft, quickly absorbed component, and hard, penetrating component.

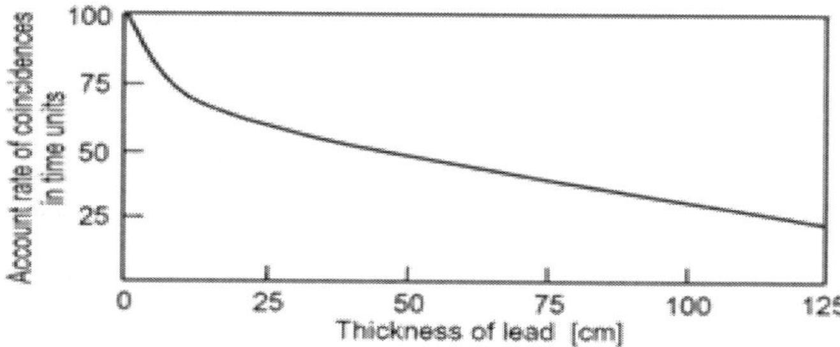

Figure 24. A curve of absorption of cosmic ray particles in lead. From Rossi (1932a).

These components were not accurately differentiated; soft were those cosmic ray particles that were absorbed by about 10 cm of lead, and what passed through this layer of lead were considered hard. It was established that the degree of absorption of the hard component by various substances is approximately proportional to the weight of the substance per cm^2, and the absorption of the soft component seemed abnormally great (as it became clear later, phenomenological separation of CR into soft and hard components mostly corresponds to the separation by nature of energetic particles).

Besides, Rossi (1932b) decided to investigate whether CR really form secondary particles in substances, as followed from the experiments of D. Skobelzyn and L. Myssowsky. For this purpose three counters were located in a triangle and surrounded from different directions by lead. Thanks to such an arrangement, simultaneous operation of counters could not be caused by one particle moving on a straight line. Nevertheless, the installation registered about 35 coincidences per hour. Coincidence could arise only when two or more charged particles simultaneously took off from lead, and B. Rossi explained the results by the generation in lead of some secondary particles under the influence of CR. At removal of the top part of the lead screen, as one would expect, the number of coincidences sharply fell, but remained equal to ten coincidences per one hour. This effect B. Rossi could not explain, and only later it became clear that it was caused by the formation of secondary energetic particles in the atmosphere. In the early 1930s was known only the Compton effect of γ-ray, the result of which was the formation of only one electron. Therefore, on the basis of theoretical representations at that time, it was impossible to explain the occurrence of a great number of the secondary particles observed during experiment: it must be a unique unknown process of interaction of photons of high energy with substance. Results of my experiment, - B. Rossi wrote, - have seemed so improbable to Editors of scientific magazine where I have presented in the beginning the article that they have refused to print it. Later article has been accepted by other magazine (Rossi, M1966, page 49).

27. Secondary CR and Energy Loss by γ-Rays on Compton Scattering and Charged Particles on Ionization

The described above experiments of B. Rossi began a variety of works on studying of secondary CR, and further research showed that the interaction of CR with substance is accompanied by a much more difficult and unusual phenomenon than it was possible to assume by the first experiments of B. Rossi. Formation of secondary particles in substance meant that the considerable part of CR observed near sea level arises in the atmosphere and strongly differs from the primary CR coming in from the border of atmosphere. In the beginning, many scientists were inclined to consider that the hard component is the primary CR reaching the surface of the Earth, and the soft component - secondary radiation. But most likely there was an assumption that only an insignificant part of primary particles could pass through the entire atmosphere without having interacted with it, and it is possible that such particles did not exist at all. Thus, all particles registered at sea level should have a secondary origin, and the structure of secondary radiation could be hard enough that even all primary cosmic ray particles would be similar.

Really, within the next two decades it was revealed that into the structure of secondary CR in the atmosphere entered not only all earlier known kinds of radiations and particles predicted by theorists, but also many new unknown particles. To know about the nature of primary CR, study at sea level and mountains of secondary particles, it was necessary to know well the mechanism of absorption of CR in air and other substances. "Discoveries of the last years, - D.V. Skobelzyn told in the report at the 1st All-Union Nuclear Conference which took place in September, 1933 in Leningrad, - have led to the problem on the nature of cosmic "ultra-radiation" and the phenomena connected with its penetration through the material environment, be put now in the forefront and occupies one of the central places among actual problems of so-called nuclear physics (Skobelzyn, 1934, page 65).

Until 1930-1932, representation was common that primary CR are photons of big energies and that absorbed during propagation in atmosphere mostly by Compton scattering. The calculations made by Klein and Nishina (1929) and Tamm (1930, 1975) lead to the following expression for probability of dispersion of a photon with energy E in substance:

$$\sigma_{Comp} = \pi n Z r_0^2 \frac{mc^2}{E} \left[\ln\left(\frac{2E}{mc^2} \right) + \frac{1}{2} \right], \tag{5}$$

where $r_0 = e^2 / mc^2$ is the "classical radius of electron", Z - the serial number of atoms of substance, n - number of atoms in 1 cm^3 of substance, m and e - weight and charge of electron, c - velocity of light. Probability of Compton scattering, according to Eq. 5, as a first approximation, is inversely proportional to the energy of a photon.

In the case in which it was considered that CR consists of charged particles, the absorption occurs mainly by energy loss on ionization. For the first time calculation of the energy loss by a moving charged particle on excitation and ionization of atoms of the environment was made by N. Bohr (1915). Applying classical electrodynamics, N. Bohr showed that energy losses quickly decrease at the increase of energy of a particle and aspire to a constant at relativistic energies. The big penetrating ability of CR could be attributed therefore their high energy. If losses of energy by a particle are caused only by ionization, as energy losses single charged particles in the air make nearby 2×10^6 eV/(g/cm^2), only a particle with energy bigger than 2×10^9 eV could pass through the atmosphere. Bohr's formula really was not applied to the phenomena connected with CR because prior to the beginning of 1930s, the opinion was standard that CR is a γ-radiation.

H. Bethe (1930) made quantum-mechanical consideration of losses of energy by the charged particle on ionization almost irrespective of the problems of the physics of CR. This results from the fact that during this period the theorists working in the field of quantum mechanics dealt with structure problems of the atom, and also were skeptical about the possibility of the application of quantum mechanics in the field of relativistic energies. According to calculations of H. Bethe (1932) and F. Bloch (1933), it was received that the average energy lost on 1 cm of a way by a particle of a charge $Z'e$ moving with speed v, is described by the following expression:

$$\left(\frac{dE}{dx}\right)_{ioniz} = \frac{2\pi n Z\left(Z'e^2\right)^2}{mv^2}\left[\ln\left(\frac{2mv^2W}{I^2(Z)\left(1-\beta^2\right)}\right)-\beta^2\right], \tag{6}$$

where $I(Z)$ is the "average potential of ionization" atoms of substance with a serial number Z, W - the maximum energy which the flying particle can transfer to electron, and $\beta = v/c$. Application of laws of conservation of energy and impulse at elastic impact for a relativistic case gives the following expression for the maximum energy W:

$$W = \frac{2mp^2}{M^2 + m^2 + 2(m/c)\left(p^2 + M^2c^2\right)^{1/2}}, \tag{7}$$

where M - weight of a particle, p - a particle impulse. From here it is visible that for the given environment of energy loss on ionization by the charged particle are functions only of its charge and speed. For the non-relativistic case at $\beta \ll 1$, ionization losses are inversely proportional to the square of speed (or energy) of a particle. In the field of big energies, ionization losses slowly grow with the energy of a particle. Thus, the theory of ionization losses well explained the behavior of the hard component of CR.

28. Secondary CR and Bremsstrahlung Radiation Losses by Energetic Charged Particles

At the passage of the charged particle near an atom kernel between them there is an interaction caused by Coulomb forces. This leads to changes in the speed of movement of the particle, and any non-uniformly moving charge, according to laws of classical electrodynamics, radiates electromagnetic waves. An example of such bremsstrahlung radiation is the continuous X-ray spectrum which is let out by electrons at their braking on the anticathode of an X-ray tube. Usually it was considered that the energy lost to electrons on bremsstrahlung radiation is small in comparison with the energy lost on ionization. However, H. Bethe and W. Heitler (1934), working in England, published results of calculations of radiating braking of electrons which led to an absolutely different conclusion. Using quantum theory and the relativistic equations, they showed that the probability of bremsstrahlung radiation by an electron with energy E of a photon with energy E' at the passage of a layer of substance of thickness of 1 cm has the following expression:

$$W_e(E,E')dE' = 4n\alpha Z^2 r_0^2 \frac{dE'}{E'}\left\{\left[1+\left(1-\frac{E'}{E}\right)^2\Phi_1-\left(1-\frac{E'}{E}\right)\Phi_2\right]\right\}, \tag{8}$$

where $\alpha = e^2/\hbar c = 1/137$ ($\hbar = h/2\pi$, and h is the Planck's constant).

The formula (3.4) is fair in the assumption that energy of an electron $E \gg mc^2$. Functions Φ_1 and Φ_2 describe the shielding of the electric field of an atom kernel by a field of orbital electrons. At full shielding

$$\Phi_1 = \ln\left(191\times Z^{-1/3}\right), \quad \Phi_2 = \frac{2}{3}\ln\left(191\times Z^{-1/3}\right)+\frac{1}{9}, \tag{9}$$

and Eq. 8 is possible to write down in the following form:

$$W_e(E,E')dE' = 4n\alpha Z^2 r_0^2 \frac{dE'}{E'}\left\{\left[1+\left(1-\frac{E'}{E}\right)^2-\frac{2}{3}\left(1-\frac{E'}{E}\right)\right]\times\ln\left(191Z^{-1/3}\right)+\frac{1}{9}\left(1-\frac{E'}{E}\right)\right\}. \tag{10}$$

It is easy to see that the expression in square brackets is close to unity at any possible ratios E'/E. Therefore, $W_e(E,E')dE'$ is approximately

$$W_e(E,E')dE' \approx 4n\alpha Z^2 r_0^2 \ln\left(191Z^{-1/3}\right)\frac{dE'}{E'}. \tag{11}$$

If we introduce the designation

$$1/t_0 = 4n\alpha Z^2 r_0^2 \ln\left(191Z^{-1/3}\right), \tag{12}$$

then instead of Eq. (3.7), there will be

$$W_e(E,E')dE' = \frac{1}{t_0}\frac{dE'}{E'}.$$

(13)

The value t_0, determined by Eq. 12, has dimensional lengths, and received the name radiating unit. The total losses of energy by electrons owing to bremsstrahlung radiation are defined by the following expression:

$$\left(\frac{dE}{dx}\right)_{bremsstr} = -\int_0^E E'W_e(E,E')dE'.$$

(14)

From Eqs. 11 and 14 follows that at the given quantity of substance on the unit of the area the loss on bremsstrahlung radiation in a substance with big serial atomic number Z is much more than in a substance with a small Z. This is easy to understand as a rejecting force which operates on a particle which is passing by an atomic nucleus, and it is proportional to a kernel charge, i.e. its serial atomic number.

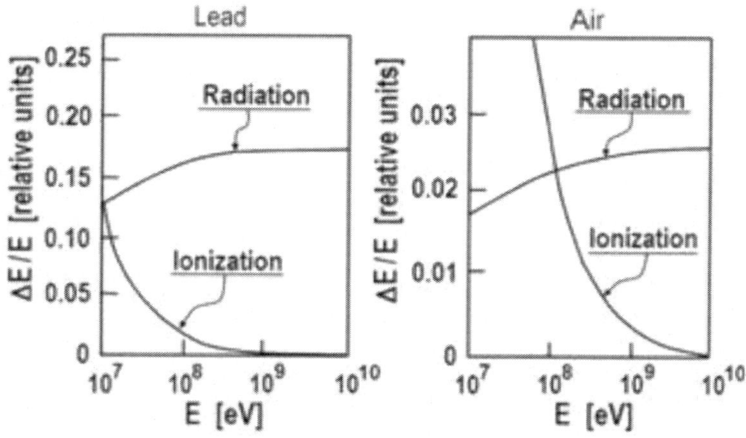

Figure 25. Ionization losses of energy and energy loss on bremsstrahlung radiation by electrons in lead and air depending on their kinetic energy E.

Losses on bremsstrahlung radiation quickly increase with electron energy and eventually surpass energy losses on ionization. The value of energy of a particle at which losses of both kinds become identical is called as critical energy. As energy losses on bremsstrahlung radiation are proportional to Z^2, and energy losses on ionization are proportional Z, the relative role of losses of energy on bremsstrahlung radiation by that more than is more serial atomic number of environment. In Figure 25 energy losses on ionization and bremsstrahlung radiation by electrons depending on their kinetic energy are shown. The absorbing environment in one case is lead, and in the other - air. From Figure 25 it is visible that critical energy for lead is equal to ~ 10 MeV, and for air ~ 100 MeV.

29. SECONDARY CR AND ENERGY LOSSES BY γ-RAYS FOR GENERATION OF ELECTRON-POSITRON PAIRS

As to absorption in a substance of CR having photon nature, the same work of H. Bethe and W. Heitler (1934) resulted also surprisingly at the first sight results of calculation of the process of formation of pairs.

The theory of P.A.M. Dirac (1926), using terminology of those years, foretold that a photon of high energy flying by near an atomic nucleus pulls out an electron from a level of negative energy and forms a usual electron with positive energy and a "hole" in the continuous background conditions with negative energy, which is a positron. In other words, the photon generates the pair of an electron-positron. According to the calculations of Bethe and Heitler (1934), the probability of formation by a photon with energy E' in the field of an atomic kernel with a charge Z pair, consisting of a positron with total energy E and electron with energy $E' - E$, carried to 1 cm of substance, is equal to

$$W_p(E',E)dE = 4n\alpha Z^2 r_0^2 \frac{dE}{E'}\left\{\left[\left(\frac{E}{E'}\right)^2 + \left(1 - \frac{E}{E'}\right)^2\right]\Phi_3 + \frac{E}{E'}\left(1 - \frac{E}{E'}\right)\Phi_4\right\}, \tag{15}$$

Functions Φ_3 and Φ_4 reflect a shielding influence. At full shielding the approached formula turns out to be

$$W_p(E',E)dE = \frac{7}{9}\frac{1}{t_0}\frac{dE}{E'}. \tag{16}$$

The parameter t_0 has the same name and value as in the formula Eq. 12. In the beginning the photon energy $2mc^2 \approx 1$ MeV at which the process of formation of pairs becomes possible, the probability of formation of pairs in a certain layer of substance quickly increases with energy of the photon, and then becomes almost constant. Besides, calculations have shown that for a photon of energy E' the probability of formation of a pair (but with the same quantity of substance on area unit) quickly increases in various substances with growth of the serial atomic number Z of substance. After the occurrence of exact calculation of the process of formation of pairs, which was made by H. Bethe and W. Heitler (1934), it became clear that only photons of small energy are absorbed in a substance mainly by Compton scattering.

At big energies of photons, the probability of Compton scattering is small, and the absorption occurs mainly by the process of formation of pairs. The probability for a photon to take the Compton-effect or to form electron-positron pairs at the passage through lead or air 1 g/cm^2 depending on the energy of the photon E' is presented on Figure 26.

From Figure 26 it can be seen that the critical energy at which the probability of both processes are equal to each other is about 5 MeV in lead and about 20 MeV in air.

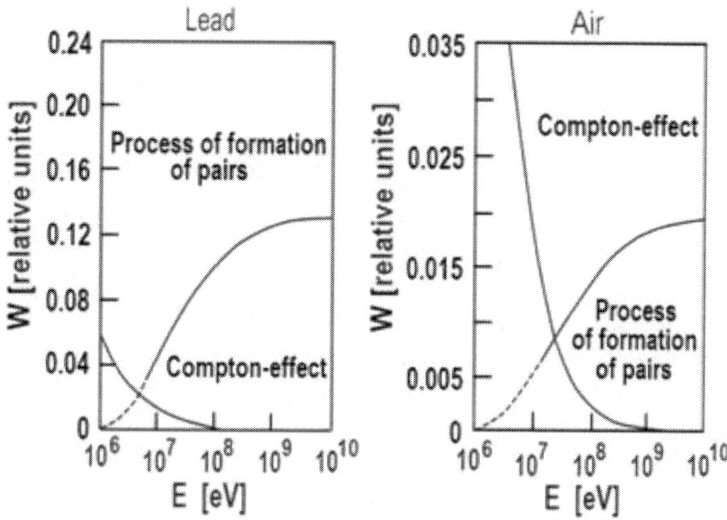

Figure 26. Probability W for a photon to test Compton-effect or to form electron-positron pair at the passage of lead (left panel) or air (right panel) of 1 g/cm^2 depending on photon energy E'.

The theory of H. Bethe and W. Heitler (1934) gave the chance to answer at last the question that very much excited physicists for a long time. Abnormal absorption of soft component CR in substances with big Z spoke to energy losses on bremsstrahlung radiation and the process of formation of electron-positron pairs that strongly increased with increasing Z. The picture of absorption of charged particles in heavy and light substances appeared essentially varied. This specific distinction between heavy and light substances was a new feature of high-energy particles.

30. DIFFICULTIES IN UNDERSTANDING THE NATURE OF PRIMARY CR

When H. Bethe and W. Heitler (1934) developed the theory described above, thanks to detection and research of the CR latitude effect, it was established that the majority of primary CR consist of the charged particles. Physicists at that time were assured that all charged particles of CR are electrons and positrons. But in this case there was the question of what energy particles of CR should possess to pass through even 1 m of lead. After all, if the theory of Bethe and Heitler (1934) were true, only electrons with absurdly big energies can pass through such a layer of an absorber. Besides, it became known that C. Anderson observed in Wilson's chamber charged particles with energy near 3×10^8 eV which was lost in the substance, a considerably smaller part of energy than was predicted by the theory. Guarded also that circumstance that theoretically calculated lengths of run electrons and positrons of big energy appeared much less experimentally certain in a run of space particles. So, for example, at passage of an electron through a thickness of the terrestrial atmosphere it should, according to the theory, lose energy equal to 10^{13} eV, whereas research of the latitude geomagnetic effect showed that primary cosmic ray particles possess energy of order 10^{10} eV, i.e. one thousand times smaller. One more difficulty consisted in that the theory was powerless to explain the mechanism of formation of the cosmic ray showers discovered in 1933.

Both the big penetrating ability of CR, and existence of cosmic ray showers represented such contradictions to the theory that H. Bethe and W. Heitler came to the conclusion that "the quantum theory is obviously unfair for electrons such high energies". However after a while, namely in 1937, there was developed a cascade theory which explained the mechanism of formation of showers in CR by the account of the process of generation of electron-positron pairs by photons and radiation of photons by electrons in the bremsstrahlung process in the Coulomb field of an atom kernel. It created a subsequently apparently incorrect representation that the circle of the questions connected with the passage of CR through any substance can be completely explained within the frames of the existing electromagnetic theory with taking into account ionization losses, Compton scattering, generation of bremsstrahlung radiation and electro-positron pairs.

31. The Problem of Primary Cosmic Ray Nature: Understanding of Necessary Research in Stratosphere

Already in the early 1930s, after it appeared that at the passage through a substance CR form secondary particles, physicists came to the conclusion that the fullest representation of CR can be received, investigating it before it will undergo the most complicated changes in the atmosphere, which on the absorbing action is equivalent 10 m of water. Making supervision at the terrestrial surface, it is possible to receive data only about a small part of the CR filtered by atmosphere." Therefore, - wrote A.F. Ioffe (1935), - the permission of a problem of studying of CR is closely connected with stratosphere research. That is the most important and characteristic for CR, exists, possibly, only in a stratosphere and only in its higher layers". Research of CR in the stratosphere was made in two ways: during lifting of devices with automatic registration on sounding balloons, and during flight of stratostat when the physicist-observer himself accompanied various measuring devices.

32. Experiments on Balloons of E. Regener in 1931-1934 up to an Altitude of 28 Km by Automatic Ionization Chambers

In 1931-1933 E. Regener (1932) organized flights of a device with automatic registration on sounding balloons to investigate intensity of CR up to a height of 28 km. For this, E. Regener changed a little the device applied to underwater measurements: the electrometer was brought in a spherical ionization chamber with a volume of 2.1 l, filled with air under a pressure of 6 atm. The appearance of the device is shown on Figure 27.

Electrometer indications through certain time intervals were photographed on a motionless photographic plate. Besides, pressure and air temperature at flight were registered automatically by means of a special adaptation consisting of a ring on which was fixed aneroid A and bimetallic plate B, bent at temperature change (see Figure 28). Both devices in Figure 28 were supplied easily by the sagittary system, which came to an end by straightedges C and D, which at change in temperature and pressure moved upwards and downwards along a photographic plate surface. The shade from the rulers crossed the shade from the electrometer thread at the bottom (straightedge C) and top (straightedge D) photographic plate parts, that gave the chance to read the value of temperature and pressure for each position of the thread of the electrometer.

Figure 27. Appearance of the device of E. Regener for high-altitude measurements.

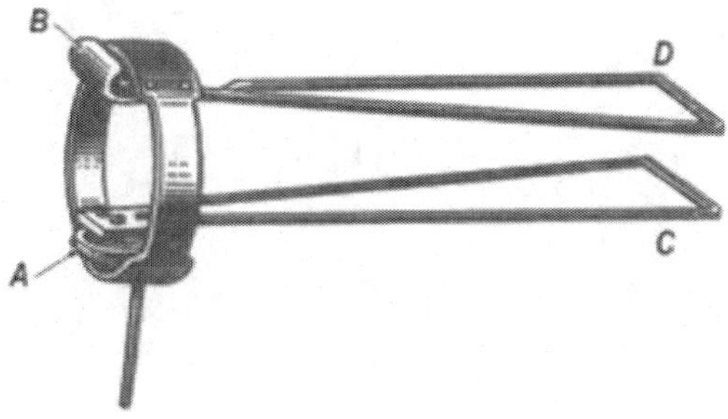

Figure 28. The device for automatic registration of pressure and temperature during the flight. From E. Regener (1932).

For protection from destruction at the moment of landing, the device, weighing only 1.5 kg, was surrounded with a rigid skeleton from above fitted by a transparent cellophane film which operated as a hotbed during the flight. The whole device was also protected from sharp fluctuations of air temperature in flight by an additional external skeleton. The equipment was suspended from several spherical balloons and rose upwards until one or two spheres, because of the expansion of the hydrogen filling them, burst. After that, the equipment smoothly returned to earth, with the remaining whole spheres playing a parachute role.

The results of E. Regener appeared unexpected: up to an altitude of 12 km (140 mm Hg), the ionization rate, i.e. the intensity of CR registered by the device, strongly increased with height growth, but above 12 km a retardation in the ionization rate increase was observed. From the height of 20 km, the increase in the rate of ionization was not registered any more, and it reached, as considered E. Regener, an almost constant limiting value. The maximum rate of ionization equaled 330 ion $cm^{-3}s^{-1}$, which almost by 150 times exceeded the rate of ionization created by CR at sea level. E. Regener understood that at such substantial growth of ionization rate the electrical potential, which at the terrestrial surface provided a saturation current, appeared insufficient at equipment lifting at big altitude (because at big cosmic ray intensity some part of ions will be recombined with electrons which will lead to some smaller value of measured electric current, i.e. – to visible smaller ionization rate; this means that the real cosmic ray intensity will be even bigger than that obtained by E. Regener).

Therefore, in 1933 and 1934 he made some flights with an open ionization chamber, which was connected through a filter with external air and worked, hence, at the pressure equal to external atmospheric pressure (Regener and Auer, 1934). The open chamber with a volume of 194 l and having a length of over one meter was made from thin aluminum of thickness of 0.3 mm (see Figure 29).

The electrometer in the open ionization chamber was located separately from the chamber and was connected to the isolated core located in the chamber, and ahead of the chamber settled down the cartridge with film and contact hours. The whole device weighed about 6 kg, i.e. was four times heavier than the device with the closed ionization chamber. Having increased the size and number of the bearing balloons, E. Regener was able to lift this bulky device to a height of over 15 km. Results are shown in Figure 30.

Figure 29. Appearance of the open ionization chamber of E. Regener and R. Auer (1934).

However, E. Regener noticed that, unfortunately, it was impossible to completely rely upon his results, as there was no confidence that the temperature, which makes a strong impact on instrument readings, was precisely considered.

In two cases, the internal part of the ionization chambers was covered by a layer of substance containing hydrogen (celluloid in one case and paraffin in other), in order to find out whether neutrons are a part of CR. E. Regener believed that if as a part of CR there are neutrons in the chamber covered with a layer of a substance containing hydrogen, the considerable additional ionization caused by fast protons, neutrons beaten out from this layer will be observed.

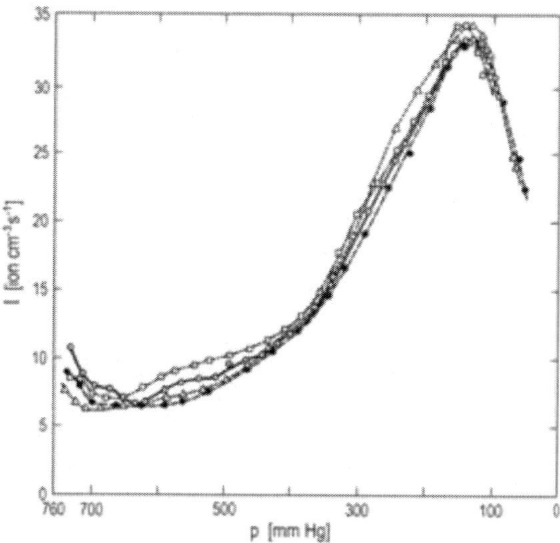

Figure 30. Results of high-altitude measurements of the ionization rate *I*, received by E. Regener and R. Auer (1934) by means of the open ionization chamber during 4 flights.

Having compared the data received with the help of ionization of the chamber covered with paraffin, and the data received in the chamber without paraffin, he showed that in the first case the ionization was higher by 2-3 %. E. Regener fairly considered that experiment errors exceed this difference, and has made the conclusion that if "neutrons exist in CR, they do not make its considerable part". Let us note that a few years later, by L. Rumbaugh and G. Locher (1936), using photo-plates covered with paraffin, they could find out the presence of neutrons in the upper atmosphere – these neutrons were part of secondary CR.

33. USING ON BALLOONS AUTOMATIC SYSTEMS OF GEIGER-MULLER COUNTERS IN COINCIDENCES AND POSSIBLE CORPUSCULAR NATURE OF PRIMARY COSMIC RAYS

In addition to measurements of intensity of CR by ionization chambers, E. Regener and his employee G. Pfotzer made for the first time at high-altitude experiments on balloons with automatically working systems of Geiger-Muller counters in coincidences (Regener and Pfotzer, 1934). In this case, impulses from the counters, through a special relay operated on the mechanism of hours and number of particles passing through the counter, it was possible to count directly under the indication of the clock's arrow. The clock was established on a rotating plate, and its dial was photographed through certain time intervals. Distinctions in the curves of absorption received by ionization chambers (see previous Section 4.2) and by Geiger Muller counters were not revealed. The absence of an increase of intensity of CR above 20 km obtained before by E. Regener (1932) explained that in its structure there are no very soft components. Comparing the data received under water at depths up to 230 м and at various altitudes in the stratosphere up to 28 km, E. Regener (1933 came to the conclusion that CR consists of at least five components, possessing various factors of absorption. Though the calculations of E. Regener concerning the structure of CR already at that time caused mistrust of many researchers, its qualitative conclusions were indisputable: first, the structure of CR at the terrestrial surface and in the stratosphere is absolutely distinguished; secondly, the CR coming in from the border of the atmosphere already has a complicated structure. The basic objection caused E. Regener to persistently consider all results received by him as confirming the photon nature of CR. On the contrary, as noticed in the review devoted to the discussion on the nature of CR, Clay (1932) wrote, "results of measurements in the stratosphere contradicted the concept about the monochromatic photon primary radiation coming on the border of atmosphere and forming in it corpuscular fluxes of high energy more likely, and testified to the corpuscular nature of primary CR".

In general, in those years, physicists aspired to deal with unequivocally the question of the nature of CR: either it is photons, or charged particles. A main role the opening of the latitude effect played was that it forced the majority of researchers to recognize the correctness of the hypothesis about the corpuscular nature of CR. Only R. Millikan with his employees and the group of E. Regener denied the existence of the latitude effect, explaining the change of intensity with meteorological effects, and firmly kept representations about the photon nature of CR almost to the middle of the 1930s.

Bewilderment was caused also by the course of the curve of the absorption received by E. Regener above 20 km, as it seemed doubtless that the ionization rate should increase up to the atmospheric border, instead of remaining constant.

34. COSMIC RAY RESEARCH IN THE STRATOSPHERE ON STRATOSTATS WITH VISUAL OBSERVATIONS

The data received by measurements on stratostats with visual observations confirmed the results of E. Regener. First of all it was measurements of Belgian physicists A. Piccard and M. Cosyns (M1933) who designed a hermetically closed gondola from aluminium and in 1931-1932 made two flights on a stratostat. The measuring equipment which could be used in this case differed in its big weight and dimensions from the equipment of E. Regener. In the first flight of A. Piccard and M. Cosyns, they used the usual device of W. Kolhörster and an ionization chamber with pressure of gas of 11 atm, in which was incorporated a one-wire electrometer. Unfortunately, the chamber during flight leaked gas that considerably lowered the accuracy of results of measurements. The maximum height which managed to reach the stratostat, 16 km, and readout were made visually each 5-10 mines (final results are shown in Figure 4.5).

The values of intensity of CR given by A. Piccard and M. Cosyns (M1933) were well coordinated with those given by E. Regener (1932) (above 9 km), though values of intensity of the CR, measured by A. Piccard and M. Cosyns have appeared on 25 % more than those of E. Regener (the reason for this has not been found out). During the flight of A. Piccard and M. Cosyns, measurements were made also by Geiger-Muller counters. The increase in the counting rate in Geiger-Muller counters with height went in parallel with an increase in the ionization rate measured by ionization chambers. During the second stratostat flight, A. Piccard and M. Cosyns included counters in the scheme of coincidences, and it appeared that the number of coincidences grew with height more slowly than the ionization rate. This divergence caused bewilderment in the experimenters, and did not receive an explanation until 1935. Besides, an unsuccessful attempt to find out the direction of the maximum intensity of CR in the stratosphere was made.

35. THE EARLIER COSMIC RAY MEASUREMENTS IN THE STRATOSPHERE ON STRATOSTATS BY VISUAL OBSERVATIONS IN FORMER USSR

On September 30th, 1933, the first Soviet stratostat "USSR" reached heights of 19 km. The stratostat was constructed by the Management of Military-Air Forces of the USSR on original designs at the Soviet factories from domestic materials (results are shown in Figure 31).

The equipment for measurement of cosmic ray intensity was prepared by A.B. Verigo (1935, 1937), and consisted of W. Kolhörster's electrometer and an electrometer of the same type that V. Hess placed in the lead filter in which top cover acted in film. He was supposed to make measurements using the device without the filter and with the filter, which would allow judging the absorption of the soft component of CR. Besides, as A. B. Verigo believed, slow rotation of the gondola of the stratostat gave the chance to study the orientation of CR at the removed cover. As removal of indications of the devices was made visually, absence of the physicist-observer on the stratostat "USSR" and command employment allowed only partial execution of the planned research. During flight, engineer K.D. Godunov removed the indications of the devices. Analyzing the obtained results, A.B. Verigo (1935) drew the conclusion that "they are close to results of Professor A. Piccard and considerably exceed the results of Professor E. Regener that proves a regular divergence of a method of measurements on sounding balloons and on stratostat. It is necessary to include finding-out of the reasons of this divergence in problems of the future flights".

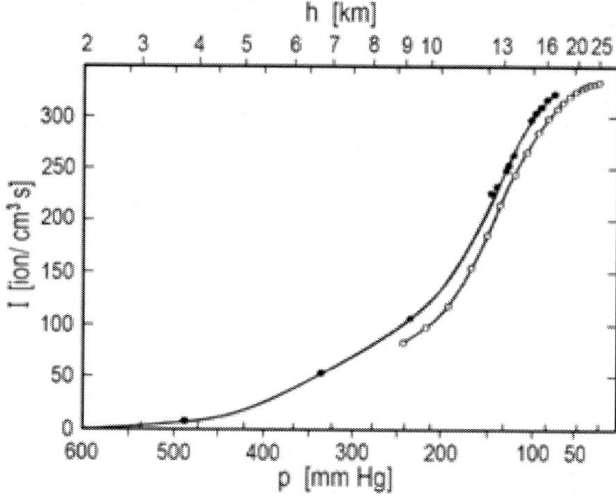

Figure 31. Results of measurements of A. Piccard and M. Cosyns (M1933), received during flights on stratostats (black points) in comparison with E. Regener (1932) on balloons (open circles).

Flight of the second Soviet stratostat "Osoaviakhim-1" occurring on January 30th, 1934, and terminated in an accident. The commander of the stratostat P.F. Fedoseenko, pilot A.B. Vasenko and the young Leningrad physicist I.D. Usyskin who was watching the equipment for research of CR were on the flight. The stratostate reached a then record height of 22 km, but during descent there was a failure, and the whole crew was lost. The ashes of the victims are buried in the Kremlin wall in Red Square in Moscow. The opening of the March 31st, 1934 All-Union conference on stratosphere study was devoted to the memories of the victims of the stratostat "Osoaviakhim-1"

The accident did not stop the Soviet physicists. During the next flight of the stratostat "USSR-1 bis" in June 1935, physicist A.B. Verigo was on board. "One of the primary goals of the flight, - he wrote (Verigo, 1937) - there was a studying of physical properties of CR". Working out of a measurement technique and manufacturing of devices was carried out by A.B. Verigo. Measurements during the flight were made simultaneously by five devices of various designs: Hess's two electrometers, two electrometers of Kolhörster's type and Wulf's electrometer. Besides, in the stratostat's gondola, two Wilson's chambers were placed. A.B. Verigo found a difference in character of change in intensity of soft and hard components of CR with height. "Owing to small absorption of the hard component, - wrote A.B. Verigo analyzing the received results, - its intensity in the stratosphere is a little more than at sea level, but the intensity of the strongly absorbed soft component of CR should increase considerably in the stratosphere, as is observed at flight". During the flight were registered considerable short-term (1-3 minute) intensity fluctuations with amplitude 10-12 %. A.B. Verigo considered them to be a characteristic of CR (let us note that actually these fluctuations were not true time variations of CR, and were caused by statistical fluctuations because of the small effective area of the used devices). During the descent of the stratostat, in order to decrease the speed, A.B. Verigo was compelled to jump out by parachute: the brave researcher was awarded the Order of Lenin.

36. Using Stratostat Observations in the USA: The Important Role of Inclined Cosmic Rays

Two flights on stratostats to which time measurements of CR were made are known. The first flight of the American stratostat "Century of Progress" took place on November 20th, 1933. Unfortunately, the received results so strongly differed from all previous measurements that they did not represent any interest. Most likely, the reason for this was the thick lead screen that surrounded the ionization chamber.

Besides, W. Swann and G. Locher (1935) made a flight on the stratostat of the National Geographical Society of the USA (because of damage to the cover, it rose only to a height of 12 km). Measurements which were made by the Geiger-Muller counters included in the scheme of coincidence confirmed the results of A. Piccard and M. Cosyns (M1933), and allowed to establish that though the intensity of CR in the vertical direction is at a maximum, it is impossible to neglect the CR coming even in directions close to the horizontal (at the height of 12 km their intensity was 20 %).

37. AUTOMATIC MEASUREMENTS OF CR BY GEIGER-MULLER COUNTERS ON SOUNDING BALLOONS: IMPORTANCE OF DATA RADIO-TRANSMISSION

Schemes of coincidence of Geiger-Muller counters represented a huge advantage over ionization chambers, as they did not differ in bulkiness and in big weight yet gave the chance to measure the intensity of CR in a particular direction. Therefore, they began to be used with success for automatic cosmic ray measurements especially on flights of sounding balloons. The method of sounding balloons had one essential lack: it was necessary to search for the device after it fell back to earth (this was not always simple) to see the photographic plate and to have the possibility to study the obtained data. Therefore, before research, there was the question of the creation of equipment which not only automatically would register intensity of CR, but also could transfer the results of measurements by means of a radio transmitter to an observer who is on the Earth. Such equipment was developed for the first time by S.N. Vernov (1934, 1935a,b). "For studying of CR, - S.N. Vernov (1938a) in the report at the 2nd All-Union Conference on an Atomic Nucleus, passing in 1936 in Moscow, told, - it is rather important to make measurements of their intensity in the stratosphere at various latitudes. However, use for these purposes of automatic devices in equatorial and polar latitudes is strongly complicated by that in these areas the probability to find devices after the flight is rather small. Therefore, I on the basis of radio-balloon-sounds of professor P.A. Molchanov developed the special installation transmitted radio-signals about intensity of CR from various heights. An installation body is the Geiger-Muller counter connected to the amplifier and a radio transmitter. Reception of signals is made on hearing by several observers". The device created by S.N. Vernov consisted of a battery of 1500 V, two Geiger-Muller counters and a three-cascade amplifier which strengthened the impulses coming from counters, and allocated from the total number of pulses only occurring in two counters simultaneously. From the amplifier a relay also worked, which switched on the transmitter. The receiver established on the Earth accepted signals from the transmitter. Installation gross weight was about 12 kg, and its size was 45 cm × 45 cm × 60 cm. Between counters was placed a layer of lead in the thickness of 2 cm, therefore coincidence could be caused only by particles with energy of more than 7×10^7 eV. From time to time the total number of pulses in one of counters was registered that gave the chance to register also photons.

For checking the reliability of the work of the installation, in July 1934 it was lifted by plane to a height of 5800 m. After the checking during the plane flight, the measurements were made by means of eight balloons filled by hydrogen (each in a volume of 5 м^3). Let us note that S.N. Vernov's offers on automation of data recording and their transfer by radio laid down the basis not only of the further measurements on sounding balloons, but also of cosmic ray research on geophysical rockets and later − on satellites. In Figure 32. is shown a young S.N. Vernov (27 years old) at work.

The flights on sounding balloons were made near Leningrad and Yerevan. Here is how the Armenian newspaper "Horurdain Ajastan" for November 10th, 1936 describes the flight of the new cosmic ray device: "On November 6th about two o'clock in the morning from the Yerevan airdrome is started up the automatic working device on radio-balloon-sounds for studying of CR which for 57 minutes has reached heights of 9 km. In this time from the device, the data about intensity of CR has been obtained by radio. Despite bad atmospheric conditions, the device worked well and broadcasted signals were accurately listened."

Figure 32. Sergei Nikolaevich Vernov at work (1937).

As in the previous experiments on stratostats, the intensity of CR measured by the method of coincidences of Geiger-Muller counters increased with height more slowly than the intensity received by ionization chambers. Originally this divergence was explained by the presence in the upper layers of the stratosphere of very soft radiation which was absorbed in the walls of Geiger-Muller counters, and it was not registered by method of coincidence. Besides, the results received by S.N. Vernov specified a reduction of the increase of cosmic ray intensity (or even maximum presence) at the height of about 10 km.

38. GROSS'S EXPLANATION OF THE DIFFERENCE IN RESULTS OBTAINED BY VERTICAL COUNTER TELESCOPE AND IONIZATION CHAMBER

Underlined above were the difference in results obtained by vertical Geiger-Muller counter telescope and ionization chamber on the dependence of cosmic ray intensity with altitude. The explanation was found by B. Gross (1933, 1934). First of all, it is necessary to take into consideration that an ionization chamber measures a high-altitude course of global intensity $I(x)$ of the CR obtained by the device registering in all directions. The method of Geiger-Muller counters included in the scheme of coincidence gives the chance to investigate the intensity of the CR coming from a vertical direction. It is natural that high-altitude curves for both cases will be essentially varied.

B. Gross (1933, 1934) showed that irrespective of assumptions of the nature of CR and the law of their absorption, knowing a high-altitude course of global intensity $I(x)$, it is possible to define a high-altitude course of intensity in the vertical direction $f(x)$. Assuming that the direction of primary CR at depth x and their connected secondary coincides with the initial direction of the primary CR on the atmospheric border, he received the following relation between $f(x)$ and $I(x)$:

$$f(x) = I(x) - x(dI(x)/dx), \qquad (17)$$

where x is the thickness of a layer of air from the border of atmosphere to the considered place (in g/cm^2). This relation also gives the connection between intensity of CR in a vertical direction and global intensity. As one would expect, the high-altitude course for global intensity and intensity in a vertical direction does not coincide. Differentiating Eq. 17, B. Gross obtained that

$$\frac{df(x)}{dx} = -x\frac{d^2I(x)}{dx^2}.$$

(18)

The high-altitude course of global intensity of CR, according to E. Regener (1932) and E. Regener and R. Auer (1934), had a distinct inflection point above 12 km. In any inflection point $d^2I(x)/dx^2 = 0$. Hence, according to Eq. 18 $df(x)/dx = 0$.

This means that at this depth is the maximum presence in a curve of a high-altitude course of vertical intensity of CR. Such a maximum can arise only in the case when primary CR, penetrating into the atmosphere, form secondary ionizing energetic particles. The assumption of E. Regener that primary CR penetrate into the atmosphere to a certain depth without absorption disappeared. Therefore, B. Gross calculated a curve $f(x)$ on a known curve $I(x)$ obtained experimentally. The inflection point in $I(x)$ should correspond to a curve maximum in $f(x)$.

39. Measurements of Cosmic Ray Vertical Intensity up to 29 km and the Discovery of a Maximum at an Altitude of about 15 km: Pfotzer's Curve

At last, G. Pfotzer (1936), an employee of E. Regener, managed to measure experimentally a high-altitude course of cosmic ray intensity in the vertical direction up to a height of 29 km. Using a vertically directed telescope consisting of Geiger-Muller counters, included in the scheme of the threefold coincidence, G. Pfotzer established the existence of an accurately expressed maximum at the height about 15 km (depth of atmosphere of 90 g/sm^2).

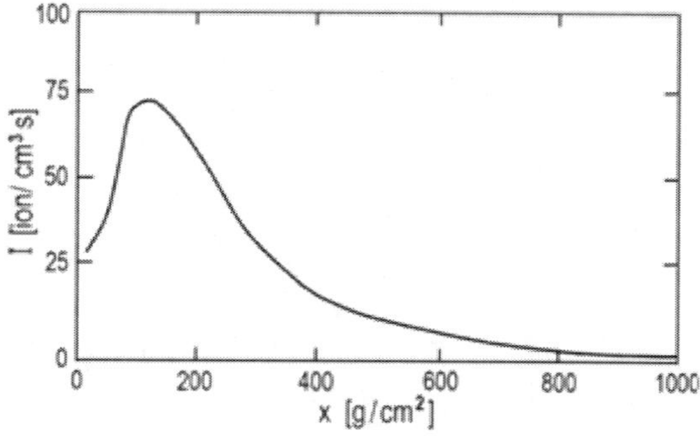

Figure 33. A curve of dependence of vertical intensity of CR on the depth of atmosphere x (in g/cm^2), obtained by G. Pfotzer (1936), - so called "Pfotzer's curve".

This maximum it was possible to explain by the presence only of the occurrence of secondary CR in the upper atmosphere. The "Curve of Pfotzer" - growth of intensity of CR with height up to a maximum near to the top border of atmosphere, and then its falling to the value of intensity of the primary stream from space (see Figure 33).

40. COSMIC RAY RESEARCH OF THE LATITUDE EFFECT IN THE STRATOSPHERE ON SOUNDING BALLOONS BY MILLIKAN'S GROUP AND BY S.N. VERNOV

Many flights with automatically working devices were realized by Millikan's group (Bowen et al., 1937, 1938) up to high altitudes in the stratosphere on sounding balloons at different latitudes (in Figure 34 are shown Robert Millikan with his group at the start of flight). At about the same time S.N. Vernov (1938, 1939) measured in the stratosphere on sounding balloons the cosmic ray latitude effect. These results of S.N. Vernov and Millikan's group will be considered in more detail below.

Figure 34. R.A. Millikan and his employees start sounding balloons. California, August 1935.

41. ON THE NATURE OF PFOTZER'S CURVE AND PRIMARY COSMIC RAYS

The course of Pfotzer's curve received a definitive explanation only after the occurrence in 1937 of the cascade theory of cosmic ray atmospheric showers. It has appeared that in the upper atmosphere (at a distance of several radiating units from the boundary) primary CR at the interaction with nucleus of air atoms generates electron-photon showers, and then with a further increase in depth of the atmosphere, the number of electron-positron pairs in showers gradually decreases because of reduction in the energy of the particles generating them.

Despite the valuable information received from research in the stratosphere, the main problem of the nature of primary CR remained still without an answer. It seemed that there was only one conclusion: lifting equipment to the border of the atmosphere, and direct measurements of the structure of CR. However, even before this method was realized, physicists managed to obtain interesting data on primary CR while studying their geomagnetic effects.

42. DISCOVERY OF THE COSMIC RAY LATITUDE EFFECT AT SEA LEVEL BY J. CLAY IN 1927, BUT WITHOUT THE CORRECT EXPLANATION

In 1927, for the first time, the latitude was found by the Dutch physicist J. Clay (1927). On the steamship "Slamat," J. Clay came back to Holland from the island of Java, where for a long time he'd studied the high-rise course and time variations of CR (J. Clay used the term ultra-radiation up to the middle of 1930s) using ionization chambers. Experiencing difficulties in defining the self radiation of ionization chambers that brought uncertainty in values of the intensity of CR, J. Clay decided to measure it in hydrochloric mines in Holland, and consequently carried with himself an ionization chamber. On the way, J. Clay made some measurements, and with surprise found that the intensity of CR decreases at the approach to the equator by 10 - 15 %. Originally, he decided, "this reduction can be explained by reduction of γ-radiation from an emanation in atmosphere". J. Clay did not manage to define the radiation of the ionization chamber itself in Holland, but, having become interested in the received results, he repeated measurements two more times, in 1928 and 1929, during voyages from Holland to the island of Java and back (Clay, 1928, 1930). J. Clay again found that around the equator, the intensity of CR decreases approximately by 14 %, but he could not offer any comprehensible explanation to the received effect and considered that "it most likely specifies in a local terrestrial origin of the cosmic ray latitude effect." The point of view of J. Clay is easy to understand if one recollects that his works appeared when the opinion was standard that CR are photons of large energy. Besides, errors of measurements were so great that the intensity change of all was some percent, it appeared, and it was hard to give this value. That fact is spoken to by the fact that the works of J. Clay did not generate a great interest, and only Bothe and Kolhörster (1929) gave the correct interpretation of results of Clay (1927), having connected change in intensity depending on latitude with the influence of the magnetic field of the Earth.

43. THE CORRECT EXPLANATION OF COSMIC RAY LATITUDE EFFECT AT SEA LEVEL

Bothe and Kolhörster (1929) for the first time noticed that it is possible to receive direct data on the nature of primary CR, i.e. on the radiation coming on the border of atmosphere, investigating changes of intensity of CR on the surface of the Earth depending on the magnetic latitude of the place of observation (so-called latitude effect). Really, if at least part of primary CR is made up of charged particles, they will deviate with the magnetic field of the Earth in such a manner that intensity at the poles will appear considerably greater than at the equator. If primary CR completely consist of photons, their intensity would be identical in an every spot on the globe (if they are isotropic in space).

44. Cosmic Ray Measurements at High Latitudes: No Latitude Effect

For the purpose of investigating the phenomenon discovered by Clay (1927) in the region of high latitudes, Bothe and Kolhörster (1930) traveled from Hamburg to Spitsbergen, but, to their big disappointment, did not notice any dependence of cosmic ray intensity on latitude around the North Sea. The problem of geographic distribution of the intensity of CR inspired a set of researchers to carry out measurements in various places around the globe. Therefore, it is known that in 1928 physicist F. Begonek measured the intensity of CR during the first Arctic flight under the direction of Umberto Nobile on the dirigible balloon "Italy" over the pole, but a cosmic ray latitude effect was not found.

45. Cosmic Ray Measurements at Different Latitudes by Millikan's Group: No Difference in Intensity, Conclusion on the Photon Nature of Primary Cosmic Rays, and Religious Aspect

R. Millikan and G. Cameron (1928) also did not notice an essential difference in the intensity of CR measured in Bolivia (19° S) and in Pasadena in the State of California (34° N). Later, R. Millikan (1930) repeated the measurements, and again did not find almost any difference between Pasadena and Churchill in Canada (59° N). It is not necessary to say that after the measurements made by Nobel Prize winner Robert Millikan that were famous for accuracy and care in the experiments, the existence of a cosmic ray latitude effect seemed very doubtful. Even if there was any change in cosmic ray intensity with latitude, it was so a little that R. Millikan and many other researchers were inclined to consider that it is caused by the difference in atmospheric conditions at various geographical points.

Summarizing all the received results, one of the largest authorities in, physics, CR G. Hoffmann (1932) wrote: "As a whole the results received until this moment, it is necessary to recognize negative. The majority of observers come to the conclusion that within an accuracy of experiences intensity of CR is constant and identical, and even those authors who observe its changes with geomagnetic latitude, ascertain them with known care".

Robert Millikan and his employees considered the absence of a considerable cosmic ray latitude effect as proof of the opinion that primary CR consist of photons. As was already written in the previous Chapter, R. Millikan in 1928 put forward a hypothesis according to which CR are the first shout of atoms, which continuously are born in world space. In the hands of R. Millikan, the reason that CR testify to continuous synthesis of atoms in interstellar space turned to be proof of the expression, "the Founder continues the work." R. Millikan tirelessly popularized the idea, acting at various meetings of unscientific character, and one from his messages, the truth, without his permission appeared on New Year's day 1931 in the Sunday supplement to the popular newspaper "The New York Times." It occupied six columns on the first page - a place that is usually reserved only for speeches of the President of the United States. Readers of the newspaper learned that (in the expression of the reporter) "Millikan gives battle to the awful second law of thermodynamics" and "the Founder continues the work." The article caused rough interest and received especially wide popularity among the religious population.

46. Arthur Compton's Objections Against the Millikan's Theory of Primary Cosmic Ray Origin

Another Nobel Prize winner, Arthur Compton, put forward serious objections to the basic postulate on which all of Millikan's cosmological theory, that CR are only high energy photons from newly created atoms in the Universe, was based. A. Compton considered the experiences of W. Bothe and W. Kolhörster and of J. Clay in cosmic ray latitude effect measurements to specify that primary CR undoubtedly consist of charged particles.

47. New Measurements of Latitude and Longitude Geomagnetic Effects, and the Origin of the Plateau at High Latitudes

Just at this time, J. Clay and H. Berlage (1932), firmly convinced in the existence of the cosmic ray latitude effect, for the fourth time repeated measurements of cosmic ray intensity on the way from Holland to the island of Java. This time, however, onboard the steam-ship "Christians Huygens," a much more sensitive big ionization chamber designed by E. Steinke was established (the statistical errors of the received data, according to the authors, was smaller than 1 %). At the approach to the geomagnetic equator, a reduction in intensity of 16 % in comparison with its value at high latitudes, was observed. Analyzing and comparing the obtained results with the results of previous expeditions, J. Clay (1932) received strongly pronounced dependence of the intensity of CR on latitude (Figure 35).

At that time, to the large bewilderment of many researchers, J. Clay has correctly explained the absence of increase in intensity of CR at latitudes above 50° by taking into account that primary CR with energy less than 4×10^9 eV are absorbed in the terrestrial atmosphere and do not reach the surface of the Earth. Therefore, the increase in the flux of these particles expected at high latitudes over the border of the terrestrial atmosphere cannot be revealed at sea level. At once, the results received at high latitudes that seemed strange became clear. Besides, J. Clay noticed for the first time that intensity of CR depends not only on latitude, but also on longitude of the place of observation. This effect, which subsequently received the name longitudinal geomagnetic effect, J. Clay connected with the fact that the centre of a magnetic dipole is displaced with respect to the centre of the Earth.

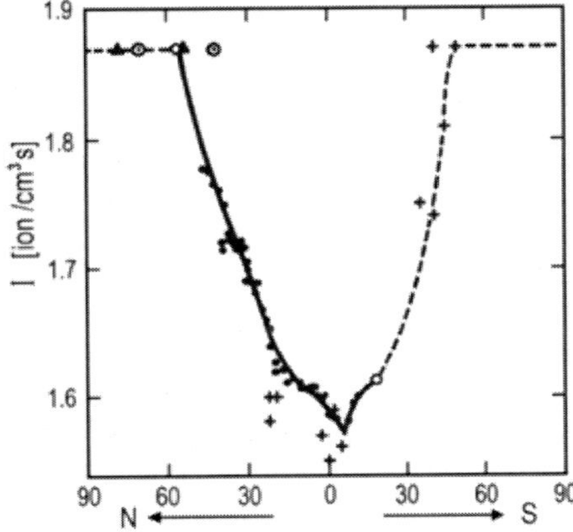

Figure 35. Dependence of intensity of CR (ionization rate I in unshielded chamber) on the geographic latitude, obtained by J. Clay (1932) according to measurements by ionization chambers (mainly during various expeditions between the island of Java and Holland).

48. EIGHT COSMIC RAY EXPEDITIONS ORGANIZED BY ARTHUR COMPTON IN 1932 IN MANY REGIONS OF THE EARTH

To be convinced definitively of the existence of the cosmic ray latitude effect and to prove the case, Arthur Compton (1932, 1933) organized in 1932 eight expeditions for CR research at various latitudes. Eight expeditions on whose work some tens of researchers participated made observations at various altitudes in 69 points scattered across the globe (their arrangement is shown on a card presented in Figure 36).

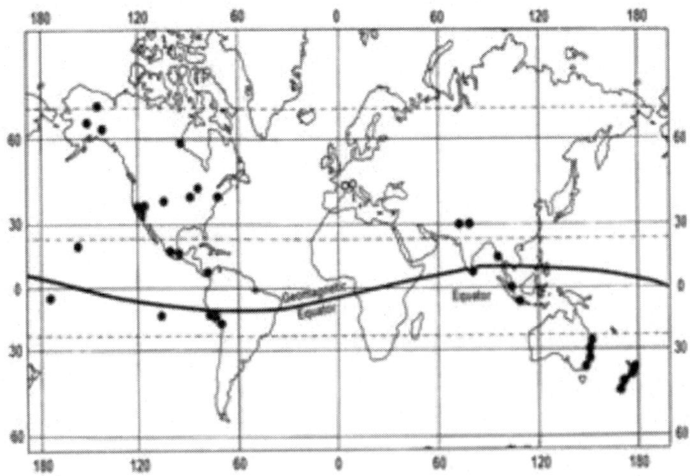

Figure 36. An arrangement of the basic points in which were made measurements of intensity of CR during the eight expeditions under the direction of A. Compton.

This collective work of physicists of various countries was then, perhaps, the largest physics expedition in history, and its high level organization, scope and the high techniques of measurement represent a great interest. The most interesting expedition headed by Arthur Compton was made in New Zealand, Australia, Northern Canada, on the Hawaiian Islands and in Illinois in USA (Compton, 1932). Members of this expedition floated 50 thousand miles, visited five continents, and five times crossed the equator. Measurements were made not only at sea level, but also, where there was a possibility presented, on high mountains. All eight expeditions (Compton, 1933) were equipped by specially made portable ionization chambers (Figure 37 and 38).

Figure 37. Arthur Compton near ionization chamber, developed by him for eight cosmic ray expeditions worked in 1932.

The steel spherical chamber in a diameter of 10 cm was surrounded with three screens of thickness 2.5 cm each,(two lead, one bronze) and filled with argon under a pressure of 30 atm thanks to which it was possible to receive ionization currents by which it was possible to measure by ordinary electrometer of type Lindeman (Figure 38).

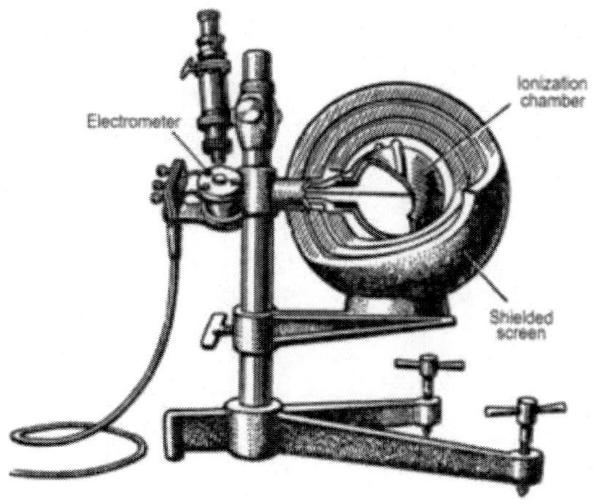

Figure 38. The device of the ionization chamber developed under the direction of A. Compton.

The ionization caused by CR was defined by comparison with the ionization made by radiation from a standard radium source, located at a certain distance from the chamber. The chamber was adapted for forwarding conditions and worked absolutely reliably even under heaving conditions.

49. MAIN RESULTS OBTAINED IN COMPTON'S EIGHT COSMIC RAY EXPEDITIONS IN 1932 AND CONCLUSION THAT MILLIKAN'S THEORY ON THE ORIGIN OF PRIMARY CR IS WRONG

Using the data obtained by all expeditions, A. Compton constructed curves of dependences of CR intensity on geomagnetic latitude and altitude $I = f(\lambda, H)$, where λ is the geomagnetic latitude and H is altitude (Figure 39).

From Figure 39 it is clearly visible that cosmic ray intensity at the equator (1.61 ion.cm^{-3}.s^{-1}), gradually increases up to the latitude 50-55° (1.84 ion.cm^{-3}.s^{-1}) by 14 % from the average size. Higher, the latitude intensity of CR does not vary. Therefore, the existence of the cosmic ray latitude effect which caused of so many discussions was not subject to more doubt. Besides, A. Compton showed that if instead of geomagnetic latitude one takes geographical latitude, the CR latitude effect comes to light not so accurately.

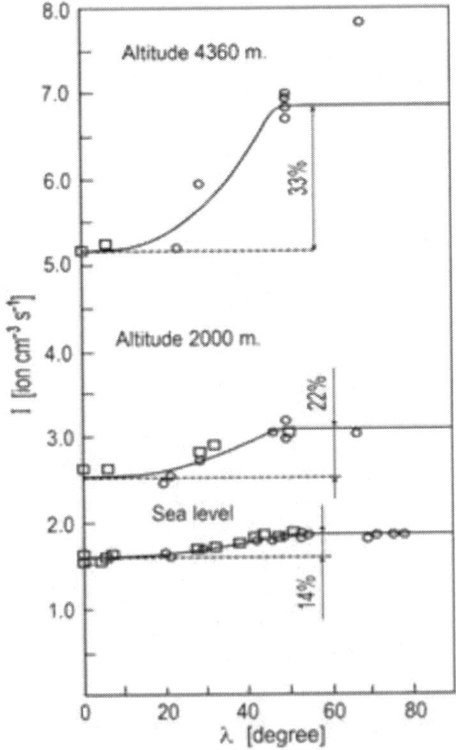

Figure 39. Change of intensity of CR I (in ion.cм$^{-3}$.s^{-1}) with geomagnetic latitude λ at various altitudes H above sea level (under the data received by eight expeditions under the direction of A. Compton).

A particular interest was caused by the increase found out during the A. Compton CR latitude effect expedition with the increase in height of a place of observation. At a height of 2000 m, the amplitude of the CR latitude effect was already 22 %, and at a height of 4360 m - 33 %. If one is to judge at the unique point which has been obtained on Alaska at latitude 68°, the amplitude of the CR latitude effect should be even more. However, at this point it is impossible to consider the great value of intensity of CR authentic as it was restored from the salvaged records of the lost La Kur's group. The expedition headed by La Kur made CR measurements at high latitudes and on Mount McKinly (Alaska), and tragically perished, having fallen in a glacial crack. Its results were restored from the records found at the site of the accident. The increase of the CR latitude effect with altitude above sea level, A. Compton explained that at large heights primary particles of smaller energies arrive, testing stronger the influence of the Earth's magnetic field on primary CR particles.

In September 1932, A. Compton, just arrived from Arctic regions, on the basis of results of his expeditions, publicly declared that CR undoubtedly consist of dressed up charged particles and that the conclusions of R. Millikan and his cosmogenic theory on primary CR origin were erroneous.

50. COSMIC RAY LATITUDE MEASUREMENTS BY MILLIKAN'S GROUP IN 1932

Robert Millikan, having refrained from answering Arthur Compton, organized his own expedition to Arctic regions, and sent his pupil H.V. Neher to South America. H.V. Neher designed a special self-recording anti-vibrating device, which combined big accuracy and sensitivity with huge stability to the concussions, allowing work in the most varied conditions. In Figure 40 is shown the photo of H.V. Neher and R.A. Millikan considering the new device. The device consisted an ionization chamber connected to an electroscope, and a photo-camera the film of which was set in motion by clockwork.

Figure 40. R.A. Millikan and its pupil H.V. Neher consider the ionization chamber designed by H.V. Neher for CR expedition measurements.

Robert Millikan again didn't find a CR latitude effect. In November 1932, H.V. Neher informed R. Millikan from Peru that his new device also did not register a change of CR intensity with latitude. R.A. Millikan immediately organized a-press conference and informed reporters on the absence of a CR latitude effect.

51. HISTORICAL DEBATE BETWEEN TWO NOBEL PRIZE WINNERS OVER COSMIC RAY ORIGIN

After the press conference of R.A. Millikan where he informed reporters of the absence of a CR latitude effect, a historical debate began between Robert Millikan and Arthur Compton about the nature of CR. This debate received wide publicity, having got on pages of newspapers (e.g., see Figure 41).

Figure 41. Newspaper page "The Pasadena Star-News" from December, 30th, 1932 with photos of A. Compton and R. Millikan.

In this Figure the newspaper page "The Pasadena Star-News" from December, 30th, 1932 where photos of A. Compton and R. Millikan are placed is presented and the entire account about "battle" between two Nobel Prize winners is given. Unfortunately, the press was not as excited with the question itself of who was in the scientific right as it was in the fact that one of the Nobel winners can be mistaken.

Unknown materials some until now from the history of this discussion resulted in the work of D. Kevles (1978).

52. ERRORS IN NEHER'S EXPERIMENT AND NEW MEASUREMENTS BY MILLIKAN'S GROUP ON AIRPLANE

In the meantime, H.V. Neher, coming back from South America to USA, registered the CR latitude effect, as it appeared that his device was simply faulty earlier on the way from the

USA to South America. The message of H.V. Neher about errors in the earlier experiment caused indignation in R.A. Millikan, because in his debate with A.H. Compton, he mainly based his rhetoric on Neher's wrong results. Nevertheless, in the spring of 1933, R.A. Millikan and his employees made a series of measurements by planes and were convinced, at last, of the reality of the CR latitude effect (Bowen et al., 1933). Results of the whole series of measurements by the planes, presented in Figure 42, showed that the increase of CR intensity with height appears much faster at the higher latitudes than on the equator.

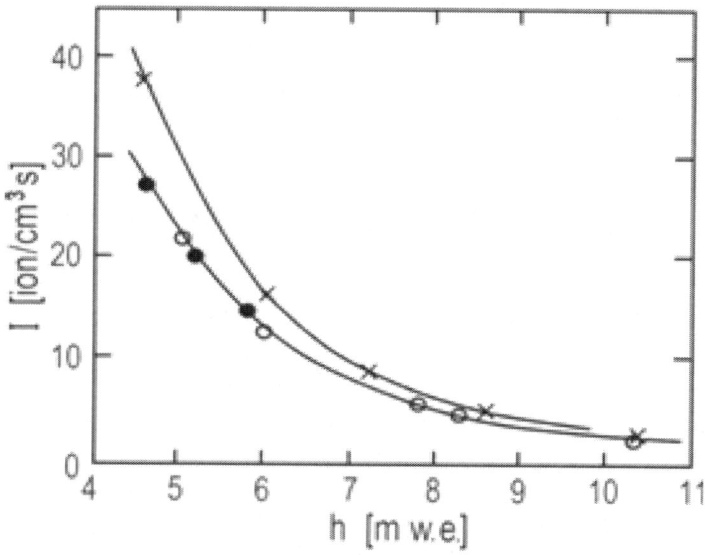

Figure 42. Dependence of CR intensity on the height, obtained by R.A. Millikan and his employees (Bowen et al., 1933) during flights by planes in Merch Fields (daggers), Panama (points), and Peru (circles).

53. FINAL APPROVING OF COSMIC RAY LATITUDE EFFECT EXISTING AND PROBLEM OF PRIMARY COSMIC RAY NATURE: ENDING OF DISCUSSION BETWEEN R. MILLIKAN AND A. COMPTON

Thus, the CR latitude effect discovered by J. Clay was investigated by many researches by means of different equipments in various points of the globe, and its existence undoubtedly proved that the considerable part of primary CR consists of charged particles. It is important that there was not confidence that all primary CR had a corpuscular nature, and there was an opinion that it could be divided on two components: the basic, having a photon nature and consequently not rejected by the magnetic field of the Earth, and an additional, consisting of charged particles which the magnetic field of the Earth influences, changing its intensity depending on geomagnetic latitude. On the equator only the photon component is observed, and its intensity is identical on the whole surface of the Earth.

Recognizing the existence of the CR latitude effect, R.A. Millikan never spoke more about "birth cries" of atoms, having given to priests to solve, "whether the Founder continues work". When in 1936 A. Compton wished to continue the discussion about the nature of CR,

R. Millikan has warned him against it, fairly considering that "public will observe of it as behind the dog fight between two Nobel Winners, and it will help nobody". With this reason A. Compton, of course, could not disagree. So was finished the well known discussion in the history of CR physics between two Nobel Prize winners, lasting several years and stimulating great development of CR research. It was shown that studying CR geomagnetic effects (latitude and azimuthal) allowed to make not only general qualitative conclusions, but also quantitative conclusions about energy and a sign on the charge of primary particles coming from space.

54. The First Theoretical Investigations by Carl Störmer of Charged Energetic Particles Moving in a Geomagnetic Field for Explanation of Aurora Phenomenon: Possible Application to Primary Cosmic Rays

As a first approximation it is possible to present the magnetic field of the Earth in the form of a field of a magnetic dipole with the moment equal 8×10^{25} Gs.cm^3. The centre of the dipole is displaced 340 km from the centre of the Earth, and its axis makes a corner nearby 11° with the axis of the Earth's rotation (therefore geomagnetic latitudes not coincide with geographical). The geomagnetic field reaches distances from the dipole centre of about ten radii of the Earth, and, despite the rather small intensity of the magnetic field, the big way which take place of primary CR particles, will lead to that even the charged particles of high energies should test it. In the early thirties representations about the possible influence of geomagnetic field on CR were very foggy. However secondary particles with big energies, formed in the terrestrial atmosphere, do not have time to deviate considerably in the field of the Earth and consequently keep the direction of movement of primary particles.

In the early 1930s representations about the possible influence of the Earth's magnetic field on CR were very foggy, though up to that time already there were theoretical calculations of movement of the charged particles in a geomagnetic field. These calculations were done by the Norwegian physicist Carl Störmer (1907), i.e. before the discovery of CR, for an explanation of polar aurora (Fredrik Carl Mülertz Störmer, -September 3, 1874 – August 13, 1957, -was a Norwegian mathematician and physicist, known both for his work in number theory and for studying the movement of charged particles in the magnetosphere and the formation of polar aurora).

Carl Störmer supposed that polar aurorae are caused by the charged particles let out by the Sun during periods of raised solar activity and getting to areas of high geomagnetic latitudes. Within several years, having executed tiresome calculations, C. Störmer calculated a set of trajectories of movement of the charged particles which, despite the influence of a geomagnetic field, got to the Earth. Subsequently it appeared that the Störmer's theory is directly inapplicable to polar aurora, as the particles causing aurora have too small of an energy to get from the outside into the zone of aurora located near the geomagnetic latitude 65°. After discovering the CR latitude effect when it became clear that the considerable part of primary CR consists of charged particles, and there was a necessity strictly to calculate dependence of intensity of CR on geomagnetic latitude, Störmer's theory developed for polar aurora, found wide application in the physics of CR. For the decision of a problem on movement of the charged particle in the magnetic dipole field of the Earth, C. Störmer (1930) started with the system of differential equations describing the behavior of an electron in a magnetic field:

$$m\ddot{x} = e\left(H_z\dot{y} - H_y\dot{z}\right),\ m\ddot{y} = e\left(H_x\dot{z} - H_z\dot{x}\right),\ m\ddot{z} = e\left(H_y\dot{x} - H_x\dot{y}\right), \tag{19}$$

where m - weight of electron, e - its charge, $\mathbf{H}$ - a vector of intensity of a magnetic field, x, y, z - coordinates of electron.

C. Störmer showed that in the field of a dipole the movement of an electron can be spread out on two components (see Figure 43): 1) movement in a meridian plane OPQ in which there is an electron, and 2) a rotation together with plane OPQ round axis of dipole OZ, on the angle $\varphi = f(\lambda, t, \rho)$, where λ - geomagnetic latitude, ρ — radius of curvature of electron's trajectory in the magnetic field.

Analyzing movements of electrons in the field of a dipole, Störmer (1930) found that the inclination of its trajectory to meridian plane defined by angle θ, is from the equation

$$-\sin\theta = \frac{\cos\lambda}{\psi^2} - \frac{2\gamma}{\psi\cos\lambda}. \tag{20}$$

Here γ - Störmer's constant of integration depending on an initial direction of movement of particle, and ψ is defined by equation

$$\psi = (r/r_E)(R/R_E)^{1/2}, \tag{21}$$

where r_E - radius of the Earth, r - distance from the centre of the Earth to electron, R - magnetic rigidity of electron, equal, by definition, product of intensity of a magnetic field on the radius of curvature of its trajectory $R = H\rho$, and R_E is a rigidity of electron which radius of curvature is equal in a plane of geomagnetic equator to radius of the Earth.

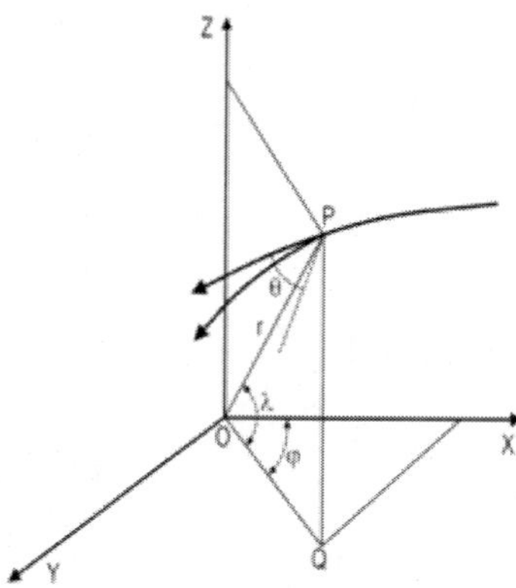

Figure 43. Movement of the charged particle in the field of magnetic dipole of the Earth: OZ - direction of axis of dipole, θ - angle between the direction of trajectory of movement and meridian plane OPQ. According to Störmer (1930).

As the intensity of magnetic field on the equator H = 0.32 Gs, and $\rho = r_E = 6.38 \times 10^8$ cm, that $R_E = 2 \times 10^8$ Gs.cm. For relativistic electrons, a relation takes place 300 Hρ = E (eV), and from here its energy E = 60 GeV. Hence, the charged particles with energy of order 60 GeV and smaller should be deviated strongly by the Earth's magnetic field.

55. DEVELOPMENT OF STÖRMER'S THEORY IN APPLICATIONS FOR COSMIC RAY TRAJECTORIES IN A GEOMAGNETIC FIELD

The considerable quantity of analytical calculations of movement of particles in the equatorial plane of a geomagnetic dipole was executed by Boguslavsky (M1929). Some trajectories of the particles arriving to the Earth and moving in a plane of geomagnetic equator, calculated by E. Brüche (1930), are shown in Figure 44. From the drawing, it is visible that not all the particles possessing identical energy behave equally.

Negative particles which in the absence of a magnetic field of the Earth should fly by from the East side of globe, actually thanks to a magnetic field, having described loop-trajectory, can come nearer to a surface of the Earth. At the same time particles, which as it seemed, are directed directly to the Earth, will abruptly turn to the East, without reaching a terrestrial surface. If one considers particles which move not in the equatorial plane, an even more difficult picture will turn out. In Figure 45 the trajectories, calculated by E. Brüche (1930) for this case are presented.

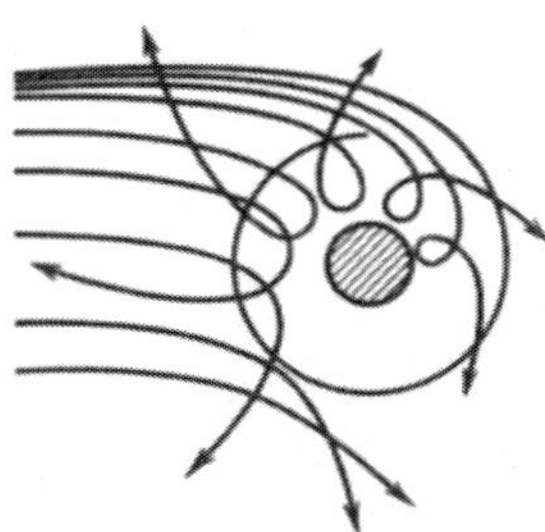

Figure 44. Trajectories of particles in the plane of geomagnetic equator calculated by E. Brüche (1930).

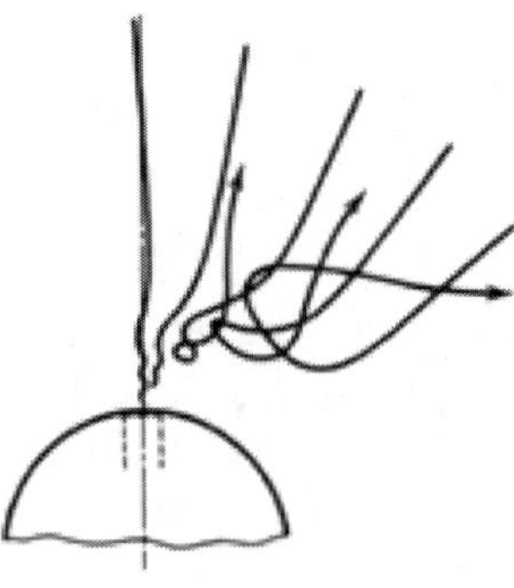

Figure 45. Trajectories of the particles coming under various angles to the axis of magnetic dipole, calculated by E. Brüche (1930).

Besides, E. Brüche (1931) developed a method of observation of trajectories of the charged particles directly in the laboratory, making a deviation of a "threadlike" bunch of electrons in a cathodic tube by the small electromagnet having the form of a sphere. In Figure 46 two trajectories, received by E. Brüche experimentally are shown. Their kind to the smallest details coincides with theoretically calculated trajectories.

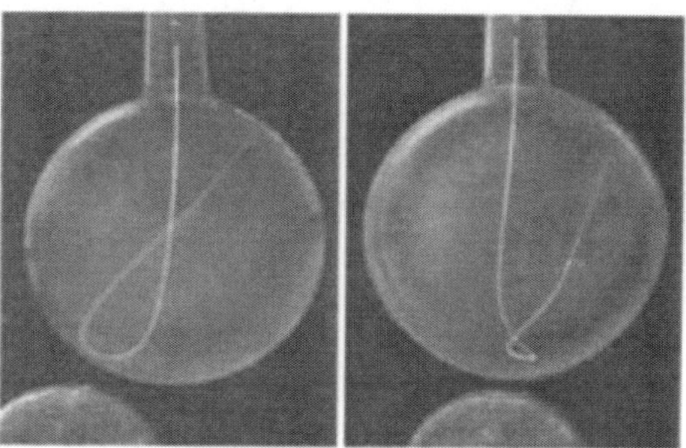

Figure 46. Two trajectories of the charged particles obtained by E. Brüche (1931) experimentally.

56. DEVELOPMENT OF STÖRMER'S THEORY IN APPLICATIONS FOR COSMIC RAYS IN GEOMAGNETIC FIELD: ALLOWED AND FORBIDDEN TRAJECTORIES

Eq. 20 obtained by Carl Störmer for the theory of polar aurora was applied to the problem of CR by Störmer (1931), and also by B. Rossi (1930) and G. Lemaitre and M.S. Vallarta (1933). Therefore, B. Rossi (1930) noticed for the first time that, using the theory of Carl Störmer, it is easy to receive results interesting for physicists of CR. Instead of considering the movement of the charged particle with the set rigidity from world space to the Earth, he suggested choosing the particle which has come from the set direction in some point P near the terrestrial surface, and then tracking its trajectory in the opposite direction through the magnetic field of the Earth. There are two possibilities: either the particle can leave in space, without crossing the surface of the Earth, or it will return back to the Earth. In the first case it meant that the particle could come from world space to point P (the resolved direction for arrival of space particles with certain magnetic rigidity), and in the other case the particle could not get to the Earth (the forbidden direction).

B. Rossi (1930) used the equations obtained by Carl Störmer, and showed that in each point on the Earth for positive particles of set magnetic rigidity in space there is a cone with an axis focused on the East, in which all directions for arrival of space particles are forbidden. This cone has received the name of "a cone of Störmer". For negative particles, there is a similar cone of Störmer with an axis directed to the West.

As the physicists studying CR were not interested in the beginning in calculations of the valid trajectories of particles, and aspired to define what directions for arrival of CR to the Earth are resolved, and what are forbidden, the theory of Carl Störmer for the permission of this question appeared very useful. C. Störmer (1931) also came to the idea that developed by him a theory for an explanation of polar aurora that can be applied to solving problems, interesting for physicists who were engaged in CR. He showed that because $|\sin\theta| \leq 1$ no energetic charged particle can get from world space into the area, for which

$$\left| -\frac{\cos\lambda}{\psi^2} + \frac{2\gamma}{\psi\cos\lambda} \right| > 1 . \tag{22}$$

Hence, if Eq. 22 is carried out, there is an area of values λ and ψ which cannot be reached by particles. The form of this area depends on the value of parameter γ. The border between the forbidden and resolved areas is defined from the condition

$$\left| -\frac{\cos\lambda}{\psi^2} + \frac{2\gamma}{\psi\cos\lambda} \right| = 1 . \tag{23}$$

Let's consider the trajectory of a particle beginning in infinity (i.e. $\psi = \infty$). In the process of particle approach to the Earth ψ will decrease and will reach the minimum value

$$\psi_E = \sqrt{R/R_E} \quad \text{at } r = r_E . \tag{24}$$

But so that the particle could reach really a surface of the Earth, it is necessary that in the process of reduction of ψ from ∞ to ψ_E the value $|\sin\theta|$ does not exceed 1. However it is easy to see that if $\gamma > 1$ and $\psi_E < 1$, that Eq. 22 will be always carried out, and consequently such a particle cannot reach a surface of the Earth. To find the border of the forbidden area, Eq. 23 was used at $\gamma = 1$. The border of the forbidden area for latitude λ_1 is defined by parameter ψ_1, satisfying the equation

$$\left| -\frac{\cos\lambda_1}{\psi_1^2} + \frac{2\gamma}{\psi_1\cos\lambda_1} \right| = 1 . \tag{25}$$

Thus any particle with parameter ψ_1 cannot reach surfaces of the Earth in the range of latitudes from $0°$ to λ_1 because for these values of λ always $|\sin\theta| > 1$. The forms of such forbidden areas in coordinates ψ and λ were calculated by Carl Störmer, and some of them are shown in Figure 47 for different values of parameter γ.

In Figure 47 dark areas correspond to forbidden zones, white - to the allowed zones. Apparent in the drawing, for $\gamma > 1$ allowed area shares on two parts between which there is a forbidden area, therefore, though the internal area is allowed, particles cannot get to it. To pass from parameter ψ_E to energies of particles, Carl Störmer calculated the Table 3 that is resulted below.

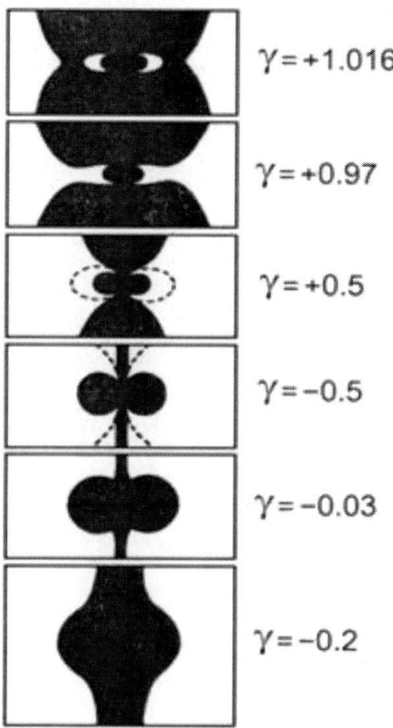

Figure 47. Forms of the forbidden areas in coordinates ψ and λ for various values of the parameter γ, calculated by Carl Störmer.

Table 3. Relation between parameter ψ_E and energy of particles

ψ_E	Energy of particles, 10^{10} eV		
	Electrons	Protons	α-particles
0.1	0.06	0.017	0.018
0.2	0.24	0.16	0.23
0.4	0.95	0.86	1.56
0.6	2.15	2.05	3.93
0.8	3.82	3.72	7.25
1.0	5.96	5.85	11.52

So, from the theory of C. Störmer followed that for the given energy of particles there is a minimum geomagnetic latitude λ_1, below which any particle cannot to get to the Earth. The most simple expression for critical rigidity of particles (so, and for energy) turns out for a vertical direction, when $\theta = 0$. Really, in this case from Eq. 20 for the cutoff value ψ_{Ec} the relation follows

$$\psi_{Ec} = \cos^2 \lambda / 2\gamma \,. \tag{26}$$

As for a surface of the Earth $r = r_E$, $\psi_{Ec} = \sqrt{R_{Ec}/R_E}$, and from Eq. 26 it follows at $\gamma=1$ that

$$R_{Ec} = \frac{R_E}{4} \cos^4 \lambda = 14.9 \cos^4 \lambda \ [10^9 \ \text{eV}]. \tag{27}$$

In Table 4 the minimum values of energy for various geomagnetic latitudes λ which should possess electrons to arrive in the vertical direction are represented.

Table 4. The cutoff energy E_c for various geomagnetic latitudes λ which should possess electrons to arrive in the vertical direction (calculations for protons and α-particles at that time, appear, did not represent the interest as has been standard to consider that primary CR are electrons)

λ	E_c, eV	λ	E_c, eV
0°	1.49×10^{10}	60°	9.31×10^{8}
20°	1.16×10^{10}	80°	1.36×10^{7}
40°	5.13×10^{9}	90°	0

57. EXPECTED PLANETARY DISTRIBUTION OF COSMIC RAY INTENSITY IN THE FRAME OF LIUVILLE THEOREM

Though Störmer (1930, 1931) found a huge number of the open trajectories on which particles could get to the Earth, there was still the unresolved a problem of the intensity of the CR coming in a certain direction. As calculation of trajectories of particles in a geomagnetic field at that time were extremely difficult, for the decision of this problem it was necessary to make, it appeared, extremely long and bulky calculations of all possible trajectories to define how their concentration from infinity varies at the approach to the Earth.

However, there was very simple solution to the problem. In B. Rossi's memoirs (Rossi, M1966, page 66) one notices that it discussed a problem of calculation of intensity of CR in a geomagnetic field with one of the largest physicists of the present, Enriko Fermi, who noticed that here it is necessary to take advantage of the general theorem of mechanics of preservation of the number of particles in the unit of phase volume known under the name of the theorem of Liuville. E. Fermi and B. Rossi (1933) showed that if CR are distributed in space isotropic, i.e. in regular intervals in all directions at large distances from the Earth, as in a constant magnetic field of the Earth Lorenz's force is perpendicular to the direction of the speed of the particle, the energy of the particle to vary not will be, and application of the theorem of Liuville leads to a conclusion about a constant intensity of CR near to the Earth in all resolved directions.

To a similar conclusion G. Lemaitre (Belgium) and M.S. Vallarta (Mexico) independently came (Lemaitre and Vallarta, 1933). They have applied a mechanical computer, the so-called differential analyzer developed by V. Bush in Massachusetts Technology Institute, for calculation of trajectories, and found that not all trajectories out of a cone of Störmer are resolved. It appeared that some of these trajectories, before they leave to infinity, cross a terrestrial surface and consequently intensity of CR particles along them should be equal to zero.

As a result of calculations taking into account the theorem of Liuville, Lemaitre and Vallarta (1933) found distribution of intensity of cosmic rays over the terrestrial atmosphere depending on geomagnetic latitude for particles of various rigidity (Figure 48).

In Figure 48 for 100% intensity of CR at infinity is accepted, and the numbers at the curves give the value $\psi = \sqrt{R/R_E}$. Each curve corresponds to a certain energy E which can be defined from Table 3 depending on the particle's nature. From the drawing it is visible that for particles of certain energy there is a latitude λ_1 below which any particle does not come to the Earth. There is also a latitude λ_2, above which particle of the given energy can come to the Earth from all directions. Between latitudes λ_1 and λ_2 particles of this energy come to the Earth only in some directions lying within a certain cone (so-called "the resolved cone of Lemaitre and Vallarta"). Dependence between energy of particles E and critical latitudes λ_1 and λ_2, according to their calculations resulted in Table 5 for the case if CR particles are electrons. G. Lemaitre and M.S. Vallarta continued the calculations for the area lying between latitudes λ_1 and λ_2 in the work Lemaitre and Vallarta (1936). In 1975 in connection with M.S. Vallarta's 75th anniversary there was published the article of S. Korff (1975) in which the historical review of development of the theory of Lemaitre - Vallarta and its experimental checking is given.

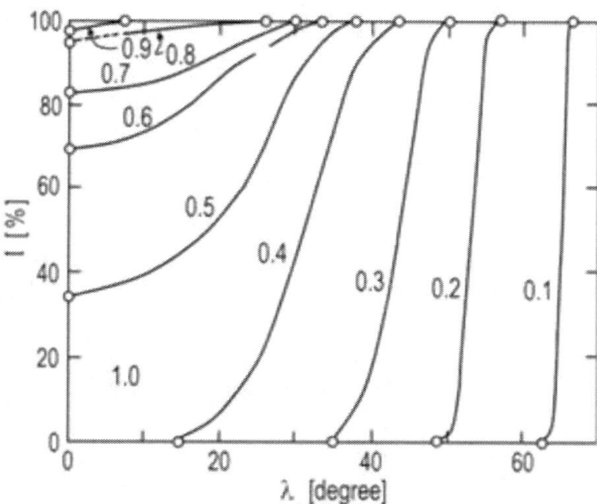

Figure 48. Dependence of intensity of CR over terrestrial atmosphere from geomagnetic latitude λ for particles of the various rigidity, received by Lemaitre and Vallarta (1933).

It is necessary to remember that all calculations were made by G. Lemaitre and M.S. Vallarta for intensity of CR at the atmosphere border, i.e. without influence of the terrestrial atmosphere. In a case that the Earth did not possess an atmosphere and particles would be homogeneous for one of curves Figure 5.16 would define energy change of intensity of CR with latitude. If the Earth did not possess an atmosphere, but particles would be non-uniform, change of intensity of CR with latitude would be defined by the sum of curves, and it is necessary to know, what percent of particles as a part of CR possesses the energy corresponding to each separate curve.

Table 5. Relation between ψ and E, from on the side, with λ_1 and λ_2, from the other side

$\psi = \sqrt{R/R_E}$	E, eV	λ_1	λ_2
0.1	6×10^8	64.4	66.7
0.2	2.4×10^9	49	57
0.3	5.4×10^9	35	50
0/4	9.5×10^9	12	44
0.5	1.5×10^{10}	-	40
0.6	2.2×10^{10}	-	36
0.7	2.7×10^{10}	-	34

Atmosphere influence is reduced by absorption of particles with energies smaller than the energy necessary for passage of all thickness of the atmosphere.

This energy, as shown by Kolhörster and Tuwim (1931) and by J. Clay (1932), equals 4×10^9 eV. G. Lemaitre and M.S. Vallarta counted up that since latitude 47.5° the magnetic field of the Earth complicates hit of particles with such energy to the Earth.

Thus, on a surface of the Earth for a curve latitude effect, according to the calculations of G. Lemaitre and M.S. Vallarta, decreasing of intensity of CR at reduction of latitude from 47.5° to the equator was expected. For latitudes from 47.5° to a pole independence of intensity of CR with latitude due to particles with energies lower 4×10^9 eV not getting to a surface of the Earth should be observed.

The calculations of G. Lemaitre and M.S. Vallarta were well coordinated with the experimental results received by A. Compton (1933), J. Clay (1932) and other authors.

So, A. Compton (1933) made comparison of the experimental results with the theory of Lemaitre and Vallarta (1933) and showed that the curve found it latitude effect can be received from the theory of G. Lemaitre and M.S. Vallarta if one were to assume that CR consist of magnetically not rejected photon components and a flux of the charged particles with energies lying in limits from 5×10^9 eV to 1.3×10^{10} eV.

The theory of G. Lemaitre and M.S. Vallarta has undergone to the sharp criticism from C. Störmer (1934) who objected the application of the theorem of Liuville for simplification of difficult calculations of intensity of CR, and considered necessary calculation of all resolved trajectories. Interesting discussion on this problem in which many scientists have taken part has begun.

58. THE USING OF EAST-WEST GEOMAGNETIC EFFECT FOR DETERMINING THE SIGN OF PRIMARY COSMIC RAY CHARGED PARTICLES

Bruno Rossi (1931) paid attention to that circumstance that if primary CR contain unequal quantities of positive and negative particles, the intensity of the radiation coming from the East and from the West will be varied. An especially big difference between the number of the particles coming from the East and the number from the West is expected in the case when all CR particles have charges of one sign. This effect received the name East-West asymmetry.

B. Rossi (1931) tried to find out East-West asymmetry by means of an installation which became known as "a telescope of CR" later. It consisted of several Geiger-Muller counters included in the scheme of coincidence and located so that their centers lay on one line named as an axis of the telescope. Simultaneous discharges in counters could cause only those particles which moved in the directions close to an axis of the telescope, i.e. the telescope allowed to define intensity of CR coming from a certain direction. In the experiments of B. Rossi by CR telescope with axis directing serially on the East and on the West concerning geomagnetic meridian were found CR intensities coming correspondingly from East and from West directions. "Any appreciable difference in counting rates in both cases, - wrote B. Rossi (M1966, page 69) - I did not manage to notice".

After the work of G. Lemaitre and M.S. Vallarta (1933), in which was developed the theory of behavior of charged particles in the magnetic field of the Earth, and once again persistently underlined that searches for the East-West asymmetry can deal with the problem of the sign of charge of primary CR particles, regular works in this direction, soon began to be crowed full successes. First of all it was T.H. Johnson (1933a) who made measurements in Mexico City (29° N, altitude 2250 m above sea level), and L. Alvarez and A. Compton (1933), working in the same place. Measurements were made by means of a telescope whose axis made 45° with the vertical, and showed that intensity of the CR coming from the West is greater than that coming from the East. The difference in intensity was about 10%.

After some months B. Rossi (1934), making measurements in Eritrea in Africa (11° N, altitude 2370 m above sea level) found that the surplus of CR intensity from the Western direction reaches 26 %, thus, it became clear that for the most part (or maybe all), primary CR consist of positively charged particles. This was an absolutely unexpected result, as among supporters of the hypothesis about the corpuscular nature of CR the opinion prevailed that primary CR consist of electrons. "If the CR East-West asymmetry had been opened few years earlier, - recollects B. Rossi (M1966, page 70), - we without fluctuating would have made the conclusion that primary CR are protons or heavier kernels as other positive particles then did not know. Now clearly that it would be the correct conclusion. However, in 1932 the new positive particle — positive electron, or positron was discovered. Therefore the question on the nature of primary CR remained uncertain before obtaining of new experimental results".

In the beginning, T.H. Johnson, A. Compton, and many other researches were inclined to consider that primary CR consist of positrons. In 1933, soon after the first measurements, T.H. Johnson (1933b) conducted more detailed research of the dependence of azimuthally asymmetry with latitude and height of a place of observation and showed that its value increases at approach to the equator. Therefore, at a height of 3000 m it made 13% on equator, 7% at latitude 25° and 2% at latitude 48°. Besides, azimuthally asymmetry increased with height increase: at measurements in Peru (on the geomagnetic equator), it was only 7% at sea level, but at the height of 4200 m reached 16%. Therefore, T.H. Johnson drew the conclusion that almost all particles coming to the Earth have the positive sign of charge.

As from the theory of geomagnetic effects it followed that on the equator the minimum energy of primary particles coming from the East makes 3×10^{10} eV, and from the West - 1×10^{10} eV, for an explanation of the results T.H. Johnson admitted that in primary CR there can be a considerable number of particles with energies from 1×10^{10} to 3×10^{10} eV. This was well coordinated with estimations of energy of the CR particles made by C.D. Anderson (1932) and P. Kunze (1933) on the basis of experiments with a Wilson's chamber placed in a strong magnetic field.

59. Measurements of Cosmic Ray Geomagnetic Effects in USSR: Latitude Effect at Sea Level at High Latitudes, Azimuthally Effect at Mountain Height, and Latitude Effect in Stratosphere

In the Soviet Union the CR latitude effect was observed for the first time in 1932 by A.B. Verigo (1938). On the ice-breaker "Malygin," he made measurements of intensity of CR from Archangelsk (65° N) to Franz Josef land (82° N). "Results made us during the expedition of measurements, - summarized A.B. Verigo the observation results, - show that the value of CR intensity on all route of expedition remained a constant within experimental errors of measurements". This result was in good agreement with many other measurements of the CR latitude effect in the region of high latitudes. Later, in 1934, azimuthally asymmetry investigated by V.M. Dukelsky and N.S. Ivanova (1935) at mountain Alagez in Armenia (35° N) found that the intensity of CR coming from the West in 9% exceeds the intensity of CR coming from the East (see Figure 49).

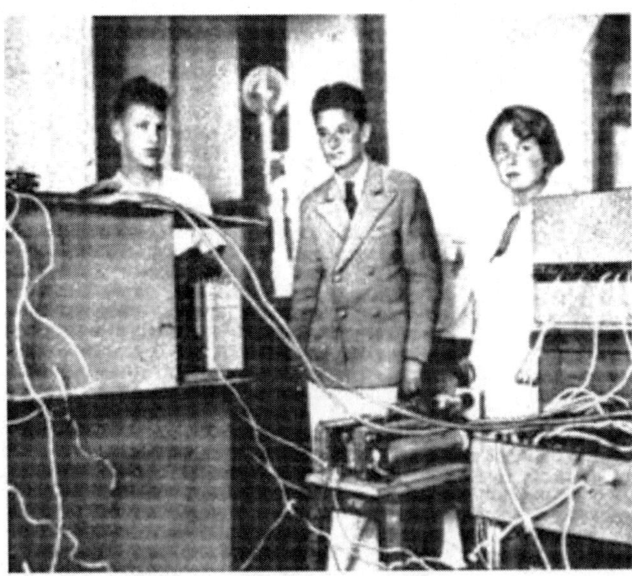

Figure 49. Equipment checking before the expedition beginning on mountain Alagez. From left to right: A.A. Maleev, N.M. Kocherjan, and N.S. Ivanova.

The following important step was made by S.N. Vernov (1937, 1938, 1939), measuring latitude effect of CR on balloons in the stratosphere. Effective CR research in the stratosphere became possible after the offering by S.N. Vernov in 1934 of data transmission method by radio. The first balloons flights of equipment designed by S.N. Vernov were carried out in

1936 near Leningrad (56° N) and in Yerevan (35° N). "Intensity of CR in the stratosphere at latitude 35° is much less than at the high latitudes, - drew the conclusion S.N. Vernov (1937), - distinction starts to affect considerably at 9 km. At 12 km (the limiting height of the flights in Leningrad) the intensity of CR at latitude 35° it appears is 2.5 times smaller. The sharp distinction in intensity of CR at various latitudes proves the presence of corpuscular fluxes with energy smaller than 5×10^9 eV". Besides, in 1937 during balloon flights from a steamship "Sergo" from Black sea to the Far East measurements of intensity of CR in stratosphere in the areas close to the magnetic equator also were made. "Received results of data flights in equatorial areas, - S.N. Vernov (1938) wrote, - will well be coordinated among themselves. In stratosphere intensity in equatorial areas in 4 times is less, than at latitude of Leningrad".

The big value of the CR latitude effect in stratosphere, defined in a range from 5° to 56°, allowed S.N. Vernov (1939) to draw the conclusion that already 90% of primary CR particles consist of charged particles. Besides, comparing the data received at three latitudes, S.N. Vernov could define a kind of a power spectrum of primary CR particles.

60. WIDE-WORLD MEASUREMENTS OF COSMIC RAY GEOMAGNETIC EFFECTS AND DETERMINING OF COSMIC RAY PLANETARY DISTRIBUTION

I.S. Bowen, R.A. Millikan and H.V. Neher (1937, 1938) in 1937 also made a series of measurements at large heights, lifting an ionization chamber on sounding balloons at points with different geomagnetic latitudes. The curve of dependence of CR intensity with height at the big latitudes appeared basically same as the curve of G. Pfotzer (1936), to within the amendment caused by that they measured global intensity. At latitude reduction the observable curve went less abruptly that specified in increase latitude effect at height increase. It has been shown that intensity of CR in the stratosphere varies with latitude change almost by 3 times. On the basis of received data, Bowen et al. (1937, 1938) also could define a power spectrum of primary CR particles.

In general, those years there were many interesting works on studying of the CR East-West asymmetry, latitude and longitudinal effects. Among them it is necessary to mention works of J. Clay (1934), T.H. Johnson and D.V. Read (1937), A.H. Compton and R.N. Turner (1937). A.H. Compton (1936) summarized all received results in the form of the diagram which is presented in Figure 50. Curves on the diagram are lines of equal intensity of CR, or as they were named by A.H. Compton, "izocosmes". In the drawing the geomagnetic equator and the geomagnetic parallels corresponding 50° northern and 50° of southern latitude are noted. From the diagram it is visible that intensity of CR changes not only with latitude, but also with longitude. Therefore, around the Indian Ocean intensity along the geomagnetic equator falls to a minimum because the centre of the terrestrial magnetic dipole does not coincide precisely with the geographical centre of the Earth, and is displaced in the direction to Indian Ocean. Existence of the longitudinal effect follows from the theory of G. Lemaitre and M.S. Vallarta (1933) and for the first time, found out by J. Clay (1932), did not cause more doubts.

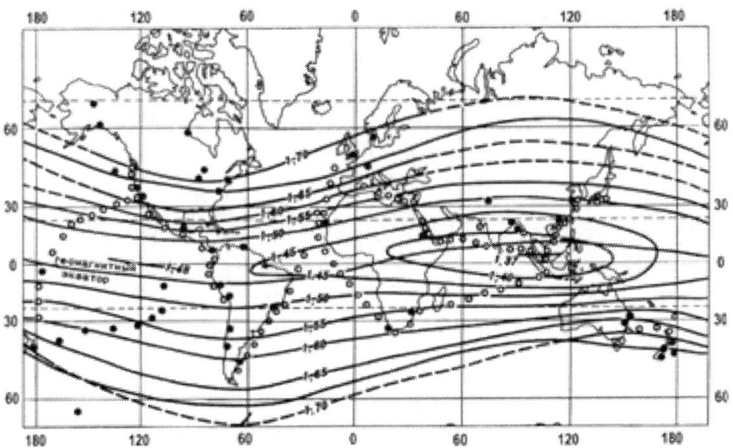

Figure 50. Curves of equal intensity of CR at the sea level, or "izocosmes", obtained by A. Compton (1936). Numbers at curves mean the intensity of CR in $ion \times cm^{-3}.s^{-1}$. Points note places where measurements were made.

Besides, izocosmes obviously followed not to geographical parallels, but to geomagnetic, that testified that change of intensity of CR falls mainly under the influence of a magnetic field of the Earth.

61. ON THE ROLE OF COSMIC RAY TEMPERATURE EFFECT IN MEASUREMENTS OF THE LATITUDE EFFECT AT SEA LEVEL

However it was impossible to completely exclude the influence of atmospheric conditions on the intensity of CR. Really, in 1937, A.H. Compton and R.N. Turner (1937) during the 12 ship routes through Pacific Ocean found that the form of an observed curve of CR intensity changes with latitude depends on the season in which the data are obtained. Though similar seasonal changes of intensity were observed even earlier, the work of Compton and Turner (1937) represents the first serious research of this effect. The further careful studying of the influence of meteorological factors led to the opening of temperature effect which consists of the fall of intensity of CR with increase of temperature of the air. The explanation of the temperature effect was given by P.M.S. Blackett (1938) on the basis of a hypothesis about spontaneous disintegration of μ-mesons.

As the average temperature of the atmosphere at the equator is much higher than at middle and high latitudes it should lead to changes of CR intensity with latitude (decreasing with increasing the latitude). Exact measurements have shown that about one third of the observed latitude effect at sea level is caused by the temperature effect, and the remaining two thirds is caused by the influence of the magnetic field of the Earth on the charged particles of primary CR.

62. Geomagnetic Effects and the Nature of Primary Cosmic Rays

Discovery and research of the CR latitude effect and the East-West azimuthally asymmetry allowed the establishment that primary CR consist of positively charged particles. If one excludes the existence of new, previously unknown positively charged stable particles, primary CR could consist either of protons or from positrons, as by then it was already known that positive μ-mesons are non-stable particles and consequently cannot be a part of primary CR. For a long time, as it was already told, scientists believed that primary CR particles are positrons.

T.H. Johnson (1938), in connection with studying of geomagnetic effects of CR, came out, however, with the assumption that primary CR can consist of protons. According to T.H. Johnson's reasons, as geomagnetic effects are observed and on a surface of the Earth, primary particles with energy smaller than 1.5×10^{10} eV, or their secondary products should reach there. However, neither positrons with such energy nor the secondary particles created by them in the terrestrial atmosphere can overcome the thickness of the atmosphere. From here it followed that basically primary CR should consist of heavy positive particles, i.e. protons, as only these particles or the secondary particles created by them are capable of reaching a terrestrial surface.

This assumption received acknowledgement after the experiences spent in the Chicago University by M. Schein, W.P. Jesse, and E.O. Wollan (1941). Within 1940 this group of researchers made measurements on sounding balloons of the number of particles at various heights up to 20,000 m, where residual thickness of the atmosphere was less than one radiating unit of length (i.e. thus primary particles were registered basically). Between counters which joined in threefold, fourfold and fivefold coincidences were located lead plates with thickness from 4 to 18 cm. The scheme of the experiment and the received results are presented in Figure 51.

Experiments of M. Schein, W.P. Jesse, and E.O. Wollan (1941) have showed that registered intensity of primary CR particles smoothly increases with increase in height up to a pressure of about 2 cm Hg, and the observable particles are not positrons. Really, in these experiments, particles which were found at very big heights as have shown the experiments made with lateral counters at their passage through lead did not form showers in such quantity as it was possible to expect from high energy positrons and electrons. Moreover, it appeared that these particles were absorbed by lead not so effectively, as positrons and electrons for the data received at different thickness of lead plates (4, 6, 8, 10, 12, and 18 cm) fell neatly down on one curve. As measurements were made at geomagnetic latitude 51°, and at this latitude only primary particles with energy of more than 3×10^{9} eV could be registered, Schein et al. (1941) came to the conclusion that positrons with such energy in primary CR are not present. Schein et al. (1941) understood also that primary CR could not be positive μ-mesons. "Therefore, - concluded Schein et al. (1941), - it is most likely that incoming primary CR consist of protons". As to the curve of G. Pfotzer (1936) which was received earlier as a result of measurements without a lead screen (a curve **B** in Figure 51), Schein et al. (1941) drew the conclusion that it represents a high-rise course of intensity of electrons and positrons completely having a secondary origin. One of the basic sources of these secondary electrons and positrons, Schein et al. (1941) considered the decay of positive and negative μ-mesons.

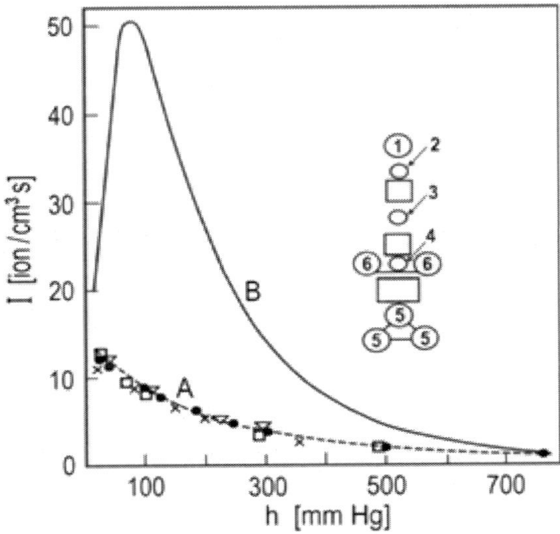

Figure 51. Results of measurements of the number of primary CR particles at the big heights, obtained by M. Schein, W.P. Jesse, and E.O. Wollan (1941) (curve A). The curve B obtained by G. Pfotzer (1936). On an insert, the experiment scheme is shown.

The experiments of Schein et al. (1941) have fundamental value, as primary CR particles for the first time were measured and were shown basically to be protons. It is difficult to overestimate the value of early studies of geomagnetic, as thanks to them, by the beginning of the 1940s it was possible to obtain directly the important information on the nature of primary CR. By other ways in those years it was impossible to receive such information.

63. RECOGNIZING OF COSMIC RAY DISCOVERY IMPORTANCE: THE NOBEL PRIZE IN PHYSICS FOR 1936

In 1936 Victor Hess received the Nobel award in physics "for discovery of space radiation" (see Figure 52). Meanwhile, under the status of Nobel Prizes, they should be awarded "for the latest achievements..., and for old works only when their value has not been found out until recently".

The Nobel Prize in Physics was shared by Victor Hess, for the discovery of cosmic rays, and Carl Anderson, for the discovery of the positron. According to Carlson (2012), Arthur Compton, in his letter nominating Hess for the prize, wrote, "The time has now arrived, it seems to me, when we can say that the so-called cosmic rays definitely have their origin at such remote distances from the Earth, that they may properly be called cosmic, and that the use of the rays has by now led to results of such importance that they may be considered a discovery of the first magnitude." The award only 24 years after the discovery of cosmic rays reflected, apparently, long existing doubts at first in the fact of existence of space radiation, and then in the value of this discovery for physics.

Figure 52. At the Nobel ceremony at December 1936 in Stockholm: Victor Hess (right) and Carl Anderson (middle) are seated beside chemistry laureate Peter Debye. From Carlson (2012).

Congratulating Victor Hess, the Chairman of the Nobel Committee in Physics of the Swedish Royal Academy of Sciences told: "As a result of purposeful researches of effects of the radioactive radiations executed with big experimental skill, you have discovered the surprising radiation coming from the depth of space, - space radiation. Existence of space radiation has put before us the new important problems, concerning formations and destructions of a matter and has opened a new area of researches ". In the Nobel lecture "Ungelöste Probleme in der Physik" Victor Hess underlined that the further research of cosmic rays, probably, will lead "to discovery of still many elementary particles, neutrino and the negative protons the existence of which was postulated by theorists in recent years". He with satisfaction noticed that the Nobel Prize was divided with Carl Anderson, who discovered positrons in cosmic rays. Already Victor Hess understood that "the great interest represents possible influence of cosmic rays on live organisms". He called for further comprehensive investigation of space radiation by all known methods.

In 1938 Victor Hess was compelled to leave his native land because of the prosecution from the Nazis, and moved to the USA where until 1956 he worked at Fordham University. In his long life Victor Hess has published about sixty papers and several books devoted to cosmic rays and adjacent problems.

In 1964, at the age of 81 years, two years after the solemn anniversary devoted to the fiftieth anniversary of the discovery of space radiation Victor Hess died. The well known American physicist, professor of the Massachusetts Institute of Technology Bruno Rossi, the active participant and the head of many researches in the field of cosmic rays, in the book "Cosmic rays" (Rossi, M1966) writes that "the fiftieth anniversary has come at a critical moment for the physicists studying cosmic rays, and may be, for the physics of cosmic rays. The physics of cosmic rays has actually ceased to exist as an independent branch of a science;

probably, future historians of science will close the head about cosmic rays by the fiftieth anniversary of discovery of Victor Hess".

It is difficult to agree with this opinion. Now, at the 100-th anniversary, cosmic ray research continues to develop in many geophysical and astrophysical aspects and in combination with accelerators – even in the aspect of elementary particle and high energy physics.

ACKNOWLEDGMENTS

One of us (Lev Dorman) is thankful to Israeli Ministry of Science for support in the frame of the project "National Centre of Knowledge in Cosmic Rays and Space Weather". Our gratitude to David Shai Applbaum for checking and improve English in this paper.

REFERENCES

Alvarez L. and A.H. Compton, 1933. "A positively charged component of cosmic rays", *Phys. Rev.*, Ser. II, 43, No. 10, 835-836.

Behounek F., 1926. "Zum Ursprung der durchdringenden Strahlung der Atmosphäre", *Phys. Ztschr.*, 27, No. 1, 8-10.

Bethe H., 1930. "Zur Theorie des Durchgangs schneller Korpuskularstrahlen durch Materie", *Ann. Phys.*, 397, No. 3, 325-400.

Bethe H., 1932. "Bremsformel für Elektronen relativistischer Geschwindigkeit", *Ztschr. Phys.*, 76, No. 5-6, 293-299.

Bethe H. and W. Heitler, 1934. "On the Stopping of Fast Particles and on the Creation of Positive Electrons", *Proc. Roy. Soc. London*, Ser. A, A146, No. 856, 83-112.

Blackett P.M.S., 1938. "On the Instability of the Barytron and the Temperature Effect of Cosmic Rays", *Phys. Rev.*, Ser. II, 54, No. 11, 973-974.

Bloch F., 1933. "Bremsvermögen von Atomen mit mehreren Elektronen", *Ztschr. Phys.*, 81, No. 5-6, 363-376.

Boguslavsky S.A., M1929. *Paths of Electrons in Electromagnetic Fields*, Moscow-Leningrad. In Russian

Bohr N., 1915. "On the Decrease of Velocity of Swiftly Moving Electrified Particles in Passing Through Matter", *Phil. Mag.*, Ser. 6, 30, No. 178, 581-612.

Bothe W. and W. Kolhörster, 1929. "Das Wesen der Hohenstrahlung", *Ztschr. Phys.*, 56, No. 11-12, 751-777.

Bothe W. and W. Kolhörster, 1930. "Vergleichende Höhenstrahlungs messungen auf nordlichen Meeren", *Berl. Ber.*, 26, 450-456.

Bowen F, R.A. Millikan, and H.V. Neher, 1933 "New High-Altitude Study of Cosmic-Ray Bands and a New Determination of Their Total Energy Content", *Phys. Rev.*, Ser. II, 44, No. 4, 246-252.

Bowen I.S., R.A. Millikan, and H.V. Neher, 1937. "The influence of the Earth's magnetic field on cosmic rays intensities up to the top of the atmosphere", *Phys. Rev.*, Ser. II, 52, No. 2, 80-88.

Bowen I.S., R.A. Millikan, and H.V. Neher, 1938. "New light on the nature and origin of the incoming cosmic rays", *Phys. Rev.*, Ser. II, 53, No. 11, 861-885 (1938).

Brüche E., 1930. "Experimente zu Störmers Polarlichtteorie", *Phys. Ztschr.*, 31, No. 22, 1011-1015.

Brüche E., 1931. "Wo erreichen kosmische Elektronenstrahlen im die Erd Kugel", *Phys. Ztschr.*, 32, No. 1, 31-37.

Carlson Per, 2012. "A century of cosmic rays", *Phys. Today*, 65, No. 2, 30-36.

Clay J., 1927. "Penetrating Radiation", *Proc. Roy. Acad. Amsterdam*, 30, 1115-1127.

Clay J., 1928. "Penetrating Radiation II", *Proc. Roy. Acad. Amsterdam*, 31, 1091-1097.

Clay J., 1930. "Ultraradiation (penetrating radiation). III. Annual variation and variation with the geographical latitude", *Proc. Roy. Acad. Amsterdam*, 33, No. 7, 711-718.

Clay J., 1932. "Earth-Magnetic Effect and the Corpuscular Nature of the Cosmic Radiation", *Proc. Roy. Acad. Amsterdam*, 35, 1282-1290 (1932).

Clay J., 1933. "Results of the Dutch cosmic ray expedition 1933", *Physica*, 1, No. 5, 363-382.

Clay J. and H. Berlage, 1932. "Variation der Ultrastrahlung mit der geographischen Breite und dem Erdmagnetismus", *Naturwissenschaften*, 20, No. 37, 687-688.

Compton A.H., 1932. "Progress of cosmic ray survey", *Phys. Rev.*, Ser. II, 41, No. 5, 681-682.

Compton A.H., 1933. "A geographic study of cosmic rays", *Phys. Rev.*, Ser. II, 43, No. 6, 387-403.

Compton A.H., 1936. "Recent developments in cosmic rays", *Rev. Sci. Instr.*, 7, 71-81.

Compton A.H. and R.N. Turner, 1937. "Cosmic Rays on the Pacific Ocean", *Phys. Rev.*, Ser. II, 52, 799-814.

Corlin A., 1934. "A New Hard Component of the Cosmic Ultra-Radiation", *Nature*, 133, No. 3350, 63-63.

Coulomb C.A., 1785. "Troisième Mémoire sur l'Electricité et le Magnetismo", *Mémoires de l'Académie Royale des Sciences*, Paris, 612-638.

Curie M., 1898. "Rayons e'mis par les composés de l'uranium et du thorium", *Comptes rendus*, Paris, 126, No. 15, 1101-1103.

Curie M., 1899. " Les rayons de Becquerel et le polonium", *Rev. Gen. Sci.*, 10, 41-50.

Dirac P.A.M., 1926. "Relativity Quantum Mechanics with an Application to Compton Scattering", *Proc. Roy. Soc. London*, Ser A, A111, No. 758, 405-423.

Dukelsky V.M. and N.S. Ivanova, 1935 "Influence of filtration on cosmic ray azimuthally asymmetry", *JETP*, 5, No. 6, 512-519. In Russian.

Elster J. and H. Geitel, 1900. "Weitere Versuche über die Elektrizitäts zerstreuung in abgeschlossenen Luftmengen", *Phys. Ztschr.*, 2, No. 38, 560-563.

Fermi E. and B. Rossi, 1933. "Azione del campo magnetico terrestre sulla radiazione penetrante", *Rend. Lincei*, 17, 346-350.

Geiger H., 1924. "Über die Wirkungsweise des Spitzenzählers", *Zeitschr. Phys.*, 27, No. 1, 7-11.

Geiger H and W. Muller, 1928. "Elektronenzählrohr zur Messung schwächster Aktivitäten", *Naturwissenschaften*, 16, No. 31, 617-618.

Geitel H., 1900. "Über die Elektrizitätszerstreuung in abgeschlossenen Luftmengen", *Phys. Ztschr.*, 2, No. 8, 116-119.

Gockel A., 1911. "Messungen der durchdringenden Strahlung bei Ballonfahrten", *Phys. Ztschr.*, 12, No. 14, 595-597.

Gross B., 1933. "Zur Absorption der Ultrastrahlung", *Zeitschr. Phys.*, 83, No. 3-4, 214-221.

Gross B., 1934. "Zur mittleren Durchdringungsvermogen der Ultrastrahlung", *Phys. Ztschr.*, 35, No. 17, 746-747.

Hess V.F., 1911. "Über die Absorption der γ-Strahlen in der Atmosphäre", *Phys. Ztschr.*, 12, No. 22-23, 998-1001.

Hess V.F., 1912. "Über Beobachtungen der durchdringenden Strahlung bei sieben Freiballonfahrten", *Phys. Ztschr.*, 13, No. 21-22, 1084-1091.

Hoffmann G., 1925. "Registrierbeobachtungen der Höhenstrahlung im Meeresniveau", *Phys. Ztschr.*, 26, No. 1, 40-43.

Hoffmann G., 1926. "Registrierbeobachtungen der Höhenstrahlung im Meeresniveau", *Ann. Phys.*, 385, No. 16, 779-807.

Hoffmann G., 1932. "Probleme der Ultrastrhlung", *Phys. Ztschr.*, 33, No. 17, 633-662.

Ioffe A.F., 1935. "Cosmic rays", *Proc. of All-Union Conf. on Stratosphere Research*, Moscow-Leningrad, Academy of Sciences USSR Press, 385-388. In Russian.

Johnson T.H., 1933a. "The Azimuthal Asymmetry of the Cosmic Radiation", *Phys. Rev.*, Ser. II, 43, No. 10, 834-835.

Johnson T.H., 1933b. "Preliminary Report of the Results of Angular Distribution Measurements of the Cosmic Radiation in Equatorial Latitudes", *Phys. Rev.*, Ser. II, 44, No. 10, 856-858.

Johnson T.H., 1938. "Cosmic-Ray Intensity and Geomagnetic Effects", *Rev. Mod. Phys.*, 10, No. 4, 193-244.

Johnson T.H. and D.V. Read, 1937. "Unidirectional Measurements of the Cosmic-Ray Latitude Effect", *Phys. Rev.*, Ser. II, 51, No. 7, 557–564.

Kevles D.J., 1978. "Physicists and the Revolt Against Science in the 1930's", *Phys. today*, 31, No. 2, 23-30.

Klein O. and Y. Nishina, 1929. "Über die Streuung von Strahlung durch freie Elektronen nach der neuen relativistischen Quantendynamik von Dirac", *Zeitschr. Phys.*, 52, No. 11-12, 853-868.

Kolhörster W., 1913a. "Über eine Neukonstruktion des Apparates zur Messung der durchdringenden Strahlung nach Wulf und die damit bisher gewonnen Ergebnisse", *Phys. Ztschr.*, 14, No. 21, 1066-1069.

Kolhörster W., 1913b. "Messungen der durchdringenden Strahlung im Freiballon in größeren Höhen", *Phys. Ztschr.*, 14, No. 22-23, 1153-1156.

Kolhörster W., 1923. "Intensitates und Richtungsmessungen der durchdringenden Strahlung", *Berl. Ber.*, 24, 366-377.

Kolhörster W. and L. Tuwim, 1931. "Die spezifische Ionisation der Höhenstrahlung" *Zeitschr. Phys.*, 73, No. 1-2, 130-136.

Kramer W., 1933. "Die Absorption der durchdringenden Strahlung in Wasser und die Analyze der Absorptionsfunktion", *Zeitschr .Phys.*, 85, No. 7-8, 411-434.

Kurz K., 1909. "Die radioaktiven Stoffe in Erde und Luft als Ursache der durchdringenden Strahlung in der Atmosph¨are", *Phys. Ztschr.*, 10, No. 22, 834-845.

Kyker G.C. and A.R. Liboff, 1978. "Absolute cosmic ray ionization measurements in a 900-liter chamber", *J. Geophys. Res.*, A83, No. 12, 5539-5549.

Lemaitre G. and M.S. Vallarta, 1933. "On Compton's Latitude Effect of Cosmic Radiation", *Phys. Rev.*, Ser. II, 43, No. 2, 87-91.

Lemaitre G. and M.S. Vallarta, 1936. "On the Allowed Cone of Cosmic Radiation", *Phys. Rev.*, Ser. II, 50, No. 6, 493-504.

Millikan R.A., 1924. "Atoms and Ethereal Radiations", *Nature*, 114, No. 2856, 141-143.

Millikan R.A., 1930a. "History of research on cosmic rays", *Nature*, 126, No. 3166, 14-16, 29-30.

Millikan R.A., 1930b. "On the Question of the Constancy of the Cosmic Radiation and the Relation of these Rays to Meteorology", *Phys. Rev.*, Ser. II, 36, No. 11, 1596-1603.

Millikan R.A., M1935. *Electrons (+ and −), protons, photons, neutrons and cosmic rays.* Chicago.

Millikan R.A., M1939. *Electrons (+ and −), protons, photons, neutrons and cosmic rays,* GONTI, Moscow-Leningrad. In Russian.

Millikan R.A. and J. Bowen, 1923. "Penetrating radiation at high altitudes", *Phys. Rev.*, Ser. II, 22, No. 2, Minutes of the Pasadena Meeting, May 5, 1923, 198-198.

Millikan R.A. and G.H. Cameron, 1926. "High Frequency Rays of Cosmic Origin III. Measurements in Snow-Fed Lakes at High Altitudes", Phys. Rev., Ser. II, 28, No. 5, 851-868.

Millikan R. and G.H. Cameron, 1928a. "New Precision in Cosmic Ray Measurements; Yielding Extension of Spectrum and Indications of Bands", *Phys. Rev.*, Ser. II, 31, No. 6, 921-930.

Millikan R. and G.H. Cameron, 1928b. "The Origin of the CR", *Phys. Rev.*, Ser. II, 32, No. 4, 533-557.

Millikan R.A. and G. Cameron, 1928c. "New on cosmic rays", *Physics Uspehi (UFN)*, 8, No. 2, 121-140 (1928). In Russian.

Millikan R.A. and G.H. Cameron, 1928d. "High altitude tests on the geographical directional and spectral distribution of cosmic rays", *Phys. Rev.*, Ser. II, 31, No. 2, 163-173.

Myssowsky L.V., M1929. *Cosmic Rays*, Gosizdat, Moscow-Leningrad. In Russian.

Myssowsky L. and L. Tuwim, 1926a. "Versuche über die Absorption der Höhenstrahlung im Wasser", *Zeitschr. Phys.*, 35, No. 4, 299-303.

Myssowsky L. and L. Tuwim, 1926b. "Versuche über die Richtung der Höhenstrahlung im Meeresniveau", *Zeitschr. Phys.*, 36, No. 8, 615-622.

Myssowsky L. and L. Tuwim, 1928. "Absorption in Blei, sekundäre Strahlen und Wellenlänge der Höhenstrahlung", *Zeitschr. Phys.*, 50, No. 3-4, 273-292.

Pfotzer G., 1936. "Dreifachkoinzidenzen der Ultrastrahlung aus vertikaler Richtung in der Stratosphäre", *Zeitschr. Phys.*, 102, No. 1-2, 23-40.

Piccard A. and M. Cosyns, M1933. *Etude du Rayonnement Cosmique*, Bruzelles, Marcel.

Regener E., 1929. "Messungen über das kurzwellige Ende der durchdringenden Höhenstrahlung", *Naturwissenschaften*, 17, No. 11, 183-185.

Regener E., 1932a. "Über das Spektrum der Ultrastrahlung I. Die Messungen im Herbst 1928", *Zeitschr. Phys.*, 74, No. 7-8, 433-454.

Regener E., 1932b. "Messung der Ultrastrahlung in der Stratosphäre", *Naturwissenschaften*, 20, No. 38, 695-699.

Regener E., 1933. "Die Absorptionskurve der Ultrastrahlung und ihre Deutung", *Phys. Ztschr.*, 34, No. 8, 306-323.

Regener E. and R. Auer, 1934. "Weitere Messungen der Ultrastrahlung in der oberen Atmosphäre mit offenen Ionisationskammern", *Phys. Ztschr.*, 35, No. 19, 784-788.

Regener E. and G. Pfotzer, 1934. "Messungen der Ultrastrahlung in der oberen Atmosphäre mit dem Zählrohr", *Phys. Ztschr.*, 35, No. 19, 779-784.

Rossi B., 1930a. "Method of Registering Multiple Simultaneous Impulses of Several Geiger's Counters", *Nature*, 125, No. 3156, 636-636.

Rossi B., 1930b. "On the Magnetic Deflection of Cosmic Rays", *Phys., Rev.*, Ser. II, 36, No. 3, 606-606.

Rossi B., 1931. "Measurements on the Absorption of the Penetrating Corpuscular Rays coming from Inclined Directions", *Nature*, 128, No. 3227, 408-408.

Rossi B., 1932a. "Absorptionsmessungen der durchdringenden Korpuskularstrahlung in einem Meter Blei", *Naturwissenschaften*, 20, No. 4, 65-65.

Rossi B., 1932b. "Nachweis einer Sekundarstrahlung der durchdringenden Korpuskularstrahlung", *Phys. Ztschr.*, 33, No. 7, 304-305.

Rossi B., 1933. "Über die Eigenschaften der durchdringenden Korpuskularstrahlung im Meeresniveau", *Zeitschr. Phys.*, 82, No. 3-4, 151-178.

Rossi B., 1934. "Directional Measurements on the Cosmic Rays Near the Geomagnetic Equator", *Phys. Rev.*, Ser. II, 45, No. 3, 212-214.

Rossi B., M1966, *Cosmic Rays*, Atomizdat, Moscow. In Russian.

Rumbaugh L. and G. Locher, 1936. "Neutrons and other heavy particles in cosmic radiation of the stratosphere", *Phys. Rev.*, Ser. II, 49, No. 11, 855-855.

Rutherford E. and H.L. Cooke, 1903. "A Penetrating Radiation from the Earth's Surface", *Phys. Rev.*, 1st Ser., 16, No. 3, Minutes of the Eighteenth Meeting, 183-183.

Schein M., W.P. Jesse, and E.O. Wollan, 1941. "The Nature of the Primary Cosmic Radiation and the Origin of the Mesotron", *Phys. Rev.*, Ser. II, 59, No. 7, 615-615.

Schindler H., 1931, "Übergangseffekte bei der Ultrastrahlung", *Zeitschr. Phys.*, 72, No. 9-10, 625-657.

Skobelzyn D.V., 1924. "Investigation of γ-rays by Wilson's method in connection with the problem on ray's energy scattering", *ZhRFHO*, 56, No. 2-3, 120-125. In Russian.

Skobelzyn D., 1927. "Die Intensitätsverteilung in dem Spektrum der γ-Strahlen von Ra C", *Zeitschr. Phys.*, 43, No. 5-6, 354-378.

Skobelzyn D., 1929a. "Die spektrale Verteilung und die mittlere Wellenlänge der Ra-γ-Strahlen", *Zeitschr. Phys.*, 58, No. 9-10, 595-612.

Skobelzyn D., 1929b. "Über eine Neue Art sehr Schneller β-Strahlen", *Zeitschr. Phys.*, 54, No. 9-10, 686-702.

Skobelzyn D.V., 1934. "Problems of CR", *Proc. 1-st All-Union Nuclear Conf.*, Moscow-Leningrad, 65-112. In Russian.

Skobelzyn D.V., M1936, *Cosmic Rays*, ONTI, Moscow-Leningrad. In Russian.

Steinke E., 1927. "Über die durchdringende Strahlung im Meeresniveau", *Zeitschr. Phys.*, 42, No. 8, 570-602.

Steinke E., 1928. "Neue Untersuchungen über die durchdringende Hesssehe Strahlung", *Zeitschr. Phys.*, 48, No. 9-10, 647-689.

Steinke E., 1929. "Wasserversenkmessungen der durchdrinkenden Hessschen Strahlung", *Zeitschr. Phys.*, 58, 183-193.

Störmer C., 1907. "Sur les trajectories des corpuscles electrises dans l'espace sous l'action du magnetisme terrestre avec application aux aurores boreales", *Arch. sci. phys. et natur.*, Geneve, Ser. 4, 24, 5-18, 113-158, 221-247, 317-364.

Störmer C., 1911. "Sur les trajectories des corpuscles electrises dans l'espace sous l'action du magnetisme terrestre avec application aux aurores boreales", *Arch. sci. phys. et natur.*, Geneve, Ser. 4, 32, 117-123, 190-219, 277-314, 415-436, 501-509.

Störmer C., 1912. "Sur les trajectories des corpuscles electrises dans l'espace sous l'action du magnetisme terrestre avec application aux aurores boreales", *Arch. sci. phys. et natur.*, Geneve, Ser. 4, 33, 51-69, 113-150.

Störmer C., 1930. "Periodische Elektronenbahnen im Felde eines Elementarmagneten und ihre Anwendung auf Brüches Modellversuche und auf Eschenhagens Elementarwellen des Erdmagnetismus, *Zeitschr. Astrophys.*, 1, No. 4, 237-274.

Störmer C., 1931. "Ein Fundamentalproblem der Bewegung einer elektrisch geladenen Korpuskel im kosmischen Raume. Erster Teil", *Zeitschr. Astrophys.*, 3, No. 1, 31-52.

Störmer C., 1934. "Critical Remarks on a Paper by G. Lemaitre and M. S. Vallarta on Cosmic Radiation", *Phys. Rev.*, Ser. II, 45, No. 11, 835-838.

Swann W. and G. Locher, 1935. "Measurements of the Angular Distribution of Cosmic-Ray Intensities in the Stratosphere with Geiger-Muller Counters", *Phys. Rev.*, Ser. II, 47, No. 4, Minutes of the Pittsburgh Meeting, December 27-29, 1934, 326-326.

Tamm Ig., 1930. "Über die Wechselwirkung der freien Elektronen mit der Strahlung nach der Diracsehen Theorie des Elektrons und nach der Quantenelektrodynamik", *Zeitschr. Phys.*, 62, No. 7-8, 545-568.

Tamm I.E., 1975. "On interaction of free electrons with radiation according to Dirac's theory on electrons and quantum electrodynamics", In Collection of Scientific Papers of Academician I.E. Tamm, Nauka, Moscow, Vol. 2, 24- . In Russian.

Verigo A.B., 1934. "Discussion on report of D.V. Skobelzyn", *Proc. 1-st All-Union Nuclear Conf.*, Moscow-Leningrad, 114-114. In Russian.

Verigo A.B., 1935. "The using of ionization method in the research of cosmic rays in the stratosphere", *Proc. of All-Union Conf. on Stratosphere Research*, Moscow-Leningrad, Academy of Sciences USSR Press, 413-421. In Russian.

Verigo A.B., 1937. "Results of cosmic ray investigations during the flight of stratostat 'USSR-1bis'", *Nature (Russian)*, 26, No. 8, 16-29 (1937). In Russian.

Verigo A.B., 1938. "Measurements of cosmic ray intensity in Arctic", *Proc. of Gosud. Radievii Institute (GRI)*, 4, 183-206. In Russian.

Vernov S.N. and N.A. Dobrotin, 1977a. "50-th anniversary of fundamental discovery in cosmic ray physics", *Physics Uspekhi (UFN)*, 123, No. 3, 531-535. In Russian.

Vernov S.N., 1934. "On the Study of Cosmic Rays at the Great Altitudes", *Phys. Rev.*, Ser. II, 46, No. 9, 822-822.

Vernov S.N., 1935a. "Radio-Transmission of Cosmic Ray Data from the Stratosphere", *Nature*, 135, No. 3426, 1072-1073.

Vernov S.N., 1935b. "The using of Geiger-Muller counters for cosmic ray research in stratosphere", *Proc. of All-Union Conf. on Stratosphere Research*, Moscow-Leningrad, Academy of Sciences USSR Press, 423-427. In Russian.

Vernov S.N., 1937. "Cosmic ray measurements in stratosphere at magnetic latitude 35°", *DAN USSR*, 14, No. 5, 263-266. In Russian.

Vernov S.N., 1938. "Cosmic ray latitude effect in stratosphere", Izvestia Ac. of Sci. USSR, Series Phys., No. 5/6, 738-740. In Russian.

Vernov S.N., 1938a. "Cosmic ray research in stratosphere by radio-transmission signals", *Izvestia of Academy of Sciences of USSR*, Series Phys., 2, No. 1/2, 121-122. In Russian.

Vernov S.N., 1938b. "The cosmic ray latitude effect in stratosphere", *Izvestia of Academy of Sciences of USSR*, Series Phys., 2, No. 5/6, 738-740. In Russian.

Vernov S.N., 1939. "Analysis of the cosmic ray latitude effect in stratosphere", *Reports of Academy of Sciences of USSR (DAN USSR)*, 23, No. 2, 141-143. In Russian.

Vernov S.N. and N.A. Dobrotin, 1977b. "Fiftieth anniversary of a fundamental discovery in cosmic ray physics" *Sov. Phys. Usp.*, 20, No. 11, 970-972.

Wilson C.T.R., 1900. "On the Leakage of Electricity Through Dust-Free Air", *Proc. Cambr. Phil. Soc.*, 11, No. 1, 32-36.

Wilson C.T.R., 1901. "On the Ionisation of Atmospheric Air", *Proc. Roy. Soc. London*, 68, No. 444, 151-161.

Wilson C.T.R., 1925. "Acceleration of β-particles in Strong Electric Fields such as those of Thunderclouds", *Proc. Cambr. Phil. Soc.*, 22, No. 4, 534-538.

Wulf T., 1909. "Über die in der Atmosphäre vorhandene Strahlung von hoher Durchdringungsfähigkeit", *Phys. Ztschr.*, 10, No. 5, 152-157.

Wulf T., 1910. "Beobachtungen über die Strahlung hoher Durchdringungs fähigkeit auf dem Eiffelturm", *Phys. Ztschr.*, 11, No. 18, 811-813.

In: Homage to the Discovery of Cosmic Rays …
Editor: Jorge A. Perez-Peraza

ISBN: 978-1-63483-084-3
© 2015 Nova Science Publishers, Inc.

Chapter 2

HISTORY OF THE FIRST ANTI-PARTICLE DISCOVERY IN COSMIC RAYS (DEDICATED TO 80 YEARS OF POSITRON DISCOVERY)

Lev I. Dorman[1,2] and Irina V. Dorman[3]

[1]Israel Cosmic Ray and Space Weather Centre with Emilio Segré Israel-Italian Observatory, affiliated to Tel Aviv University, Golan Research Institute, and Israel Space Agency, Israel
[2]IZMIRAN of Russian Academy of Sciences, Moscow, Russia
[3]Institute of History of Science and Technology of RAN, Moscow, Russia

ABSTRACT

We describe in detail the dramatic history of how the positive electron (positron) was discovered in 1932. This discovery influenced the development of the fundamental basis of elementary particle physics very much, and in 1936 it was recognized by the Nobel Prize in Physics.

1. CONNECTION OF THE PROBLEM OF COSMIC RAY NATURE WITH THE PROBLEM OF PARTICLES EXISTING IN NATURE

The problem of the nature of CR has been closely connected with the problem of what particles exist in nature and how high energy particles behave. For many years, it was necessary to solve these two problems simultaneously. There were not yet particle accelerators, and CR were for many years a unique source of high energy particles. The physics of CR and the physics of high energy particles were synonymous.

Research of CR in the basic area of energies resulted in many remarkable discoveries, first and foremost - discovery of new particles. The positron was the first such particle.

2. Relativistic Quantum Equation for Electrons of P.A.M. Dirac as the Theoretical Basis for Existence of Positrons

In 1928 the young English physicist-theorist P.A.M. Dirac (he was at that time only 26 years old) made an attempt to combine the main principles of quantum mechanics with the special theory of relativity and developed the relativistic quantum equation for electrons. This equation consistently considered the presence at electron's spin, and explained the thin structure of the spectrum of the hydrogen atom. However, according to the majority of theorists known at that time, Dirac's theory was lacking in that it predicted the existence of particles with negative energy ($E < 0$). A little later, P.A.M. Dirac (1930, 1931) stated an ingenious guess that actual conditions with negative energy are filled, and a "hole" (empty seat) in this filled background corresponds to a particle with positive energy, but having a charge opposite to a charge of electron, i.e. P.A.M. Dirac postulated the existence of a positively charged antiparticle (antielectron). In the beginning, this assumption seemed absolutely unreal, and Dirac tried to identify the antiparticle necessary to it with a proton as other positively charged particles at that time were not yet known. However, according to the theory, the weight of an antiparticle should equal precisely the weight of electron, i.e. actually it was a problem of the existing anti-electron - a positive electron (positron).

The discovery in 1932 by C.D. Anderson of a positron in CR was a triumph of the theory of P.A.M. Dirac. The connection between a rather narrow area, - physics of CR,- and fundamental problems of all physics became obvious. The discovery of the first antiparticle in cosmic rays started from using Wilson's chamber inside a strong magnetic field. In the summer of 1930, R.A. Millikan and his young employer Carl Anderson designed in the laboratory of the Californian Institute of Technology a vertical Wilson's chamber in which, unlike all previously used chambers, the piston moved upwards and downwards, instead of horizontally. Let us note that the fact that the positron was found for the first time in CR caused later, already after positron discovery, surprise to many experimenters. So, P.M.S. Blackett (M1935, page 64) wrote about it: "positron discovering in cosmic radiation is represented for the first time casual because the positron could be without difficulty observed and in laboratories at least about five years before". By this P.M.S. Blackett meant the laboratory experiments with γ-radiation of radioactive sources.

3. Using Wilson's Chamber inside of a Strong Magnetic Field for Measuring of Particle Energy

Nevertheless, the positron was discovered in CR, and the history of this discovery starts from using Wilson's chamber inside of a strong magnetic field. D. Skobelzyn (1929), as a result of the experiments with Wilson's chamber placed in a magnetic field of intensity 1000 Gs came to the conclusion that the bottom limit of energy of particles of cosmic radiation is 20 MeV. However, it seemed doubtless that energy of primary and even secondary particles of CR possessing huge penetrating ability should be much more. Therefore it was of great interest to create an experimental installation for measurement of energy of particles in cosmic radiation. Several groups of experimenters in different countries were engaged to solve this problem.

As P.S. Epstein (1948) in his article, devoted to the Robert Millikan's 80 anniversary from the date of its birth wrote that, R.A. Millikan with intuition inherent in it has seen that the method offered by D.V. Skobelzyn, opens "new promising possibilities for the decision of a problem on the nature of CR and the mechanism of their absorption".

Figure 1. C.D. Anderson with Wilson vertical chamber in the strong magnetic field, designed in the Laboratory of R.A. Millikan.

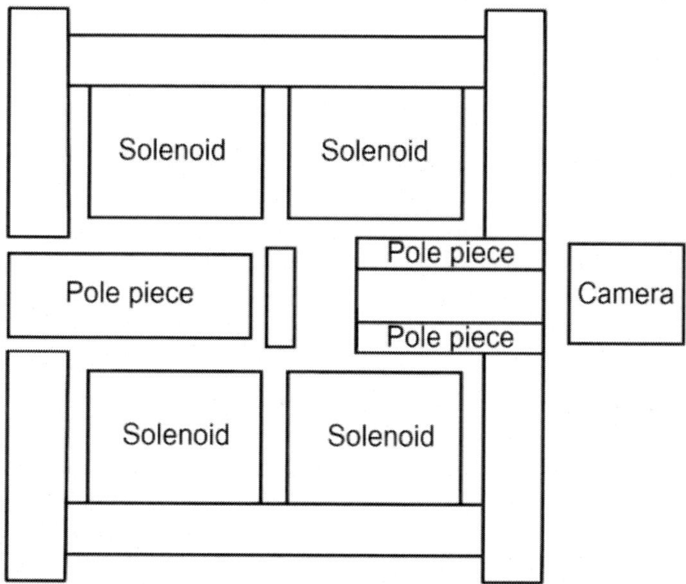

Figure 2. The scheme of installation for measurement of energy and charge of particles of the cosmic radiation, constructed in the Californian Institute of Technology.

In the summer of 1930, R.A. Millikan and his young employee Carl Anderson designed in the laboratory of the Californian Institute of Technology a vertical Wilson's chamber in which unlike all previously used chambers the piston moved upwards and downwards instead of horizontally. As it was supposed that the basic part of incoming radiation comes from above, vertically flying particles should leave in the chamber traces bigger lengths, and this gave the chance to find out and approximately to measure even their insignificant curvature. The chamber in the size $17 \times 17 \times 4$ cm^3 was located between poles of a powerful magnet on whose coil the current in 2000 A was passed (the generator was necessary for production of such current almost of 1000 horsepower). The scheme and appearance of the installation created by R.A. Millikan and C.D. Anderson, are shown in Figures 1 and 2.

At the maximum intensity of magnetic field H = 24,000 Gs it was possible to measure curvature of trajectories of particles in the installation with radius up to 7 m that corresponds to magnetic rigidity of Hρ = 1.7×10^7 Gs.cm. The energy of such particles (if their weight is equal to weight of electron) makes E = 300×Hρ eV= 5×10^9 eV = 5,000 MeV that in 250 times exceeds values of energies, accessible to measurements before.

4. Discussion on Two Photos Obtained by C. D. Anderson at 1931

In the summer of 1931, C.D. Anderson received the first results, described in Anderson (1932). He obtained trajectory photos of many particles in a wide energy range up to several GeV. Besides, C.D. Anderson noticed that approximately half of the trajectories of particles in the magnetic field deviate in one direction, and the second half - in another. Assuming the charged particles of CR fly from top to bottom, C.D. Anderson drew the conclusion that half of particles have a negative charge, and half - positive. Two photos, received in 1931, on which the traces of two particles with charges of different signs starting from one centre are clearly visible, are presented in Figures 3 and 4.

"These photos were shown by me at lectures in Cambridge and in Paris at Poincare's Institute in November, 1931, - recollected Robert Millikan (M1935; M1939, page 222). - The first of them (see Figure 3), as I considered, showed that CR beat out from an atomic kernel a proton whose trace is bent downwards and to the right, and a negative electron, the left trace going downwards and to the left. This extraordinary photo was published for the first time in December, 1931 together with a picture of C.D. Anderson under the name: "CR destroy nuclear kernels". Therefore, in the beginning, R.A. Millikan and C.D. Anderson decided that positively charged particle is a proton. Traces of particles in the first photo (Figure 3) confirmed their opinion as ionization along the trace of a positive particle with energy of 130 MeV was much greater than ionization along a trace of a negative particle with energy of 120 MeV. There is an opinion that this photo delayed positron discovery for at least eight months, as only later it became clear that out of several thousand photos investigated by C.D. Anderson, only in one photo was there a trace definitely belonging to a proton. In general, it appeared that photos on which traces of electrons and protons are visible from one point simultaneously were extremely rare.

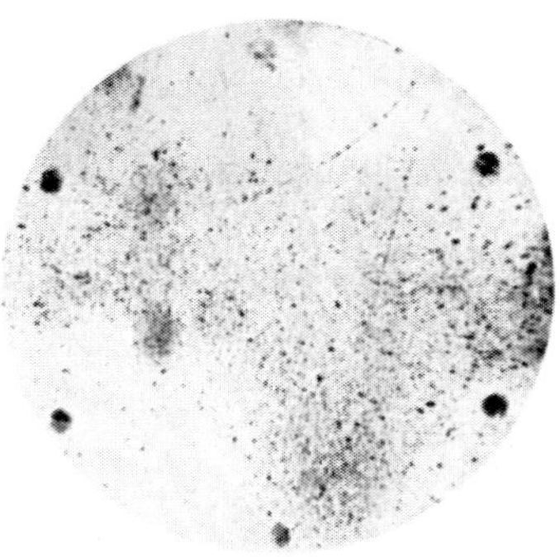

Figure 3. Photo received by C.D. Anderson (1932) in the summer of 1931 on which are visible traces of two particles. More thin trace belongs to electron with energy of 120 MeV, and thicker - to proton with energy of 130 MeV. Magnetic field intensity – 17,000 Gs.

Figure 4. One of the first photos received by C.D. Anderson (1932). Traces of two particles are clearly visible. The particle which has left a trace of the big curvature, is electron with energy of 27 MeV. The trace of smaller curvature is left then still by the unknown particle bearing a positive charge. Intensity of magnetic field was 12,000 Gs.

The second photo received almost at the same time (Figure 4), delivered to C.D. Anderson, according to R.A. Millikan, a lot of anxiety and caused long discussions. The left trace, according to R.A. Millikan and C.D. Anderson, belonged to an electron with an energy of 27 MeV, and the second trace, to a proton with an energy of 450 MeV. Surprise was caused by the fact that ionization along both traces was identical though already then it was clear that a proton of such energy should make on the way considerably more ionization than an electron. For an illustration, in Figure 5 it is shown how, according to H. Bethe (1932) calculations, ionization along a trajectory of electron and proton depends upon their energy, which was defined on curvature of trajectory of particle in the magnetic field.

In Figure 5 on the axis of ordinates is the loss of energy by a particle in water, in MeV per g/cm^2, and on an axis of abscissas $\log_{10}(H\rho)$. From the drawing it is visible that electrons and protons of big energy $E > 10^9$ eV possess identical ionizing ability, but at energy reduction the ionizing ability of a proton is much stronger than that of an electron. If the value of the product of $H\rho$ is less than 10^6 Gs.cm, protons possess ionizing ability in 6 times exceeding that of electrons, and on ionization made in Wilson's chamber the proton trace can be distinguished easily from a trace of the electron.

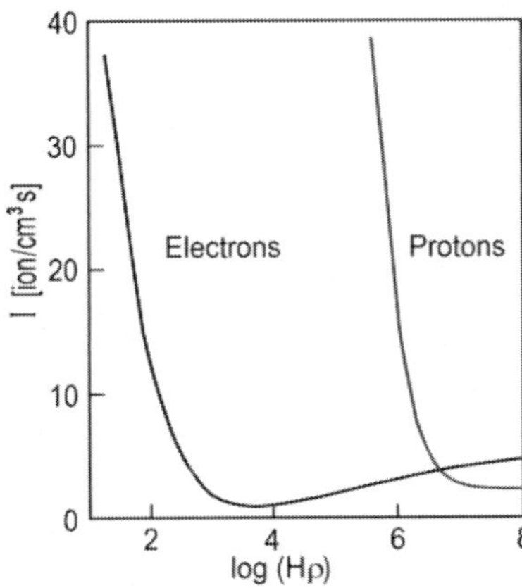

Figure 5. Dependence of ionizing ability of electrons and protons from their energy, according to H. Bethe (1932) calculations.

Hence, if the positive particle in photo Figure 4 is a proton, the proton of such energy should make ionization several times bigger than that of an electron. Meanwhile, ionizing ability can be estimated reliably enough on number of droplets per unit of the way formed by a particle along a trajectory in Wilson's chamber. The careful analysis made by C.D. Anderson showed that ionization along both traces was absolutely identical. "We could not find other exit from this difficulty, - recollected R.A. Millikan (M1935; M1939, page 233), - how to make the assumption that for areas such extraordinary high energies which until now anybody had not to face yet, the theory giving dependence of ionization from energy of protons, is in something incorrect. Besides, representations about the proton as about the basic unit of a positive electricity so have deeply taken roots that any hypothesis seemed at that time comprehensible if only with its help it was possible to keep usual representations".

5. DISCOVERY OF ELECTRON WITH A POSITIVE CHARGE (POSITRON) IN 1932

It was necessary to collect additional data which could confirm or deny the assumption that the positive particle observed in the experiences of Anderson is a proton, and at that time it was not so simple as it can seem now. We will recollect that operation and photographing in Wilson's chamber occurred casually and to photograph a particle trace, it was necessary for it to have passed through the chamber just before expansion. As a result, only in one photo out of 50 was there a trace of a particle of big energy, and only a small part of these photos showed a pair of particles starting with the general centre. Single traces were obviously insufficient, as never is it precisely known in what direction the particle moved through the chamber. Particles that were considered positive and moving from top to bottom actually could appear as normal negative electrons, coming from below.

C.D. Anderson found a brilliant way out. In Wilson's chamber, a horizontal lead plate of 6 mm in the thickness was placed. Having passed the plate, the particle lost a part of the energy, i.e. the value of $H\rho$ decreased, and it allowed the definition of its direction. It appeared that fears of C.D. Anderson were not in vain: in the Wilson's chamber particles, moving both from top to bottom, and from below upwards were observed.

On August 2, 1932, C.D. Anderson received a photo in which the existence of a positive particle with weight close to the weight of an electron was proved. This photo, having big historical significance resulted on Figure 6.

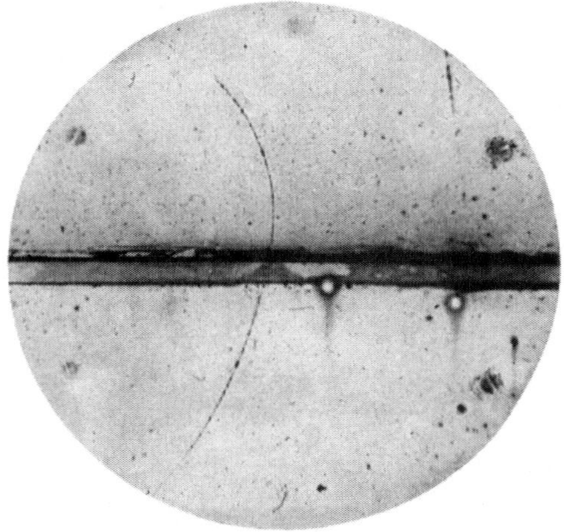

Figure 6. C.D. Anderson obtained this photo of great scientific and historical significance, at August 2, 1932. This photo proved the existence of the first antiparticle – the positron.

The trace is left by a particle, moving from below upwards. Without a lead plate, this trace certainly would be wrongly accepted for a trace of a normal electron, going from above. The direction of movement of the particle in this case was obvious, as the trace over the plate has considerably smaller radius of curvature, than under a plate, i.e. energy of a particle decreased after plate passage. Therefore, the particle has come from below, and on a deviation in a magnetic field, it was possible to conclude that it has a positive charge. If the particle had the weight of an electron, its energy in the beginning was 63 MeV ($H\rho = 2.1 \times 10^5$ Gs.cm), and after plate passage equalled 23 MeV ($H\rho = 7.5 \times 10^4$ Gs.cm).

If the particle in Figure 6 was a proton, after passage of the lead plate a proton with magnetic rigidity of $H\rho = 7.5 \times 10^4$ Gs.cm would have had the kinetic energy of only 0.3 MeV, and the density of ionization along its trace should be many times more than observed. Besides, such a proton would have run in the chamber gas only 5 mm, while the observable length of a trace of the particle in the photo is more than 5 cm, i.e. in 10 times more. "Having received and having shown a picture, C.D. Anderson has found out it at once relevancy, - R.A. Millikan (M1935; M1939, page 220) recollected, - and together with S.H. Neddermeyer has spent all night long in vain attempts to explain the phenomenon photographed by it, remaining within the limits of former representations. It has appeared impossible ". Therefore, C.D. Anderson, after night reflexions, came to the conclusion that the trace should belong to a "free positive electron". In September 1932, in the magazine "Science," the short message on discovery of a new particle - a positive electron was published (Anderson, 1932).

From September, 1932 until March, 1933 C.D. Anderson received a considerable quantity of photos confirming the existence of a positive electron. Doubts did not remain anymore! "From 1300 investigated photos of traces of space particles, - C.D. Anderson (1933a) in the paper which was sent to the magazine "The Physical Review" on February 28, 1933 wrote, - on 15 positive particles passing through lead any of which cannot have the same big weight as the weight of a proton are visible. Thus the existence of a positive particle of unit charge and weight a lot smaller than the weight of a proton is established". C.D. Anderson (1933b) showed that the specific ionization made by the new positive particle differs from the specific ionization of electron no more than by 20%. As the specific ionization made by the charged particle is proportional (at equality of other parameters) to a charge square, hence, the discovered positive particle charge could not differ from the charge of an electron by more than 10%. A little later, this new particle with the positive charge and mass of the electron was recognized as the antiparticle for electron and began being called a positron. For this discovery C.D. Anderson received the Nobel Prize in 1936 (together with Victor Hess, who received this prize for discovery of CR).

6. THE CHECKING OF THE POSITRON DISCOVERY BY OTHER RESEARCHES

Soon after the discovery by C.D. Anderson of a positive electron, other researchers published photos of similar traces of positive particles. One of the best installations for studying of energy of CR was constructed in Rostock (Germany) by P. Kunze (1933). Wilson's chamber was located in the solenoid weighing more than one ton through which winding the current in 1000 A. Inside of the solenoid was passed a magnetic field equal of 18,000 Gs for which it was required to use capacity of the whole electrical-power station. Therefore the installation was located inside the building of this station and worked only in the morning time of an underload of the power station. The current was passed through the solenoid directly ahead of chamber expansion of all within 2 - 3 seconds. During one day it was possible to make only 18 experiments, then the coil should cool down within 24 hours. The magnetic field which was created during passage electrical current through the solenoid was in all premise so great that small iron subjects flew around the room, and devices gave ridiculous readings. The Wilson's chamber in experiences of P. Kunze, as well as at C.D. Anderson, was located vertically. P. Kunze received some tens of photos of high quality on which traces of particles were clearly visible also their curvature it was possible to measure precisely. From the calculations of P. Kunze it followed that particles of CR have positive and negative charges, and their energy can reach 10 Gev.

7. EXPERIMENTS WITH WILSON'S CHAMBER IN THE MAGNETIC FIELD OPERATED BY GEIGER-MULLER COUNTERS AND CONFIRMING OF POSITRON EXISTENCE

While C.D. Anderson worked in the USA, P.M.S. Blackett and G.P.S. Occhialini (1933) in Cavendish Laboratory of the Cambridge University in England designed in 1932 a Wilson's chamber in a magnetic field operated by two Geiger-Muller counters. P.M.S. Blackett to those time was already well-known researcher in radio-activity by means of

Wilson's chamber, and G.P.S. Occhialini only began scientific activity." Usually photographing was made at random, - recollected subsequently P.M.S. Blackett (M1935, page 15), - only the small percent of pictures gave the ways connected with CR. However it is possible to make so that fast particles photographed themselves, having passed through two counters and influencing through these counters on the relay connected with the mechanism, making expansion".

P.M.S. Blackett and G.P.S. Occhialini arranged two Geiger-Muller counters as follows: one - over, and another - under the Wilson's chamber in such a manner that in the chamber expansion was made while the particle passed through both Geiger-Muller counters simultaneously. The probability of reception of pictures on which traces of particles of high energy were observed, in this case considerably increased. Using this witty method of connection of the counters included in the scheme of coincidence, and Wilson's chamber Blackett and Occhialini (1933) received in the spring of 1933 (at this time article of the C.D. Anderson was already printed) photos, in 75% of which were fixed traces of particles of high energy. Despite the fact that what researchers used was a rather weak magnetic field (from 2000 to 3000 Gs), they managed to establish that approximately half of traces were left by the particles having a positive charge with weight of an electron. Therefore the discovery of new particle made several months before by C.D. Anderson was confirmed.

8. ON THE CONNECTING OF THE POSITRON WITH DIRAC'S THEORY

In the same laboratory where the experiments were made, P.M.S. Blackett and G.P.S. Occhialini worked. P.A.M. Dirac, P.M.S. Blackett, and G.P.S. Occhialini, for whom the quantum-relativistic theory of electrons by P.A.M. Dirac and its difficulties was well known, identified the opened by C.D. Anderson positive electron with a particle postulated by P.A.M. Dirac. Moreover, at once all has risen on the places.

According to the theory of P.A.M. Dirac, the positron has small time of a life in substance as at a meeting with ordinary (negative) electron both particles disappear, annihilated and formed at annihilation energy $2mc^2$ + kinetic energy of positron, is transferred to two photons. For this reason positrons so are difficult to find. In turn high-energy photons possessing energy more than $2mc^2$ (i.e. than 1 MeV), under the Dirac's theory are capable of generating a pair of electron-positron.

9. EXPERIMENTALLY CHECKING OF DIRAC'S THEORY CONSEQUENCE ON THE GENERATION OF ELECTRON-POSITRON PAIRS BY HIGH-ENERGY PHOTONS

Experimentally the consequence followed from the Dirac's theory about generation of electron-positron pairs by high-energy photons has been checked up by J. Chadwick, P.M.S. Blackett and G.P.S. Occhialini (1934) in England, I. Curié and F. Joliot (1933) in France and C.D. Anderson (1933c) in USA. It was established that γ-radiation from the sample of a radioactive element ThC" falling on a lead plate, leads to the formation of electron-positron flux. In photos in Wilson's chamber traces electron and a positron, starting with one centre were clearly visible. Moreover, it appeared that positrons can be obtained at the bombardment by α-particles from a radioactive substance of polonium of a target from aluminum or a pine forest. In other words, the new kind of radioactive transformation - positron radioactivity and simultaneously - the phenomenon of artificial radioactivity was discovered. The calculations, which were made by H. Bethe and W. Heitler (1934) showed that at enough big energies photons are absorbed in substance at the expense of the process of formation of pairs, instead of for the account Compton dispersion as believed earlier. Now it began to be possible to explain the strong absorption noticed by earlier many experimenters in substance of γ-beams at great energies.

10. FUNDAMENTAL ROLE OF POSITRON DISCOVERY FOR DEVELOPMENT OF EXPERIMENT AND THEORY OF ELEMENTARY PARTICLE PHYSICS

Positron discovery excited both theorists and experimenters studying CR. The first understood that research of CR can promote understanding of fundamental problems of elementary particle physics, and the second realized the necessity of application of relativistic quantum mechanics for explanation of experimental data.

"Thus discovery of one more, before an unknown particle was essential at all, - underlined W. Heisenberg (1976a, 1977a), - the set of particles without some serious consequences for the physics bases has been discovered still. By discovering of positron it was essential that really it was also discovered new symmetry, the association of particles-antiparticles closely connected with Lorenz group of the special theory of relativity, and also the transformation of kinetic energy of interacting particles in weight of rest in new particles and back".

The discovery of the positron was extremely important for physics of elementary particles because before, it was considered that there are only two fundamental particles: electrons and protons. Analysing the situation created in physics in the 1930s, W. Heisenberg (1976b, 1977b) recollected that "the matter was represented constructed finally from electrons and protons. Experiences of the C.D. Anderson and P.M.S. Blackett with all definiteness have proved that this hypothesis about a matter structure was incorrect. Electrons could be born and destroyed; their number could vary, and they could not be 'elementary' in initial understanding of this word. There was a necessity to reconsider the definition of concept 'simplicity'."

11. EIGHTY ANNIVERSARY OF THE "YEAR OF MIRACLES" FOR PHYSICS OF COSMIC RAYS AND ELEMENTARY PARTICLE PHYSICS

In the history of physics of atomic nucleus, CR, and elementary particles the year 1932 left an indelible mark. It is enough to recollect that in this "Year of Miracles" as it was named by physicists, the positron and neutron were discovered, and the proton-neutron model of atomic nuclei was created. On February 27, 1932 in "Nature," the first message of J. Chadwick (1932) was published on the discovery of the neutron. Neutron discovery, i.e. a neutral variant of the proton, was unclearly forecast by E. Rutherford in 1920, and sufficiently changed the situation. It became possible to refuse definitively the assumption of the presence in atomic nuclei of electrons. Really, presence in nuclei of electrons seemed rather strange, first, because of the unclear nature of big width of β-spectra, and secondly, because of loss by them in a nucleus of their spin, and the problem with statistics (so-called "nitric catastrophe"). Creation of the proton-neutron model of atomic nuclei put in the forefront a problem of an origin of nuclear forces and stimulated big jump in CR physics in the middle and end of 1930s and later, in the 1940s.

ACKNOWLEDGMENTS

One of us (Lev Dorman) is thankful to Israeli Ministry of Science for support in the frame of the project "National Centre of Knowledge in Cosmic Rays and Space Weather". Our gratitude to David Shai Applbaum for checking and improve English in this paper.

REFERENCES

Anderson C.D., 1932a. "Energies of Cosmic-Ray Particles", *Phys. Rev.*, Ser. II, *41*, No. 4, 405-421.

Anderson C.D., 1932b. "The Apparent Existence of Easily Deflectable Positives", *Science*, *76*, No. 1967, 238-239.

Anderson C.D., 1933a. "The positive electron", *Phys. Rev.*, Ser. II, *43*, 491- 494.

Anderson C.D., 1933b. "Cosmic ray positive and negative electrons", *Phys. Rev.*, Ser. II, *44*, 406-416.

Anderson C.D., 1933c. "Free positive electrons resulting from the impact upon atomic nuclei of the photons from Th C", *Science*, *77*, No. 2001, 432-432.

Bethe H., 1932. "Bremsformel für Elektronen relativistischer Geschwindigkeit", *Ztschr. Phys.*, *76*, No. 5-6, 293-299.

Bethe H. and W. Heitler, 1934. "On the Stopping of Fast Particles and on the Creation of Positive Electrons", *Proc. Roy. Soc. London*, Ser. A, *A146*, No. 856, 83-112.

Blackett P.M.S., M1935. *Cosmic Radiation*, ONTI, Kharkov. In Russian.

Blackett P.M.S. and G.P.S. Occhialini, 1933. "Some Photographs of the Tracks of Penetrating Radiation", *Proc. Roy. Soc. London, Ser. A*, *A139*, No. 839, 699-720, 722, 724, 726.

Chadwick J., 1932. "Possible Existence of a Neutron", *Nature*, *129*, No. 3252, 312-312.

Chadwick J., P.M.S. Blackett, and G.P.S. Occhialini, 1934. "Some Experiments on the Production of Positive Electrons", *Proc. Roy. Soc. London*, Ser. A, *A144*, No. 351, 235-249.

Curié I. and F. Joliot, 1933. "Sur l'origine des électrons positives", *Comptes rendus*, Paris, *196*, No. 21, 1581-1583.

Dirac P.A.M., 1930. "A theory of electrons and protons", *Proc. Roy. Soc. London*, Ser A, *A126*, No.801, 360- 365.

Dirac P.A.M., 1931. "Quantized singularities in the electromagnetic field", *Proc. Roy. Soc. London*, Ser A, *A133*, No. 821, 60-72.

Epstein P.S., 1948. "Robert Andrews Millikan as Physicist and Teacher", *Rev. Mod. Phys.*, *20*, No. 1, 10-25.

Heisenberg W., 1976a. "Cosmic radiation and fundamental problems in physics", *Naturwissenschaften*, *63*, No. 2, 63-67.

Heisenberg W., 1976b. "The Nature of Elementary Particles", *Phys. today*, *29*, No. 3, 32-39.

Heisenberg W., 1977a. "Cosmic radiation and fundamental problem in physics", *Physics Uspekhi (UFN)*, *121*, No. 4, 669-677. In Russian.

Heisenberg W., 1977b. "The nature of elementary particles", *Physics Uspekhi (UFN)*, *121*, No. 4, 657-668. In Russian.

Kunze P., 1933. "Magnetische Ablenkung der Ultrastrahlen in der Wilsonkammer", *Zeitschr. Phys.*, *80*, No. 9-10, 559-572.

Millikan R.A., M1935. *Electrons* (+ *and* □), *protons, photons, neutrons and cosmic rays.* Chicago.

Millikan R.A., M1939. *Electrons* (+ *and* □), *protons, photons, neutrons and cosmic rays,* GONTI, Moscow-Leningrad. In Russian.

Skobeltzyn D., 1929. "Über eine Neue Art sehr Schneller β-Strahlen", *Zeitschr. Phys.*, *54*, No. 9-10, 686-702.

Vernov S.N. and N.A. Dobrotin, 1977. "50-th anniversary of fundamental discovery in cosmic ray physics", *Physics Uspekhi (UFN)*, *123*, No. 3, 531-535. In Russian.

In: Homage to the Discovery of Cosmic Rays …
Editor: Jorge A. Perez-Peraza

ISBN: 978-1-63483-084-3
© 2015 Nova Science Publishers, Inc.

Chapter 3

HISTORY OF THE FIRST MESON DISCOVERY IN COSMIC RAYS (DEDICATED TO THE 75TH ANNIVERSARY OF MUON DISCOVERY)

Lev I. Dorman[1,2] *and Irina V. Dorman*[3]

[1]Israel Cosmic Ray and Space Weather Centre with Emilio Segré Israel-Italian
Observatory, affiliated to Tel Aviv University,
Golan Research Institute, and Israel Space Agency, Israel
[2]IZMIRAN of Russian Academy of Sciences (RAN), Moscow, Russia
[3]Institute of History of Science and Technology of RAN, Moscow, Russia

ABSTRACT

The discovery of μ-mesons, unlike the discovery of positrons, was not the result of one individual observation, but a conclusion from a whole series of experimental and theoretical research projects between 1933-1937. In 1933, cosmic radiation was phenomenologically divided into two components with various factors of absorption in substances. Particles in the "soft" component were strongly absorbed in lead, thus forming cascade showers. Particles in the other, "hard," component passed through large thicknesses of substance (up to 1 m of lead), and it was established that their absorption degree in various substances is approximately proportional to the mass of the substance. From the analysis of the traces left by particles of the "hard" component in a Wilson's chamber, it then became clear that they should have the same charge as electrons, but with large mass. At first, the penetrating particles were interpreted as protons, but it appeared that run, degree of ionization along a trace, and radius of curvature in a magnetic field were not coordinated in any way with this assumption. Only in 1937, Anderson and Neddermeyer obtained a photo with a penetrating particle of this new type. The particle in the photo had a run of 4 cm and a magnetic rigidity of about 10^5 Gs.cm. Only one conclusion could be made: the observable particle was neither a proton nor a positron. What was discovered was a particle with intermediate mass, between the mass of an electron and the mass of a proton, which received name "mesotron." Later, they were called μ-mesons or muons.

1. On the Nature of Soft and Hard Secondary Cosmic Ray Components

The experiences begun with Anderson and Neddermeyer (1934), studying the losses of energy by particles of cosmic radiation in lead plates placed across a Wilson's chamber, being in a magnetic field, specified the existence of two components because some of the particles tested as having much smaller losses than the others. Later, after Bethe and Heitler (1934) published calculations of the probability of bremsstrahlung radiation generation by fast electrons and the process of the birth of pairs by high-energy photons, there was the possibility to explain the absorption of the soft component, observed by B. Rossi, and properties of strongly absorbed particles in the experiences of C.D. Anderson and S.H. Neddermeyer. On the other hand, it was necessary to recognise that if the hard component, in the experiences of B. Rossi, and the less absorbed particles, in the experiments of C.D. Anderson and S.H. Neddermeyer, are electrons, they test much smaller losses in substance, which follows from the theory of Bethe and Heitler (1934). In connection with this, in the mid-thirties, even the possibility of the inapplicability of quantum electrodynamics for electrons of large energy was discussed. There was a choice between two possibilities: either the calculations of H.A. Bethe and W. Heitler cease to be fair at energies exceeding some critical value, or the hard component consists of particles harder than electrons, as, according to the theory, energy losses to radiation are inversely proportional to the square of the mass. Each of these assumptions had supporters among the physicists studying CR. In the beginning it seemed to be the first possibility, but as the value of energy loss for electrons of higher and higher high-energies was measured, it was more probable that there were no indications on an inconsistency in the existing theory. Then, when Weiszäcker (1934) and Williams (1934), on the basis of theoretical reasons, came to the conclusion that the calculations of H.A. Bethe and W. Heitler concerning bremsstrahlung radiation electrons should be fair, and at energies in some GeV, they started to discuss the second possibility seriously. Besides, from experimental research, it was already known that some observable CR showers are so great that the energy generating their electrons should make some GeV, while the majority of penetrating particles had energy of only 1 GeV, which is less.

Williams (1934) came out with the assumption that penetrating particles of CR probably possess the mass of a proton. One of the difficulties connected with this hypothesis consisted in the necessity of the existence not only positive, but also negative protons, because experiments with a Wilson's chamber in a magnetic field showed that penetrating particles of CR have charges of both signs. In the meantime, continuing the experiences on measurements of loss of energy by particles of CR, Anderson and Neddermeyer (1936) found in many photos traces of particles with magnetic rigidity Hρ in limits from 10^5 to 10^6 Gs.cm which passed through a lead plate 3.5 mm thick without creating any appreciable secondary particles and losing much less energy than was predicted by theory for electrons. From the analysis of the traces left by these particles in the Wilson's chamber, it became clear that they should have the same charge as an electron, but with large mass. At first Anderson and Neddermeyer tried to interpret the penetrating particles also as protons, but it appeared that run, degree of ionisation along a trace, and radius of curvature in a magnetic field were not coordinated in any way with this assumption.

In Figure 1 is one of the first photos received by Anderson and Neddermeyer that specified the existence of a particle of the new type. The particle in the photo had a run of 4 cm and magnetic rigidity about 10^5 Gs.cm. The energy of a proton with such a run should be only 1.5 MeV. But a proton of such energy would have a radius of curvature equal to 20 cm (in an applied magnetic field), three times more than the value of the radius observed in the picture. Besides, at such magnetic rigidity protons should make ionization at least three times more strongly than electrons. However, in the photos, the density of ionization along the traces of penetrating particles was approximately the same as along traces of electrons, from which followed that these particles cannot be protons. Only one conclusion could be made: the observable particle was neither a proton nor a positron. One more photo of a penetrating particle was soon received; it is interesting that it, too, had a positive charge. Only later were penetrating particles with a negative charge found.

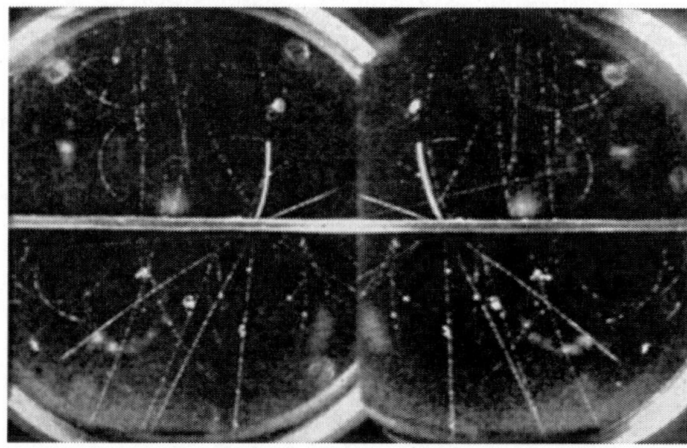

Figure 1. One of the first photos received by Anderson and Neddermeyer (1936) on which the trace of the particle having mass between the mass of an electron and the mass of a proton was found.

Anderson and Neddermeyer (1936) considered that the curvature of a trace can influence, besides the magnetic field, other factors, and drew the conclusion that "if observable curvature has been entirely created at the expense of a magnetic deviation, it is necessary to conclude that this trace corresponds to a massive particle with the relation e/m much bigger than for a proton." The definitive judgement about the particle nature was postponed until then, and additional data was not yet received. Thus, by the end of 1936 it seemed almost obvious that in CR there are particles, of previously unknown type, with a mass between the mass of an electron and the mass of a proton.

2. RESEARCH OF PENETRATING PARTICLES OF UNKNOWN TYPE WITH MASS BETWEEN THAT OF AN ELECTRON AND PROTON: DIFFICULTIES IN DETERMINING THE MASS OF THE PARTICLE

The existence of particles with intermediate mass was definitively proved by the experiments of Neddermeyer and Anderson (1937) and Street and Stevenson (1937), on the basis of research of traces of particles in a Wilson's chamber and of studying loss of energy by penetrating particles in substances. In the spring of 1937, Neddermeyer and Anderson

(1937) replaced the lead plate with platinum with a thickness of 1 cm (platinum was used because of its larger density) and made measurements of losses of energy for separate particles by defining the curvature of a trace of particle before plate passage. About 6000 photos were received, the research of which allowed the drawing of important conclusions. In Figure 2, results of measurements of loss of energy for 55 particles are presented. In Figure 2 on the abscissa is presented the energy of particle E_1 before plate passage in MeV, and on the axis of ordinates – the loss of energy by a particle in platinum in MeV/cm. Measurements were made for particles of three types: shower, generated shower, and single. Negative values of loss of energy for some particles, available in Figure 2 as S.H. Neddermeyer and C.D. Anderson marked, are connected with discrepancies in measurements. From the drawing it is visible that particles concerning losses of energy tested by them definitely break up into two groups: single particles with both big and small energies which lose a small share of energy at passage through a plate, and of a shower and generated shower, the particles losing a large share of energy. As losses of energy for particles of the second group were coordinated with settlement values of these values for electrons in the assumption that they lose energy on ionisation and bremsstrahlung radiation, particles of this group undoubtedly were electrons and positrons.

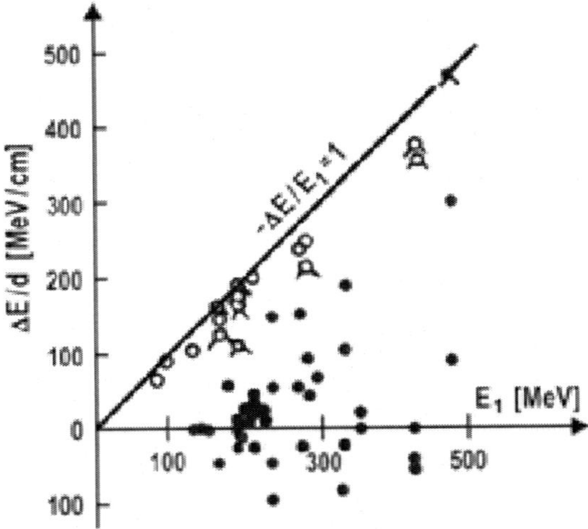

Figure 2. Losses of energy by particles of cosmic radiation in a platinum plate, measured by Neddermeyer and Anderson (1937). Open circles – particles of shower, open circles with two small lines – particles generated showers, and black circles – single penetrating particles.

Neddermeyer and Anderson (1937) drew, thus, the important conclusion that the big penetrating ability of particles is not simply a function of the energy of a particle. Particles of the same energy behaved differently only depending on whether they were single, shower, or shower-generated particles. From here it was possible to conclude that electrons cannot behave as absorbed particles at energies below some critical value and as penetrating particles at energies above the critical value. The conclusion that penetrating particles are not electrons arose. For penetrating particles of loss of energy it was possible to explain completely by only ionization losses, i.e. losses to bremsstrahlung radiation were very small. This once again confirmed the assumption that observable penetrating particles possess mass much bigger than the mass of an electron. From the previous experiments, as it was already told, it was clear that these particles cannot be protons. Existence of a new type of particle with intermediate mass did not raise the doubts more. Having made the assumption of existence of an unknown particle of intermediate mass, S.H. Neddermeyer and C.D. Anderson could not, however, define its mass with little comprehensible accuracy. This results from the fact that at magnetic rigidity of an order of 10^6 Gs.cm and above charged particles with mass up to several hundred of electron masses it is impossible on density of ions along a trace to distinguish them from electrons.

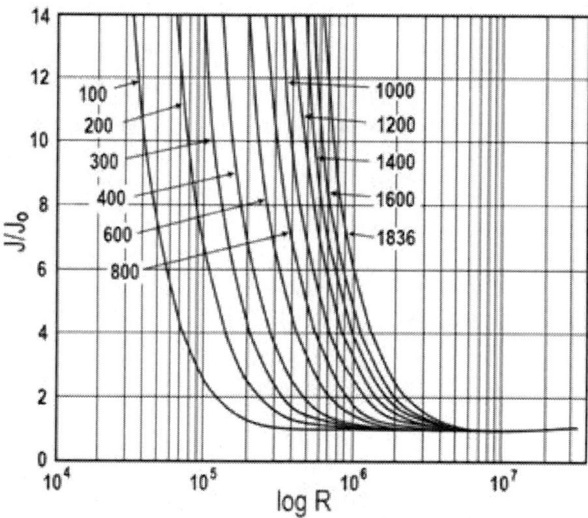

Figure 3. The nomogram, connecting relative density of ionization along a trace J/J$_0$ with magnetic rigidity R of particles with different masses from 100 to 1836 m$_e$. The value of R is presented in logarithmic scale.

For an illustration in Figure 3 a nomogram is presented, connecting relative density of ionization J/J$_0$ (where J — observable density of ionization, and J$_0$ - the minimum density of ionization) with magnetic rigidity of a particle R. Different curves in Figure 3 correspond to particles with unit charge, but with various masses which values (in electronic mass) are noted near each of the curves. Knowing magnetic rigidity and density of ionization along a trace, it is possible, using the nomogram, to define mass of a particle. It is well visible that at small values of magnetic rigidity curves are divided from each other, but at rigidity increase ($R > 10^6$ Gs.cm) they practically merge. Therefore, it is possible to measure reliably the mass of only those penetrating particles which possess small enough energy.

3. EXPERIMENTS WITH PENETRATING UNKNOWN PARTICLES OF SMALL ENERGIES AND ESTIMATION OF THEIR MASS

The problem of investigating penetrating particles which possess enough small energy was solved by Street and Stevenson (1937), working at Harvard University in the USA. As penetrating particles of small energy seldom meet, Street and Stevenson decided to increase the number of useful photos in the chamber, choosing for registration only those particles which stop in the chamber. The scheme of installation of Street and Stevenson for reception of photos penetrating particles of small energy in Wilson's chamber is presented in Figure 4.

The Wilson's chamber C worked only from those particles which caused the discharge in counters G_1, G_2, and G_3, but did not cause the discharge in one of counters G_4, i.e. from the particles which have stopped in the chamber. To register only penetrating particles, a lead block with the thickness 11 cm was placed in front of the chamber. As the method of measurement the of density of ionisation in Wilson's chamber is based on calculation of a number of drops along a trace to make the best estimation of density of ionization, expansion in the chamber automatically was delayed by approximately one second after the particle passed through the chamber. For this time interval ions diffused on distances, sufficient that drops condensed on them were separated considerably from each other which much facilitated their accounting. In the autumn of 1937 Strit and Stevenson received the first photos, one of which, representing a great interest, is shown on Figure 5.

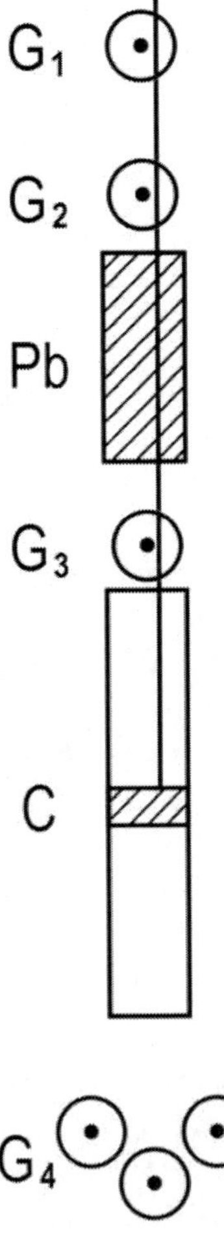

Figure 4. The scheme of installation of Street and Stevenson (1937) for obtaining in Wilson's chamber of photos of penetrating particles of only small energy.

In the photo of the particle trace the density of ionisation along which exceeds the ionization density of an electron by approximately six times is visible. The magnetic rigidity of a particle defined on curvature of a trace makes 9.5×10^4 Gs.cm, and its run is equal 7 cm. The particle, obviously seen, cannot be a proton, as a proton with such rigidity should have an energy of only 0.44 MeV and run 1 cm. The mass of the particle defined by J.C. Street and E.C. Stevenson (1937), was about 130 electronic mass.

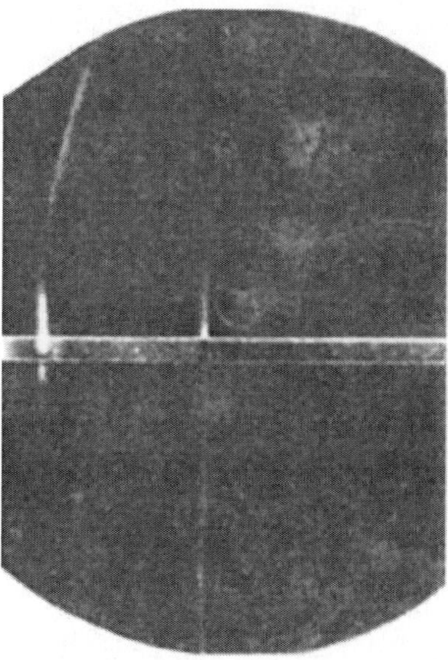

Figure 5. The photo of meson trace, on which J.C. Street and E.C. Stevenson (1937) for the first time have defined approximately its mass.

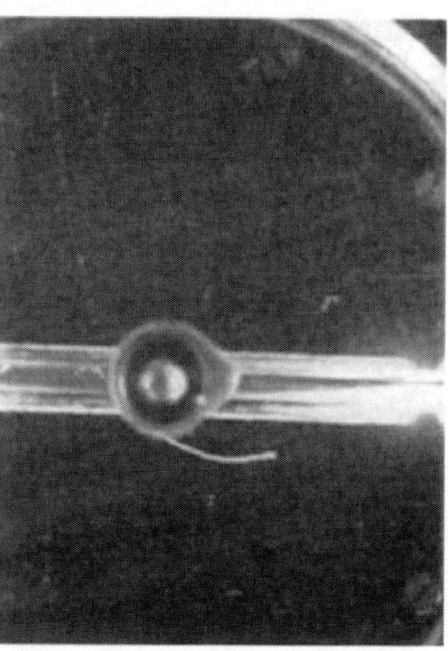

Figure 6. A photo received by Neddermeyer and Anderson (1938) which is considered as the most convincing proof of meson existence.

In the same year, Nishina et al. (1937), using a similar method, found that the mass of an intermediate particle is more and makes 180 - 260 electronic masses. Let's mention once again one photo received by the Neddermeyer and Anderson (1938) which gave, apparently, the most convincing proof of existence of a particle with intermediate mass (Figure 6). In Figure 6 a wide white horizontal strip in the middle of a picture - is one of the operating counters placed in the working volume of the chamber. The chamber worked only in the case when the particle passed through the counter placed over the chamber, and the counter placed inside of the chamber that considerably increased probability of observation of the particles stopping in the chamber. In the picture is visible the trace of a particle moving from top to bottom and possessing considerable energy, but which lost almost all of that energy in the walls of the counter and the chamber which stopped it in the bottom part. The particle is positively charged, but could not be a positron because the density of ionisation along the trace in the top part of the chamber appears obviously more than the density and the ionization made by a positron. The particle could not be a proton as in this case it would have energy of 1.4 MeV in the top part of the chamber. A similar proton should create a trace with much bigger density of ionization than is revealed in the photo. The run of the particle after counter passage was 1.5 cm, whereas an electron or a positron having given curvature of a trace, should have run much more, and a proton run should be less than 0.02 cm. Judging by calculations, the mass of the new particle was about 200 electronic masses. At the beginning the new particle named variously: baritron, yukon, mesotron, and, at last, meson (from the Greek word " $\mu\varepsilon\sigma o\zeta$ " that means intermediate). The last name - meson - became standard. Already after the discovery of mesons it was recollected that elsewhere Kunze (1933) published a photo of an unclear trace of the same type as the trace of a particle of intermediate mass. But then nobody paid attention to this photo, considering it simply an error of experiment.

4. More Precise Experiments for Determining of Meson's Mass

After 1937 many physicists tried to measure the mass of a meson more precisely. So, Corson and Brode (1938) at the Californian University in Berkeley defined the mass of a particle not by magnetic rigidity and ionization along a trace as it was made earlier, but rather on magnetic rigidity and particle run. The scheme of the installation which was used for defining the mass of a meson by Corson and Brode is shown in Figure 7. The installation consisted of two Wilson's chambers C_1 and C_2, located one over another. Chamber C_1 was located in a magnetic field, and on curvature of a trace of a particle in this chamber magnetic rigidity was defined. Chamber C_2 contained 15 lead plates in the thickness of 0.63 cm each, and particle run was measured in this chamber. Both chambers worked from the particles causing the discharge in Geiger-Muller counters G_1, G_2, and G_3. For mass definition of mesons D.R. Corson and R.B. Brode made a nomogram (Figure 8), connecting mass of a particle (in electronic masses) with various observable characteristics of the trace left by a particle in a Wilson's chamber on magnetic rigidity of particle of Hρ (in Gs.cm), and value of run in air R (at temperature 15° C and pressure 750 mm Hg), and pressure of 750 mm Hg), and by the ratio D of observable ionization to the minimum ionization.

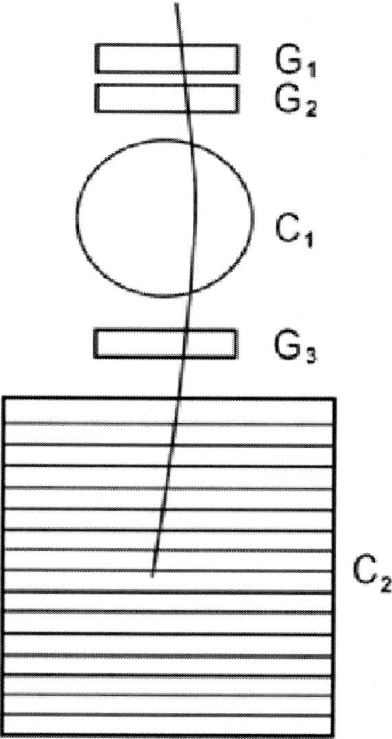

Figure 7. The scheme of installation of Corson and Brode (1938) for mass definition of meson.

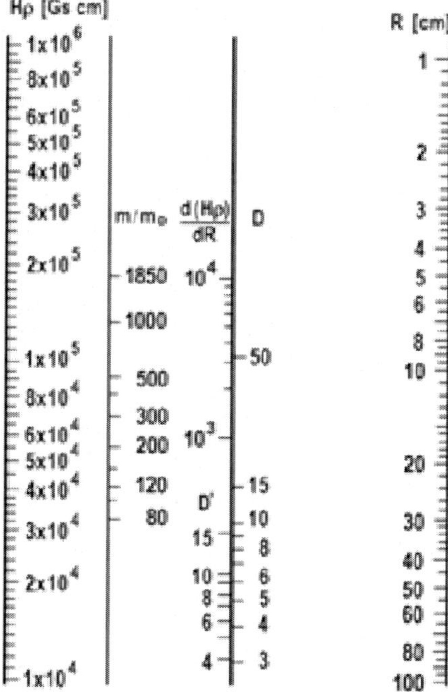

Figure 8. The nomogram, constructed by Corson and Brode (1938) and connecting mass of a particle with various observable characteristics of the trace left by a particle in the Wilson's chambers C_1 and C_2.

The nomogram was constructed in such a manner that any straight line, connecting two characteristics received experimentally, crossed a scale of masses, specifying required mass of particles. Corson and Brode (1938) showed that if all measured sizes are known with an identical degree of accuracy, the most exact value of mass turns out at the measurement of magnetic rigidity and particle run. According to their measurements, the mass of meson is about 250 electronic masses. The mass of a meson was measured in the next years by various methods of the times, but the received values so strongly differed from each other that it was possible to tell with confidence only that it exceeded electronic mass by approximately 200.

The idea was often come up with even that the particle can not have one certain size of mass, and there is a set of the masses lying in enough wide limits. The description of various methods of measurement of the mass of mesons and comparison of the received results resulted in the work of D.I. Blokhintsev and P.E. Nemirovsky (1947). It appeared, however, that no disorder of masses exists and the value of mass accepted now of the μ-meson, defined in many experiments with them artificial received, is equal to 206.8 electronic masses.

The discovery of mesons by Neddermeyer and Anderson (1937) allowed the establishment at last of the nature of the secondary hard cosmic radiation observed at sea level: the penetrating particles are mesons, and strongly absorbed shower particles - electrons and positrons.

5. Discovery of Mesons and Theory of Nuclear Forces

It is necessary to recollect that even before the discovery of meson, the existence of a particle of intermediate mass was predicted for theoretical reasons. After E. Fermi created the remarkable theory of β-decay, it became clear at once to some physicists that the possibility of the new type of interaction between the nucleons, carried out by particles of final mass opens. Tamm (1934), by analogy that photons are exchange objects of an electromagnetic field, assumed that nuclear forces arise at the expense of an exchange between the nucleons, carried out in steams of particles — electron plus neutrino[1], and created the quantitative theory of electron-neutrino nuclear forces, so-called β-forces.

However, the ideas taken as a principle of the theory of β-forces, were used to construct the further theory of nuclear forces. As the β-forces of Tamm (1934) found are extremely weak, Yukawa (1935), keeping the basic idea of carrying over of forces by particles of final mass, carried out nuclear interactions at the expense of existence of the new hypothetical particles strongly interacting with nucleons. These quanta of a field of nuclear forces had mass defined by the radius of action of nuclear forces and equal, according to estimations of H. Yukawa, to approximately 200 electronic masses. Fantastic as it was represented at that time, the idea of H. Yukawa was extraordinarily courageous. Even simple communication of

---

[1] The existence of neutrinos (a neutral particle with zero or very small mass) was postulated by V. Pauli in 1931 for an explanation of that fact that electrons at β-decay of radioactive elements possess a wide spectrum of energies. In the beginning, the new particle was named neutron, but after the discovery in 1932 by J. Chadwick of a heavy neutral particle with the mass approximately equal to mass of a proton, E. Fermi suggested naming the particle participating in β-decay neutrino, which in Italian means "a small neutron". To big disappointment, it was found that the theory of β-forces appeared insolvent, as these forces were extraordinarily weak in comparison with forces which could explain the stability of a nucleus.

mass of field particles with the radius of the forces, noticed by H. Yukawa and presently considered so obvious, it was difficult to realise.

Years were required before the majority of theorists and experimenters all over the world accepted this idea. In the meantime, H. Yukawa developed the theory and to explain weak β-decay, foretold that the particle connected with a field of nuclear forces should be unstable and break up on electron and neutrino (hence, particle's spin should be an integer). Average time of a life of "a particle of Yukawa" before decay, according to estimations, was 0.5×10^{-8} seconds. Though the introduction of a hypothetical particle was connected with quantization of a field of nuclear forces, H. Yukawa at the end of the work noticed that it, possibly, is a part of cosmic radiation where energy exists, sufficient for generation of such particles in a substance. It is not surprising that after the discovery of CR mesons some physicists identified it with the quantum of a field of nuclear forces predicted by H. Yukawa. Later, it appeared that it was not so. Nevertheless idea that the mesons found in CR and the particles predicted by H. Yukawa are the same promoted experiments into the expected decay of mesons.

6. DISCOVERY OF MESON'S DECAY AND CR TEMPERATURE EFFECT

Already in the mid-1930s, during measurements of intensity of CR in the atmosphere depending on height, research done by groups in different countries found some anomalies that were difficult to explain from the point of view of loss of energy by particles in the atmosphere. There was an impression that particles of CR are absorbed in the air more strongly than in dense substances if one compares layers to identical mass on a unit area. Moreover, it appeared that in the rarefied layers of air at big heights, the intensity of cosmic radiation decreases more strongly than in dense beds at low heights, i.e. the rarefied layers of air absorb cosmic particles more effectively, than the dense layers. In other words, absorption depends not only on through what mass of substance there passed cosmic particles, but also on what is the time required to pass through the substance. In 1938, Kulenkampff and Böhm (1938) showed that it is possible to explain these anomalies if one were to admit that mesons are unstable and have a life time comparable to the due course of their passage through atmosphere. Really, if this assumption is fair, some mesons should decay spontaneously earlier, before they will reach the end of the run.

In dense absorbers the decay of mesons will play an insignificant role thanks to the smaller length it will take for the mesons lose the energy and stop. In the air, spontaneous decay is added to the usual absorption of mesons, and air appears an especially effective absorber the more it is rarefied. "The idea of H. Kulenkampff and K. Böhm, - B. Rossi (M1966, page 108) recollected, - many physicists, including W. Heisenberg and H. Euler in Germany and P.M.S. Blackett in England, which considered numerous consequences from this hypothesis picked up. Both in the end of 1938 and in the beginning of 1939 the question on radioactive decay of mesons caused heated discussions among the physicists studying cosmic rays". It is valid that Euler and Heisenberg (1938), at the creation of a consecutive picture of interactions in CR, considered a hypothesis about spontaneous decay of mesons and used this in the cascade theory of showers. According to their representations, primary CR were positrons, having an energy power spectrum. Positrons form the cascade of showers in which at a depth of atmosphere of approximately in 100 g/cm^2 the set of photons already is born. These photons, at interactions with atoms of air, form mesons[2].

Electrons and positrons are formed in a shower, and also as a result of the decay of mesons at high altitude is the soft component, and this explains the presence in G. Pfotzer's curve of a maximum. Un-broken-up mesons penetrating up to small heights form the hard component, the absorption of which is described by the "tail" of Pfotzer's curve (in those times it seemed that all processes occurring at the passage of CR through substance could be completely explained on the basis of electromagnetic interaction of particles). That same year P.M.S. Blackett (1938), on the basis of a hypothesis about spontaneous decay of mesons, offered a graceful explanation of temperature effect in CR. Upon heating, the atmosphere expands, and mesons should spend more time in passage through same layer of air (in g/cm^2), than in the case of a colder atmosphere. As a result of this, the quantity of broken up mesons will be greater in hot air, hence, the measured intensity will be less than at the lower temperature.

7. EXPERIMENTAL CHECKING OF MESON'S DECAY AND ITS DEPENDENCE FROM ENERGY

In the spring of 1939, B. Rossi, who worked at that time under the invitation of C.D. Compton at the Chicago University, decided to check a hypothesis about instability of mesons by comparing the absorption of mesons in the air and in graphite. B. Rossi and two employees of C.D. Compton - N. Hilberry and J.B. Hoag (B. Rossi et al., 1939) made measurements in Chicago (180 m above sea level), in Denver (1600 m), on Eho Leik (3240 m) and on Maunt Evans (4300 m). The installation, a scheme of which is shown in Figure 9, consisted of three counters included in the scheme of coincidence and surrounded with a layer of lead of thickness 10 cm for shielding from soft component.

In all points except Chicago, measurements were made with the graphite block (about 87 g/cm^2) over counters, and without it. Results of these measurements are presented in Figure 10. On an axis of abscissas the total thickness of substance (air + graphite) over counters in g/cm^2 is presented, and on an axis of ordinates – the number of coincidences in one minute is presented. Apparent in the drawing, absorption in graphite appeared much less than in air. So, for example, on Eho Leik (depth of atmosphere of 699 g/cm^2) addition of graphite of 87 g/cm^2 reduces the number of mesons by 10% whereas addition of air of 87 g/cm^2 reduces the number of mesons by 20%. This appreciable additional absorption spoke to decay of mesons in the atmosphere's way, and measuring it, B. Rossi et al. (1939) could estimate the average time of a meson's life.

---

[2] Subsequent researches have shown that in primary cosmic radiation there is less than 1% of electrons and positrons. Besides, it has appeared that mesons are actually generated at nuclear interactions of protons with the nuclei of air atoms.

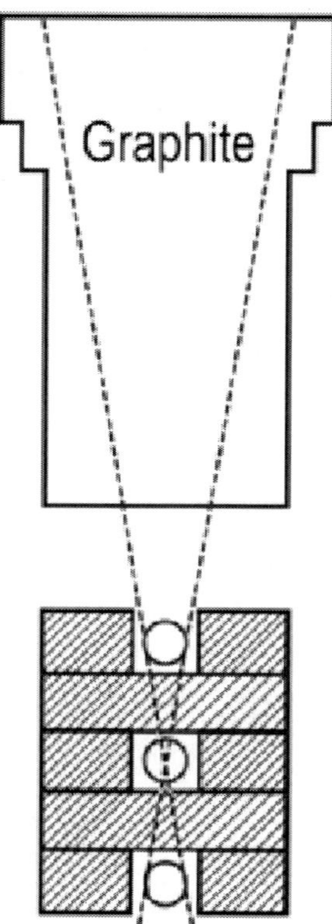

Figure 9. The scheme of installation of B. Rossi, N. Hüberry and J.B. Hoag (1939) for comparison of absorption mesons in the air and in graphite.

Thus it was necessary to consider the fact that according to the theory of relativity the definition of an interval of time depends on the readout system in which there is a measurement. Let us suppose that the average time of a meson's life, measured in the system of readout moving together with meson, is τ_0. Thanks to relativistic effect of delay of a course of moving clock, the average time of a life of meson, measured by the "motionless" observer (i.e. in this case in the system of readout connected with a terrestrial surface), will be equal

$$\tau = \frac{\tau_0}{\sqrt{1-\left(v^2/c^2\right)}} = \frac{\tau_0 E}{mc^2},$$

(1)

where v – speed of meson, E - a total energy of meson, and m - mass of rest meson. In other words, the fasted a meson moves, the more its time of a life. Therefore, for defining of average time of a life of rest mesons according to measurements of moving mesons it is necessary to know their energy. Having estimated average energy of observed mesons, B. Rossi et al. (1939) received for the average time of life of rest mesons a value of order 2 µs. In the following years similar measurements would be made by many physicists, and as a whole confirmed the results of Rossi et al. (1939). Then, Rossi and Hall (1941) measured average time of life of moving mesons of different energies, and showed that it varies depending on energy in full conformity with predictions of the theory of relativity according to Eq. 1. The observable effect was great enough - average time of life for the fastest mesons, registered in the experiment, was three times more than for the slowest.

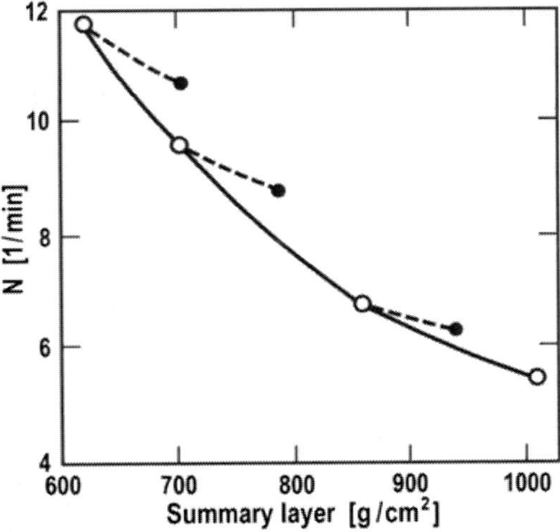

Figure 10. Results of measurements of absorption mesons in air (open circles) and in graphite (black points) received by Rossi et al. (1939); N - number of coincidences in one minute.

8. DIRECT OBSERVATIONS OF MESON'S DECAY IN WILSON'S CHAMBER

The direct observation of decay of mesons in the photos received in a Wilson's chamber, and a method of late coincidence was the definitive proof of justice of a hypothesis about spontaneous decay of mesons.

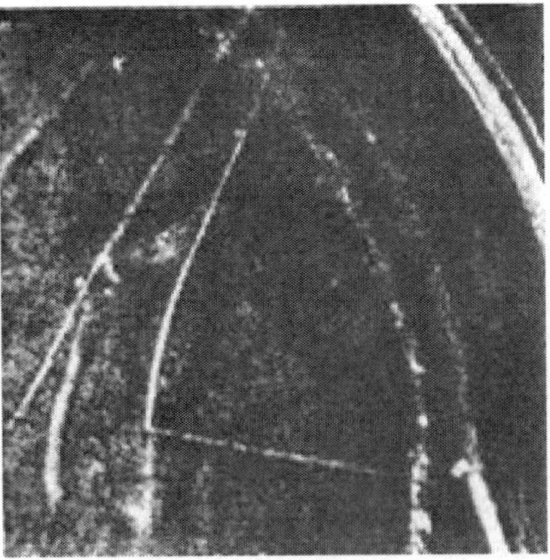

Figure 11. The first photo of decay of meson in Wilson's chamber, received by Williams and Roberts (1940).

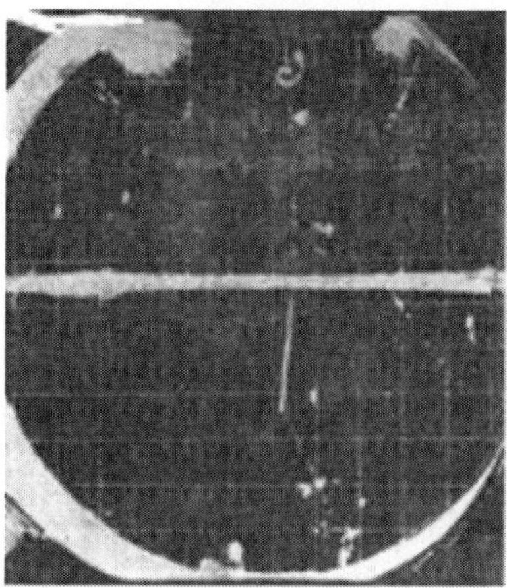

Figure 12. The photo of meson decay received by R.W. Thompson (1948).

The first photo of a meson's decay in a Wilson's chamber was received by Williams and Roberts (1940). In this photo (Figure 11) the trace of a slow meson, stopped in the chamber's gas, and the trace of a positron is visible leaving under a big angle from the end of a trace of meson. In the next years there were many similar photos.

In Figure 12 an especially successful photo of decay of meson which was received by Thompson (1948) in the Laboratory of Massachusetts Technological Institute is presented (for a long time up to the end of the 1940s it was a commonly accepted wrong opinion that μ-meson decays on two particles: electron or positron and neutrino.

9. USING TIME DELAY COINCIDENCES SCHEMES FOR DETERMINING DECAY TIME OF REST MESON

The following important step was made by Rasetti (1941). The Williams and Roberts (1940) photo of a meson's decay, made in all energies and stopped in substance was fixed. However, the stopped meson decays up not at once, and after a while, equal to about its average time of a life, i.e. electron, let out at decay of meson, will take off from an absorber in which there was a decay, to a delay in some μsec. To observe late emission of electrons from mesons which have stopped in substance, Rasetti (1941) made the experiment the scheme of which is shown in Figure 13.

Mesons which of 10 cm stopped in the block of lead T with thickness 10 cm and caused the discharge in one of three counters A and in single counters B, C, and D, but did not cause the discharge in any one of counters F were selected by means of the auxiliary electronic scheme. Along each party of an absorber settled down on three counters E which registered electrons let out at decay of stopped mesons. The scheme of late coincidence worked only in the event that the discharge in any of counters E arose less than through 1 μs after passage meson. Two other schemes of late coincidence worked with a delay in 2 and 15 μs. F. Rasetti found that most of the electrons had been let out with a delay of more than 2 μs, but less than 15 μs.

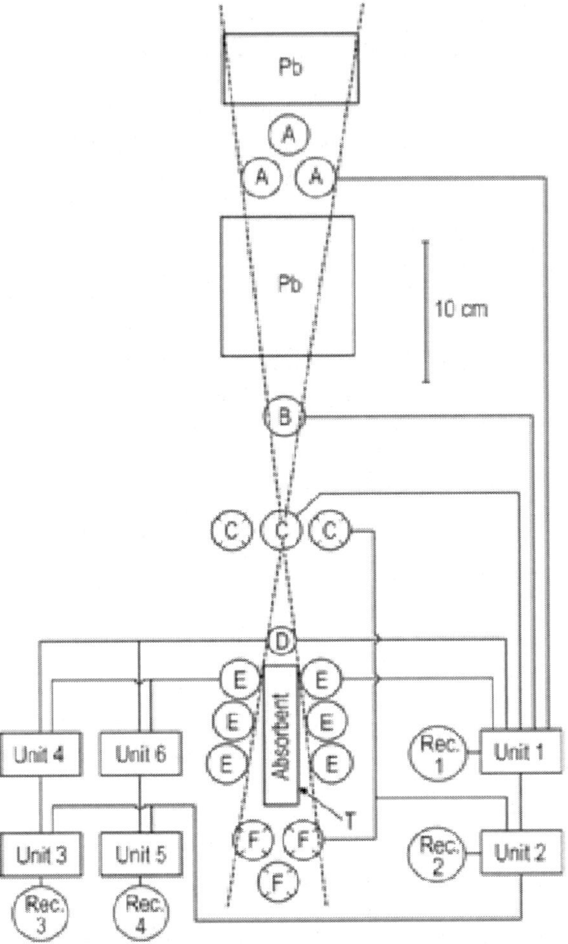

Figure 13. The installation scheme of Rasetti (1941) which help to observe decay of stopped mesons.

From relations of speed of the account in three channels (corrected on casual coincidence) F. Rasetti received that the average time of life of rest mesons τ_0 is equal to about 1.5 μs. The year after Rossi and Nereson (1942) made exact quantitative research of decay mesons with the installation similar to the installation of F. Rasetti (1941). The use of an "electronic clock" was a main feature of their experiment. The "electronic clock" allowed to measure a time lag between passage of a meson and emission of a decay electron or positron. Impulses from three counters registering stopping of absorber mesons, and from the counters registering decay electrons or positrons, moved on to two separate inputs of the measuring scheme. This scheme gave the target impulse whose amplitude was proportional to the time interval between two entrance impulses, and that allowed determination with big statistical accuracy a curve of dependence $N(t)$ of numbers of mesons whose decay occurred after a delay on t μs in the range of t from 0 to 10 μs. The received curve of decay mesons, stopped in the lead, is presented on Figure 14. It appeared that decay of mesons just as radioactive decay of atoms submits to the exponential law, i.e. has spontaneous character. The time of life of a meson τ_0 found from experimental data appeared equal to 2.15 ± 0.1 μs.

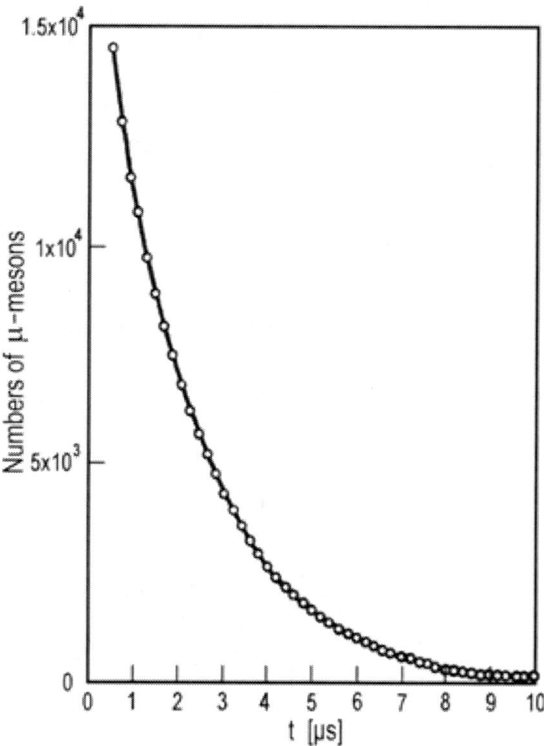

Figure 14. A decay curve of mesons, stopped in the lead (Rossi and Nereson, 1942).

A year later Nereson and Rossi (1943) made again similar measurements for various absorbers and showed that in all cases the found time of a meson life was identical. Similar measurements were made in the following years by different groups of physicists and confirmed the results that F. Rasetti and B. Rossi and N. Nereson made in 1939 - 1943. "That on advantage to estimate the contribution of the European scientists, it is necessary to consider - marked B. Rossi (M1966, page 115) - in what severe conditions it was necessary to them to work in days of the war. The Italian physicists, for example, carried out the most part of the works, hiding in cellars where they had to hide equipment from the Germans occupying Rome in 1943". In detail, the problem of meson decay is considered in a review paper by Feinberg (1947). It is necessary to take into account that the article of E.L. Feinberg (1947) was written before the discovery of the continuous character of a power spectrum electrons decay from stopped mesons, and calculations are made in the assumption that mesons decay to up to two, instead of to three particles.

10. DISCOVERY OF MESOATOMS FORMATTED BY NEGATIVE MESONS

After the decay of mesons was established, it became clear that they cannot be a part of primary cosmic radiation as believed earlier, and should be created in the atmosphere by any other primary particles. Having compared the number of mesons stopped in the iron block to the number of decay electrons or positrons left of the block, Razetti (1941) found out that only half of mesons break up with emission electrons or positrons. As experiences with Wilson's chamber showed that at sea level there is present about identical numbers of positive and negative mesons, F. Rasetti drew the conclusion that positrons are formed at the decay positive mesons, and negative mesons are grasped by kernels of atoms according to Tomonaga's and Araki's (1940) theoretical predictions. Really, in 1940 the Japanese physicists S. Tomonaga and G. Araki (1940) specified that positive and negative mesons after stoping in a substance should behave in essentially various ways, namely: a positive meson never can approach very close to an atomic nucleus (as he tests repelling from outside by the positively charged kernel) and after a while tests spontaneous decay.

A negative meson, on the other hand, is drawn by a kernel and consequently can be grasped before it will have time to decay. If one were to assume that interaction between mesons and kernels is strong enough, almost all negative mesons will be grasped by kernels and only the few will decay by electrons. If interaction is weak, part of the mesons will be absorbed by kernels, and a part will decay. In any case expected time of a life negative mesons in a substance should be less than for positive mesons. Besides, upon capturing the negative meson, the kernel allocates additional energy that should lead to atomic kernel explosion. The probability of capture of a negative meson by an atomic kernel was estimated by Migdal and Pomeranchuk (1940). It appeared that the probability of capture by an atomic kernel of a fast moving meson is insignificantly small and only a slow negative meson can be grasped by atomic kernel.

11. EXPERIMENTAL DETERMINATION OF FORMATION MESON-ATOMS IN DIFFERENT SUBSTANCES WITH SMALL AND BIG Z

The direct experimental proof of the formation of meson-atoms was received by Conversi et al. (1945). These experiments were similar to the earlier ones of F. Rasetti and B. Rossi and N. Nereson described in Section 9, except for an added device, capable of allocating mesons of one sign (a so-called magnetic lens). The similar magnetic lens was used earlier by B. Rossi (1931) in experiments on studying CR. The scheme of experimental installation used in the experiments of Conversi et al. (1945), is shown in Figure 15.

The magnetic lens consisted of two iron blocks F_1 and F_2, magnetised in opposite directions and nearby to each other. Over a lens and under it two Geiger-Muller counters A and B, included in the scheme of coincidence, impulses from which just as impulses from counters C, placed under absorber T, moved on the scheme of delayed coincidence settled down. Thus, this installation registered mesons, stopped and decayed, up with formation electrons or positrons through time t (the value of t varied from $t_{min} = 1$ μs to $t_{max} = 4.5$ μs. Depending on the direction of the magnetisation of both blocks F_1 and F_2 the lens could collect mesons of a certain sign and reject mesons of opposite sign (Figure 16). It was shown that, as one would expect, only positive mesons stopped in iron and tested spontaneous decay, and negative mesons disappeared, without giving electrons decay.

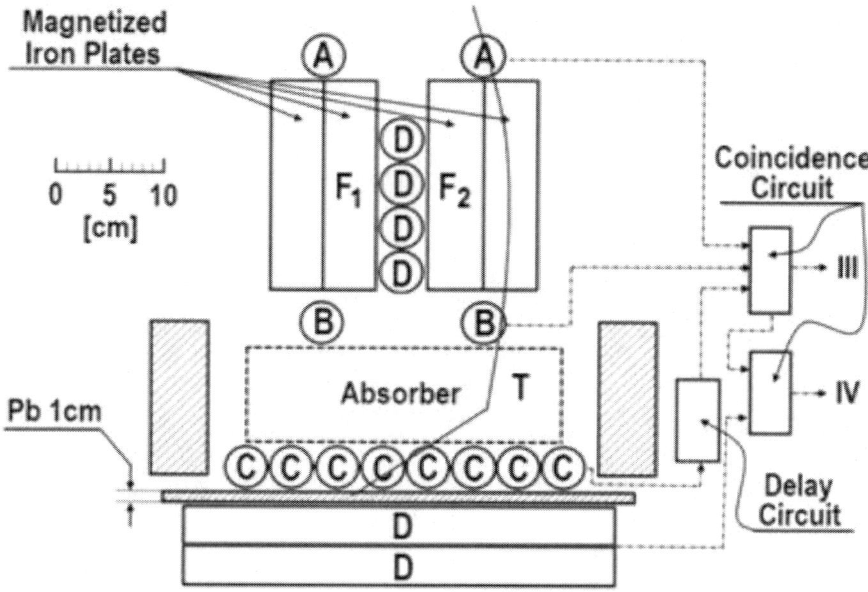

Figure 15. The scheme of experimental installation of Conversi et al. (1945) for research of behaviour positive and negative mesons, stopped in an absorber T.

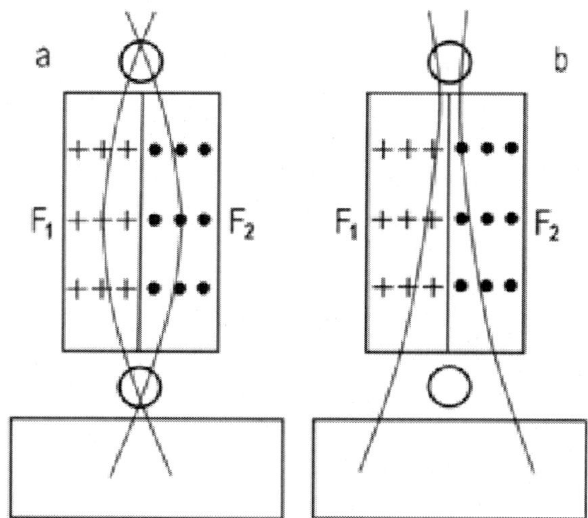

Figure 16. The scheme illustrating action of a magnetic lens. According to Conversi et al. (1945).

The received results contradicted those given by P. Auger, according to whom almost all negative mesons, stopped in aluminium, show decay. Though the experiences of P. Auger were considered erroneous, Conversi et al. (1947) decided to make measurements for a case when the absorber is a substance with a small serial atomic number Z. As such, the substance graphite (Z = 6) was chosen. It unexpectedly appeared that in this case spontaneous decay test both positive, and negative mesons. From here the important conclusion was drawn that negative mesons, stopped in an absorber with big serial number Z, are absorbed by atomic nuclei, while in an absorber with small Z they often test spontaneous decay.

12. The Sharp Contradiction Between Experiments with Mesons and Predictions from H. Yukawa's Theory

The results received by M. Conversi, E. Pancini, and O. Piccioni (1945, 1947), have been repeated and specified in many works by different authors. Using various absorbers, it was possible to establish that the probability of nuclear capture of negative mesons gradually increased with an increase in a serial atomic number Z of an absorber. It is clear, as heavy atomic nucleus possess big positive charge and more strongly draw negative mesons. It was surprising that negative mesons could not be absorbed by atomic nucleus in such substances as, for example, carbon (Z = 6), while on the basis of the theoretical predictions of S. Tomonaga and G. Araki (1940) that considered that mesons observed in CR are none other than "quanta of a field of nuclear forces", it was possible to expect with confidence that all negative mesons, stopped in carbon, should be grasped by atomic nuclei before they have time to decay. Really, average time of life of mesons is equal to 2×10^{-6} s, exceeding the time which by theoretical estimations was required for nuclear capture by at least 20 million times. For an explanation of the results of the experiments described above, it was necessary to assume that actually the interaction value between cosmic mesons and atomic nucleus is much weaker than was supposed in the theory of H. Yukawa (1935). This sharp contradiction between the theory of H. Yukawa and experiment resulted in the physicists being in deadlock.

Discovery of a "semi-heavy" particle was only the first part of the scheme offered by H. Yukawa. For justifying the whole picture it was required first that this particle break up into an electron plus a neutrino, secondly, that it have whole spin, and thirdly, that it be strongly co-operative with nucleons, providing stability of an atomic nucleus.

After the decay of mesons was established, it appeared that the life time of meson was a hundred times more than it is necessary for an explanation of β-decay under the theory of H. Yukawa. This divergence was inexplicable, but nevertheless did not create the big complications. The problem of spin was much more essential. For an explanation of spin dependence of nuclear forces it was necessary that the spin of meson be equal to 1, but in this case calculation of bremsstrahlung radiation and other processes led to ridiculously strong growth of the cross-section of interaction with energy. Moreover, experiments with ionization chambers showed that frequency of the showers resulting in bremsstrahlung radiation of mesons with spin 1 would be more than in 10 times above observed values. Finally, there were serious bases to consider that the mesons, opened in CR, do not interact strongly with nucleons.

13. Supposition on Two Types of Mesons

It is necessary to notice that elsewhere, before the experiments of Conversi et al. (1945, 1947) with a magnetic lens, Japanese physicists, working in days of the Second World War in conditions of full isolation, came to conclusions, which were absolutely correct. They established that mesons in CR do not interact with substances so strongly as postulated in Yukawa's (1935) theory and for this reason pass through big thicknesses of substance instead of absorbed owing to strong nuclear interactions. To overcome this difficulty, Sakata and Inoue (1946) and Tanikawa (1947) supposed about the existence of two various types of mesons, one of which is identical to the meson of Yukawa, and the other of which is observed in hard component of CR. Without knowing about the work of Japanese scientists, Marshak and Bethe (1947) also stated a hypothesis about the existence of two mesons. Moreover, they assumed that a "nuclear" meson breaks up into the meson observed in the hard component of CR, and into a neutral particle. Lattes et al. (1947) received experimental proof of the existence of two types of mesons, using just appeared emulsions for nuclear researches. They named the mesons, observed in hard component of CR, μ-meson (or muon), and the meson, responsible for nuclear interactions, π-meson (or pion).

The nuclear-active π-meson discovered by C.M.G. Lattes, G.P.S. Occhialini, and C.F. Powell satisfied all the conditions formulated by H. Yukawa (1935) for particles carriers of strong interactions.

W. Heisenberg recollected about one interesting case when mesons helped the decision of rather fundamental problem, in (1976, 1977): "Here, in Germany, before the war the government did not approve the relativity theory, in particular relativistic delay of time in moving bodies about which it has been told that is absurd, purely theoretical gamble. Business has reached even proceedings concerning an admissibility of teaching of the theory of a relativity at Universities. At one of such trials I had a possibility to express that decay time of muon should depend from its speed: muons which moved almost with a velocity of light, should decay more slowly that muons which moved with smaller speeds, - according to prediction of the theory of relativity. Experimental results have confirmed such prediction: time delay could be observed on experience directly, and this fact has opened doors for the relativity theory in Germany. Therefore I always feel gratitude to muons". Therefore, in the 1930s in CR two unknown particles - a positron and μ-meson were found. Discovery of these particles which arise at the interaction of particles of high energy with substance was the beginning of a major stage in the physics of elementary particles. Until accelerators of billions of eV were built, CR remained a unique known source of particles with such high energy, and the basic discoveries in the physics of elementary particles therefore are closely connected with research in CR. "Discovery of muons has not caused changes in bases of physical laws, similar themes to changes which were caused by positron discovering, - wrote W. Heisenberg (1976, 1977), - However detection of the muon has opened interesting feature of a spectrum of particles. This spectrum has broken on two subsystems supposing only weak ability to a combination, hadrons and leptons".

ACKNOWLEDGMENTS

Lev Dorman is thankful to Israeli Ministry of Science for support in the frame of the project "National Centre of Knowledge in Cosmic Rays and Space Weather". Our gratitude to David Shai Applbaum for checking and improve English in this paper.

REFERENCES

Anderson C.D. and S.H. Neddermeyer, 1934. "Energy-loss and the production of secondaries by cosmic ray electrons", *Phys. Rev.*, Ser. II, 46, No. 4, Minor Contributions, 325-325.

Anderson C.D. and S.H. Neddermeyer, 1936. "Cloud Chamber Observations of Cosmic Rays at 4300 Meters Elevation and Near Sea-Level", *Phys. Rev.*, Ser. II, 50, No. 4, 263-271.

Bethe H.A. and W. Heitler, 1934. "On the Stopping of Fast Particles and on the Creation of Positive Electrons", *Proc. Roy. Soc. London, Ser. A*, A146, No. 856, 83-112.

Blackett P.M.S., 1938. "On the Instability of the Barytron and the Temperature Effect of Cosmic Rays", *Phys. Rev.*, Ser. II, 54, No. 11, 973-974.

Blokhintsev D.I. and P.E. Nemirovsky, 1947. "The mass of meson", in *Meson*, Ed. I.E. Tamm, Gostekhizdat, Moscow-Leningrad, 56-71. In Russian.

Conversi M., E. Pancini, and O. Piccioni, 1945. "On the Decay Process of Positive and Negative Mesons", *Phys. Rev.*, Ser. II, 68, No. 9-10, 232-232.

Conversi M., E. Pancini, and O. Piccioni, 1947. "On the Disintegration of Negative Mesons", *Phys. Rev.*, 71, No. 3, 209-210.

Corson D.R. and R.B. Brode, 1938. "The Specific Ionization and Mass of Cosmic-Ray Particles", *Phys. Rev.*, Ser. II, 53, No. 10, 773-777.

Euler H. and W. Heisenberg, 1938. "Theoretische Gesichtspunkte zur Deutung der kosmischen Strahlung", *Erg. Exakt. Naturwiss.*, 17, 1-69.

Feinbeg E.L., 1947. "Decay of meson", in *Meson*, Ed. I.E. Tamm, Gostekhizdat, Moscow-Leningrad, 80-113. In Russian.

Heisenberg W., 1976. "Cosmic radiation and fundamental problems in physics", *Naturwissenschaften*, 63, No. 2, 63-67.

Heisenberg W., 1977. "Cosmic radiation and fundamental problems in physics", Physics Uspekhi (UFN), 121, No. 4, 669-677. In Russian.

Kulenkampff H. and K. Böhm, 1938. "Über die azimutale intensitätsverteilung der Röntgen Bremsstrahlung", *Phys. Ges.*, 19, No. 1, 5- .

Lattes C.M.G., G.P.S. Occhialini, and C.F. Powell, 1947. "Observations on the Tracks of Slow Mesons in Photographic Emulsions", *Nature*, 160, No. 4066, 453-456.

Marshak R.E. and H.A. Bethe, 1947. "On the Two-Meson Hypothesis", *Phys. Rev.*, Ser. II, 72, No. 6, 506-509.

Migdal A.B. and I.Ya. Pomeranchuk, 1940. "On the end of mesotron's track in Wilson's chamber", *DAN USSR*, 27, No. 7, 652- . In Russian.

Neddermeyer S.H. and C.D. Anderson, 1937. "Note on the Nature of Cosmic-Ray Particles", *Phys. Rev.*, Ser. II, 51, No. 10, 884-886.

Neddermeyer Seth H. and Carl D. Anderson, 1938. "Cosmic-Ray Particles of Intermediate Mass", *Phys. Rev.*, Ser. II, 54, No. 1, 88-89.

Nereson N. and B. Rossi, 1943. "Further Measurements on the Disintegration Curve of Mesotrons", *Phys. Rev.*, Ser. II, 64, No. 7-8, 199-201.

Nishina Y., M. Takeuchi, and T. Ichimiya, 1937. "On the Nature of Cosmic-Ray Particles", *Phys. Rev.*, 52, No. 11, 1198-1199.

Rasetti F., 1941. "Evidence for the Radioactivity of Slow Mesotrons", *Phys. Rev.*, Ser. II, 59, No. 9, 706-708.

Rossi B., 1931. "Magnetic Experiments on the Cosmic Rays", *Nature*, 128, No. 3225, 300-301.

Rossi B., 1933. "Über die Eigenschaften der durchdringenden Korpuskularstrahlung im Meeresniveau", *Zeitschr. Phys.*, 82, No. 3-4, 151-178.

Rossi B., M1966, *Cosmic Rays*, Atomizdat, Moscow. In Russian.

Rossi B. and D.B. Hall, 1941, "Variation of the Rate of Decay of Mesotrons with Momentum", *Phys. Rev.*, 59, No. 3, 223-228.

Rossi B. and N. Nereson, 1942. "Experimental Determination of the Disintegration Curve of Mesotrons", *Phys. Rev.*, Ser. II, 62, No. 9-10, 417-422.

Rossi B., N. Hilberry, and J.B. Hoag, 1939. "The Disintegration of Mesotrons", *Phys. Rev.*, Ser. II, 56, No. 8, 837-838.

Sakata S. and T. Inoue, 1946. "On the Correlations between Mesons and Yukawa Particles", *Progr. Theor. Phys.*, 1, No. 4, 143-150.

Street J.C. and E.C. Stevenson, 1937. "New Evidence for the Existence of a Particle of Mass Intermediate Between the Proton and Electron", *Phys. Rev.*, Ser. II, 52, No. 9, 1003-1004.

Taketani M., 1971a. "Meson theory of Yukawa in Japan", *Progr. Theor. Phys. Suppl.*, No. 50, 18-24.

Taketani M., 1971b. "Methodological Approaches in the Development of the Meson Theory of Yukawa in Japan", *Progr. Theor. Phys. Suppl.*, No. 50, 12-24.

Tamm I.E., 1934. "Exchange Forces between Neutrons and Protons, and Fermi's Theory", *Nature*, 133, No. 3374, 981-981 (1934).

Tanikawa Y., 1947. "On the Cosmic Ray Mesons and the Nuclear Meson", *Progr. Theor. Phys.*, Kyoto, 2, No. 4, 220-221.

Thompson R.W., 1948. "Cloud-Chamber Study of Meson Disintegration", *Phys. Rev.*, Ser. II, 74, No. 4, 490-491.

Tomonaga S. and G. Araki, 1940. "Effect of the Nuclear Coulomb Field on the Capture of Slow Mesons", *Phys. Rev.*, Ser. II, 58, No. 1, 90-91.

Weizsäcker C.E.V., 1934. "Ausstrahlung bei Stößen sehr schneller Elektronen", *Zeitschr. Phys.*, 88, No. 9-10, 612-625.

Williams E.J., 1934. "Nature of the High Energy Particles of Penetrating Radiation and Status of Ionization and Radiation Formulae", *Phys. Rev.*, Ser. II, 45, No. 10, 729-730.

Williams E.J. and G.E. Roberts, 1940. "Evidence for Transformation of Mesotrons into Electrons", *Nature*, 145, No. 3664, 102-103.

Yukawa H., 1935. "On the Interaction of Elementary Particles" *Proc. Phys. Math. Soc. Japan. Ser 3*, 17, No. 48, 139-148.

In: Homage to the Discovery of Cosmic Rays ...
Editor: Jorge A. Perez-Peraza

ISBN: 978-1-63483-084-3
© 2015 Nova Science Publishers, Inc.

Chapter 4

THREE CYCLE QUASI-PERIODICITY IN AP, IMF AND COSMIC RAYS: IMPLICATIONS FOR GLOBAL CLIMATE CHANGE

H. S. Ahluwalia

Department of Physics and Astronomy,
University of New Mexico, Albuquerque, New Mexico, US

ABSTRACT

We describe the discovery of a three-cycle quasi-periodicity (TCQP) in the planetary indices Ap/aa, interplanetary magnetic field (IMF) intensity (B) and galactic cosmic rays (GCRs) [Ahluwalia, 1997, 1998, 1999]. It led to the development of a heuristic methodology for predicting the peak smooth sunspot numbers (SSNs) and the rise time for a new solar activity cycle. The scheme was used to predict key parameters for the SSN cycle 23. The predictions were right on the mark, one of the very few predictions to have achieved this status. The smoothed SSNs for cycle 24 have been increasing slowly since its onset in December 2008. Ahluwalia and Jackewicz [2011]- -hereafterAJ11- -have predicted that cycle 24 will be only half as active as cycle 23 and reach its peak in May 2013 ± 6 months. We report on the present status of the ascending phase of cycle 24 and compare its timeline with those of the previous ten cycles (10-23) of the twentieth century; cycle 24 is rising the slowest. Drawing from the past history, AJ11 note that we may be on the eve of a Dalton-like minimum. We compare the ascending phase of cycle 24 with those of cycles 5 and 14 that led to the start of two recent Grand Minima, namely the Dalton and the Gleissberg Minima respectively. We discuss the physical significance of our findings and speculate about their implications for the global climate change and the probable socioeconomic and political future happenings.

INTRODUCTION

The solar variability is best expressed in terms of the sunspot number (SSN) cycles. The sunspots form and dissolve on the solar disc (photosphere) with an average period of 11.2 yr for a (Schwabe) cycle; the actual periods have ranged from 9 to 13.5 yr. In 1908, George Ellery Hale discovered that sunspots have intense magnetic fields embedded in them. The spots tend to occur in associated pairs, the magnetic polarity of the spots of the pair are of opposite sign in a solar hemisphere; if the preceding spots in the northern hemisphere have positive polarities, the preceding spots in the southern hemisphere will have negative polarities. The polarities of the preceding spots shift from one cycle to the next as the solar polar field reverses polarities. Therefore, our sun is a magnetic variable star with ~22 yr period (Hale cycle). The actual evolution of the solar magnetic field is complex [Wang et al., 2005].

SSN SERIES: 1700 -2012

A plot of the annual mean SSNs for 1700-2012 is shown in Figure 1, every fourth cycle is boldly labeled. One notes that beginning with cycle 10, there is a pattern where even cycles of the even-odd pairing are less active; it disappears after cycle 21. Also, every third cycle is less active (14, 17, 20, and 23). Moreover, SSNs seem to exhibit a period ~80-100 years (Gleissburg, 1958). At present, we have little understanding of the physical cause(s) of these solar features.

The thick black line in Figure 1 shows the 11yr mean of SSNs (long-term trend). It suggests that sun regularly undergoes periods of enhanced activity interspersed with noticeable periods of lower activity. The positions of three recent Grand Minima are boldly labeled in Figure 1, namely the Maunder Minimum (1645-1715), the Dalton Minimum (1790-1830), and the Gleissburg Minimum (1989-1902); they are labeled with the acronyms: MM, DM, and GM respectively. We emerged from a Grand Maximum recently and are sliding into a period of lower activity. Ahluwalia and Jackewicz [2011a] - -hereafter AJ11- - predict that we are at the advent of a Dalton-like minimum.

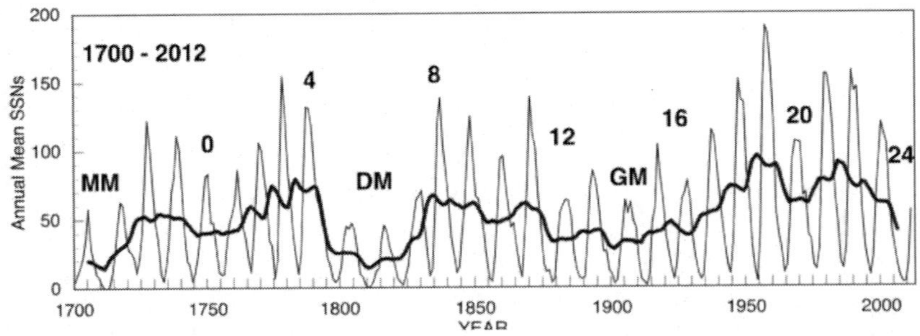

Figure 1. Plot of the annual mean SSNs for 1700-2012 is shown in Figure 1, every fourth cycle is boldly labeled. The thick black line in Figure 1 shows the 11yr mean of SSNs.

PLANETARY INDICES AP AND AA

The solar variability manifests at earth through the interaction of the magnetized solar wind plasma with the terrestrial magnetosphere. Bartels [1962] designed the planetary index

Ap to measure the geo-effectiveness of the solar wind. The index has a linear scale with a range between 0 (quietest day) and 400 (most disturbed day); the maximum value of Ap has never been observed to-date, the data go back to 1932. The aa index bears a linear correlation to Ap and covers a longer period to 1868. These indices are proportional to the amplitude of the solar wind electric field at earth orbit [Ahluwalia, 2000a], given by the product of solar wind speed (V) and interplanetary magnetic field (IMF) intensity (B); in-situ measurements of B and V started in October 1963 [Snyder, Neugebauer, and Rao, 1963]. The data are readily available at the National Geophysical Data Center website SPIDER of the National Oceanic and Atmospheric Agency (NOAA) at Boulder, CO. Ahluwalia [2011] derived the following fit parameters between B and Ap observations for 1963-2009, with a correlation coefficient (cc) 0.83, at a confidence level (cl) > 99%:

$$B = 0.23Ap + 3.3, nT \tag{1}$$

THREE CYCLE QUASI-PERIODICITY

A severe form of interaction between the solar wind and the terrestrial magnetosphere occurs via the interplanetary remnants of the coronal mass ejections (ICMEs). It leads to the visual aurora in the polar regions of earth, often accompanied by great geomagnetic storms. The CME frequency increases in step with an increase in the activity of a SSN cycle [Gosling et al., 1992; Webb and Howard, 1994; Manoharan, 1997; Ahluwalia, 2000, and references therein].

Silverman [1992] presents the power spectrum of the monthly auroral occurrence for 1500-1948 for about 45, 000 observations, showing significant power at periods of 83 yr, 55 yr, 33.3 yr, and 11.1 yr. Similarly, Alicia et al. [1993] present the power spectra of the monthly Ap data for 1932-1982, showing significant power at 10.3 yr and 32.1 yr; they also present the power spectra for the SSNs for the same period, showing significant power at 10.4 yr and 33.3 yr. Strangely, the 22 yr Hale period is absent in both analyses (auroral as well as Ap data). Alicia et al. note that 33 yr line in SSN data is very unstable, being absent in the SSN power spectra for 1920-1982. Therefore, we call it a 'three cycle quasi-periodicity (TCQP).'

TCQP IN AP AND B

Figure 2 shows a plot of the annual mean values of Ap and B for 1963-2009. The positions of the SSN maxima (M) and minima (m) are marked. The vertical dashed lines are drawn through the intervals of the solar polar field reversals. In addition to the Schwabe cycle in the two time series, the following additional features are also seen.

1. The annual mean value of B reached the highest value (9.1 nT) in 1991 (cycle 22); in 2009 B declined to ~ 2/3 of its value in 1963, the lowest value ever since the measurements in space began in 1963. The Wilcox Solar Observatory (WSO) data

indicate that solar polar field strength for cycle 23 is significantly lower (~ 50%) than for the previous 3 cycles.

2. TCQP in the limited B data is highlighted in Figure 2 with dashes, a similar trend is seen in the Ap data. The slope of the trend-line changes from positive to a steeper negative value after 1996 SSN minimum (cycle 22), as is also the case for aa data (see Figure 5, below). This implies that sun is in a transition phase to a new state of lower activity whose characteristics may unfold in the near future.

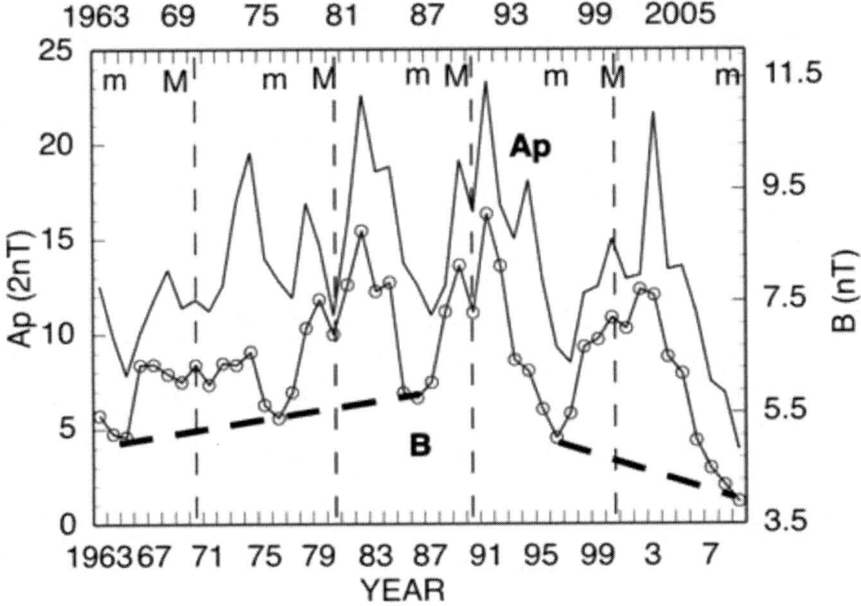

Figure 2. Plot of the annual mean values of Ap and B for 1963-2009; positions of the SSN maxima (M) and minima (m) are marked. TCQP in the limited B data is highlighted.

TCQP IN GCR DATA

Ahluwalia [1997a] extended the shielded ion chamber (IC) data string, by combining the Yakutsk (1953–1994) and Cheltenham/Fredericksburg (1937- 1972) hourly data sets; Yakutsk IC data are erroneous after 1988. The data are normalized to 100 % in 1965. Both sites are located at sea level at high latitudes with an atmospheric cutoff (Ra) <4 GV which is greater than the geomagnetic vertical cutoff (Ro) of 1.7 GV at Yakutsk and 2.2 GV at Cheltenham/Fredericksburg. The median rigidity of response (Rm) of IC to the GCR differential rigidity spectrum is 67 GV; 50% of IC counting rate lies below Rm [Fujimoto et al., 1984; Ahluwalia and Fikani, 2007]. The extended IC data string is depicted in Figure 3. For a comparison, the annual mean hourly rates for the Climax neutron monitor (CL/NM) are also plotted in Figure 3 for 1951-2006, normalized to 100 % in May 1965; Rm = 11 GV for CL/NM. The tracking between the two strings, a factor <6 apart in Rm values, for four cycles (17– 21) is truly remarkable.

Ahluwalia [1997b] discovered the TCQP in the extended IC data string and noted its correspondence to TCQP present in the solar and geophysical data over a much longer time interval. Figure 3 shows that the cosmic ray intensity (CRI) minimum in 1939 (cycle 17) is distinctly shallower than CRI minima for cycles 18 and 19 and CRI minimum for cycle 20 is shallower than for cycles 21 and 22. This pattern is extended to cycle 23 with the help of CL/NM data also plotted in Figure 3. One notes the following features:

1. The TCQP (shown by dashed lines at the top of Figure 3) in the residual modulation for the two data strings stand out, indicating that GCR intensity during 1933 minimum should be higher than the residual modulation in 1965, an expectation confirmed by the measurements made by Neher [1971] at balloon altitudes in Bismarck, ND.

2. The recovery during an even cycle (18, 20) is to a lower level than for a negative cycle (17, 19, 21), as noted earlier by Ahluwalia [1994] as well as Webber and Lockwood [1988]; the latter authors note that GCR intensity measured by NMs for the broad maximum is ~1.5% lower (for ICs the difference is ~0.3%). The recovery for cosmic ray modulation for cycle 23 (not shown, CL/NM data are not available after 2006) is to the highest level ever observed in the instrumental era [Ahluwalia and Ygbuhay, 2010; Mewaldt et al., 2010]. The physical significance of this feature is not clear yet but it supports the assertion that sun is in a transition phase to a new state of lower activity, as noted above.

Kota and Jokipii [1983] show that charged particle drifts in inhomogeneous IMF, with the warped heliospheric neutral current sheet (HCS) included, lead to a " pointy" and a " broad" maximum in the recovery phase of the 11 yr modulation for a 1.6 GeV proton, the pointy peak is higher than the broad peak (see their Figure 6). These patterns are expected to recur during alternate SSN cycles (Hale cycle). Potgieter and Moraal [1985] are unable to reproduce this effect in a self-consistent simulation for a NM data (at an order of magnitude higher Rm) that incorporates drifts; they get the same profile as Kota and Jokipii do but intensity is higher for broad maximum than for the pointy peak (see their Figure 19) which is reverse of what is observed for CL/NM. They conclude, ". . . long-term NM count rates present the first (and so far only) problem of internal self-consistency for these models." Lockwood and Webber [1997] suggest that difference in intensities on recovery for alternate cycles may result from a modulation in the helio-sheath region (see their Figure 7); earlier, Webber and Lockwood [1987] argued similarly (see their Figure 2). Ahluwalia [2005] emphasizes that heliosheath related modulation effects are minimal at earth orbit, at rigidities >3 GV; the geomagnetic vertical cut-off for CL/NM is 3 GV. In short, the present situation is that drift hypothesis is able to account for the 22 yr (Hale cycle) profile of NM observations but not the difference in intensities for the two peaks during alternate SSN minima.

Figure 4 shows a plot of the annual mean values of Helium fluxes measured on board IMP 8 (it stopped operating in October 2006) in the three channels of the University of Chicago cosmic ray nuclear composition (CRNC) instrument, namely: >250 MeV, 160–220 MeV, and 110–130 MeV, spanning 1972–2004 (cycles 21, 22 and parts of cycles 20, 23); the curves for the last two data strings very nearly overlap.

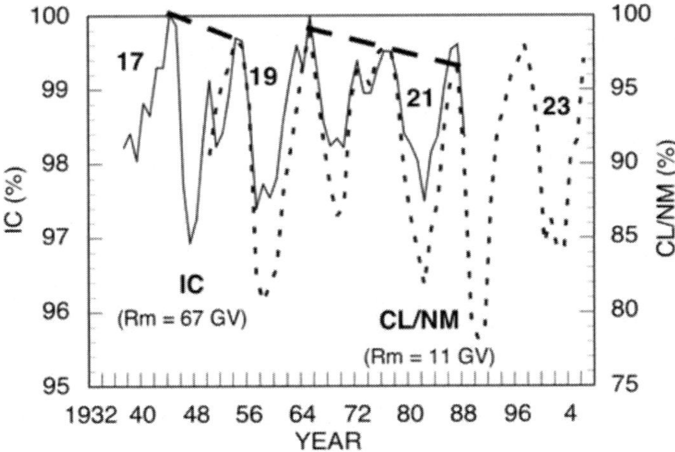

Figure 3. Cosmic ray data are plotted for 1937- 2006 for seven SSN cycles (17-23); TCQP is shown by dashed lines at the top. See text for more details.

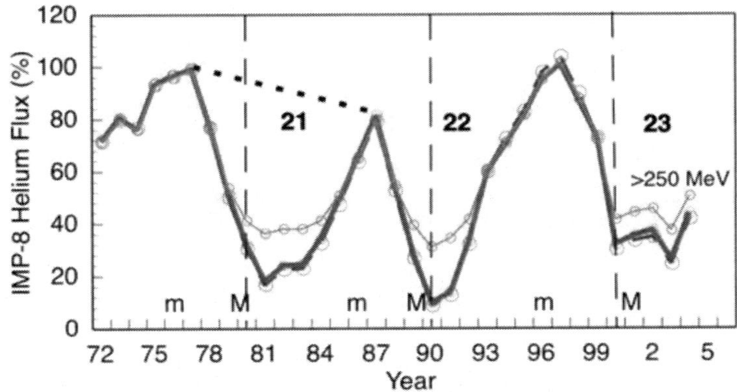

Figure 4. Plot of the annual mean values of Helium fluxes measured on board IMP 8 in three channels of the University of Chicago cosmic ray nuclear composition instrument, namely: >250 MeV, 160–220 MeV, 110–130 MeV, for 1972–2004 (cycles 21, 22 and parts of cycles 20, 23).

Epochs of solar activity maxima (M) and minima (m) are marked; vertical dashed lines are drawn through the middle of the solar polar field reversal intervals. One notes that all channels exhibit similar temporal variations; in particular note that the annual mean values in three channels recover to a lower level in 1987 compared to the recovery in 1976 for cycle 20 and 1997 for cycle 22; a similar behavior is seen for the annual mean hourly rates at higher GCR rigidities (NM and IC data) in Figure 3. Lockwood et al. [2001] present a plot of 26-day average of the count rate of >70 MeV ions as well as fluxes of 121–230 MeV protons and 169-458 MeV Helium from IMP 8 in their Figure 2. The data behavior is similar to CRNC Helium data plotted in Figure 4. An energy loss through adiabatic cooling (in an expanding solar wind) is the most important modulation mechanism for these particles [Zhang, 1997]; adiabatic cooling is most significant closer to the sun.

The data depicted in Figures 3 and 4 confirm that the TCQP in GCR residual modulations for 1944-2006 span a wide rigidity range (4 GV < Rm < 67 GV) covering a variety of energetic charged particle species. Furthermore, Ahluwalia [2002, and references therein] shows that CRI is inversely proportional to B. So, TCQP in CRI implies the presence of TCQP in B data string for the period when in situ measurements of B are not available i.e. prior to 1963. None of the currently available solar dynamo theories are able to account for the TCQP observed in the annual mean B values. Furthermore, the annual mean value (8.7 nT) of aa in 2009 is similar to its value in 1912 (8.9 nT), indicating that we may be returning to the geomagnetic and solar conditions near the early part of the 20th century (GM), in terms of the interplanetary electromagnetic state at earth orbit. In 2009, CRI reached the highest value observed since the instrumental (IC) measurements began in 1935 [Ahluwalia and Ygbuhay, 2010; Mewaldt et al., 2010]. Once again, these facts support the notion that sun is undergoing a transition to a new less active state, beginning at the minimum of SSN in 1996 (cycle 22). Perez-Peraza and Velasco-Herrera reported the discovery of 30 yr cycle in cosmogenic ^{14}C and ^{10}Be radio-nuclide data (A-11-0142-08) at the 37th COSPAR Scientific Assembly, Montreal, Canada, 2008.

PREDICTING SSNs

Figure 5 shows a plot of the annual mean values of the aa index and SSNs for 15 solar cycles (1843–2009); the aa data for 1844–1868 are from (Nevanlinna and Katja, 1993), they may not be as reliable as the data after 1868 derived by Mayaud [1972]. We use only the aa data from 1901 (cycle 14) onwards, covering the 20th century. In addition to the 11 yr (Schwabe) cycle in the two time series, one notes the presence of the TCQP in aa index (high-lighted by dashes); for every third sunspot cycle, the annual mean aa value is the lowest. The index seems to be riding a line of positive slope after the 1901 minimum; no such trend is seen in SSN minima. After the 1996 aa minimum, the trend-line changes slope to a steeper negative value.

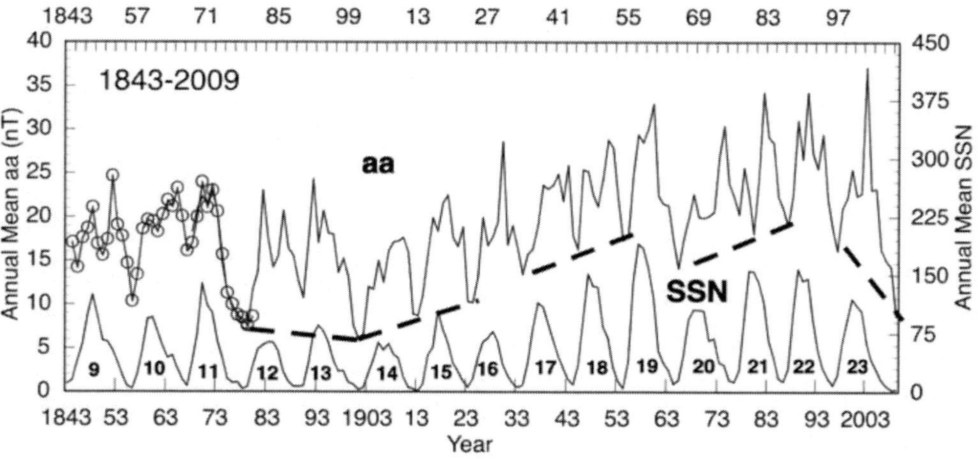

Figure 5. Plot of the annual mean values of aa index and SSNs for 15 solar cycles (1843–2009); TCQP in aa index is high-lighted by dashes. See text for more details.

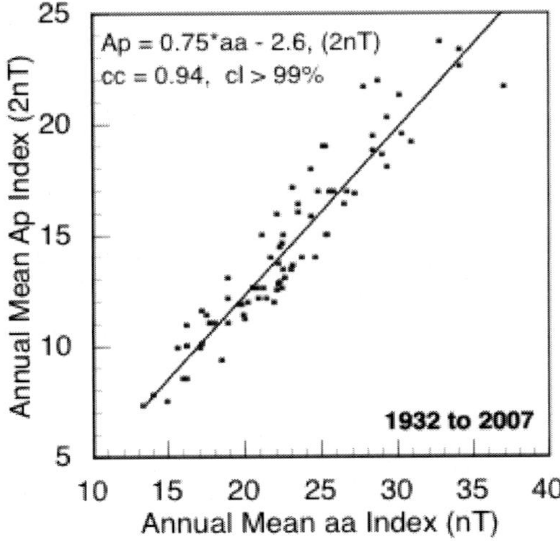

Figure 6. Plot shows the linear correlation between Ap and aa for 1932-2007.

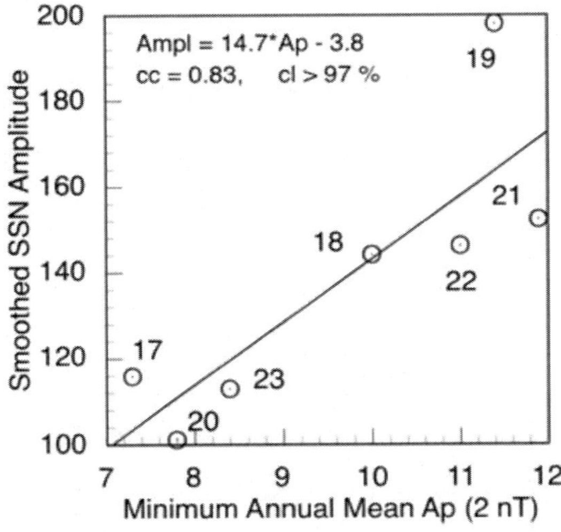

Figure 7. Correlation between the annual mean Ap, one year after the minimum of the previous SSN cycle and the amplitude of the new cycle.

The annual mean aa value (8.7 nT) in 2009 is comparable to its value (8.9 nT) in 1912, near cycle 14 minimum, and the slope of the trend-line before 1901 is negative. So, aa/Ap data may have a period ~200 yr (de Vries/Suess cycle) embedded in them. Figure 6 shows the linear correlation between Ap and aa for 1932-2007, with cc = 0.94 at cl > 99 %.

The empirical procedure for computing smooth SSNs at the peak of a new SSN cycle derives from the discovery by Ahluwalia [1998] that there is a linear correlation between the amplitude of a SSN cycle and the minimum value of Ap at the start of the cycle; the amplitude (Ampl) of a SSN cycle is defined as the monthly mean of the smooth SSN at cycle maximum minus the monthly mean smooth SSN at the preceding minimum.

Figure 7 shows the correlation between the annual mean Ap, one year after the minimum of the previous SSN cycle and the amplitude of the new cycle. The data are plotted for the last seven cycles (17–23) of the 20th century. The datum point for cycle 19 deviates the most from the regression line; it is the most active cycle observed since1600s. The fit parameters for the linear correlation are given below:

$$SSN\ Ampl = 14.7 * Apmin - 3.8 \tag{1}$$

The 2009 annual mean value for Ap = 3.9 ± 0.3 nT, and the annual mean smooth SSN for 2008 = 2.9 ± 0.2, giving a monthly mean smooth SSN at cycle 24 peak 56.4 ± 4.4; the ± indicates the propagated error, the mean standard deviation of the points (excluding cycle 19) is ± 32. The predicted smooth SSN at cycle 24 peak has nearly half the value obtained at the peak for cycle 23 (120.8).

Ahluwalia [2003] set up a regression relation to determine the rise time (Tr) for a SSN cycle, defined as the number of months between smooth SSNs at cycle minimum and maximum. The regression relation uses aa index data beginning with cycle 9 minimum (1855). The fit parameters for the updated linear correlation (including data for cycle 23) are given by:

$$Tr\ (mos) = 61.\ 6 - aamin\ (nT) \tag{2}$$

The maximum deviation of the points from the regression line is ± 6 months (mos). Equation (2) supports Waldmeier [1955] hypothesis that lower activity cycles rise to peak later in time (see his Figure 4). Also, it predicts that maximum value of Tr (when aamin = 0) is ~62 mos. The annual mean value for aa index in 2009 = 8.7 ± 0.3 (nT), giving Tr = 52.9 ± 6 months. Since the minimum smooth SSN (1.7) occurred in December 2008, cycle 24 will attain its peak in May 2013 ± 6 mos. So, Ahluwalia technique predicts that SSN cycle 24 will reach a value of 56.4 ± 4.4 smooth SSNs in the month of May 2013 ± 6 mos. If TCQP endures in the latter half of the de Vries cycle in the aa/Ap index, we may see the next two SSN cycles (25 and 26) to be successively less active; at this time, we cannot predict their activity nor can we state whether they will be long or short cycles. This may be one of the consequences of the new state that sun has been transitioning into since 1996.

TIMELINE COMPARISONS

We use smooth SSNs (to suppress the transients) to describe the profiles of several SSN cycles of the last century. Figures 8a,b show a comparison between the timelines of cycle 24 activity and ten prior cycles (14–23) for 35 mos after the onset. The cycles are normalized at the origin by subtracting the smooth SSN at the onset from those for the subsequent months. The observed timelines fall into two distinct groups. The cycles 14-17, 20, and 23 are slow risers like cycle 24; they are depicted in Figure 8a, the slow rise cycles have <100 smooth SSNs at their peak. The cycles 18, 19, 21, and 22 rise rapidly (compare scales on the left of Figures 8a and 8b); they have >100 smooth SSNs at their peak. They are plotted in Figure 8b; cycles 18, 19, 21, 22 are very active (see Figure 1 above), 19 being the most active cycle ever observed. The four cycles also define the recent Grand Maxima in solar activity that we emerged out of recently.

Additional points to be noted are as follows:

- For the 35 mos period shown in Figure 8a, the timeline for cycle 23 overlaps with that for cycle 20 since the onset for 21 mos (the profiles are barely distinguishable). Early on, cycle 24 follows cycle 17 timeline closely but deviates significantly later on. After 27mos, cycle 24 became more active than cycle 14 and after 30 mos it surpassed the activity of cycle 15 and appears to be leveling off. An interesting question is whether cycle 24 may end up being more like cycle 5 that led to the onset of DM. The ascending phase of cycle 24 observed to-date appears to be consistent with the AJ11 prediction [Ahluwalia and Ygbuhay, 2012].

- Figure 8b shows that cycle 22 started out to be more active than cycle 19, but became less so after 18 mos. The even-odd symmetry in sunspot cycles broke down with cycle 22. Figure 2 shows that the annual mean value of B reached the highest value in cycle 22 and the lowest value in cycle 23. Furthermore, the data at the Wilcox Solar Observatory indicate that the solar polar field strength for cycle 23 is lower ($\sim 50\%$) than for the previous three cycles. The peculiar solar behavior during cycles 22 and 23 is not understood at present but it is consistent with the inference that sun is in transition to a new state of lower activity.

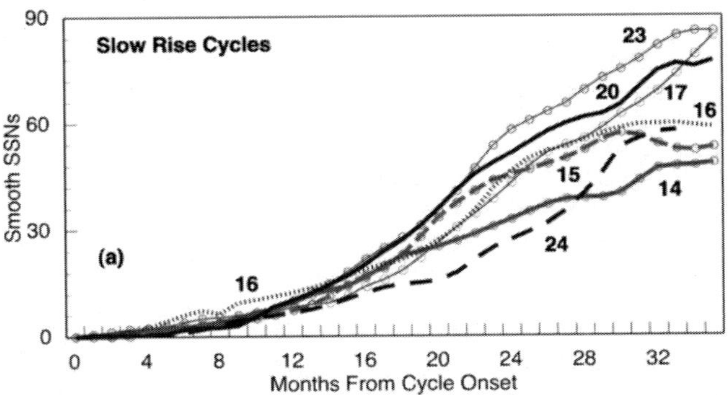

Figure 8a. Slow rise cycles (14-17, 20, 23) have < 100 smooth SSNs at their peak. See text for more details.

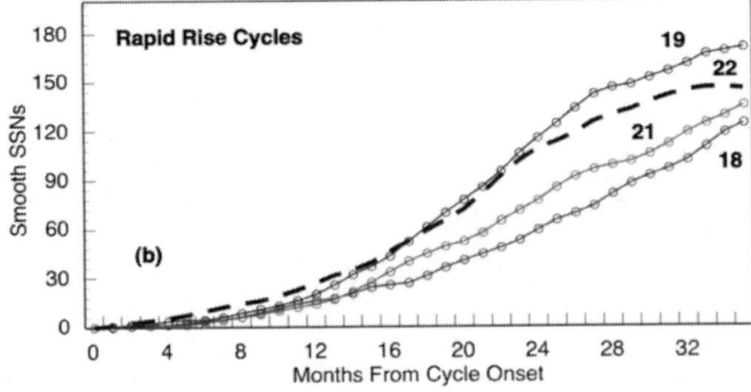

Figure 8b. Rapid rise cycles (18, 19, 21, 22) have > 100 smooth SSNs at their peak. See text for details.

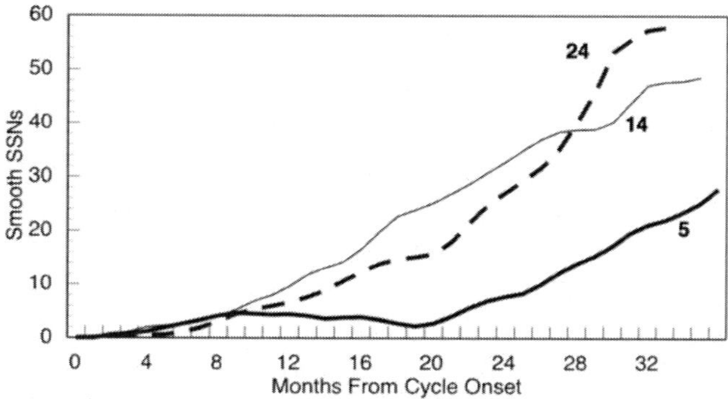

Figure 9. Comparison between ascending phases of cycles 5, 14, 24 for 35 mos after the onset.

Figure 9 shows a comparison between the ascending phases of cycles 5, 14 and 24 for 35 mos after the onset. Clearly, cycle 5 is rising the slowest; at its peak, the smooth SSN 49.2 may be compared to 56.4 predicted for cycle 24. Also, cycle 24 seems to follow the timeline of cycle 14, it led to the onset of the GM (at its peak, the smooth SSN reached a value 64.2). At this point in time, it is not clear whether the timeline of cycle 24 will lie betwixt those of cycles 5 and 14 or it would level off. We shall follow closely the differences between the three timelines as cycle 24 reaches its peak in 2013.

Probable Future Happenings

The change in the slope of B (in Figure 2) for cycle 23 may portend an increasing trend for the GCR flux and a decreasing trend in the global temperatures. One is reminded of the prolonged descending phase of cycle 4 leading to DM (see Figure 1) when earth was cooler. Therefore, our heuristic analysis (not based on any model) cannot rule out the premise that cycle 24 may herald the onset of a DM-like minimum in the present century. This inference contradicts the conclusion drawn in IPCC [2000] report. We note that Feulner and Rahmstorf [2010] use a coupled climate model to explore the future global temperatures. They expect a temperature off set of no more than $-0.30°$ C in 2100, much smaller than the published estimates for anthropogenic global warming according to IPCC [2001]. A careful analysis is needed to estimate how the anticipated decrease in global temperature may compare with an increase in global warming due to an increase in concentration of the greenhouse gases (such as CO_2) in the atmosphere and other causes described in the IPCC reports.

It is interesting to note that the interval of the DM coincided with a significant rise in inflation in England; for example, the price of wheat rose by a factor < 3 [Menzel, 1959]. Considering that human population is projected to reach 10 billion in the 21st century, the higher food prices (if they occur) may cause social and political upheaval in the overpopulated countries of Asia and Africa; people living in the towns of sub-Saharan Africa spend a bigger share of their income on food than do urban residents almost anywhere else in the world. Laborers use up over half their wage just to eat. So, they will spend less on school fees, sanitation, and health. The food, fuel, and climate crises are by themselves serious issues, but taken together, their impact could be disastrous for the world economy and intranational and international conflicts, threatening the security of individuals and nations.

Optimistically speaking, we may hope that advances in sciences and the agricultural technology in the 21st century may meet the challenge of food shortage and mitigate the adverse circumstances forecast by Thomas Robert Malthus FRS (1766–1834) nearly two centuries ago. In this context, we note that the agricultural development begun in Mexico by Norman Borlaug in 1943 led to a new disease resistant high yield varieties of wheat and the founding of the International Maize and Wheat Improvement Center. In 1961 India was on the brink of mass famine because of its rapidly growing population. With the advice from Bourlag, the new wheat and rice seeds were tried in the State of Punjab. It led to a dramatic increase in the agricultural output and India became one of the most successful rice producers and a major rice exporter. Similar results were obtained in the Philippines and a 'Green Revolution" was born [Borlaug, 2000]. Its success spread technologies that had already existed but not used outside industrialized nations; such as: modern irrigation projects, pesticides, synthetic nitrogen fertilizer, alongside improved high yielding crop varieties (they were domesticated plants bred specifically to respond to fertilizers), significantly increasing the amount of calories per acre.

ACKNOWLEDGMENTS

I am grateful to the providers of the solar, interplanetary, cosmic ray, and geomagnetic data used in this study and to Roger C. Ygbuhay for technical help. This research is supported by NASA EPSCoR award NNX09AP76A-Q01378.

REFERENCES

Ahluwalia, H. S. Repetitive patterns in the recovery phase of cosmic ray 11–year modulation. J. Geophys. Res., 99, 11, 561–11, 567, doi:10.1029/94JA00720, 1994.

Ahluwalia, H.S. Galactic cosmic ray intensity variations at a high latitude sea level site: 1937-1994. *J. Geophys. Res.*, 102, 24229-24236, 1997a.

Ahluwalia, H.S. Three activity cycle periodicity in galactic cosmic rays? *Proc. 25th Int. Cosmic Ray Conf.*, Durban, South Africa, 2, 109-112, 1997b.

Ahluwalia, H.S. The predicted size of cycle 23 based on the inferred three cycle quasi-periodicity of the planetary index Ap. *J. Geophys.* Res., 103, 12103–12109, 1998.

Ahluwalia, H. S. Ap time variations and interplanetary magnetic field intensity. J. Geophys. Res., 105, 27481–27487, doi:10.1029/2000JA900124, 2000a.

Ahluwalia, H.S. Prediction of galactic cosmic ray modulation by CMEs at solar cycle 23 maximum. *Adv. Space Res.*, 26, 857-860, 2000b.

Ahluwalia, H.S. IMF intensity and galactic cosmic ray modulation. *Adv. Space Res.*, 29, 439-444, 2002.

Ahluwalia, H.S. Meandering path to solar activity forecast for cycle 23. *Solar wind Ten: Proc. Tenth Intern.* Solar Wind Conf. Velli, M., Bruno, R., Malra, F. (Eds.) AIP: CP679, pp.176–79, 2003.

Ahluwalia, H.S. Cycle 20 solar wind modulation of galactic cosmic rays: understanding the challenge. *J. Geophys. Res.* 110, A10106, doi:10.1029/2005JA011106, 2005.

Ahluwalia, H. S., and M. M. Fikani. Cosmic ray detector response to transient solar modulation: Forbush decreases, *J. Geophys. Res.,* 112, A08105, doi:10.1029/ 2006JA011958, 2007.

Ahluwalia, H.S., and R.C. Ygbuhay. Current status of galactic cosmic ray recovery from sunspot cycle 23 modulation. *AIP Conf. Proc. 1216,* Twelfth Int. Solar Wind Conf. Eds. Maksimovic, M. et al.,pp. 699-702, 2010.

Ahluwalia, H.S. Timelines of cosmic ray intensity, Ap, IMF, and sunspot numbers since 1937. *J. Geophys. Res.,* 116, A12106, 9 pages, doi:10.1029/2011JA017021, 2011a.

Ahluwalia, H. S., and J. Jackiewicz. Sunspot cycle 23 descent to an unusual minimum and forecasts for cycle 24 activity. *Adv. Space Res.,* in press, 2011b.

Ahluwalia, H.S. and R.C. Ygbuhay. Sunspot cycle 24 and the advent of a Dalton-like minimum. *Adv.Astron.,* v.2012, Article ID 126516, 5 pages, doi:10.1155/2012/126516, 2012.

Alicia, L. C. G., W.D. Gonzalez, S.L.G. Dutra, and B.T. Tsurutani. Periodic variation in the geomagnetic activity: A study based on the Ap index. *J. Geophys. Res.,* 98, 9215–9231, 1993.

Bartels, J. Collection of geomagnetic planetary indices Kp and derived daily indices, Ap and Cp for the years 1932 to 1961. North Holland, New York. 1962.

Borlaug, N. The green revolution revisited and the road ahead. Lecture to Norwegian Nobel Institute, Oslo. 2000.

Feulner, G. and S. Rahmstorf. On the effect of a new grand minimum of solar activity on the future climate on Earth. *Geophys. Res. Lett.,* 37, 5 pages, L05707, doi:10.1029/ 2010GL042710, 2010.

Fujimoto, K., A. Inoue, K. Murakami, and K. Nagashima. Coupling coefficients of cosmic ray daily variations for muon telescopes, *Rep. 9, Cosmic Ray Res. Lab.,* Nagoya, Japan, 1984.

Gleissberg, W. The eighty year sunspot cycle. J. Br. Astron. Assoc., 68, 148-152, 1958.

Gosling, J.T., D.J. McComas, J.L. Phillips, and S.J. Bame. Counter-streaming solar wind halo electron events: Solar cycle variations. *J. Geophys. Res.,* 97, 6531- 6535, 1992.

Kota, J., and J.R. Jokipii. Effects of drift on the transport of cosmic rays: VI. A three-dimensional model including drifts. *Astrophys. J.* 265, 573–581, 1983.

Lockwood, J.A., and W.R. Webber. A comparison of cosmic ray intensities near earth at sunspot minima in 1976 and 1987 and during 1995 and 1996. *J. Geophys. Res.,* 102, 24221–24228, 1997.

Lockwood, J.A., W.R. Webber, and H. Debrunner. Differences in the maximum intensities and the intensity-time profiles of cosmic rays in alternate solar magnetic field polarities. *J. Geophys. Res.,* 106, 10635-10644, 2001.

Manoharan, P.K. The solar cause of interplanetary disturbances observed in the distance range 0.24-1 AU. *Geophys. Res. Lett.,* 24, 2623-2626, 1997.

Mayaud, P.N. The aa indices: A 100 year series characterizing the magnetic activity. *J. Geophys. Res.,* 77, 6870–6874, 1972.

Menzel, D.H. *Our Sun,* Harvard University Press, Cambridge, Mass, USA, 1959.

Mewaldt, R.A., et al. Record setting cosmic ray intensities in 2009 and 2010. *Astrophys. J.,* 723, L. 1-6, 2010.

Nevanlinna, H., and Katja, E. An extension of the geomagnetic activity index series aa for two solar cycles (1844–1868). *Geophys. Res. Lett.*, 20, 2703–2706, 1993.

Neher, H. V. Cosmic rays at high latitudes and altitudes covering four solar cycles, *J. Geophys. Res.*, 76, 1637–1651, doi:10.1029/JA076i007p01637, 1971.

Potgieter, M.S., and H. Moraal. A drift model for the modulation of galactic cosmic rays. *Astrophys. J.* 294, 425–440, 1985.

Silverman, S.M. Secular variations of the aurora for the past 500 years. *Rev. Geophys.*, 30, 333-351, 1992.

Snyder, C.W., M. Neugebauer, and U.R. Rao. The solar wind velocity and its correlation with cosmic ray variations and with solar and geomagnetic activity. *J. Geophys. Res.*, 68, 6361–6370, 1963.

Waldmeier, M. *Ergebnisse und Probleme der Sonnenforschung*, Leipzig, Germany, 2nd edition, 1955.

Wang, Y.M., J.L. Lean, and N.R. Sheeley. Modeling the sun's magnetic field and irradiance since 1713. *Astrophys. J.*, 625, 522-538, 2005.

Webb, D.F., and R.A. Howard. The solar cycle variation of the coronal mass ejections and the solar wind mass flux. *J. Geophys. Res.*, 99, 4201- 4220, 1994.

Webber, W.R., and J.A. Lockwood. Interplanetary radial cosmic ray gradients and their implication for a possible large modulation effect at the heliospheric boundary. *Astrophys. J.* 347, 534–542, 1987.

Webber, W. R., and J. A. Lockwood. Characteristics of the 22-year modulation of cosmic rays as seen by the neutron monitors, *J. Geophys. Res.*, 93, 8735–8740, doi:10.1029/JA093iA08p08735, 1988.

Zhang, M. A linear relationship between the latitudinal gradient and 26 day recurrent variation in the fluxes of galactic cosmic rays and anomalous nuclear components: I. Observations. *Astrophys. J.* 488, 841–853, 1997.

In: Homage to the Discovery of Cosmic Rays …
Editor: Jorge A. Perez-Peraza

ISBN: 978-1-63483-084-3
© 2015 Nova Science Publishers, Inc.

Chapter 5

COSMIC RAY FLUCTUATIONS

I. Ya. Libin[*]

International Academy of Appraisal and Consulting, Moscow, Russia

ABSTRACT

For the first time in space physics, by using power spectrum analysis on the μ meson intensity records of a high counting rate instrument (106 counts/min) operated at Chacaltaya, Bolivia, it has been possible to identify the presence of cosmic ray fluctuations of 15 to 19 and 25 to 28 cycles per hour (cph). These frequencies were found at a 99% level of significance in the average of 1034 three-hour samples extending from November 1965 to June 1966. Comparison with Pioneer 6 measurements (Ness et al., 1966) indicates that the magnetosheath field, at a distance of about 10 earth radii (RE), and the interplanetary magnetic field show peaks in spectral density that correspond closely (though not identically) with peaks in muon intensity.

It has been suggested, that these cosmic ray fluctuations are caused by the fluctuations of geomagnetic field corresponding to an amplitude of about 20 γ in the dipole magnetic field. Magnetic field measurements conducted in the magnetosphere by Explorer 6 (Judge and Coleman, 1962) and Explorer 12 (Patel and Cahill, 1964) show similar oscillations. Recent observations by ATS 1 confirm the presence of magnetic oscillations of similar frequencies at about 6 RE . These frequencies are also found in the geomagnetic micropulsations observed at the surface of the earth.

The integrated power of cosmic ray fluctuations in the frequency range of 6 to 30 cph has been studied at various periods, and its solar-terrestrial relationship is examined. The association of observed frequencies in cosmic rays with frequencies in the solar photosphere and the interplanetary magnetic field, as well as the resonance frequencies of the magnetosphere, are discussed.

The origin and propagation mechanisms of fluctuations in the cosmic rays observed by instruments on the ground, in the stratosphere, and in space are discussed.

Recent experimental results on fluctuations at frequencies from $10{-}3$ to $10{-}6$ Hz are discussed. The primary cause of the statistically significant fluctuations in the cosmic rays is the scattering of charged particles of the cosmic rays by random in homogeneities of the interplanetary magnetic field.

---

[*] Email: libin@bk.ru

As a result of numerous experiments and calculations (1978-2008), the method of short-term prediction and diagnostics of shock waves by means of ground-based observations of cosmic-ray scintillations, is elaborated. The study of scintillation phenomena in cosmic rays is divided into a number of problems connected with the interaction of charged particles of cosmic radiation with the matter and fields which they encounter in the entire length of their propagation. It is shown that high-frequency scintillations greater than or equal to 0.00001 Hz are able to sense an approaching shock wave and that scintillations with periods of the order of 10-20 and 40-50 min are most sensitive to interplanetary medium disturbances near the earth. Since cosmic rays of different energies are sensitive to different processes in interplanetary space at different distances from the earth, the conditions in the interplanetary medium can be studied by measuring particle fluxes at different energy ranges.

Power spectral analysis of cosmic-ray intensity recorded by eight stations was carried out over a wide range of frequencies from 2.3×10^{-8} Hz to 5.8×10^{-6} Hz (2–500 days) during the period 1964–2005. Spectrum results of large-scale fluctuations have revealed the existence of a broad peak near 250–285 days and a narrower peak at 45–50 days during the studied epochs as a stable feature in all neutron monitors covering a wide rigidity range. The cosmic-ray power spectrum displayed significant peaks of varying amplitude with the solar rotation period (changed inversely with the particle rigidities) and its harmonics.

The amplitudes of 27-day and 13.5-day fluctuations are greater during the positive-polarity epochs of the interplanetary magnetic field ($qA > 0$) than during the $qA < 0$ epochs. The comparison of cosmic-ray power spectra during the four successive solar activity minima have indicated that at the low-rigidity particles the spectrum differences between the $qA > 0$ and $qA < 0$ epochs are significantly large.

Furthermore, the spectrum for even solar maximum years are higher and much harder than the odd years. There are significant differences in the individual spectra of solar maxima for different cycles.

Are discussed, the variations in the form of the cosmic-ray fluctuation power spectrum as an interplanetary shock wave approaches the Earth have been calculated for different values of cosmic ray anisotropy. The relevant experimental estimates of the power spectra are inferred from the data of cosmic ray detection with the ground-based neutron monitors at cosmic-ray stations. A comparison between the theoretical and experimental estimates has demonstrated an important role of the cosmic ray anisotropy spectrum in the generation of the power spectrum as the latter is rearranged before the interplanetary medium disturbances.

In the present chapter to use spectral autoregressive methods in order to improve the detection of variations in the spectral composition of the cosmic ray intensity and in the B_z component of the interplanetary magnetic field. By using this methodology, some regularities in the variation of the power spectra of the cosmic ray intensity are determined immediately before the arrival of a shock wave to Earth. So, the shock wave arrival is preceded by the appearance of a wave with period of 4 to 8 hours in the power spectra of the B_z component. The results confirm the possibility of the use of the cosmic rays and the geomagnetic activity as a tool for the diagnosis of the interplanetary environment and also for the prediction of the arrival of powerful shock waves to Earth, 1-2 days in advance.

We study the time evolution of power spectra of galactic cosmic ray fluctuations during the last three solar cycles (1968-2002) using data of 5-min count rates from two far spaced high-latitude neutron monitors, Tixie Bay (Russia) and Oulu (Finland).

It is shown that the power spectrum of cosmic ray fluctuations is a subject of a regular long-term periodic 11-year modulation in phase with solar activity, in accordance with variations of the inertial part of the interplanetary magnetic field turbulence power spectrum. These results present a new kind of modulation of cosmic ray intensities.

Keywords: Cosmic rays, interplanetary space, scintillation, Forbush decreases, interplanetary magnetic fields, interplanetary medium, shock wave propagation, solar flares

I. INTRODUCTION

The solar wind and the magnetic fields frozen in it pass by the Earth at a velocity on the order of 300-700 km/s and largely determine the "Space Weather" in the space environment of our planet: It is the variations of the solar wind, which is interacting with the magnetic field of the Earth,- which are the primary causes of such events as geomagnetic storms and all their manifestations. Systematic observations of the-solar wind are accordingly necessary for research in geophysics and astrophysics.

In particular, a study of the turbulence in the solar wind is one of the central problems of modern astrophysics. The propagation of shock waves in a turbulent medium in space gives rise to many physical processes, in particular, the scattering and acceleration of the charged particles of the cosmic rays.

Sensitive probes of the solar wind are the cosmic rays, which can furnish information on not only the large-scale processes in the solar wind, but also the small-scale structure of the wind in a study of short-period changes (variations, fluctuations, oscillations) in the cosmic-ray intensity at the earth's surface, underground, in the atmosphere, in the stratosphere, and directly in space. In the modern scientific literature, the term "short-period variations" in cosmic-ray intensity is replaced by "fluctuations" or "scintillations" of cosmic rays, which mean rapid changes (with periods ranging from a few minutes to a few hours) in the cosmic-ray intensity detected with respect to the average background intensity.

Cosmic-ray fluctuations were first observed by Dhanju and Sarabhai in the late 1960s (Dhanju and Sarabhai, 1967).

Although today the study of these variations is a well-established field in space physics, the main question-the nature of the observed cosmic-ray fluctuations-remains unresolved in many studies.

The problem is that it is a complicated and laborious process to distinguish and study these short-period changes in the cosmic-ray intensity. The only methods which were available until recently for studying the spatial and temporal variations in the flux density, spectrum, and composition of the cosmic rays ran into serious difficulties in attempts to study fluctuations, since in order to distinguish the fluctuations and to determine their nature it was necessary (on the one hand) to deal with time-varying processes with continuously varying amplitudes, phases, and frequencies and (on the other hand) to distinguish and study the many correlations between the fluctuations themselves and various geophysical, heliophysical, and astrophysical processes.

Spectral methods have made it possible not only to detect oscillations with various periods in data on the cosmic rays (and to evaluate the reliability of these oscillations), but also to carry out qualitative and quantitative comparisons of various processes.

Study of the dynamics of cosmic-ray fluctuations and analysis of their appearance and development, in association with various geophysical and heliophysical phenomena and their parameters, has made it possible to extract several curious results: sharp increases in the amplitudes of the oscillations stem from flare activity on the sun, the structure of the solar wind, and perturbations of various types near the Earth.

It is here that we can see the enormous possibilities for space physics which are hidden in data on the short-period changes in the cosmic rays-possibilities which even today allow us to use cosmic-ray fluctuations as a forewarning of space perturbations.

As cosmic rays propagate through interplanetary space, several fluctuational phenomena occur in connection with the scattering of charged particles by random inhomogeneities of the interplanetary magnetic field. From the published data it is possible to draw several quite general conclusions about the nature and sources of the cosmic-ray fluctuations (Dorman and Libin, 1984a, 1984b, 1985; Ginzburg, 1988):

1. fluctuations in acceleration processes;
2. fluctuations during propagation in the solar corona and in the escape from the solar quasiconfinement;
3. fluctuation processes during propagation in interplanetary space and in the local galaxy;
4. fluctuations during passage through the earth's magnetosphere;
5. fluctuations due to short-period variations in meteorological factors.

Only in observations of the various secondary components of the cosmic rays at the earth do all five of these sources operate.

In observations on low-orbit satellites, geophysical rockets, and balloons, the fifth of these sources can be completely ignored. In observations outside the earth's magnetosphere, only the first three sources of fluctuations are important. It should be noted that the fluctuational phenomena caused by source in the galactic and solar cosmic rays are identical, so that separate information on the fluctuations of each type is important for drawing detailed conclusions from observational results.

For a thorough analysis of the factors and processes-which give rise to statistically significant fluctuations or, more precisely, to the formation and development of statistically significant fluctuations, it is necessary to identify as many as possible of the existing periodicities in the cosmic ray intensity, working from all the information available on the various components; to follow the time evolution of these periodicities; to discard all unimportant effects; and, finally, to determine the relationships among the important processes and phenomena (and to do this analytically, to the extent possible).

Governing factors here are not only the quality of the raw data but also the choice of analysis methods and the testing of the reliability of the results on the basis of some physical model or other.

II. Observation QF Periodicities in the Cosmic Rays

Previous methods for studying the spatial and temporal variations of the intensity, spectrum, and composition of the cosmic rays ran into serious difficulties in attempts to study fluctuational phenomena in the cosmic rays, since in order to distinguish these fluctuations and to determine their nature it was necessary, on the one hand, to deal with time varying processes with continuously varying amplitudes, phases, and frequencies and, on the other, to distinguish and study the numerous correlations among the fluctuations themselves and various geophysical, heliophysical and space-physical processes (Millican et al., 1934, Dorman and Luzov, 1967a, 1967b; Bazilevskaya et al., 1970; Kozlov et al, 1973; Gulinsky and Libin, 1979; Kozlov, 1980, 1981, 1985, 1986, 1996, 1997, 1999; Blokh et al., 1984; Gulinsky et al., 1988; Dorman et al., 1987a, 1988a; Nonaka et al., 2003; Mulligan et al, 2009; Mishra and Agarwal, 2011; Kozlov and Kozlov, 2011).

The standard correlation methods and the method of statistical periodogram analysis proved insufficiently informative because of the restricted size of the data files which could be used. Studies of fluctuations on the basis of large data files resulted in a significant smoothing of the results, as is particularly obvious in several studies.

For all the usefulness and versatility of the Fourier method, the corresponding calculations yield the amplitudes, periods, and phases of only nonrandom functions of the time, and thus are pertinent only when these amplitudes, phases, and periods are fixed. As Dorman (Jenkins and Watts, 1972; Bendat and Pearsol, 1983; Max, 1983; Dorman and Libin, 1985) has shown, the advantages of spectral analysis lie in the reliable and unambiguous monitoring of a comparison of different records, since the functions which are being compared are functions of the same parameter: the frequency.

In addition, the number of analyzed processes are a substantial restructuring of its statistical characteristics, ie, are non-stationary processes. In this case, is not defined the concept of the spectrum, and the classical transformation of data, based on fast Fourier transformation and the Blackman-Tukey (Jenkins and Watts, 1972), often gives incorrect results.

The usual method in this situation - the allocation of plots of quasi-stationary faces several difficulties. Such areas (if any) may be short, and the small amount of data the Fourier method gives poor results and does not allow to separate close frequencies. At the same time, the problem of separation of frequencies is one of the main points of the investigated problem, since each of the frequencies may be associated with different physical mechanisms of interaction of Space Physics and meteorological parameters.

For the purpose of separation of frequencies for short segments of data recently used autoregressive methods, which essentially consists in the introduction of the additional assumption that all the investigated process can be described by an autoregressive model of order p is unknown (for t = 0,1,2, ...)

$$x_{t+1} = \sum_{i=0}^{p} a_{i+1} x_{t+1}$$

Under this assumption are evaluated, one way or another, the coefficients of the autoregressive and selected the best order, and only on these factors are uniquely calculated spectrum. This approach, using different algorithms (such as methods of Berg, the Levinson-Durbin, Pisarenko, Prony and their modifications), was implemented and Gulinsky-Yudakhin (Dorman and Libin, 1984, 1985; Gulinsky et al., 1988) and in some cases gave good results.

1. Spectral Analysis

Essentially no serious restrictions are imposed on the behavior of the random process, and the spectral approach presents the extremely important opportunity of not only detecting prominent periods in the data but also testing their reliability, although it must be kept in mind that no mathematical model of any sort is capable of reflecting all the complexities of fluctuational phenomena in the cosmic rays.

There is accordingly no need to go into detail here on the various approaches in spectral analysis. We will simply mention the basic ideas, the basic steps of the analysis, the algorithms for finding spectra, and certain aspects of their use which are pertinent to the study of fluctuational phenomena in the cosmic rays.

In practice, the procedure for evaluating spectra consists of several steps: a preliminary analysis, the calculation of sampling correlation functions and spectral estimates, and the interpretation of the results.

We used the following scheme of calculations for the analysis of fluctuations of cosmic rays (Dorman and Libin, 1985):

a) *The data to be analyzed.* The data consist of the data files from measurements of the cosmic-ray intensity by neutron monitors and telescopes at various stations (or at a single station), $N_{ik}(t)$; the results of measurements of the pressure, $P_{ik}(t)$, and the temperature $T_{ik}(t)$; and the values of the coupling coefficients $W_{ik}(R)$, where i specifies the particular instrument, and k the station.

b) *Corrections for meteorological conditions.* The operations of correcting for the pressure and the temperature.

c) *Choice of optimum analysis interval.* The idea here is to introduce an interval of frequencies to be studied, $f_1 < f < f_2$ where $f_1 \geq 10^{-5}$ and $f_2 < 10^{-2}$ Hz for the cosmic-ray fluctuations caused by scattering by random inhomogeneities of the interplanetary magnetic field, and $f_1 \geq 10^{-7}$ and $f_2 < 10^{-5}$ Hz for fluctuations caused by the sector structure of the interplanetary magnetic field.

The data digitization interval is determined:

1 min for	$10^{-3} < f < 10^{-2}$ Hz,
5 min for	$10^{-4} < f < 10^{-3}$ Hz,
1 hour for	$10^{-5} < f < 10^{-4}$ Hz,
12 hours for	$10^{-6} < f < 10^{-5}$ Hz,
24 hours for	$10^{-7} < f < 10^{-6}$ Hz.

d) *Test for a steady state.* The record is broken up into n equal intervals; mean values of the data and the dispersion are calculated for each interval; and these sequences are checked for the presence of trends or other time variations which cannot be attributed exclusively to the sampling variability of the estimates. The convergence of autocorrelation functions of the intervals is tested.

e) *Reduction of the processes to a stationary (or quasi-stationary) form.* This step means eliminating trends, filtering the data, etc.

f)   *j) Trial analysis.* This step involves a preliminary evaluation of sampling power spectra, deciding on the need of some particular filter, and deciding whether to study the spectrum over the entire frequency range or only a part of it.

g)   *Spectral analysis.* Autocovariation and autocorrelation functions and sampling power spectra are determined. The appropriate cutoff point is chosen (the spectral window is narrowed), and the results in one-dimensional data files are reported.

h)   *Test for periodicity.* Theoretically, the presence of quasiperiodic components in a random process will be manifested as δ-functions in its spectral density.

In practice, the spectral density is found to contain sharp peaks, which may be erroneously attributed to a narrow-band random noise, so that the analysis of periodicity is carried out on the basis of calculations of autocorrelation functions and sampling power spectra, the results of the trial analysis, and the physical interpretation of these results.

Information on the frequencies identified and on the amplitudes at these frequencies is extracted as the product of a visual control. The test for periodicity is actually the initial stage of preparation for two-dimensional analysis.

i)   *Determination of the probability distribution.* I) According to (Jenkins and Watts, 1973), the probability distribution of a random, quasiperiodic process X_n ($n = 0, 1, 2, \dots, N$) with a zero mean deviation x can be estimated from

$$\hat{P}(x) = \frac{N_x}{NB},$$

where B is a narrow interval positioned symmetrically with respect to x, and N_x is the number of values of the realization which fall in the interval $x \pm B$.

The value of P(x) is found by partitioning the entire file of values of X_n to be analyzed into the appropriate number of subfield of equal size, tabulating the values of x by subfield, and dividing by the width of the subfield, B, and the sample size N in such a manner that N values of the sequence {x} will satisfy the condition $N = \sum N_i$ The determination of the probability distribution is a completely optional step; it is used as an auxiliary step for testing the resulting distributions for normality.

j)   *Test of normality.* The probability distribution found for the values of the process under study is compared with a theoretical normal distribution. One of the most convenient tests of normality is the χ^2 test, in which the distribution is used as a measure of the discrepancy between the theoretical and observed distributions;

$$\chi^2 = \sum_{i=1}^{k} \frac{(f_i - F_i)^2}{F_i},$$

here k is the number of intervals into which the data are grouped, and f_i is the number of observed values in interval i (Dorman et al., 1979; Dorman and Libin, 1984).

k)   *Test of correlation of realizations.* The question of whether there is a correlation between the files of input data is resolved. In many cases it is sufficient to visually estimate the basic physical properties of the original processes. In general, the test involves calculating mutual correlation functions, a mutual correlation coefficient, and coherence functions. The shift between realizations is estimated in this step.

l)   *Test of equivalence of uncorrelated realizations.* This test combines the results of the analysis of the individual realizations for which statistical equivalence has been established. This operation is performed by calculating the corresponding mean weighted values from the estimates found in the analysis of the individual realizations: While estimates of the spectral density G_l (f) and G_2 (f) with the corresponding degrees of freedom V_1 and V_2 (the condition $V_1 = V_2$ is generally assumed) are found from the individual realizations, the resultant estimate of spectral density is found in the form

$$\mathcal{G}_p\,(f) = \frac{\mathcal{G}_1\,(f)\,v_1 + \mathcal{G}_2\,(f)\,v_2}{v_1 + v_2},$$

where the number of degrees of freedom of this estimate is $V_p = V_1 + V_2$.

m)   *Determination of mutual correlation functions and mutual spectra.* This step, carried out in accordance with the algorithms of (Dorman and Libin, 1984a, 1984b, 1985), involves knowledge of the relationships which can be observed between different realizations of a process under study or between different processes.

n)   *Determination of coherence functions.* This step is required for estimating the correlation of the realizations, for estimating the accuracy of the estimates of the frequency characteristics, and, finally, for directly solving several applied problems for the program (shifting processes with respect to each other, determining the statistical delay between processes, etc.).

The determination of coherence functions precedes the determination of the mutual power spectra of the amplitude and phase functions. The determination of the frequency characteristics is carried out only when information on the spectrum as a whole is not required, and all that we need are the amplitudes and phases at the prominent frequencies, i.e., where the coherence function is at a maximum.

2. Wavelet Analysis

A full analysis of the limitations and disadvantages of the Fourier methods were given in (Stehlik and Kudela, 1984; Dorman et al., 1979; Kay and Marpl, 1981; Dorman and Libin, 1984a, 1985).

In contrast to these standard methods, one of the most powerful modern tools for working with non-stationary processes today is the so-called Wavelet spectral analysis, allows us to investigate the temporal evolution of the fundamental frequencies in non-stationary series (Morlet-Wavelet) for analyzing localized variations of power (Torrence and Compo, 1998).

The concept of wavelet (wavelet in a literal translation - a small wave) was introduced into scientific Grossman (A. Grossman) and Morlet (J. Morle) in the mid 80s. (Grossman and Morlet, 1984) by the need to analyze the properties of seismic and acoustic signals (Torrence and Compo, 1998), which are inherently unsteady. (In the various translations of foreign articles in the Russian language there are still terms "surge," "wavelet function", etc.).

Currently, digital signal analysis based on wavelet transform, become applicable to the problems of pattern recognition and video compression, the very different nature (pictures of minerals, satellite images of clouds or the surface of planets, galaxies, pictures of different turbulent processes, etc.); processing and synthesis of various signals such as speech, vibro-acoustic, seismic signals, to study the properties of turbulent fields, etc.

Wavelets are special functions in the form of short waves (bursts) with zero integral value, and the localization of the axis of the independent variable (t or x), are able to shift on this axis, and scaling (stretching / compression). Any one of the most commonly used types of wavelets generates a complete orthogonal system of functions. VT is widely used to study non-stationary signals, non-uniform fields and non-stationary time series.

As a result of this analysis, we obtain information about when and what frequency the same time, what is their nature, the existence of direct or non-linear relationships between the processes (such as solar and terrestrial phenomena), the presence of phase characteristics between the two.

Wavelet analysis includes Morlet Wavelet analysis, Cross-Wavelet analysis of coherent, Quadratic Wavelet Transform of a coherent analysis, Wavelet transform of the coherent signal/noise analysis, Global Vayvlet Range.

3. Autoregressive Methods

3.1. Autoregressive Moving-Average Model

Autoregressive moving-average model (ARMA) - one of the mathematical models used for the analysis and prediction of stationary time series in statistics. ARMA model generalizes two more simple time series model - a model of autoregressive (AR) and the model moving average (MA).

Model ARMA (p, q), where p and q - are integers that specify the order of the model is called the next generation process time series: $\{X_t\}$:

$$X_t = c + \varepsilon_t + \sum_{i=1}^{p} \alpha_i X_{t-i} + \sum_{i=1}^{q} \beta_i \varepsilon_{t-i}.$$

where c — constant, $\{\varepsilon_t\}$ — white noise,
$\alpha_1, \ldots, \alpha_p$ and $\beta_1, \ldots, \beta_q$ — real numbers, the coefficients of the autoregressive and moving average coefficients, respectively..

Such a model can be interpreted as a linear multiple regression model, in which as the explanatory variables are the past values of the dependent variable itself, but as a regression balance - moving averages of the elements of white noise. ARMA-processes are more complex compared with similar in behavior or the AR-MA-processes in a pure form, but the ARMA-processes are characterized by fewer parameters, which is one of their advantages..

If we introduce the lag operator $L: \ Lx_t = x_{t-1}$, then ARMA-model can be written as follows

$$(1 - \sum_{i=1}^{p} \alpha_i L^i)X_t = c + (1 + \sum_{i=1}^{q} \beta_i L^i)\varepsilon_t$$

Introducing the shorthand notation for polynomials of the left and right sides finally be written

$$\alpha(L)X_t = c + \beta(L)\varepsilon_t$$

For the process to be stationary, it is necessary that the roots of the characteristic polynomial of the autoregressive part $\alpha(z)$ lie outside the unit circle in the complex plane (in modulus strictly greater than one).. Stationary ARMA-process can be represented as an infinite MA-process

$$X_t = \alpha^{-1}(L)c + \alpha^{-1}(L)\beta(L)\varepsilon_t = c/a(1) + \sum_{i=0}^{\infty} c_i\varepsilon_{t-i}$$

For example, the process ARMA (1,0) = AR (1) can be represented as MA-process of infinite order with coefficients decreasing geometric progression:

$$X_t = c/(1 - a) + \sum_{i=0}^{\infty} a^i\varepsilon_{t-i}$$

Thus, ARMA-processes can be considered as MA-processes of infinite order with certain restrictions on the structure coefficients. The small number of parameters to describe the processes they enable rather complex structure. All stationary processes can be arbitrarily approximated by ARMA-model of a certain order with considerably fewer parameters than just the use of MA-models.

3.2. No Stationary (Integrated) ARMA

In the presence of unit roots of the polynomial autoregressive process is no stationary. The roots of less than unity in practice, are not considered, since these are processes of explosive character. Accordingly, to test the stationary of a time series of basic tests - tests for unit roots. If the tests confirm the presence of unit root, then analyzed the difference between the original time series and a stationary process of the differences of some order (usually the first order is sufficient, and sometimes the second) ARMA-based model.

Such models are called ARIMA-models (integrated ARMA) models or Box-Jenkins. The model ARIMA (p, d, q), where d-the order of integration (the order of differences of the original time series), p and q - the order of AR and MA - parts of the ARMA-process differences d-th order can be written in the operator form

$$\alpha(L)\,\Delta^d\,X_t = c + \beta(L)\varepsilon_t\,, \qquad \Delta = 1 - L$$

The process of ARIMA (p, d, q) is equivalent to the process ARMA (p + d, q) with d unit roots.

For the ARMA model construction on a series of observations necessary to determine the model order (numbers p and q), and then the coefficients themselves.

To determine the order of the model can be applied investigation of such characteristics of time series, as its autocorrelation function and partial autocorrelation function.

To determine the coefficients used methods such as the method of least squares and maximum likelihood method.

III. DETECTION AND MONITORING OF FLUCTUATIONS OF COSMIC RAYS ON EARTH

Fluctuational phenomena in the cosmic rays according to data obtained with various instruments.

The first detailed observations of cosmic-ray intensity, carried out by Dhanju and Sarabhai (Dhanju and Sarabhai. 1967) by means of a cosmic-ray scintillation apparatus with an area of 60 m^2 at Chacaltaya (Bolivia), resulted in the discovery of short-period variations in the intensity which turned out to be quite regular over a prolonged time. Dhanju and Sarabhai's apparatus, with an average count rate of the order of 10^6 counts/min, consisted of three identical instruments which were installed at a height of about 17 200 feet. The data on the cosmic rays were recorded every 12 s, and the pressure was measured each minute.

The power spectrum of the fluctuations was analyzed on the basis of data files each containing 180 values, with an overall delay of the order of 30 and with a digitization interval of about 1 min. Spectral estimates -found from considerable observational material show that the peaks in the power spectra of the fluctuations are permanent features in all the 3-h intervals studied by the investigators. When the spectral estimates are combined, it is found that the peaks are smeared only slightly in the interval of spectral frequencies from 1 to 6 cycles per hour, largely due to imperfections of the data analysis. Nevertheless, peaks are observed in the high-frequency region (at 15-16, 18, and 25 cycles per hour) which are essentially immune to smearing.

A major research effort on fluctuations of the cosmic-ray neutron component has been carried out for a long time now by means of the neutron supermonitors of the worldwide network of cosmic-ray stations.

The hourly and 5-min data on the cosmic-ray intensity over the years 1966-1969 at the stations at Alert, Deep River, Sulphur Mt., Chacaltaya and Calgary (the range of geomagnetic cutoff rigidities is 0.1-15 GV) have yielded cosmic-ray fluctuations over the broad frequency range 10^{-7} - 10^{-3} Hz. (Dhanju and Sarabhai. 1967; Libin, 1979, 1983a, 1983b; Kozlov, 1983).

After the peaks due to the rotation of the earth and the sun are eliminated, the spectrum of cosmic-ray fluctuations becomes a very smooth power-law function f$^{-\nu}$ (where 1.65<ν< 1.95) with peaks at certain basically fixed frequencies.

Our understanding of the fluctuations in the nucleon component of the cosmic rays has benefited substantially from studies by V.Kozlov, G.Krymskii, and N.Chirkov (Kozlov and

Krimsky, 1983; Kozlov et al., 1984, 2003; Kozlov and Tugolukov, 1984; Kuzmin, 1984; Kozlov, 1999, Kozlov and Kozlov, 2008) on the basis of cosmic-ray measurements at the Earth by the neutron supermonitors at the Tixie Bay, Yakutsk, and Khabarovsk stations over the years 1969-2009.

This analysis used observation intervals with a substantial range of environments in interplanetary space and in the earth's magnetosphere.

Comparison of the results of fluctuational researches for the different intervals revealed that stable fluctuations appear in the cosmic rays in both the low-frequency range (with periods ~7 h) and the high-frequency range (with typical periods of 15-20 and 40-50 min) during intense disturbances of the interplanetary medium and in the earth's magnetosphere.

The appearance and existence of these fluctuations are characteristic of only disturbed intervals. Power spectra calculated from the mutual correlation function of the observational data from the Tixie Bay and Irkutsk stations, the Tixie Bay and Alert stations, and several other stations separated in longitude show that the 7-h variation, at least, is common to all the stations during disturbances of the interplanetary medium and the geomagnetic field.

It is difficult to overestimate the value of this research, since it not only involves a study of a phenomenon which previously was essentially unknown but also raises the hope that we will acquire an effective instrument for predicting processes of various types in interplanetary space.

Kozlov et al. (Kozlov, 2002; Kozlov et al., 1984, 2003) are studying the relationship between the change in the power spectrum of the cosmic-ray fluctuations calculated from data acquired by the neutron supermonitors and the passage of intense shock waves.

The increase in the amplitude of the fluctuations in a rather narrow spectral region (we are dealing here with fluctuations with typical periods of about 20 and 40 min) before the arrival of the shock front, followed by a significant decrease in this amplitude after the passage of the front, gives us information about the dynamics of the propagation of shock waves. The very fact that an increase occurs in a certain part of the spectrum "is a sort of forerunner of the approaching shock wave, and the absence of these fluctuations allows us to say that we have discovered a 'wake' behind the front of the propagating shock wave.

This effect is analyzed in detail below; here we simply wish to stress that a study of the cosmic-ray fluctuations makes it possible to identify reliably the fact that a shock wave is passing, even in the essentially total absence of explicit manifestations of the influence of this shock wave on the global cosmic-ray intensity (i.e., even in the absence of Forbush-decreases).

For example, the disturbances of the interplanetary medium observed on 22 September and 22 November 1978 could barely be seen in data on the cosmic ray intensity at high-latitude stations.

During September 1978 no effect at all of interplanetary perturbations on the observed cosmic ray intensity was seen (or, at least, nothing could be seen visually). Nevertheless, the appearance of fluctuations with periods of the order of 20 and 40 min in the cosmic-ray measurements was perfectly clear (Dorman and Libin, 1984a, 1984b, 1985).

3.1. Cosmic-Ray Fluctuations during Forbush Decreases

Research on fluctuations in the intensity of the neutron component of the cosmic rays on the basis of data from the stations at Moscow, Tixie Bay, Yakutsk, and Khabarovsk (Libin, 1981, 1983; Bezrodnikh et al., 1982; Dorman et al., 1980, 1985, 1986; Kozlov and Markov, 2007) (at low geomagnetic activity time) it was shown that the power distribution of the fluctuations is quite flat over the entire frequency range studied, with fluctuation amplitudes which do not go beyond even the 95% confidence interval. Fluctuations at the "noise" level are therefore typical of unperturbed conditions in the solar wind.

During Forbush-effect caused by unsteady disturbances, on the other hand, we observed a "pumping" of the power toward higher frequencies in the spectrum (to fluctuations with periods of the order of a few tens of minutes or less) before the onset of the Forbush-decreases, and we observe a pumping downward along the frequency scale (to fluctuations with periods of 40-50 min and more) during and after the Forbush decrease.

Spectral estimates calculated from data from the scintillation telescope of the Institute of Terrestrial Magnetism, the Ionosphere, and Radio Wave Propagation over the years 1977-1997 showed that, regardless of whether there is a direct, obvious effect of interplanetary perturbations of a time-varying type on cosmic rays, there are statistically significant fluctuations (exceeding the 95% confidence interval) with periods of 17-20 and 40-60 min in the data from observations of the cosmic rays if there is a perturbation of the type near the earth. The sensitivity of the fluctuations (i.e., the distance from the earth to the perturbation of the interplanetary medium) is determined exclusively by the dimensions of the inhomogeneities in the interplanetary magnetic field and thus the energy of the particles which are detected. It should simply be noted that, particularly for the scintillation telescope of the Institute, fluctuations with periods ranging from about 20-40 to 165 min (the corresponding energies are 15-50 GeV) are detected most efficiently; this range of periods corresponds to the maximum of the coupling coefficients of the particular instrument.

Analysis of the cosmic ray fluctuations in April and May 1978 on the basis of the 5-min data on the cosmic-ray intensity obtained from the scintillation telescopes at the IZMIRAN and also at Bologna revealed a similar picture: the power spectra of the fluctuations over the intervals 10-13 April and from 29 April to 9 May showed fluctuations exceeding the 95% confidence interval with periods of the order of 17-20, 20-45, 95-115, and 165 min (Figure 1a and Figure 1b). The spectral estimates showed that the prominent peaks were permanent features in all the individual samples of I-day intervals.

Comparison of all the spectra calculated for the interval from 28 April to 1 May (spectral estimates on the total ionizing component of the vertical and inclined telescopes of the Institute and at Bologna and on the muon component of the azimuthally telescope of the IZMIRAN were used) reveals that there is an essentially complete agreement of all the estimates: The composite spectrum suffers essentially no blurring, and there are stable peaks corresponding to fluctuations with periods of 12-13, 16, 20, 45, and 165 min. The spectrum found experimentally for the frequency interval $5.10^{-7} < f < 2.10^{-4}$ Hz was compared with the spectrum expected in interplanetary space (expected on the basis of the kinetic theory of cosmic-ray fluctuations with allowance for data on the power spectrum of the fluctuations of the interplanetary magnetic field, data on Forbush-decreases, and data on cosmic-ray variations associated with the passage of active regions across the central meridian of the sun).

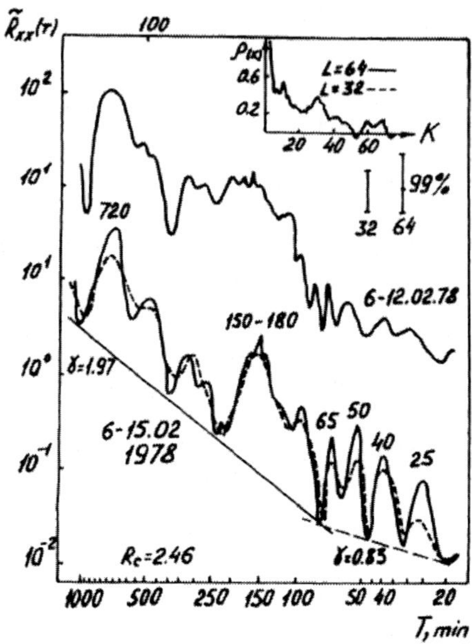

Figure 1a. The power spectrum of cosmic ray fluctuations for the February, 1978.

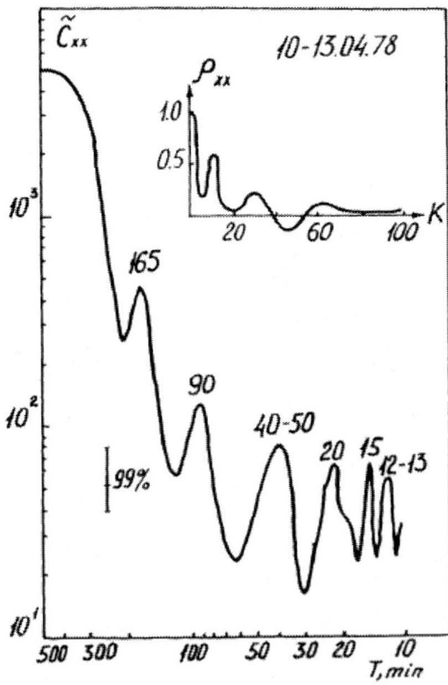

Figure 1b. The power spectrum of cosmic ray fluctuations for the April, 1978.

These results confirmed the conclusions of the theory for a nonlinear interaction of high-energy cosmic rays with perturbations of the interplanetary magnetic field, and they confirmed the theory for a diffusive propagation of cosmic rays at rigidities up to a level of the order 250 GV, with a correction for modulation of the cosmic rays by active processes on the sun.

Studies of fluctuations in the frequency interval $2.10^{-7}<f<10^{-4}$ Hz on the basis of data obtained at the stations at Washington, Resolute Bay, Kiel, Deep River, and Pic du Midi showed that during disturbances the spectral index γ of the fluctuations varies over the interval $2.0< \gamma<3.0$, corresponding to an increase in the number of inhomogeneities 10^{11} cm in size and larger.

A study of the fluctuations of the interplanetary magnetic field during the passage of shock waves by the earth on the basis of data on cosmic-ray fluctuations completely confirmed the increase in the power in the high-frequency part of the spectrum a day before the beginning of the perturbation, followed by a pumping of power down the frequency scale.

The spectrum recovers the shape typical of geomagnetically quiet conditions about a day after the beginning of the perturbation. A joint analysis of the fluctuations in the interplanetary magnetic field and in the cosmic-ray intensity reveals the simultaneous existence of oscillations with periods of 40-45 and 80-90 min, which may indicate a turbulent generation mechanism $(l= r+ 2)$.

Spectral-temporal analysis has been used to study the dynamics and frequency spectra of fluctuations of the interplanetary magnetic field and of the cosmic rays at times preceding and following the onset of perturbations in the interplanetary medium near the earth for the different intervals from 1968 – 2007 (Perez Peraza et al., 2011).

The results demonstrate both an essentially complete agreement of the observed fluctuations and the simultaneity of the changes in the spectra of the cosmic rays and the interplanetary magnetic field. The observed simultaneous intensification of the Changes in (al the index r and the power of the spectral estimates for the events) power spectra began 12 hours before the beginning of the perturbation (Dorman and Libin, 1985).

3.2. Observations of Cosmic-Ray Fluctuations during Solar Flares

Analysis of the cosmic-ray fluctuations during solar flares is a topic of independent interest. Transky (Perez Peraza et al., 1978; Filipov, 1981; Dorman and Libin, 1985) detected some additional peaks (with relative amplitudes of the order of 15%) in the time evolution of the intensity of solar cosmic rays from the flare of 12 November 1960 by working from observations of cosmic rays by the neutron monitors at the stations at Nederhorst, London, Lindau, Leeds, Kiel, and Climax.

These additional peaks appeared a few hours after the maximum was reached. Similar results were found for a flare on 4 August 1972 (with a relative amplitude of the order of 5%). Observations by the neutron monitors at Tixie Bay and Deep River revealed fluctuation frequency spectra with some typical stable peaks corresponding to cosmic-ray fluctuations with periods ranging from 30 to 150 min. At a solar wind velocity of the order of 5.10^7 cm/s, these peaks correspond to inhomogeneities of the order of 10^{11}-10^{12} cm in size.

These estimates are confirmed by those found by Libin *et al.* (Libin et al., 1979a, 1979b; Libin, 1980; Dorman and Libin, 1984a, 1985) who worked from ground-based observations of the cosmic-ray intensity by means of the scintillation telescope of the IZMIRAN on 7 May

1978 to estimate the characteristic dimensions of inhomogeneities in the interplanetary magnetic field.

They found dimensions of the order of 10^{11}-$0.5.10^{12}$ cm. A similar result was observed for the flare of 25 September 1978 (Libin, 1980): a detailed analysis of the fluctuational phenomena in the cosmic rays shows that although the direct "inspirations" for the cosmic-ray fluctuations during flares are inhomogeneities of the random interplanetary magnetic field (particularly shock waves) the actual sources of at least some of the fluctuations simply could not be identified in the comparison with various geophysical and heliophysical parameters) are intense chromo spherical flares of importance 2 and higher, which cause interplanetary and magnetosphere disturbances. In fact, it is probably not isolated flares but series of flares which are responsible.

3.3. Fluctuations in the Cosmic Rays in Front of the Interplanetary Shock Waves

The observed intensity of cosmic rays fluctuates at frequencies $f < 10^{-3}$ Hz (T > 20 min). The origin of these fluctuations is the turbulence of the interplanetary magnetic field (Perz Peraza et al., 1978; Dorman et al., 1983; Dorman and Libin, 1984a, 1984b, 1985).

Owens (Owens, 1974a,b) shows that the power spectrum of fluctuations of the cosmic rays $P_I(f)$ is related to the spectrum of the interplanetary magnetic field $P_B(f)$ in the frequency region $f > 10^{-5}$ Hz (T < 30 hrs), for particles with energy greater than a few GeV, through the following relationship:

$$P_I(f)/(n_0)^2 = A(f)P_B(f)\, d_p / (B_0)^2$$

where n_0 is the mean flux of cosmic rays, B_0 is the intensity of the interplanetary magnetic field and d_p is the projection of the anisotropy of the cosmic rays on the interplanetary magnetic field. The power spectra of cosmic ray fluctuations in the presence of perturbations of the interplanetary environment were analyzed in a number of papers (see (Dorman and Libin, 1985) and bibliography cited therein), where variations of the power spectrum of cosmic rays in low frequencies (Gulinsky and Libin, 1979; Gulinsky et al., 1988) and high frequencies (Kozlov, 1986) 1 ~ 2 days before the arrival of the shock wave to Earth, are reported.

Those results are basically obtained by means of the Fourier transform and they are based on the hypothesis of the stationary of the transformed functions, which diminishes the confidence in the results.

In the present section, an attempt is made to improve the methodology used in previous studies. The methods used here to estimate the variations in the power spectra are described in detail in Gulinsky (Gulinsky et al., 1986, 1988) and may be applicable to short non stationary time series, as is the case in cosmic ray and interplanetary magnetic field data.

Autors use several series of 5-min and hourly values of cosmic ray intensity (neutron and ionization components). The data are taken from the following stations: Moscow (1984-2006), Tiksi (1980-1986), Baksan (1984), Apatiti (1984-2006), Utrecht and Kerguelen (1977-1999).

Hourly measurements of the parameters of interplanetary plasma V, n, T and of interplanetary magnetic field $|B|$, B_x, B_y and B_z for the 1977-2006 period. The correlation function and the power spectrum are the basic instruments of analysis employed here (Kay and Marpl, 1981). The methodology used here, unlike the standard one (Bendat and Pearson 1983) can be applied to the analysis of short length time series when the steady-state condition does not hold.

The spectral analysis is based on the approximation of the corresponding time series $\{x_t\}$ by means of autoregressive models with constant coefficients, in the stationary case, or time dependent coefficients in the non stationary one, namely:

$$x_t = \sum_{i=1}^{p} a_i(t)\, x_{t-i} + \zeta_t$$

where p is the order of the autoregressive model, $\{a_i\}$ are the autoregressive coefficients and x_t is the noise.

The order p of the model varies depending on the length of the series and on the character of the processes: it grows with the length and the stationarity of the series. For the non stationary case, the coefficients $\{a_i\}$ are expanded with the aid of cubic spline functions:

$$a_i = \sum_{s=-1}^{N-1} a_{is} B_s(t)$$

where s = -1, 0, 1 and i = 1,2,...p. The coefficients $\{a_i\}$ are estimated by minimizing $\{S(\dot{x}_t - x)\}^2$, where

$$\dot{x}_t = \sum_{s=1}^{N} \sum_{i=1}^{p} \alpha_{is} B_s(t)\, x_{t-i},$$

selecting adequately the parameters $\{a_{is}\}$. Thus, the problem is reduced to solving a system of linear equations. In order to calculate the power spectrum at any given time t, the values $\{a_{is}\}$ are substituted in the expression

$$P(f,t) = (2\pi)^{-1} \sigma^2 \sum_s \sum_i \alpha_{is} B_s(t) \exp(if(p-y))$$

The selection of data was based on the available information about solar activity, geomagnetic activity, interplanetary plasma and magnetic fields. We selected data sets of 5-min and hourly values for the 1977 and 1980-1994.

The power spectra of cosmic ray fluctuations for 5- minutes and hourly intervals are almost insensitive to the choice of the data interval. In the frequency interval of fluctuations with periods of 1 to 24 hours the nature of the spectrum is basically determined by the 24 hours oscillation which is related to cosmic ray anisotropy and terrestrial rotation (Stehlik and

Kudela, 1984; Kudela and Langer, 1994; Kudela and Venkatesan, 1996; Kudela et al., 1996), and by the resonance frequency of the 8 hour period, as in the case of Tiksi and Appatity stations, and by the 6-7-hour period, in the case of Moscow, Utrecht and Baksan stations (Dorman and Libin, 1984a, 1984b, 1985).

The power spectra of the interplanetary magnetic fields were calculated on the basis of hourly data, that is, for short series. In order to test the stability of the spectra, the calculations were done for different p-orders. For additional reliability, the power spectra of the interplanetary magnetic field components were also constructed on the basis of Prognoz-7 satellite data, from January 1978 to April 1979, with resolutions of 1 hour and 5 minutes. In these frequency ranges, the results show a high level of consistency with those of other authors (see for example (Bloch et al., 1984)).

Even in quiet periods the time-dependent series of the interplanetary magnetic field components are essentially non stationary during the 27-days corotational period. However, within the sectorial structure of the interplanetary magnetic field it is possible to find steady-state intervals of 2-3 days (Kuzmin, 1984).

As in previous observations (Obridko and Shelting, 1983, 1985, 2009), we find that power spectra for different steady-state intervals may differ significantly.

In non-perturbed intervals, the power spectrum of the B_z - component shows a power law shape for fluctuations with periods greater than 10 hours, with a secondary maximum around T=5 hours.

During the day immediately preceding the shock wave arrival, the 5-hr wave increases and the low frequency oscillations decrease, or a second maximum appears near the 10-hrs period. In the two analyzed cases, the power spectrum of the perturbed oscillations rises in the high frequency region (T < 3 hrs). The cosmic ray spectra immediately before the shock wave arrival from 5-min data are illustrated in (Gulinsky et al., 1988) (Tiksi station on 22.04.82 and Moscow station on 07.04.86). They exhibit a power law shape for T > 20 min with some small peaks in the high frequency region (T < 20 min).

The main feature of these spectra is the relative depression in the low frequency region (T > 20 min).

Dotted lines in (Dorman and Libin, 1985) correspond to spectra obtained with high p-orders; they show systematic peaks in the low frequency region, coinciding with the results of Sakai (Sakai ett al., 1985), though we do not observe a significant power amplification at high frequencies. Sometimes the oscillation can be observed directly in the Bz - time series without the help of spectral analysis.

Before the arrival of a shock wave, the cosmic ray power spectrum suffers significant changes, displayed by oscillations with periods in the range of 20 min. to 24 hrs. The presence of peaks is not always a precursor of perturbations.

High frequency peaks may or may not occur during quiet periods or before a shock wave arrival. However, in quiet periods no significant power amplification is observed in the high frequency domain.

Before the arrival of a shock wave the power spectra of cosmic rays differ significantly from the spectra of the B_z component of the interplanetary magnetic field. The variation in B_z spectra before the arrival of the shock does not have a regular character. However, sometimes the arrival of the shock wave can be associated with the appearance of a sharp wave with a period of 6 to 8 hours which stays clear during nearly 24 hours.

A more detailed explanation of the influence of anisotropy on cosmic ray spectra is given in Dorman et al. (Dorman et al., 1986), where the different nature of particle reflexions according to the particle rigidity range is taken into account. The appearance of a wave with T > 28 hrs ahead of the shock front and the stochastic acceleration process that generates the additional particle flux are discussed in (Obridko and Shelting, 1985, Dorman and Libin, 1984a, 1984b, 1985, Bezrodnikh et al., 1982), and references cited in those papers. The mechanism of generation of the wave with a period of 6 - 8 hours is not clear and requires further analysis. Sometimes, the generation of the wave takes place between two shock waves and, on other occasions one of the shocks is a reverse shock wave. Regularities in the behavior of cosmic ray fluctuations may be a useful tool for the diagnosis of perturbations in the interplanetary medium, helping to predict the arrival of powerful shock waves to the earth and geomagnetic perturbations, from 1 to 2 days in advance, in agreement with the results of Kozlov (Kpzlov, 1985).

In spite of methodological differences, this appears to confirm the applicability of cosmic ray fluctuations for monitoring and predicting the state of the interplanetary medium in the Earth's neighborhood.

IV. FURTHER DEVELOPMENT OF THE STUDY OF FLUCTUATIONS OF THE COSMIC RAYS TODAY

In science, it happens that the work done many years ago, only to find themselves in demand today. Since the work (Dorman and Libin, 1984a, 1984b, 1985; Perez Peraza et al., 2011) in recent years have become the basis for the development of methods for forecasting the state of the interplanetary medium near the Earth, the basis of methods of early diagnosis of large-scale magnetic disturbances.

4.1. A New Index of Solar Activity - Cosmic Ray Scintillation Index ("Blink")

The basis of method of early diagnosis of large-scale magnetic "stoppers" in the interplanetary magnetic field on the effect discovered in (Kozlov et al., 1973, Kozlov and Markov, 2007, 2008; Kozlov and Kozlov, 2008; Kozlov, 2008, 2011). The essence of the phenomenon lies in the fact that a day before the registration of large-scale disturbances on the Earth solar wind high-latitude neutron monitors record the "blinking" in cosmic rays.

By means of compilation of alternating (for the noise-like process) indicator of the frequency spectrum of fluctuations, the author was introduced by the spectral-temporal scintillation index (Kozlov et al, 1984).

As a result, three-dimensional dynamic spectrum P (f, t) is reduced to the usual numerical sequence - the GCR scintillation index. This allowed the authors to apply to the scintillation index, all known methods of quantitative analysis of time series (Kozlov and Tugolukov, 1992, Kozlov, 1999).

In contrast to the traditional approach, the index of the frequency spectrum of the cosmic ray fluctuations, researchers measured a total number of inversions ("anomalous" deviations) of the amplitudes of the spectrum with respect to the natural series of numbers. Significant index change of the fluctuation spectrum - the GCR scintillation index, are the desired useful signal quantitatively describe the dynamics of the fluctuations of the GCR intensity in front of a large-scale "magnetic mirror" before its registration in Earth orbit.

In the article (Kozlov and Markov, 2007) devoted the no stationary transient oscillatory process of sign reversal of the general solar magnetic field was previously detected in the work to studying GCR intensity fluctuations. To quantitatively describe the GCR fluctuation dynamics during the geoeffective phases of the 11-year cycle, authors introduced a new parameter of cosmic ray fluctuations, which indicates the degree of magnetic field in homogeneity. This was done to confirm (or deny) the conclusion (Kozlov and Markov, 2007) on the detection of the no stationary transient oscillatory process of sign reversal of the general solar magnetic field drawn based on studying cosmic ray fluctuations.

The article of Kozlov stated that the Forbush effects are distributed as groups or series rather than randomly during the solar cycle geoeffective phase. After averaging of the 5-min data on the GCR intensity for each solar rotation, the series of Forbush decreases will be observed as a sharp and deep minimum in the GCR intensity, as is registered at the beginning of the 11-year cycle decline phase. After a similar averaging of the cosmic ray scintillation index, the characteristics dynamics of GCR fluctuations are also revealed on this larger scale (Kozlov and Kozlov, 2008), which indicates that these dynamics have a scaling, selfsimilar, or fractal character.

As pointed out by Kozlov (Kozlov and Kozlov, 2011): "It is customary to consider that the time series in terms of a unit variable gives rather limited information. Nevertheless, the time series has information content: it bears traces of all other variables participating in the description of the system dynamics. The procedure of reducing the initial no stationary (in the sense of average values) series to the no stationary form by eliminating a trend in data is a weak but necessary link of the spectral-time approach in analyzing time series.

However, the eliminated trend can also bear useful information "accumulated" during the latent phase of activity source formation. From this it follows that a usual (frequency or integral) histogram of initial data bears the most complete information". Thus, the problem is to extract only the possible periodicities (signal) from noise.

In (Kozlov and Kozlov, 2011) shows the results of calculations of the annual, 27-day and daily values of the cosmic rays fluctuations, which use different methods of time series analysis: linear regression, wavelet transform, the method of "superposed epoch" and cross-correlation analysis.

As a result, established that the introduced GCR fluctuation parameter is an inverse factor relative to the degree of magnetic field regularity. In other words, this means that the fluctuation parameter indicates the degree of magnetic field inhomogeneity in the vicinity of inter planetary waves and during geoeffective phases of the 11-year cycle.

As pointed out by Ayvazian (Ayvazian et al., 1983), the desired signal may be contained in the second and subsequent moments of the distribution function of the GCR intensity. This is shown by evaluation of the asymmetry coefficient. In this case, instead of the asymmetry coefficient is better to use a two-parameter Weibull distribution function (Ayvazian et al., 1983). These results are also confirmed by checking the statistical hypotheses (Dorman and Libin, 1985; Libin and Perez Peraza, 2009). In this case, the hypothesis of normal distribution of the intensity of galactic cosmic rays is rejected at a significance level of 99%. The hypothesis of Weibull the distribution points to the possibility of its adoption at a significance level of 90%.

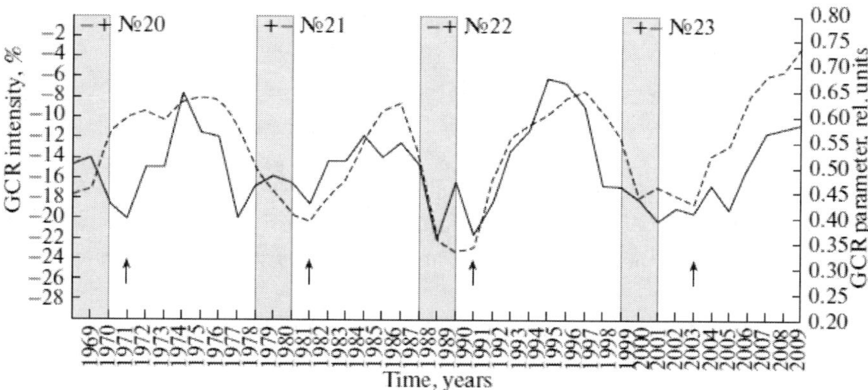

Figure 2. Calculated annual values of GCR fluctuation parameter (the solid curve, right scale) and GCR intensity (the dashed curve, left scale) (Kozlov and Kozlov, 2011).

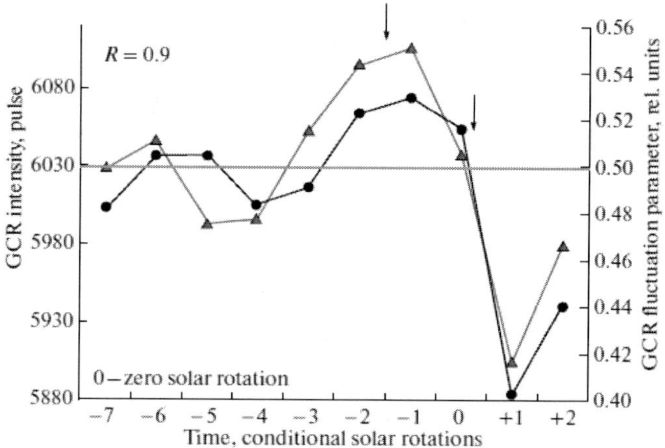

Figure 3. The average values of the fluctuation parameter and GCR for 23 cycle of SA (Kozlov and Kozlov, 2011).

According to the 5-minute observations of fluctuations of cosmic rays at the station of Oulu, has been calculated annual values of the parameter fluctuations (four 11-year solar activity cycle, № 20-23).

The correlation between the results of calculations and observations (the coefficient of their linear regression relation is R = 0.9), probably due to the 11-year variation. This is confirmed by wavelet analysis of the 11-year variation in the parameter fluctuations of galactic cosmic rays detected quite clearly (Figure 2 (Kozlov and Kozlov, 2011)).

Attention is drawn to the low-frequency drift of the period of 11-year variations in the low-frequency region, i.e. towards larger periods than during the 11-year cycle.

Why is the fact of detection of a similar drift so important? The point is that the appearance of an low-frequency drift according to the terminology of Kozlov) can precede a prolonged failure of the 11-year cyclicity (Frik, 2003). The record high present-day GCR intensity is possibly related to weakening of the global solar dipole field; such data has already appeared. Obridko (Obridko and Shelting, 2009) indicated that in 2008 the solar dipole magnetic moment decreased to values typical of the beginning of the 20th century.

Local fields are also anomalously low (Below and Gaidash, 2009). Such a prolonged period with the complete disappearance of sunspots was observed only at the beginning of the last century.

After exclusion of 11-year variation is high-frequency component, which includes 27-day variation of the calculated parameters. The authors of (Kozlov and Kozlov, 2011) defined it as a parameter fluctuations in the GCR. The envelope of the 27-day variations of the cosmic rays fluctuations parameter (in all 4 solar cycles, 20-23), reaching a maximum at the stage of complete reversal of the general solar magnetic field, i.e., in the beginning of the descending limb of the solar activity cycle.

As a result of Kozlov (Kozlov and Kozlov, 2011) research found that a significant maximum in the fluctuation parameter exists on average one-two solar rotations before the beginning of a deep minimum in the GCR intensity (Figure 3).

This points to the prognostic possibilities of the fluctuation parameter. However, it is necessary to additionally use the quantitative criterion for making decision concerning the prediction of the geoeffective period in order to perform a medium term prediction for the period about one solar rotation. According to the epoch superposition method, we deal with a conditional average "event"; therefore, quantitative criterion is averaged.

Proceeding from the probabilistic interpretation of the fluctuation parameter, we can select a probability level of $P \geq 0.5$ as a limiting level above which the occurrence probability of the solar cycle geoeffective phase can be considered significant (see http://www.forshock.ru). Has been proved a high correlation the introduced parameter (the index) with Wolf numbers ($R \geq 0,74$) and a number of major geomagnetic storms with Dst <-150 nT ($R \geq 0,65$).

It should be noted, that the established relationship between the GCR fluctuation parameter and the degree of magnetic field inhomogeneity is the decisive factor in deciphering the GCR fluctuation dynamics.

The nonrandom non Gaussian character of the GCR fluctuation parameter is caused by no stationary semiannual variation that reflects the transient no stationary oscillatory process of sign reversal of the general solar magnetic field. Precisely this transient oscillatory process is responsible for the maximal geoeffectiveness and duration of the polarity reversal phase, which manifests itself in a sharp and deep minimum in the GCR intensity during the final stage of field sign reversal.

On the basis of studies for early detection of large-scale "magnetic mirrors" in the interplanetary medium was established predictive expert system "FORSHOCK" (in IKFIA), consisting of Database "RECORD" (Turpanov et al., 2001) and Knowledge Base "CODE" (Corpuscular-line diagnostics) (Kozlov and Krimsky, 1993, Kozlov, 2002).

The system allows a continuous calculation of the cosmic rays fluctuations index (for a five-minute data of the global network of polar stations of cosmic rays). Based on the results of this calculation (real time) is determined by the precursor approaching the Earth's orbit front of large-scale disturbances. The results of monitoring and operational forecasting daily exposed to the global Internet http://ikfia.ysn.ru/fluctuations/index.php (see also the site of IKI - http://spaceweather.ru/datasets/).

V. ORIGIN OF THE COSMIC RAY FLUCTUATIONS

Detailed experimental studies (Libin, 1980; Dorman and Libin, 1984a, 1984b, 1985; Kozlov et al., 2003a, 2003b; Kozlov and Kozlov, 2008, 2011) have shown that the short-period oscillations of atmospheric properties (the pressure, the temperature, and the humidity) could not be the sources of the observed fluctuations in the cosmic rays.

As for fluctuations of the earth's magnetic field and magnetosphere as sources of the observed cosmic-ray fluctuations, we note that although these fluctuations and their amplitudes do correlate with fluctuations of the magnetosphere and the geomagnetic field the appearance of significant fluctuations and the simultaneous increase in the index K_p probably have a common source: perturbations of the interplanetary medium near the earth.

This seems particularly obvious since a study of the fluctuations at stations with different geomagnetic cutoff rigidities reveals that the fluctuation amplitude increases with decreasing rigidity. This effect stems from, first, an increase in the fraction of relatively low-energy particles and, second, the insignificant contribution of the short-period changes in the geomagnetic and magnetosphere properties to the observed cosmic-ray fluctuations (according to the measurement data, the rigidity dependence of the amplitude of the cosmic-ray fluctuations is of the order of R^{-1}).

Calculations show that the maximum amplitudes of the cosmic-ray fluctuations which could be caused by fluctuations of the magnetosphere and geomagnetic properties as a series of fluctuations are a few tenths of 1%. We should point out that everything we have said refers to particles with energies up to 1 GeV. As for fluctuations of the high-energy particles of the cosmic rays which are detected by neutron monitors and telescopes, we note that the effects on these fluctuations of variations in the geomagnetic field and the magnetosphere would be substantially lower (less than 0.01-0.03%).

In summary, a detailed study of the cosmic-ray fluctuations and a joint analysis of these fluctuations with various properties of the atmosphere, the magnetosphere, and the geomagnetic field show that the decisive role in the appearance and growth of fluctuational phenomena in the cosmic rays is played by processes in the first three of the five groups defined above.

In other words, at least most of the fluctuations in the cosmic rays are of extraterrestrial origin, so that all the theoretical calculations are justified.

According to (Libin, 1985, Perez Peraza et al., 1988; Starodubtsev et al. 2003, 2006a, 2006b, 2007; Kozlov and Kozlov, 2008, 2011), the power index of the cosmic-ray fluctuations varies between 1.1 and . 3.4 over a broad frequency range with a change in the nature of the time intervals being analyzed (from quiet to disturbed). The total power in the spectral band 10^{-6}-10^{-4} changes by a factor of more than three in 3-4 months (Libin and Perez Peraza, 2009).

A study of the variability of the power spectra of the cosmic ray fluctuations over the frequency range $10^{-5} \leq f \leq 10^{-2}$ Hz can yield valuable information on the modulation mechanism (Libin, 1980; dorman and Libin, 1985).

The time evolution of the frequency spectrum of the cosmic-ray fluctuations observed during Forbush decreases accompanied by magnetic storms suggests an extraterrestrial source-interplanetary shock waves-as one of the most likely mechanisms for the formation of the fluctuations.

Comparison of the power spectra of the cosmic-ray fluctuations and measurements of the interplanetary magnetic field reveals a good correspondence in the periods of the observed fluctuations. In turn, this agreement explains the appearance of fluctuations in the cosmic-ray intensity.

They are caused by quasiperiodic oscillations of the magnetic field in shock waves, as has been pointed out by Kozlov (Kozlov and Kozlov, 2011): "The appearance of fluctuations in the cosmic-ray intensity primarily in shock waves of the piston type can be explained by the circumstance that it is in piston shock waves that we have a relatively pronounced compression of the medium and simultaneously a decrease in the diffusion coefficient".

Estimates of the magnitude of the fluctuations expected in the cosmic rays agree well with calculations of the cosmic ray intensity spectra during the passage of flare-associated shock waves, so that we can draw several conclusions regarding the role played by oscillatory structures in the magnetic field of the shock waves in the formation of fluctuations. The amplitude of the field oscillations in piston shock waves frequently exceeds the amplitude B_o by a factor of tens, and the charged particles of the cosmic rays entering the effective range of the magnetic perturbation of a piston shock wave begin to be reflected. This reflection gives rise to fluctuations at the same frequencies.

Furthermore, the pronounced regularization of the field which is observed during the passage of intense shock waves is seen as a significant shift of the peak in the fluctuation power spectrum down the frequency scale, meaning that there are essentially no small-scale inhomogeneities in the interplanetary magnetic field.

Comparison of the measured spectra of fluctuations in the cosmic rays and the field completely confirms the conclusion (reached in a study of the dynamic spectra of cosmic-ray fluctuations that there is an increase in the scale dimension of the inhomogeneities (of the turbulence). In turn, the regularization of the field should lead to a hardening of the cosmic-ray spectrum (Libin, 1980).

During quiet periods (without any significant perturbations of the interplanetary medium), on the other hand, the spectra of cosmic-ray fluctuations have a broad set of peaks in essentially all frequency ranges, because the interplanetary magnetic field has become very turbulent, i.e., because of the presence of inhomogeneities of essentially all scale dimensions.

Most of the inhomogeneities (with scale dimensions of the order of 10^{10}-10^{12} cm) are tangential shocks separating fibers of the solar wind. The force tubes of the interplanetary magnetic field intertwine, forming a complicated structure whose field is directed perpendicular to the earth-sun line. The spectrum of inhomogeneities is of power-law form, k^{-r}, where k is the wave vector, and r is the spectral index, which varies under actual conditions (in a quiet medium) from 1.1 to 2.0 because of variations in the slope and power of the spectrum of inhomogeneities for various conditions in the interplanetary medium (Dorman and Libin, 1985).

The scattering of charged particles by single inhomogeneities changes the direction of the particles by small angles. Using the dependence of the transport range on the rigidity and the nature of the field, and determining the scale dimensions of the inhomogeneities from data on cosmic-ray fluctuations, we can determine not only the frequency distribution of the sizes of the inhomogeneities of the interplanetary field but also the frequency distribution of distances between these inhomogeneities.

Analysis of the power spectra of cosmic rays under quiet conditions (according to data from the neutron supermonitor and the scintillation telescope from the stations of the

IZMIRAN) shows that at average values of all the parameters the value of B_o and the power spectrum of the interplanetary magnetic field agree well with .both the available data on the field and the general understanding of the spectrum of inhomogeneities of the interplanetary magnetic field under highly turbulent conditions.

The studies which have been carried out show that the source primarily responsible for the appearance of the observed statistically significant fluctuations in the cosmic-ray intensity is the scattering of charged particles by random inhomogeneities of the interplanetary magnetic field.

The regularization of the field (during intense solar flares, interplanetary shock waves, etc.) leads (first) to the appearance of significant peaks in the spectra, exceeding at least the 90% confidence interval, and (second) a shift of these peaks down the frequency scale because of an energy pumping from high-frequency oscillations of the field down the spectrum.

Since the change in the spectra begins at least a few hours before the onset of the perturbation of the interplanetary medium near the earth, the fluctuations of the cosmic rays can serve as a forewarning of a shock wave approaching the Earth.

As for the second source of cosmic-ray fluctuations (oscillation and acceleration processes on the Sun), we note that data on its operation have been acquired both from a study of fluctuations of the electromagnetic component of the solar radiation and from a study of the fluctuations of solar activity and of the magnetic fields in the solar atmosphere.

Observations of the temporal variations in the magnetic fields, sunspots, and the intensity of hydrogen floccules at Ussurirsk (Vitinsky et al., 1986, Kozlov and Kozlov, 2011) have revealed statistically significant oscillations with a period of the order of 20-40 min similar to the oscillations observed in data on the radio emission at $\lambda = 3$ cm (Veselovsky, 2002).

The sensitivity of solar-measurement methods has now been refined to the level that it has no N become possible to measure the natural oscillations of the Sun. Data on two oscillation modes, with periods of the order of 40-50 and 160 min, which agree well with the results of a spectral analysis of the intensity of solar cosmic rays during flares, are reported in (Tugolukov and Kozlov 1991; Turpanov, 2001).

Fluctuations with periods of the order of 160 min are present in the power spectra essentially whenever the changes involve only the amplitudes of these fluctuations, in indirect support of (Vitinsky, 1986). Study of intense chromospheres flares has shown that the overwhelming majority of proton flares are preceded by an intense buildup of radio fluctuations, which begins a day before the flare and which causes an increase in the amplitude of the 20-60-min fluctuations by a factor of 10-30 by the time of the flare.

A similar picture (change of the form of the spectrum) is observed in the solar and galactic cosmic rays. If the changes which occur in the spectra during Forbush decreases (perturbations of the interplanetary medium near the earth which begin a significant time after the flare on the sun) can be explained in a quite simple way, in the case of flare events this advance feature in the cosmic rays can be explained in terms of a significant increase in the transport . range for scattering of particles under conditions of a highly regularized field.

Indirect data indicate such an increase (Dorman and Libin, 1985). It should be noted that the observed fluctuational effects (effects occurring in advance of events and changes in spectrum) retain their properties even when flares (increases) and Forbush decreases are not seen clearly in the cosmic rays themselves.

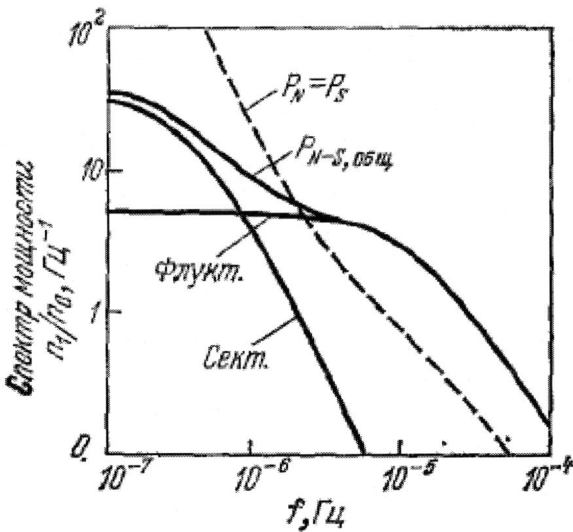

Figure 4. The power spectra of cosmic ray fluctuations (Y-axis) and frequency (X-axis) (Dorman and Libin, 1985).

A study of fluctuations thus provides a remarkable tool for diagnostics of the interplanetary medium at distances of 10^{11}-10^{13} cm from the Earth or even farther, according to recent studies (Dorman and Libin, 1985). Thus, the study of fluctuational phenomena in the cosmic rays is actually broken up into several completely independent problems involving the appearance and development of these fluctuations as a function of various processes in interplanetary space.

Figure 4 shows the overall power spectrum of the fluctuations, $P(f) = P_{sc}(f) + P_{spect}(f)$, over a broad frequency range. This figure clearly illustrates the twofold nature of the observed spectrum.

The existence of this twofold nature presents some extensive opportunities in calculations of the spectra from ground-based data. By measuring the high-frequency part of the spectrum of cosmic-ray fluctuations, one can study the spectrum of inhomogeneities of the interplanetary field. Data on the spectra in the low-frequency region can yield information on the sector structure of the field.

CONCLUSION

Extensive experiments and calculations have resulted in the development of methods for short-term predictions and diagnostics of shock waves through observations of ground-level fluctuations in the cosmic rays.

As discussed above, it is easy to explain why the effects in the cosmic rays occur in advance of the perturbation of the interplanetary medium. The cosmic rays sense inhomogeneities at a distance equal to their transport range for scattering, i.e., at a distance of the order of 10^{12}-10^{13} cm for the particles which are detected by supermonitors and scintillation telescopes at the Earth's surface.

The cosmic rays travel at nearly 1000 times the velocity at which a perturbation propagates toward the Earth, so that information on the approaching perturbation is carried to the earth essentially instantaneously by the cosmic rays-long hours before the perturbation completes its comparatively leisurely voyage to the earth. Since the energies of the cosmic rays detected at the earth stretch over a rather broad range, the distances at which the perturbations approaching the Earth will be sensed by the cosmic rays will also span a broad range: By detecting particles of various energies, one can detect inhomogeneities out to several astronomical units. This effect does not exhaust the possibilities of studying fluctuations. Studies of the power spectra of cosmic-ray fluctuations have revealed that the perturbation level of the interplanetary medium is related not only to the prominent oscillations at certain frequencies but also to the overall spectrum: the slope of the spectrum of fluctuations in the cosmic rays gradually increases to a maximum several hours before the arrival of a perturbation of the interplanetary medium at the Earth, remains at this maximum during the perturbation, and then decreases after the perturbation has passed the earth. The slope of the spectrum is at a minimum in a quiet solar wind. The possibility of investigating fluctuations in the cosmic rays have not been exhausted. Today it is possible o solve several important problems of the quantitative diagnostics and prediction of events (of importance to space research and to its applications on the Earth) such as intense solar flares, interplanetary shock waves, and perturbations propagating through interplanetary space.

Decisive steps in this direction have been taken in the works by A.Owens, D.Jokipii, G.Krymskiy, L.Dorman, M.Katz, L.Bergamasko, V.Ptuskin, A.Dolginov, I.Toptygin, V.Kozlov, A.Starodubtsev, K.Kudela, J.Perez Peraza, M.Steghlik, among others. They have established direct relationships between spectral estimates of the cosmic ray fluctuations and fluctuations of the interplanetary magnetic field. Continue experimental observations of cosmic rays fluctuations in the world. Observation data of CR fluctuations (obtained by neutron supermonitors and muon telescopes in Japan, Russia, China, Canada and other countries) are now available on the Internet.

What is the status are the study of fluctuations of the cosmic rays today?

- Introduced the cosmic ray fluctuation parameter, which is an indicator of the magnetic field inhomogeneity degree in the vicinity of shocks and during 11-year cycle geoeffective phases.
- The nonrandom non Gaussian character of the GCR fluctuation parameter is related to no stationary semiannual variation reflecting the transient nonstationary oscillatory process of sign reversal of the general solar magnetic field. This transient oscillatory process is responsible for the maximal geoeffective and duration of the polarity reversal phase, which manifests itself in a sharp and deep minimum of the GCR intensity during the final stage of the field sign reversal.
- The introduced parameter of cosmic ray fluctuations is a theoretically prognostic factor, which is very important for a medium-term prediction of geoeffective 11-year cycle periods with a lead time of ~1 solar rotation and for an online prediction of shocks with a lead time of ~1 day.
- Established and operates the system of early prediction of interplanetary shock waves in the vicinity of the Earth.

When Victor Hess launched his celebrated balloon flight on 7 August 1912 and thereby founded modern space physics, did he imagine that seventy years later the rays which he discovered would serve as forewarnings of cosmic storms?

REFERENCES

Aivazyan, S. A., I. S. Enyukov, and L. D. Meshalkin, Applied Statistics. Simulation Backgrounds and Primary Data Processing, Finansy Statistika, Moscow, 1983.

Bazilevskaya G.A., Krasotkin A.F., Okhlopkov V.P. The search for short-period fluctuations in the intensity of cosmic rays in the stratosphere // Preprint № 166. M.: FIAN USSR. 1970. 20 p.

Belov A.V., Guschina R.T., Obridko V.N., Shelting B.D., Yanke V.G. Relationship of long-term modulation of cosmic rays with the characteristics of the global solar magnetic field / / Geomagnetism and Aeronomy. T. 42. № 6. p. 727-735. 2002.

Belov, A.V. and Gaidash, S.P., Anomalously Low Solar and Geomagnetic Activities in 2007, *Geomagn. Aeron.*, 2009, vol. 49, no. 5, pp. 595–602 [*Geomagn. Aeron.*,(Engl. Transl.) 2009, vol. 49, pp. 566–573].

Bendat, J. and A. Pearsol, 1983. Application of correlation and spectral analysis, Moscow, Mir Publishing House, 312 p.

Bezrodnikh, I. P., V. A. Kuzmin and V. I. Kozlov, 1982. Dynamics of frequencial spectrum of interplanetary magnetic field fluctuations and of cosmic rays. *Geomagnetism and Aeronomy*, 22- 6, 1016-1018.

Blokh, G. M, B. M. Kujzkevsky, I. Ya. Libin and A. M. Lemberger, 1984. Spectral characteristics of low-energy solar cosmic rays obtained on the basis of data from "Prognos-6" satellite. *Geomagnetism and Aeronomy*, 24-1, 11-15.

Dorman L.I., Libin I.Ya., Blokh Ya.L. Scintillation method for studying the variations of cosmic rays. M.: *Nauka*, 1979. 108 p.

Dhanju, M. S. and V. A. Sarabhai. Short-Period Variations of Cosmic-Ray Intensity. *Phys. Rev. Lett.*, 19, 5, 252–254 (1967).

Jenkins G., Watts D. Spectral Analysis and Its Applications. M.: *Mir*, 1972. T.2. p. 47-54.

Dorman L.I., Luzov A.A. Variations of cosmic rays as a set of causal effects and random processes // Cosmic Rays. M.: *Nauka*, 1967a. No 8. p. 275-284.

Dorman L.I., Luzov A.A. Receiving information about the parameters of the variations of cosmic rays using the theory of stochastic processes. // Cosmic Rays. M.: *Nauka*, N8. 1967b. p. 285-304.

Dorman, L. I., I. Ya. Libin and O.V. Gulinsky. Study of intensity fluctuations in cosmic rays during Forbush-decreases on the basis of the data obtained with the IZMIRAN scintillation supertelescope. *Astrophysics and Space Science,* Volume 73, Number 1, 1980. p.337-347,

Dorman, L. I. and I. Ya. Libin, 1984a. Fluctuation phenomena in cosmic rays observed at Earth, in the stratosphere and in the cosmic space. *Izv. A. Sc. USSR, phys. series,* 48-11, 2143-2145.

Dorman, L. I. and I. Ya. Libin, 1984b. Cosmic ray scintillations and dynamic processes in space. *Space Sci. Rev.,* 39, 91-152.

Dorman, L. I. and I. Ya. Libin, 1985. Short-term variations of cosmic ray intensity. *Uspekhi fizicheskikh nauk,* 145-3, 403-440.

Dorman, L. I., I. Ya. Libin and K. F. Yudakhin. The role of the energy spectrum anisotropy in the variations of the cosmic-ray fluctuations power-spectrum before interplanetary medium disturbances. *Astrophysics and Space Science,* Volume 123, Number 1, 53-58, 1985. DOI: 10.1007/BF00649123.

Dorman, L. I., I. Ya. Libin and K. F. Yudakhin, 1986. Probable interpretation of changes in the form of power spectrum of cosmic ray fluctuations in the presence of perturbances of the interplanetary environment. *Geomagnetism and Aeronomy,* 26-2, 204-208.

Dorman L.I., Gall R., Gulinsky O.V., Kaminer N.S., Kudela K., Libin I.Ya., Mymrina N.V., Otaola J., Steglik M., Yudakhin K.F. The spectral characteristics of large-scale fluctuations of the cosmic rays by neutron and ionizing components and their relation to the anisotropy of the cosmic radiation. Cosmic Rays № 24. M.: VINITY. 1987a.

Dorman L.I., Gulinsky O.V., Gall R., Kudela K., Kaminer N.S., Libin I.Ya., Mymrina N.V., Otaola J., Prilutsky R., Perez Peraza J., Steglik M., Yudakhin K.F. Spectral characteristics of large-scale fluctuations of the cosmic rays by neutron and ionizing components and their relation to the anisotropy of the cosmic radiation (II). Cosmic Rays № 25.. M.: MGK AN USSR. 1988a. p.39-48.

Filippov, A. T.; Krivoshapkin, P. A.; Transky, I. A.; Kuzmin, A. I.; Krymsky, G. F.; Niskovskikh, A. S.; Berezhko, E. G. Solar Cosmic Ray Flare on September 29, 1989 by data of the Yakutsk Array Complex. Proceedings of the 22nd International Cosmic Ray Conference. 11-23 August, 1991. Dublin, Ireland. Under the auspices of the International Union of Pure and Applied Physics (IUPAP), Volume 3, Contributed Papers, SH Sessions. Dublin: The Institute for Advanced Studies, 1991., p.113.

Frik P.G. Turbulence: Approaches and models / / Moscow-Izhevsk: Institute of Computer Science. 2003.

Ginzburg V.L. Astrophysical aspects of cosmic ray research (first 75 years and prospects for the future). *Uspekhi Fiz. Nauk,* M.: AN USSR, 155 185–218 (1988).

Grossman A., Morlet J. Decomposition of Hardy functions into square integrable wavelets of constant shape //.*SIAM Journal Mathematics Analysis,* 1984. Vol. 15. P. 723-736.

Gulinsky, O. V. and I. Ya. Libin, 1979. Fluctuation phenomena in cosmic rays as observed by the multidirectional telescope of IZMIRAN. Report IZMIRAN No. 30 (258), Moscow, IZMIRAN, 24 pg.

Gulinsky O. V., I. Ya. Libin and R. Ye. Prilutsky, 1986. Fluctuations of cosmic rays during periods of perturbation. *Geomagnetism and Aeronomy,* 264, 529-534.

Gulinsky, O. V., L. I., Dorman, I. Ya. Libin, R. E. Prilutsky, J. Otaola, V. Yu. Belashev, K. M. Steglik and J. Perez Peraza, 1988. Analysis of the small-scale spectrum inferred from the ground-based cosmic ray observation data. Geofís. Int., 27-1, 3-36.

Hachiro Takahashi. Change in the Rigidity Spectrum During Short-Term Variation of Cosmic Rays. *Phys. Rev. Lett.,* 10, 3, 116–117 (1963).

Jenkins G.M. and Watts D.G. Spectral analysis and its applications< M.: Mir, 1972. 242 p.

Kay, S. M. and S. L. Marpl, 1981. *Modern Methods in Spectral Analysis. TIIER,* 69-11, 5-48.

Kozlov V.I., Kuzmin A.I., Krymsky G.F., Filippov A.T., Chirkov N.P. Cosmic ray variations with periods less than 12 hours // Proc. 13 ICRC. Denver: USA. V. 2. P. 939–942. 1973.

Kozlov V.I.. On the turbulent fluctuations of the magnetic field and cosmic rays in shock waves. // Abstracts of the VII European Symposium on Cosmic Rays. Leningrad. 1980. p. 23.

Kozlov, V. I., 1981. Turbulent pulsations of magnetic fields in presence of shock waves. *Geomagnetism and Aeronomy,* 21-6, 1115-1117.

Kozlov V.I., Borisov D.Z., Tugolukov N.N. The diagnostic method of interplanetary disturbances on the study of fluctuations of the cosmic rays and its implementation in the automation of scientific research PGO Tiksi // Izv. AN USSR. Ser. fiz. T.48. № 11. p. 2228–2230. 1984.

Kozlov, V. I., 1985. Early diagnosis of interplanetary perturbances by means of registration of scintillation modulation of cosmic rays. Preprint, IKFIA, Yakutsk, 20 pp.

Kozlov, V. I., 1986. Fluctuations of cosmic rays and dynamical processes in solar wind. In: Modulation of cosmic rays in the solar system, Yakutsk, 80-95.

Kozlov V.I., Tugolukov N.N. Scintillation intensity of cosmic rays / / *Geomagnetism and Aeronomy,* T. 32. № 3. p. 153–159. 1992.

Kozlov V.I., Krimsky P.F. Physical bases of catastrophic forecast of geophysical phenomena // Yakutsk: YNC SO RAN. 163 p. 1993.

Kozlov V.I Prediction of the solar activity cycle phases by variations of galactic cosmic ray scintillation index // Proc. Solar-Terr. Predictions Workshop. Japan. Hitachi. C. 204-206. 1996.

Kozlov V.I., Markov V V Scale-invariant features of cosmic ray fluctuation dynamics in a solar cycle // Proc. 25-th ICRC. Durban - South Africa. SH. Session 1-3. P. 425-428. 1997.

Kozlov V.I. Scaling dynamics of fluctuations of cosmic rays on the geoeffective solar cycle phases // Geomagnetizm and Aeronomy, T. 39. № 1. p. 95–99. 1999.

Kozlov V.I. Evaluation of the scaling properties of the dynamics of fluctuations of cosmic rays in the solar activity cycle / / Geomagnetism and aeronomy, T.39. № 1. p. 100–104. 1999.

Kozlov V.I. Forecast of space weather in real time according to the ground stations of cosmic rays / / Solnechnozemnaya physics. Irkutsk: ISZF SO RAN. 2. № 115. p. 96–98. 2002.

Kozlov V.I. Ksenofontov L A, Kudela K, Starodubtsev S A , Turpanov A , Usoskin I, Yanke V Real-time cosmic ray database (RECORD) // Proc. 28-th ICRC. Nsukuba. Japan. V.6/7. SH Sessions 1.1-2.3. Universal Academy Press. Inc - Tokyo. Japan. P. 3473-3476. 2003a.

Kozlov VI, Kozlov V V, Markov V V Effect of Polarity Reversal of Solar Magnetic Field in Cosmic Ray Fluctuations // Proc. ISCS-2003 Simposium "Solar variability as an input to the Earth's environment". Tatranska Lom nica. Slovakia. P. 117-120. 2003b.

Kozlov V. I. and V. V. Markov. Wavelet image of the fine structure of the 11-year cycle based on studying cosmic ray fluctuations during cycles 20–23. *Geomagnetism and Aeronomy,* Volume 47. № 1. C. 47–55. 2007.

Козлов В.И., Марков В.В. Вейвлетобраз гелиосферной бури в космических лучах // Геомагнетизм и аэрономия. Т. 47. № 1. С. 56–65. 2007.

Kozlov V. I. and V. V. Kozlov. A new index of solar activity: An index of cosmic ray scintillation. Geomagnetism and Aeronomy. Volume 48, Number 4, 463-471, 2008. DOI: 10.1134/S0016793208040063.

Kozlov V.I. The anomalous "solar activity in the" weak "cycles 20 and 23 as a manifestation of an invariant 11-year cycle / / Solar-Terrestrial Physics. Irkutsk.: ISTP. 12, T. 1. p. 32—33. 2008.

Kozlov, V., V. Kozlov. Galactic cosmic ray fluctuation parameter as an indicator of the degree of magnetic field inhomogeneity. *Geomagnetism and Aeronomy*, Vol. 51, No. 2. 2011, pp. 191-201.

Kravtsov Yu.A., Etkin V.S. Nonlinear and Turbulent Processes in Physics. // Harwood Acad. Publ. New York. 1984. Vol.3. P. 1411.

Kudela, K. and Langer, R.: On the index of cosmic ray fluctuations at neutron monitor energies. *J. Astronomical Institute of Slovak Academy*, 1994. 24 p.

Kudela, K., Venkatesan, D., and Langer, R.: Variability of cosmic ray power spectra, *J. Geom. Geoelectr.*, 48, 1017–1024, 1996.

Kudela, K., Fluchiger E.O., Torsti J., Debrunner, H.: On the character of cosmic ray variations at f$\geq$2 x 10^{-5} Hz., *Nonlinear Processes in Geophysics*, 3, 1996. p.135-141.

Kuzmin, V. A., 1984. Fluctuations of the tension of interplanetary magnetic field and those of the flux of cosmic rays in the sectorial border lines, Proceedings of 92 Interplanetary shock waves the 4th Symposium KAPG on Solar-terrestrial Physics, Sochi, 83-84.

Lander A.V., Levshin A.L., Pisarenko V.F. On the spectral and temporal analysis of the fluctuations. / / Computational and statistical methods for interpreting seismic data. Computational Seismology. Moscow: *Nauka*, In 1973. Issue 6. S. 236.

Libin I.Ya., Dorman L.I., Starkov F.A. Fluctuation phenomena in cosmic rays. Proceedings of the II All-Union Conference "Fluctuation phenomena in physical systems." Vilnius: VGU. 1979a. p.56-60.

Libin I., Dorman L.I., Lemberger A.M. Preliminary analysis of cosmic ray fluctuations on may 7, 1978. Proc. XYI-th International Cosmic Ray Conference (ICRC), Kyoto, Japan. v.4. 1979b.

Libin I. Application of scintillation telescopes to studying the cosmic ray variations. Proc. XYI-th International Cosmic Ray Conference (ICRC), Kyoto, Japan. v.4. 1979.

Libin I.Ya. The shape of ground of the power spectrum of fluctuations of CR intensity. In the book.: *Proceedings of the IZMIRAN*, 1980. p.34-40.

Libin I. Sampled spectra of the cosmic ray fluctuations power during the disturbances in interplanetary space. Proc. XYII-th International Cosmic Ray Conference (ICRC), Paris, France. SH. v.4. 1981. p.13-16.

Libin I. Fluctuations of low-energy solar cosmic ray as inferred from the observations data for September-December 1977. Proc. XYIII-th International Cosmic Ray Conference (ICRC), Bangalore, India. MG. MG.SP. v.10. 1983a. p.78-81.

Libin I. The study of fluctuations of cosmic rays during Forbush decreases. In the book. "Cosmic Rays" № 22. M.: Radio and Svyaz. 1983. p.21-44.

Libin I. The spectral characteristics of fluctuations of the cosmic rays Cosmic Rays № 23. M.: Radio and Svyaz. 1983b. p.14-20.

Libin I., Perez Peraza J. Helioclimatology, M.: MAOK, 2009. 252 p.

Max J. Methods and techniques for processing signals during physical measurements, M.: Mir, 1983. 240 p.

Millican, R,A., Anderson C.D., Neher N.V. The Three Types of Cosmic-Ray Fluctuations and Their Significance, *Phys. Rev.*, Vol. 45, No 3, 1934. p.141-143.

Mishra, R. K., R. Agarwal. 10.7-cm solar radio flux and cosmic ray fluctuations. Proc. 32ND International Cosmic Ray Conference, Beging, 2011.

Mulligan T., J. B. Blake, D. Shaul, J. J. Quenby, R. A. Leske, R. A. Mewaldt, M. Galametz. Short-period variability in the galactic cosmic ray intensity: High statistical resolution observations and interpretation around the time of a Forbush decrease in August 2006. *Journal of Geophysical Research,* Vol. 114, A07105, 12 pp., 2009.

Nonaka, T., Y. Hayashi, Y. Ishida, N. Ito, S. Kawakami, A. Oshima, S. Tamaki, H. Tanaka, T. Yoshikoshi K. Fujimoto, H. Kojima, S.K. Gupta, Atul Jain, A.V. John, D.K. Mohanty, P.K. Mohanty, S.D. Morris, K.C. Ravindran, K. Sivaprasad, B.V. Sreekantan, S.C. Tonwar, and K. Viswanathan. Study of Cosmic Ray Short Term Variations Using GRAPES-3 Muon Telescopes. Proc. 28th International Cosmic Ray Conference, 2003. pp. 3569–3572.

Obridko, V. N. and B. D. Shelting, 1983. Characteristics of the spectrum of fluctuations of interplanetary magnetic field and the level of solar activity. Solar Data, 9, 80-83.

Obridko, V. N. and B. D. Shelting, 1985. Fluctuation characteristics of the interplanetary magnetic field in the region of 3x10-10 Hz. *Geomagnetism and Aeronomy,* 25- 6, 881-885.

Obridko, V. N. and B. D. Shelting, Some anomalies in the evolution of global and large-scale fields on the Sun as the harbingers of the coming of several cycles of low // *Astron. Journal Letters,* T. 35. № 4. p. 279—285. 2009.

Owens, A. J.,1974a. Cosmic Rays Scintillations 2: General theory of interplanetary scintillations. *J. Geophys. Res.,* 75- 7, 907-910.

Owens, A. J., 1974b. Cosmic Ray Scintillations 3: The lowfrequency limit and observations of interplanetary scintillations. *J. Geophys. Res.,* 79-7, 907-910.

Pérez-Peraza, J., A. Leyva-Contreras, I. Ya. Libin, V. Ishkov, K. Yudakhin and O. Gulinsky. Prediction of interplanetary shock waves using cosmic ray fluctuations. *Geofísica* Internacional, 1998, Vol. 37, Num. 2, pp. 87-93.

Pérez-Peraza, J., L. Dorman, I. Ya. Libin. Space Sources of Earth's Climate: Natural science and economic aspects of global warming. Padova-Moscow: Euro Media, 2011. 380 p.

Sakai, T., M. Kato and E. Takei, 1985. Periodic variations of cosmic ray intensity with period of 37 minutes observed on April 25th, 1984. *J. Geomagn. And Geolbec.,* 37-6, 659-666.

Starodubtsev, S. and Usoskin I. Galactic Cosmic Ray Fluctuations: Long-term Modulation of Power Spectrum. Proc. 28th International Cosmic Ray Conference, 2003. pp. 3905–3908.

Starodubtsev, S. A., I. G. Usoskin, A. V. Grigoryev, and K. Mursula. Long-term modulation of the cosmic ray fluctuation spectrum. *Ann. Geophys.,* 24, 779–783, 2006a.

Starodubtsev, S. A., I. G. Usoskin, A. V. Grigoryev, and K. Mursula. Spectrum of rapid cosmic ray fluctuations and its long-term changes. *Geophysical Research Abstracts,* Vol. 8, 00542, 2006b.

Starodubtsev, S. A., A. V. Grigoryev, V. G. Grigoryev, I. G. Usoskin, and K. Mursula. Флуктуации космических лучей и межпланетного магнитного поля в окрестности фронтов межпланетных ударных волн. Известия РАН, Серия физическая, 2007, Том. 71, № 7, с. 1022–1024.

Stehlik, M. and K. Kudela, 1984. Power spectra of cosmic ray variations in the region of 3x10^{-10} Hz. *Acta. Phys. Slov.,* 34- 2/3, 75-85.

Torrence Ch. and Compo J.P. A Practical Guide to Wavelet Analysis. Bulletin of the American Meteorological Society: Vol. 79, No. 1, pp. 61–78.

Tugolukov N.N., Kozlov V.I. Contact the scintillation intensity of cosmic rays with solar wind parameters / / *Geomagnetism and Aeronomy,* T. 31. № 4. p. 715–716. 1991.

Turpanov A.A., Starodubtsev S.A., Grigoryev V.G., Kozlov V.I., Nikolaev V.S., Prikhodko A.N. The automatized system for the collection, treatment and analysis of neutron monitor data in real time // Proc. 27-th ICRC. Hamburg: OG–1.6. V. 6. P. 2325–2328. 2001.

Veselovsky I.S. The heliosphere and the solar wind at the maximum of cycle 23 / / "Solar-Terrestrial Physics." ISTP. Irkutsk. V. 2. p. 50-53. 2002.

Vitinsky Yu.I., Konetsky M., Kuklin G. Statistics of the sunspot. Moscow: Nauka. 201 p.. 1986.

In: Homage to the Discovery of Cosmic Rays …
Editor: Jorge A. Perez-Peraza

ISBN: 978-1-63483-084-3
© 2015 Nova Science Publishers, Inc.

Chapter 6

ON COSMIC RAYS AND SPACE WEATHER IN THE VICINITY OF EARTH

Karel Kudela[*]
Institute of Experimental Physics,
Slovak Academy of Sciences, Kosice, Slovakia

ABSTRACT

There are two types of relations between the observations of low energy cosmic rays and the effects of space weather. One is related to the interaction of cosmic ray particles including also energetic electrons with various materials, as those of satellite systems, atmosphere, ionosphere, airplane systems, human body especially at high altitudes on the ground and in space. Another one is checking the usefulness of cosmic ray measurements in creating eventual alerts of space weather effects. We attempt to present a short review of the few aspects of the second type of relations with highlighting some of the recent studies, namely on (a) fluxes of relativistic electrons and transmissivity of the magnetosphere; (b) forecasts of geoeffective events and short term alerts of the radiation storms; (c) some of the quasi-periodicities in cosmic rays useful for space weather forecasts. The chapter is an update version of the paper (Kudela, 2013).

1. INTRODUCTION

In the past decade there were published review articles, books and presentations in electronic form on relations between energetic particles in space and Space Weather research, e.g. : (Flückiger 2004; Storini 2006 and 2010; Mavromichalaki et al, 2006; Kudela et al, 2000; 2009; Singh et al., 2010) – related to cosmic ray studies; (Panasyuk, 2001; 2004) – related mainly to magnetospheric populations; (Miroshnichenko, 2003) – on radiation hazard in space; (Scherer et al., 2005; Bothmer and Daglis, 2007)) – on space weather effects and on physics behind. There exist several models of galactic cosmic rays (GCR) useful for radiation

---

[*] Email: kkudela@kosice.upjs.sk

hazard estimates. Recently e.g. paper (Mrigakshi et al., 2012) is comparing several models and test their applicability for the exposure assessment of astronauts. In that paper also references to the original model descriptions can be found. Other recently published models providing forms of energy spectra near Earth parametrizing it under different physical conditions, and comparing it with experiments for various phases of solar activity, are presented e.g. in papers (Buchvarova and Draganov, 2012; Buchvarova 2013; Matthia et al., 2013). Models of solar energetic particle (SEP) fluxes are mostly presented as probabilistic models (e.g. Tylka et al., 1997; Nymmik, 2007) or e.g. at http://smdc.sinp.msu.ru/doc/ Intact_model.pdf. Solar particle fluence as well as flux models with relevant references can be found e.g. at http://www.spenvis.oma.be/help/ background/flare/flare.html.

Here only few selected results on connection between cosmic rays and space weather effects, especially those obtained during the recent period, are mentioned. The selection cannot be taken as exhaustive on the subjects. We do not discuss the impact of CR on atmosphere and human health. One of the reviews on the topics can be found e.g. in (Singh et al., 2011) and other results obtained recently e.g. in paper (Papailiou et al., 2012c). Link of CR to clouds and climate is a controversial issue. Laken et al. (2012) deduce that there is no robust evidence of a widespread link between the cosmic ray flux and clouds. Paper by (Sloan, 2013) indicates that cosmic rays and solar activity cannot be a very significant underestimated contributor to the global warming seen in the twentieth century. For the radiation hazards in space the estimates of radiation shielding is important. A review of current transport codes and introduction to a probabilistic risk assessment approach for the evaluation of radiation shielding can be found in paper (Kim et al., 2013). Radiation hazard during mission to Mars is discussed recently e.g. in paper (McKenna-Lawlor et al., 2012).

2. ELECTRONS OF HIGH ENERGY

Much higher penetration ability of electrons into various materials in comparison with that of protons at comparable kinetic energy, is dangerous especially for satellite systems, for example due to the deep dielectric charging (e.g. Baker et al., 1987; 1998). Energetic electrons also cause the variations in crystallinity, defects and dangling bonds at the interface, which eventually results in variation of electrical properties (e.g. Laha et al, 2012; Mahapatra et al., 2006). The single event upset events (SEU) have been observed at low altitude satellites with higher probability especially in two regions, namely at South Atlantic Anomaly and at high geomagnetic latitudes. In addition, the nighttime injection of electrons to magnetosphere during the geomagnetic storms correlates with the satellite errors.

Rather extensive statistical studies of satellite anomalies indicated e.g. by (Belov et al, 2004; Dorman et al, 2005a; Iucci et al., 2006) found that characteristics for quiet and dangerous days with satellite anomalies show clear difference in high energy electron fluence. Such type of effect, studied by superimposed epoch analysis, was reported both at geosynchronous orbits as well as at low altitude satellites of the type Cosmos.

There are observed strong variations of high energy electron flux in the magnetosphere including the observations at low altitudes. Strong changes of the electron flux in connection with the geomagnetic storms in July – September 2004 have been recently reported by (Lazutin et al., 2011a; Lazutin 2012). Measurements at low altitude polar orbiting satellites

CORONAS-F at 500 km and SERVIS-1 at 1000 km at L = 3.5 over South Atlantic Anomaly have shown the changes in the intensity by a factor >100. The variability is different at different energies if checked at constant nominal McIlwain's L shell values. It is indicated that substantial part of the radiation belt particle dynamics during magnetic storms can be explained by the adiabatic effects and by the geomagnetic field reconfiguration due to variable external current systems overlapping the internal geomagnetic field. Those effects, in combination with the non-adiabatic radial diffusion, yield into the radial displacement of the outer trapping zone rather than the large losses or total disappearance of that zone. Studies of such type, even with using the older data sets available, can contribute both as an input for electron flux models important for space weather investigation, but also to understanding the changes of the geomagnetic field in the whole magnetosphere. Up to date status of the diagnostics of magnetosphere using the relativistic electron data from measurements as a tool is summarized e.g. in recently published extensive review (Tverskaya, 2011). It should be mentioned that also solar cosmic ray boundary of penetration can be used for testing the geomagnetic field model validity as it was recently shown e.g. in paper (Lazutin et al., 2011b).

High energy electron entry into the inner magnetosphere is described in terms of shock and storm injection. The shock injection, reported rather rarely, is typically of very short duration. That effect was explained originally by Tverskoy due to the drift in E and B fields in the situation when sudden short time bipolar pulses are originated (Pavlov et al., 1993). The storm injection events with longer duration are observed more frequently.

At geosynchronous orbit there are observed relativistic electrons with some delay after the increased solar wind speed observed near the orbit of Earth. The time behavior of relativistic electrons and of high energy protons in the vicinity of Earth, e.g. at geosynchronous orbit, is in some events very different. One such case, without any clear geomagnetic storm, namely that on February 18, 2011, shown by (Kudela, 2013) is specific: after the strong multiple sudden increases of solar wind speed on February 18, 2011 (observations on ACE), the electron flux >2 MeV observed on GOES reached the level of flux by > 2 orders higher than that before the event. At the same time rather strong Forbush decrease was observed at neutron monitors with various nominal vertical geomagnetic cut-offs (Athens, Rome, Lomnický štít and Oulu) and also by high energy proton channel on GOES. The event, with sharp increase of solar wind speed, was not accompanied by any clear sharp decrease in Dst.

Paper by (Reeves et al., 2011) confirmed that the geosynchronous relativistic e flux (1.8-3.5 MeV) is best correlated with the solar wind velocity observed 2 days earlier. The dependence between the relativistic electron flux and solar wind speed observed before, is not linear: while low solar wind velocity can be followed by high as well as by low fluxes of relativistic electrons, with the increase of solar wind speed the probability of high electron flux in the following interval (1-2 days), decreses. Paper by (Balikhin et al., 2011) is stressing importance of high speed, low density solar wind for the following high flux of electrons. The low altitude polar orbiting satellite perform another possibility to check relation between prehistory of solar wind and IMF (interplanetary magnetic field) respectively, and flux of relativistic electrons at various L shells through which the satellite passes. Although such study is limited to fixed (L,B) point and thus the time resolution of the study is much worse than that at geosynchronous orbit, this type of analysis is useful not only for progress in forecast of high electron fluxes in the inner magnetosphere, but also for understanding the processes of transport and acceleration of electrons. Now the analysis is in progres with SERVIS-1 and CORONAS-F data. Data from Themis mission (preliminary analysis) during the solar minimum period in the outer magnetosphere, namely during 2007-2009, indicate that the maximum correlation coefficient of the electron flux versus solar wind speed and density occurs with the delay depending both on the energy and on the L value.

Models of energetic electron distribution important for space weather studies require to include also high latitude regions. Recently (Antonova et al., 2011; Riazantseva et al., 2012) found that also auroral oval can be considered as a region of intense acceleration of energetic electrons which is important for the analysis of processes leading to the filling of the outer radiation belt and appearance of relativistic electrons. According to their studies, at high latitudes there are regions with trapping-like structure of the magnetic field. High fluxes of energetic electrons have been observed on three subsequent southern passes of CORONAS-photon satellite at L > 20. The observations of clearly enhanced electron flux were done during geomagnetically quiet periods. During these periods no significant electrons have been observed in interplanetary space outside magnetosphere. Data from such type of observations, added by revisited study of earlier measurements by low altitude polar orbiting satellites, are worth to be included in the updated models of the high energy electron flux models.

Variability of energetic electron precipitation to the atmosphere may cause changes in the ionosphere which is another link of energetic electrons to space weather. Suvorova et al. (2013a,b) have shown that fluxes of >30 keV electrons at low and middle latitudes during the storms exceeded quiet-time level by 4-5 orders of magnitudes and thus total electron content can be significantly changed.

3. USING NETWORKS OF GROUND BASED COSMIC RAY MEASUREMENTS AND OF SATELLITE DATA FOR ALERTS OF SPACE WEATHER EFFECTS

There are two types of applications of ground based and satellite measurements of CR and of suprathermal particles for providing alerts of space weather effects, namely (i) longer times (up to 20 hours) indications of consecutive geomagnetic disturbance, and (b) short time (minutes to tens of minutes) alerts of radiation storms.

3.1. Geomagnetic Disturbances

Long time ago there were reported cases with precursors in neutron monitor time series records (called as pre - decreases, pre - increases) several hours ahead of arrival of interplanetary (IP) shock to the Earth's orbit and ahead of onset of Forbush decrease (e.g. Dorman, 1963). Forbush decreases (FD) and geoeffectiveness have different profiles during individual events and their relation depends on geometry and motion of CME in interplanetary space with respect to the Earth position, as well as on interplanetary magnetic field (IMF) when CME starts to interact with the magnetosphere of Earth. The evolution of geomagnetic storms as measured by Dst and of the FD as measured by the depression of CR intensity observed by neutron monitors (NM) and muon detectors (MD) on the ground, are different (e.g. Kane, 2010; Kudela and Brenkus, 2004). When geomagnetic activity increases, both the magnetospheric transmissivity change as well as the CR anisotropy is evolving in interplanetary space. Thus the two effects have to be deconvoluted. Large geomagnetic cut-offs changes during strong storms have been reported (e.g. Tyasto et al., 2009; 2011, 2013). Nowadays there exist various geomagnetic field models involving different approaches to external current systems in magnetosphere (e.g. Alexeev et al., 1996; Tsyganenko and Sitnov, 2005 among others). Contrary to particles of lower energies in magnetosphere whose trajectory can be described by three independent cyclic motions, for CR energies there is known the only one technique for determination of the tmagnetospheric transmissivity – numerical integration of equation of motion of particle in given geomagnetic field model (reviews with references can be found e.g. in papers by Smart and Shea, 2009; Desorger et al., 2009). Different geomagnetic field models forecast the different changes of geomagnetic cut-offs and of the shifts of the asymptotic directions (e.g. Kudela et al., 2008). Changes of geomagnetic cut-offs can be parametrized by the IMF, solar wind and geomagnetic activity characteristics (Tyasto et al., 2011). Combination of both, IP anisotropy and of the cut-off change leads to estimate of rather high anisotropy for one event (Sdobnov, 2011).

Since CR particles have much higher speed in comparison with that of large-scale IMF inhomogeneity motion (moving CME "obstacle") and parallel mean free path of CR particles (λ_{par}) is large, the indication about precursory CR anisotropy originated in IP space, is transmitted quickly to distant sites, also to Earth. The decrease of CR intensity is observed down to $0.1 . \lambda_{par} . \cos(\Phi)$, Φ – IMF cone angle (Ruffolo, 1999). Using the network of ground based detectors allows to observe CR flux at Earth from various asymptotic directions simultaneously. Thus the IP anisotropy onset is observed several hours before the geomagnetic storm is starting. Theoretically, the precursors to FDs are proposed in the frame of pitch-angle transport near oblique, plane-parallel shock. By numerical simulations of CR interactions with a CME shock performed by (Leerungnavarat et al., 2003) for various shock-field angles and various values of the power-law index of magnetic turbulence of IMF provided, led to estimates that loss cone precursors (decrease of low pitch-angle particles) to FD should typically be observed by NMs about 4 hr prior to IP shock arrival. For MDs characteristic time of precursor is estimated ~15 hr prior to shock arrival.

In recent years there were published several case as well as statistical studies indicating the potential possibilities as well as the limitations of using ground based CR measurements for alerting the geoeffective events. We mention only small selections of the results. Paper (Fushishita et al., 2011) used GMDN (Global Muon Detector Network) data, by subtracting the diurnal anisotropy from many directional measurements and constructing hourly snapshots of anisotropy in the plane of asymptotic directions with contours of constant pitch angle, analyzed in detail the anisotropy before the FD on December 14, 2006. The authors show that indication of loss cone distribution appears almost 24 hours before the onset of FD and connected geomagnetic storm. Thus the first signal of anisotropy is seen by GMDN just ~ 7 hours after the CME is released from the Sun. At that time the shock was at the distance only ~0.4 AU from Sun.

Using recently completed the European project NMDB (Neutron monitor data base, http://nmdb.eu), which continues by updating and archiving of the data from many neutron monitors, various studies of CR variations can be done. Analyzing many NM records (Mavromichalaki et al., 2011) illustrated the anisotropy evolution of CR before the sudden storm commencement and geomagnetic storm on September 15, 2005. For that case a clear onset of anisotropy almost a day before the storm is reported.

Fruitfulness of the anisotropy alerts by CR observed on ground before the geomagnetic storms is step by step clarified. The probability appearance using GMDN before the geomagnetic storms of various intensity (2001 - 2007) is analyzed by (Rockenbach et al., 2011). The authors differentiate two types of anisotropy, namely loss cone and of enhanced variance one. Higher probability of CR precursors is for strong geomagnetic storms. However the statistics is not high. For the superstorms (Dst < -250 nT) the probability to observe the precursor is about 0.9, with the decreasing of the "strength" of geomagnetic storms quantified by minimum Dst, the percentage of the events accompanied by the precursors prior to the SSC decreases with decrease of |Dst|. GMDN precursors before the weaker geomagnetic storms are not appearing frequently. There are, however, events when before the weaker geomagnetic storms the anisotropy signature is reported too. (Braga et al., 2012) showed that a pitch angle distribution was changing ~5 hours before the onset of the geomagnetic storm on November 24-25, 2008.

Extensive data set from NM network allows to provide statistical studies. (Papailiou et al., 2012a) analyzed FDs in 1967 – 2006 with anisotropy Axy > 1.2% (93 events). 27 different FDs, out of 93, were chosen based on their common behavior in the asymptotic longitudinal CR distribution diagrams. Three groups are recognized, namely (1) pre-decrease in the longitudinal zone 90° – 180° noticed almost 24 h before the shock arrival; (2) pre-increase in the longitudes around and above 180° and lasts almost 12 hours until the FD and (3) pre-decrease in different longitudes and of different duration observed. The increase in the first harmonic of CR anisotropy before the shock arrival is a good tool in searching for predictors of FDs and magnetic storms and can also serve as one of the indices that characterize the occurrence of precursors. Group 1 is consecutively analyzed in detail in paper (Papailiou et al., 2012b). A long pre-decrease up to 24 hours before the shock arrival in a narrow longitudinal zone 90° to 180° is found. Specifics of the FDs and geomagnetic storms are fond for events related to western solar sources (Papailiou et al., 2013).

Increased amplitude of diurnal wave in CR intensity and the deficit of CR intensity have been identified as precursors to FDs in paper (Badruddin, 2006). (Mustajab and Badruddin, 2011) analyzed in detail the differences in geoeffectiveness due to different structures and features in IP space, with distinct plasma/field characteristics. Flux of solar energetic particles (SEP) is also used for the prediction of geoeffective events (e.g. Valach et al., 2009). The large data set collected recently (Watanabe et al., 2012) can be used for future extensive studies of precursors before the geoeffective events. The data are collected since the NM started to provide continuous measurements.

For long time period another approach to eventual forecast of geoeffective events is checked. It is connected with the search of changes in time series characteristics of CR flux records with higher temporal resolution. Short term fluctuations (T <1 h) have been probably first studied by (Dhanju and Sarabhai, 1967). Significant changes in the power spectra of rapid fluctuations are often observed about a day before and during large-scale IMF disturbances (e.g. Kozlov et al., 1973; Dorman and Libin, 1985; Kudela et al., 1996; Starodubtsev et al., 2004, 2006). Different distribution of the CR indices for 24 h before the sharp Dst decreases in comparison with that for geomagnetically quiet periods, as well as better relation of Dst to "prehistory" of CR fluctuations than to the actual fluctuations was reported in paper (Kudela and Storini, 2005). (Kozlov and Kozlov, 2011) introduced CR fluctuation parameter - indicator of the IMF inhomogeneity degree in the vicinity of shocks. This parameter is important for a medium-term prediction of geoeffective 11-year cycle periods with a lead time of ~1 solar rotation and for an online prediction of shocks with lead time of ~1 day. The parameter was shown to exceed clearly the significance interval already on October 23, 2003, well before the strong disturbances appearing by the end of October and early November 2003.

CR fluctuations have been studied in recent years also with higher statistics. In addition to ground NM and MD, the measurements of high energy particles from satellite detectors with large geometrical factors are important for checking the fine structure of CR fluctuations before, during and after geomagnetic storms (and/or FD). Such possibility gives e.g. INTEGRAL measurements (Mulligan et al., 2009). Due to high statistics (more than 1 order higher than NMs at mountains, direct measurements) the authors revealed fine structure of CR within a 3-day interval from 19.8. to 21.8. 2006 with many intensity variations in the GCR on a variety of time scales and amplitudes. Another potential possibility for CR fluctuations analysis provides the Pierre Auger project – its part Scaler. The full array was completed in 2008, with a collecting area of more than 16 000 m ^{2} and a Scaler counting rate is 2×10^8 counts.min^{-1} (Dasso et al., 2012). Data are now available with 15 min resolution. If 1 min data are available, its statistical accuracy is significantly higher than that by NMs.

During past years there started measurements by several new detectors/systems with their high potential to contribute to space weather studies, in particular to the forecasts of geoeffective events. We mention only few of them. SEVAN detector network developed in Armenia (Chilingarian et al., 2009) is now installed in several laboratories. The muon hodoscope Uragan described by (Barbashina et al., 2011 and references therein; Yashin et al., 2012) provides the detailed information about the flux, energy and angular distribution of CR primaries at energies above those to which NMs are mainly sensitive. Muon detector system named MUSTANG (Hippler et al., 2007), another one at high altitude of Mussala (Mishev and Stamenov, 2008; Angelov et al., 2008), CARPET in Andes (De Mendonca et al., 2011), KACST (Maghrabi et al., 2011) are providing continuous information about the secondary CR and its variations on the ground. Recently a new neutron monitor started to work in Spain (Medina et al, 2012). Network of NMs requires inter-calibrations. Plans for a network of mini neutron monitors, as well as progress of calibration effort of NMs are shortly presented by (Krüger and Moraal, 2012).

3.2. Irregular Increases of Low Energy Cosmic Rays

During radiation storms the most important effects are those having consequences on various element failures on satellites, on communications, on health of the astronauts etc. are high fluxes of ions with energy of several tens to hundreds MeV. Before the massive arrival of high energy ions accelerated either in solar flares or at shocks in IP space to the vicinity of Earth, NMs, if they have good temporal resolution and high statistical accuracy, and are in real-time bound in the network, can provide useful short-term alerts (several minutes to tens of minutes) in advance of the alerts prepared on basis of satellite particle measurements (Dorman, 2005b). CR ground based measurements along with the satellite measurements are useful tools for improvement of short-term forecasts of radiation storms.

Even NM at a single specific site (high latitude, high statistics) allows to obtain real time energy spectrum of SEP. South Pole combination of NM64 and that lacking usual lead shielding was illustrating that probably for the first time in the GLE 69 on January 20, 2005 (Bieber, 2006). Paper (Oh et al., 2009) reported the potential of South Pole NM data for prediction of radiation storm intensity measured by GOES. The energy spectrum was estimated (Oh et al., 2010). 31 SPEs have been associated with GLEs. Fluences and peak intensities of SPEs have good correlation with % increases in GLEs, best at channels > 350 MeV). For > 350 MeV the threshold values for GOES fluence and peak intensity are found: most SPEs above threshold are associated with GLEs, almost none below the thresholds. Ground level enhancement real-time alarm based on 8 high latitude NMs including those at high mountain is described by (Kuwabara et al, 2006). Three level alarm system is used. Out of 10 GLEs in 2001-2005 archived data the system produced 9 correct alarms. GLE system gives earlier warning than satellite (SEC/NOAA) alert in the range of 9 to 34 minutes. Including NM at various cut-offs improves in few cases the timing of the alert. Several steps of GLE alert algorithm using NM network is described by (Mavromichalaki et al., 2009). Paper (Anashin et al, 2009) describes the development of alert signal for GLEs. This tool is available online at http://cr0.izmiran.ru/GLE-AlertAndProfilesPrognosing. The first GLE in the 24th solar activity cycle has been observed on May 17, 2012, GLE 71 (Klein and Bütikofer, 2012). The highest signal was observed at South Pole, on both detectors. It was not seen at NMs with vertical geomagnetic cut-off > 3 GV. Report from the Athens group distributed (Mavromichalaki et al., 2012) informs the operational real-time Alert Code of the Athens Neutron Monitor via NMDB issued an Alert signal on 17.05.2012 at 02:13 UT, i.e. 39 min in advance from GOES. The alert was based on measurements on the three high-latitude NMs, namely Apatity, Oulu and FSMT. NOAA issued an Alert based on the recordings of the proton channel at 100 MeV when exceeding 1 pfu. This Alert was issued for the event under investigation at 17.05.2012 at 02:52 UT. For that event the NMDB alert constructed and put online in real time by the Athens Center preceded by almost 40 min the NOAA alert. Still before the NMDB alert there was observed also relatively fast increase of energetic electrons on ACE.

Another possibility of short-term forecasting of the appearance and intensity of solar ion events is that by using data of relativistic electrons measured on satellites (Posner, 2007). For the SEP and GLE 69 event (January 20, 2005) with sharp onset, the electron precursor signal was detected 20-25 minutes in advance. Relativistic Electron Alert System for Exploration (REleASE) provides the real time forecast of proton flux in the range of several MeV to ~ 50 MeV and its comparison with the SOHO energetic particle measurements (available at http://costep2.nascom.nasa.gov/). As a part of integrated Space Weather Analysis System at the site http://iswa.gsfc.nasa.gov , under Heliosphere and UMA proton flux forecast, can be found in real time the prediction of onset and intensity of first hours of well magnetically connected SEP events and the onset of poorly connected SEP events. (Nuñez, 2011), using X rays and higher energy protons describes the forecast scheme of SEP protons with E > 10 MeV. During some of the solar flares the protons accelerated to very high energies can produce by nuclear interactions with the material of solar residual atmosphere the neutrons as well as high energy photons from decay of neutral pions. At the site http://cr0.izmiran.ru/SolarNeutronMonitoring a tool for real time monitoring of solar neutron events can be found. Very high energy photons observed from several solar flares e.g. by CORONAS-F satellite can serve as a tool for identification of onset time of proton acceleration to high energies and they are in some cases seen as a precursor to GLE from few tens to several hundreds of seconds (Kurt et al., 2010; 2011). (Laurenza et al., 2009) developed a technique to provide short-term warnings of SEP events that meet or exceed the Space Weather Prediction Center threshold of J (>10 MeV) = 10 # $cm^{-2}s^{-1}sr^{-1}$. The method is based on flare location, size, and evidence of particle acceleration/escape as parameterized by flare longitude, time-integrated soft X-ray intensity, and of type III radio emission 1 MHz, respectively. By this technique, the warnings are issued 10 min after the maximum of ≥M2 soft X-ray flares. (Valach et al., 2011) used the ANN method to forecast SEP using data on X ray flares (class, position), on radio emissions (type II or IV radio bursts) and on CME (position angle, width of the CME, linear speed). The output is the forecasted flux of energetic protons (> 10MeV). There are limits on SEP forecast when only GLE are used as its signature. (Veselovsky and Yakovchuk, 2011) report that analysis and comparison to the 2001–2006 observations indicate more than 50% of SEP were omitted if only NM warning is used for forecast. Higher reliability requires using additional data on the state of solar and heliospheric activity. For the eventual forecast of GLEs and/or SEP events the change of the power spectrum density of time series not only of GCR but also in solar activity characteristics are important to check.

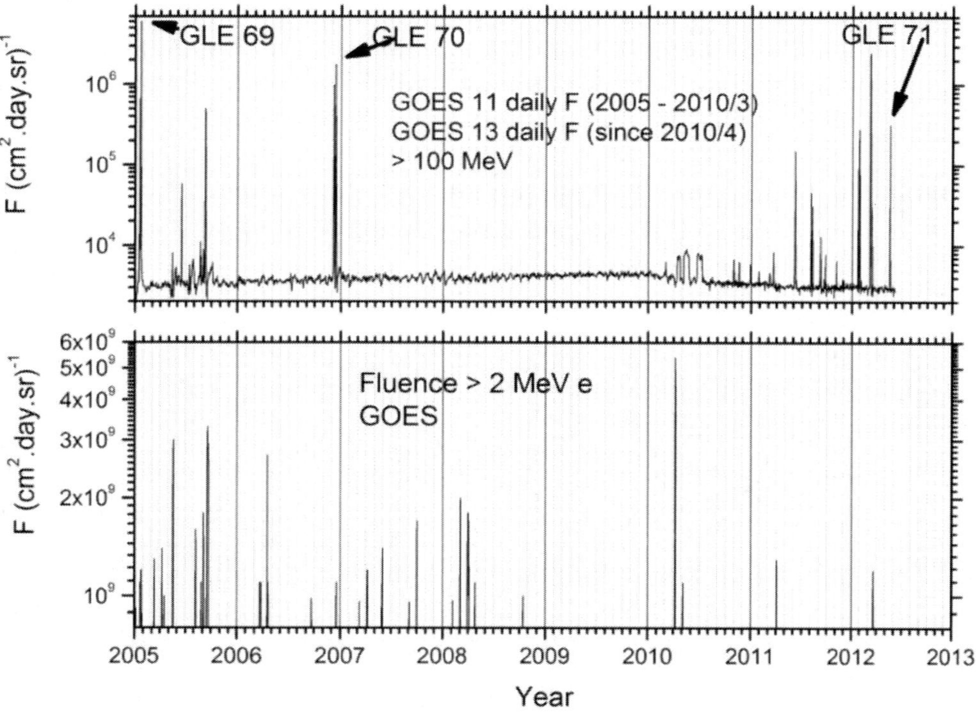

Figure 1. Daily fluences of protons > 100 MeV and of electrons > 2 MeV, based on the data downloaded from site prepared by the U.S. Dept. of Commerce, NOAA, Space Weather Prediction Center, for the period January 2005 until June 2012. Usually the highest daily fluences of protons are connected with GLEs. However, on some days there are observed high fluences due to long time duration of SEP events (e.g. 7-8 March 2012) without GLEs. The energy spectra for GLEs 69 and 70 can be found e.g. in (Vashenyuk et al., 2011; Adriani et al., 2012). The peaks of daily fluence of electrons do not coincide with those of high energy protons.

Such analysis is done e.g. by (Pérez-Peraza et al., 2011). We checked the daily fluences of protons > 100 MeV and of electrons > 2 MeV, published by the U.S. Dept. of Commerce, NOAA, Space Weather Prediction Center, for the period January 2005 until June 2012. Usually the highest daily fluences of protons are connected with GLEs. However, on some days there are observed high fluences due to long time duration SEP events (e.g. 7-8 March 2012) without GLEs. The peaks of daily fluence of electrons do not coincide with those of high energy protons (Figure 1). During radiation storms when enhanced fluxes of high energy particles accelerated either in solar flares or at the plasma discontinuities in IP space appear, one more type of space weather effect is related to energetic charged particles, namely ionisation of the atmosphere with consequences to propagation of electromagnetic waves. While galactic CR effects are assumed e.g. in papers (Velinov et al., 2011; 2012a), the anomalous component of CR is examined e.g. in paper (Velinov et al., 2012b). Velinov et al. (2013) report results of numerical calculation of galactic CR ionisation profiles at altitudes 35 – 90 km. Results of the model are important for electric parameters of the atmosphere as well as for the chemical processes. During solar flares the ionisation of the middle atmosphere is increasing. The ionisation effects of large GLEs have been analyzed e.g. in papers (Mishev et al., 2012; 2013; Usoskin et al., 2011).

For the estimates of ion production rate and of dosimetric characteristics in atmosphere during GLE events, the energy spectra of GLEs are important. They are constructed in wide energy range from satellite and ground based data. Reviews of recent GLEs and associated phenomena can be found e.g. in papers (Shea and Smart, 2012; Gopalswamy et al., 2012).

4. Quasi-Periodic Variations Observed in Cosmic Rays: Information for Space-Weather Related Effects in the Neighborhood of Earth

Forecast models used in space weather studies require also the knowledge of periodic and quasi-periodic variability of solar, interplanetary and geomagnetic effects. The same is valid also for CR. Deconvolution of the solar and cosmic ray forcing on the terrestrial climate was discussed e.g. by (Fichtner et al., 2006). We mention shortly some results on selected quasi-periodicities in CR, especially those observed on the ground.

Quasi-periodicity of ~1. 7 years in CR was reported first by (Valdes-Galícia et al., 1996). Later it was checked by wavelet technique (e.g. Kudela et al, 2002). Such wave was observed also in data from outer heliosphere in measurements on Voyager (Kato et al., 2003). Using NM data Calgary and Deep River (Kudela et al., 1991) indicated that at ~ 20 month the power spectrum density of CR is changing and rough estimate of modulation region of CR was estimated (110-130 AU).

In paper (Okhlopkov, 2011) there is indication that length of the quasi - 2 year periodicity in the even and odd numbered cycles differs by ~2 months. In cycles 20 and 22, T = 22–23.5 months, while in the 21st and 23rd T = 20.2–20.8 m. (Mendoza et al., 2006) in analysis of solar magnetic flux in the period 1971–1998 found that ~ 1.7 year is the dominant fluctuation for all the types of fluxes analyzed (total, closed, open, low and high latitude open fluxes) and that it has a strong tendency to appear during the descending phase of solar activity. (Rouillard and Lockwood, 2004) relate a strong 1.68-year oscillation in GCR fluxes to a corresponding oscillation in the open solar magnetic flux and infer CR propagation paths confirming the predictions of theories in which drift is important in modulating the CR flux. Spectral analysis of surface atmospheric electricity data (42 years of Potential Gradient, PG at Nagycenk, Hungary) showed also ~1.7 year q-per (Harrison and Märcz, 2007).

Such periodicity in the PG data is present in 1978 – 1990, but absent in 1963 – 1977. Using the monthly data of modulation parameter of CR available since 1936 from (Usoskin et al., 2011) the wavelet analysis with Morlet base function provided the profile of the spectrum density (Figure 2). The epoch without the ~1.7 year wave in atmospheric electricity measurements was found to have corresponding wavelet density of CR of lower level than that for the 1978-1990. It would be of interest to perform wavelet analysis of both time series (CR, PG) over the large interval including also the epoch after 1990. 20 month periodicity found recently by (Babic et al., 2013) in the absorbed dose monitored in Croatia is most probably linked to variations in the atmospheric production of ^{7}Be by cosmic rays. The quasi-periodicities in the range 1-2 years are observed also in other solar activity characteristics. In sunspot groups and flare index the periodicities in that range have shown differences in the solar hemispheres (Mendoza and Velasco-Herrera 2011). Recently (Vecchio et al., 2012) in detailed analysis of different components of heliomagnetic field for 1976-2003 found that quasi-biennial oscillations (QBO) are also identified as a fundamental timescale of variability of the magnetic field and associated with poleward magnetic flux migration from low latitude around the maximum and descending phase of solar activity cycle.

Quasi-periodicities in CR below ~11 years have been reported by using different methods from data 1953-1996 e.g. by (Mavromichalaki et al., 2003). Periodicities ~11 yr and ~22 yr are discussed e.g. by (Venkatesan and Badruddin, 1990). Lomb-Scargle Periodogram of Climax NM data (1953 – 2006) we computed indicates several quasi-periodicities at very low frequencies (like ~5.5, ~6.4, ~ 8.2 years). However their significance has to be tested in more detail.

The analysis done in (Pérez-Peraza et al., 2008) using the cosmogenic ^{10}Be nuclide as an indirect proxy of CR over the long time period, has shown common frequency of 30± 2 yrs in the time series of Atlantic Hurricanes. From direct measurements of CR (Ahluwalia, 1997) for the first time reported the three solar cycle trend in the data. Longer time series of CR, solar activity and IMF were examined recently (Ahluwalia, 2011). Using the direct stratospheric measurements of CR provided routinely over long time period (Stozhkov et al., 2007; 2011) the periodogram (not shown here) indicates that the three-cycle trend is present in the data.

The same indication (not shown here) was obtained from the data on modulation parameter of CR used from paper (Usoskin et al., 2011). Recently (Perez-Peraza et al., 2012) by using different techniques indicated the evidence of the existence of cosmic ray fluctuations with a periodicity of around 30 years. Detailed discussion about presence of three solar cycle quasi-periodicity in various data sets characterizing the physical status of outer space can be found in paper (Ahluwalia, 2012).

Wave of ~27 days in the ground based CR recordings has been discussed in detail by several papers (e.g. Gil et al., 2005; Richardson, 2004 among others, and references therein). (Agarwal et al., 2011) indicate that ~27 day CR variation correlates with B, Bz, v, and B(v x B). Wavelet method provides fine structure of appearing quasi-periodicities in the signal as inspected over long time.

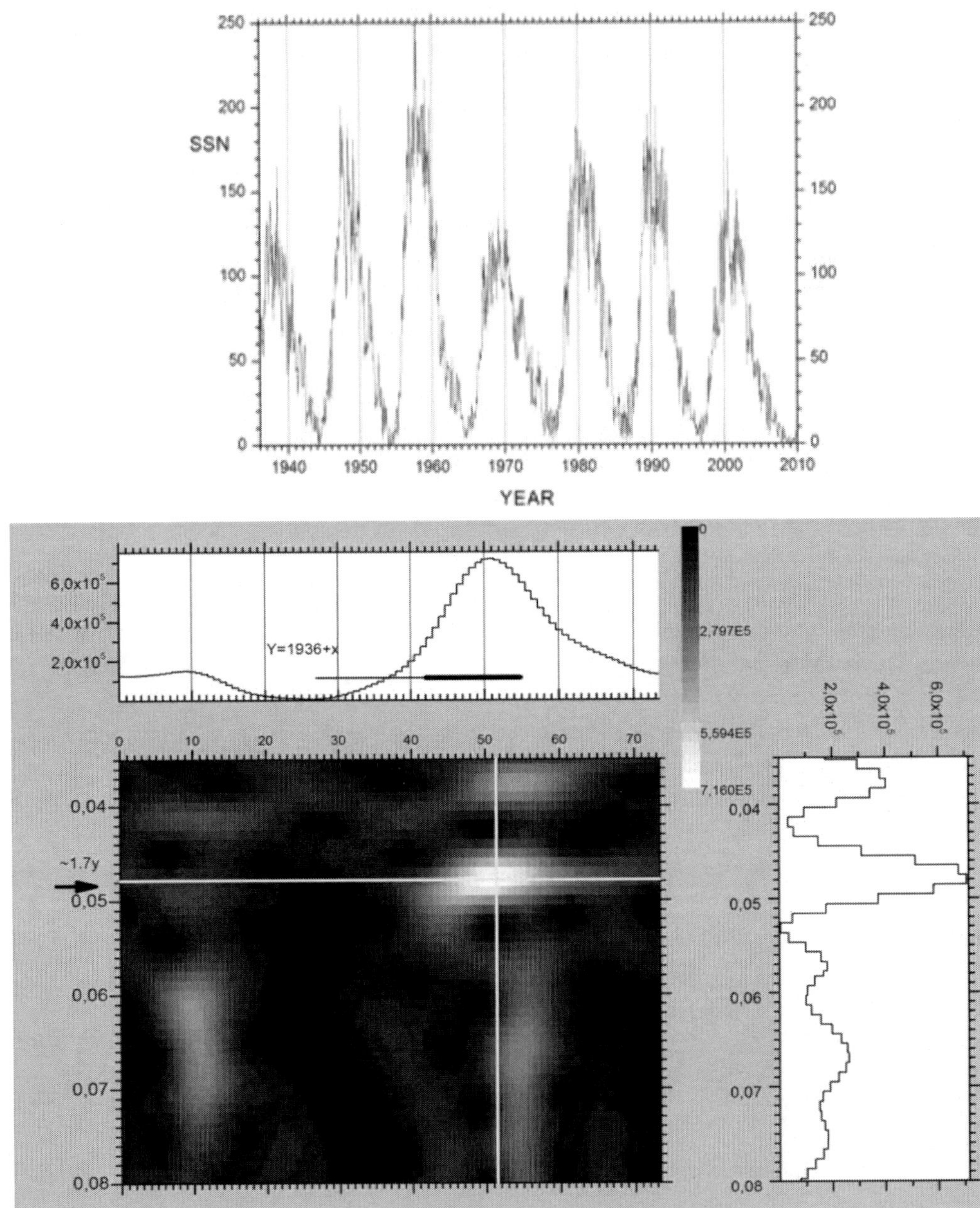

Figure 2. Wavelet spectrum density of CR modulation parameter based on data (Usoskin et al., 2011) in the plot time (years) versus frequency (in month^{-1}). Uppermost panel: Sunspot numbers. The "cross-sections" are for 1.7 year quasi-periodicities and spectral density at the maximum indicating along with ~1.7 year also ~1.3 yr and ~ 2.2 yr enhancements. The epoch with enhanced and depressed quasi-periodicity reported by (Harrison and Märcz, 2007) are marked as thick and thin horizontal lines, respectively. The time evolution of the ~ 1.3 year q-per, reported by (Mursula and Zieger, 2000) in solar wind speed and geomagnetic activity as well as by (Obridko and Shelting, 2007) in solar magnetic fields since 1915 inferred from H-alpha filament observations, is in CR different from that at ~ 1.7 years one. The 2.2-2.3 yr enhancement indicated earlier by (Mavromichalaki et al., 2005) using coronal index from coronal stations (Rybanský, 1975), is seen in CR as enhancement especially in the 22nd solar cycle. It may be the edge of QBO discussed recently by (Laurenza et al., 2012).

For example the profile of wavelet spectrum density using Climax NM data over 1953-2006 indicates the double structure during year 1986, with two peaks, one around ~27 days and another ~30-31 days. The two peak structure is similar to that reported by (Dunzlaff et al., 2008) for GCR from measurements at lower energies, EPHIN on SOHO. (Gil and Alania, 2011; 2012) reported the 3 – 4 cycling structure of ~ 27 day q-per amplitude in NM data with cut-off rigidities below 8 GV. The ~3 Carrington rotation quasi-periodicity is significant even at higher energies of primaries (Kudela, 2013). (Modzelewska and Alania, 2011) discussed the 3D model of ~ 27 day CR variations and indicated this variation of the GCR intensity for different polarity periods of the solar magnetic cycle are compatible with the NM data. By checking the linear correlation of the amplitude of ~27 day wave obtained from Climax NM (data by Sabbah, 2012, personnal communication) indicate relatively high correlation to the magnitude of IMF B, to solar activity measured by sunspot number, to tilt angle of heliospheric current sheet and to coronal hole area as deduced from the green corona line (Rybanský et al., 2001). Averaging over 1 year provides the linear relation of the amplitude of 27 day wave to those characteristics with the correlation coefficient 0.93 for the period 1977 – 2003. Similar type of analysis was done using the wavelet spectrum in the vicinity of ~13.5 days. Also here the wave is far from being monochromatic with various temporary features. The result (not shown here) is consistent with that one reported earlier by (Krymsky et al., 2008) indicating that the power spectrum of 13.5- and 27-day variations repeats the power spectrum change of the number of sunspots and tilt angle of the current sheet. Checking the dependence on solar magnetic field polarity requires more detailed study. Recently (Vieira et al., 2012) found a double structure in power spectrum density near ~ 13.5 days at muon detector. (Chowdhury and Dwivedi, 2012) reported results from wavelet analysis of GOES high energy electron data during solar cycle 23 and a part of cycle 24 and found a number of quasi-periodic oscillations in both data sets. We made a spectral analysis of the daily fluence of relativistic electrons and high energy protons measured by GOES and found the clear difference of ~ 27 day wave: while for electrons the increase around that periodicity is very clear (probably with fine structure), there is no indication about ~ 27 day recurrency in the proton data. Solar, geomagnetic and IMF parameters recently analyzed by (Katsavrias et al., 2012) by wavelet and the Lomb-Scargle periodogram identified the ~27 day per. (with ~13.5 days being its harmonic) in solar wind parameters, in Bx, By, and in the geomagnetic indices.

Paper (Rieger et al., 1984) indicated the presence of quasi-periodicity in occurrence of solar flares with very high energy photons. There are several papers dealing with this range of periods analyzed in solar activity parameters. Solar flares and consecutive CMEs along with plasma discontinuities in interplanetary space modulate the low energy CR flux near Earth. On the other hand the flares themselves contribute to SEP and GLE events. Thus temporal behaviour of CR time series is of interest around that periodicity too. (Chowdhury et al., 2010) found several intermediate-term q-per in solar activity characteristics and in CR. Period 150-160 days was found prominent during ascending phase of cycle 23 in both galactic CR and solar indices. Detailed studies of ~156 d q-per in various time series of solar activity was recently published (Akimov and Belkina, 2012). There may be more complicated structure in the time series of CR measured by NM. Wavelet analysis is needed to use in more detailed analysis of CR time series around that quasi-periodicity.

CONCLUSION

Study of relativistic electron flux variability near Earth, its relation to solar wind and IMF characteristics, is important for the forecasts of high flux of relativistic electrons having impact on satellite systems, and having different characteristics in comparison with those of high energy proton flux. The analysis of satellite measurements of electrons at various orbits, especially during and after geomagnetic and interplanetary enhanced activity intervals, are important not only for checking the geomagnetic field models and better understanding the dynamics of electrons in magnetosphere, but also as inputs for update of radiation models in the vicinity of Earth.

In recent period the new approaches in construction the alerts for the events of enhanced geomagnetic activity as well as in short time warnings before the radiation storms, with using networks of ground based CR detectors, is reported both from case and statistical studies. New measurement devices with their potential to study geoeffective events have been installed. One of the important points for the future is to estimate the efficiency of alerts based on anisotropy of CR measured at Earth, for the geoeffective events, based on long term existing data bases. Fluctuations of CR are also one of important tools for checking IMF inhomogenities in IP space and for alerting the geoeffective events. Progress can be expected if the analysis of fluctuations observed by NMs is done jointly with high geometrical factors satellite measurements and with high statistical accuracy of other ground based measurements. Joint study of CR dynamics is important to be done with solar physicists as well as with the specialists working with satellite energetic particle measurements and its analysis.

In addition to studies of irregular effects, the detailed description of quasi-periodic variations of CR along with the interplanetary, solar and geomagnetic activity physical characteristics, examined by the same methods, can contribute to discriminating of CR from other "outer-space drivers" of Space Weather and Space Climate effects.

ACKNOWLEDGMENTS

This article was supported by the realization of the VEGA grant agency, project 2/0040/13.

REFERENCES

Adriani, O. et al., Observations of the December 13 and 14, 2006, Solar Particle Events in the 80 MeV/n - 3 GeV/n range from space with PAMELA detector, *Astrophys. J.*, 742 102 doi:10.1088/0004-637X/742/2/ 102, 2012.

Agarwal, R., R.K. Mishra, S.K. Pandey and P.K. Selot, 27-day variation of cosmic rays along with interplanetary parameters,, *Proc. 32nd ICRC Beijing*, paper icrc0129, 2011.

Ahluwalia, H.S. Three activity cycle periodicity in galactic cosmic rays? Proc. 25th Int. *Cosmic Ray Conf.*, Durban, South Africa, 2, 109-112, 1997.

Ahluwalia, H.S. Timelines of cosmic ray intensity, Ap, IMF, and sunspot numbers since 1937, *J. Geophys. Res.*, 116, A12106, doi:10.1029/ 2011JA017021, 2011.

Ahluwalia, H.S., Three-cycle quasi-periodicity in solar, geophysical, cosmic ray data and global climate change, *Indian J. of Radio and Space Physics,* Vol. 41, pp. 509-519, 2012.

Akimov, L.A. and I.L. Belkina, Rieger QuasiPeriodicity in Solar Indices, *Solar System Research*, Vol. 46, No. 3, pp. 243–252, 2012.

Alexeev, I.I., E.S. Belenkaya, V.V. Kalegaev, Y.I. Feldstein and A. Grafe, Magnetic storms and magnetotail currents, *J. Geophys. Res.*, 101, A4, 7737-7747, 1996

Anashin, V., A. Belov, E. Eroshenko et al., The ALERT signal of ground level enhancemens of solar cosmic rays: physics basis, the ways of realization and development, *Proc. 31st ICRC*, Lodz, icrc1104, 2009.

Angelov, I.I., E. S. Malamova, J. N. Stamenov, Muon Telescopes at Basic Environmental Observatory Moussala and South-West University – Blagoevgrad, *Sun and Geosphere*, 3(1): 20 – 25, 2008.

Antonova, E.E., I.M. Myagkova, M.V. Stepanova et al., Local particle traps in the high latitude magnetosphere and the acceleration of relativistic electrons. *J. Atmos. Sol. Terr. Phys.* 73, 1465–1471, doi:10.1016/ j.jastp.2010.11.020, 2011.

Babic, D., J. Sencar, B. Petrinec, G. Marovic, T. Bituh and B. Skoko, Fine structure of the absorbed dose rate monitored in Zagreb, Croatia, in the period 1985-2011, *J. Environm. Radioactivity*, Vol. 118, 75-9, Apr. 2013.

Badruddin, Transient Perturbations and their Effects in the Heliosphere, the Geomagnetosphere, and the Earth's Atmosphere: Space Weather Perspective, *J. Astrophys. Astr.* 27, 209–217, 2006.

Baker, D.N., J.H. Allen, S.G. Kanekal, and G.D. Reeves, Disturbed space environment may have been related to pager satellite failure, Eos, Transactions, *AGU*, 79(40), 477, Oct 6, 1998.

Baker, D.N., R.D. Belian, P.R. Higbie, R.W. Klebesadel and J.B. Blake, Deep dielectric charging effects due to high energy electrons in the outer magnetosphere, *J. Electrostatics*, 20, 1, 3-19, 1987.

Balikhin, M.A., R.J. Boynton, S.N. Walker et al., Using the NARMAX approach to model the evolution of energetic electrons fluxes at geostationary orbit, *Geophys. Res. Lett.*, 38, L18105, 5 PP., doi:10.1029/2011GL048980, 2011.

Barbashina, N.S., I. I. Astapov, V. V. Borog et al., Investigating the Energy, Angular, and Temporal Characteristics of Forbush Effects Registered by the URAGAN Muon Hodoscope, *Bull. Russian Acad. Sci., Physics*, 75, 6, 814–817, 2011.

Belov, A.V., L.I. Dorman, N. Iucci, O. Kryakunova and I. Ptitsyna, The relation of high- and low-orbit satellite anomalies to different geophysical parameters, in Effects of Space Weather on Technology Infrastructure edited by Ioannis A. Daglis, Kluwer, 147-164, 2004.

Bieber, J.W. et al, AOGS 3rd Ann. Meeting, Singapore, July 2006.

Bothmer, V. and I.A. Daglis, Space Weather - Physics and Effects, Springer, Praxis Publ. Co., Chichester, UK, 2007.

Braga, C.R., A. Dal Lago, M. Rockenbach et al., Precursor signatures of the storm sudden commencement observed by a network of muon detectors, paper SH 550, ECRS Moscow, July 2012 (abstract).

Buchvarova, M. and D. Draganov, Cosmic-Ray Spectrum Approximation Model: Experimental Results and Comparison with Other Models, *Solar Phys.*, DOI 10.1007/s11207-012-0157-8, online 16 October 2012.

Buchvarova, M., Approximated model of primary cosmic radiation, *J. Phys.: Conf. Ser.* 409 012170, 2013.

Chilingarian, A. , G. Hovsepyan, K. Arakelyan et al., Space Environmental Viewing and Analysis Network (SEVAN), Earth Moon Planet, DOI 10.1007/s11038-008-9288-1, 2009.

Chowdhury, P., M. Khan, P.C. Ray, Evaluation of the intermediate-term periodicities in solar and cosmic ray activities during cycle 23, *Astrophys. Space Sci.*, 326, 191-201, 2010.

Chowdhury, P. and B.N. Dwivedi, Time evolution of short and intermediate term periodicities of electrons at geosynchronous orbit during solar cycle 23 and 24: A wavelet approach, *Planetary and Space Sci.*, 67, 1, 92-100, 2012.

Dasso, S., H. Asorey for Pierre Auger, The scaler mode in the Pierre Auger observatory to study heliospheric modulation of cosmic rays, *Adv. Space Res.* 49, 11, 1563-1569, 1 June 2012.

De Mendonça, R.R.S., J.-P. Raulin, F.C.P. Bertoni et al., Long-term and transient time variation of cosmic ray fluxes detected in Argentina by CARPET cosmic ray detector , *J. Atmos. and Solar-Terrestrial Physics*, 73, 11–12, 1410-1416, July 2011.

Desorgher, L., K. Kudela, E. O. Flückiger, R. Bütikofer, M. Storini, V. Kalegaev, Comparison of Earth's magnetospheric magnetic field models in the contect of cosmic ray physics, *Acta Geophysica* vol. 57 issue 1, 75 – 87, 2009.

Dhanju, M.S. and V.A. Sarabhai, Short-period variations of cosmic – ray intensity, *Phys. Rev. Lett.*, 19, 5, 1967.

Dorman L.I., 1963. Geophysical and Astrophysical Aspects of Cosmic Rays. North-Holland Publ. Co., Amsterdam (*In series "Progress in Physics of Cosmic Ray and Elementary Particles"*, ed. J.G. Wilson and S.A. Wouthuysen, Vol. 7), pp 320., 1963.

Dorman, L.I., and I.Y. Libin, Short-period variations in cosmic ray intensity, *Soviet Phys. Uspekhi*, 28, 233-256, 1985.

Dorman, L.I., A.V. Belov, E.A. Eroshenko et al., Different space weather effects in anomalies of the high and low orbital satellites, *Adv. Space Res.*, 2530-2536, 2005.

Dorman, L.I. Monitoring and forecasting of great radiation hazards for spacecraft and aircrafts by online cosmic ray data, *Ann. Geophys.* 23, 3019–3026, 2005.

Dorman, L.I., Cosmic rays and space weather: effects on global climate change, *Ann. Geophys.*, 30, 1, 9-19, 2012.

Dunzlaff, P., B. Heber, A. Kopp et al., Observations of recurrent cosmic ray decerases during solar cycles 22 and 23, *Ann. Geophys.*, 26, 3127-3138, 2008.

Fichtner, H., K. Scherer, and B. Heber, A criterion to discriminate between solar and cosmic ray forcing of the terrestrial climate, *Atmos. Chem. Phys. Discuss.*, 6, 10811–10836, 2006.

Flückiger, E.O., ECRS 2004, http://www.fi.infn.it/conferenze/ecrs2004/Pages/ Presentazioni/03%20Flueckiger.pdf.

Fushishita, A., T. Kuwabara, C. Kato et al., Precursors of the Forbush decrease on 2006 December 14 Observed with the GMDN, *Astrophys. J.*, 715 1239 doi:10.1088/0004-637X/715/2/1239, 2011.

Gil, A., K. Iskra, R. Modzelewska and M.V. Alania, On the 27-day variations of the galactic cosmic ray anisotropy and intensity for different periods of solar magnetic cycle, *Adv. Space Res.*, 35, 687-690, 2005.

Gil, A. and M.V. Alania, Cycling Changes in the Amplitudes of the 27-Day Variation of the Galactic Cosmic Ray Intensity, *Solar Phys.*, 278:447–455, DOI 10.1007/s11207-012-9944-5, 2012.

Gil, A. and M.V. Alania, The rigidity spectrum of the harmonics of the 27-day variation of the galactic cosmic ray intensity in different epochs of solar activity: 1965–20022011, *J. Atmos. Sol. Terr. Phys.*, 73, 294-299, 2011.

Gopalswamy N, H. Xie, S. Yashiro, S. Akiyama, P. Mäkelä and I.G. Usoskin, Properties of Ground Level Enhancement Events and the Associated Solar Eruptions During Solar Cycle 23, *Space Sci. Rev.*, vol. 171, Issue 1-4, pp. 22-30, 2012.

Harrison, R.G., and F. Märcz, Heliospheric timescale identified as in surface atmospheric electricity, *Geophys. Res. Lett.*, 34, L23816, 2007GL031714, 2007.

Hippler, R., A. Mengel, F. Jansen et al., "First Space weather Observations at MuSTAnG — the Muon Space weather Telescope for Anisotropies at Greifswald", in: *Proceedings of the 30th ICRC*, Merida, Mexico, 2007, v. 1, pp. 347-350, 2007.

Iucci, N., L.I. Dorman, A.E. Levitin et al., Spacecraft operational anomalies and space weather impact hazards, *Adv. Space Res.*, 37, 184–190, 2006.

Kane, R.P., Severe geomagnetic storms and Forbush decreases: interplanetary relationships reexamined, *Ann. Geophys.*, 28, 479-489, 2010.

Kato, C. K. Munakata, S. Yasue, K. Inoue and F.B. McDonald, A ~1.7-year quasi-periodicity in cosmic ray intensity variation observed in the outer heliosphere, *J. Geophys. Res.*, 108, 1367, 7 pp, doi:10.1029/ 2003JA009897, 2003.

Katsavrias, Ch., P. Preka-Papadema, X. Moussas, *Wavelet Analysis on Solar Wind Parameters and Geomagnetic Indices*, arXiv:1205.2229v1, in press, Sol. Phys. 2012.

Kim, M.H.Y., J.W. Wilson and F.A. Cucinotta, Description of transport codes for space radiation shielding, *Health Phys.*, 103, 5, 621-639, Nov. 2012.

Klein, K.-L. and R. Bütikofer, The first relativistic solar particle event of the present activity cycle observed by the network of neutron monitors! http://www.nmdb.eu/?q=node/480, 2012.

Kozlov, V.I. , A.I. Kuzmin and G.F. Krymsky, Cosmic ray variations with periods less than 12 hours, in *Proc. 13th ICRC*, 2, Denver, 939-942, 1973.

Kozlov, V.I. and V.V. Kozlov, Galactic Cosmic Ray Fluctuation Parameter as an Indicator of the Degree of Magnetic Field Inhomogeneity, *Geomagn. Aeron.*, 51, 2, 187–197, 2011.

Krüger, H., and H. Moraal, Neutron monitor calibrations: progress report, paper SH 481, ECRS 2012, Moscow, 4 p., 2012.

Krymsky, G.F., V.P. Mamrukova, P.A. Krivoshapkin, S.K. Gerasimova, S.A. Starodubtsev, Recurrent variations in the high-energy cosmic ray intensity. *Proc. 30th ICRC*, Mexico, v.1, 381-384, 2008.

Kudela, K., A.G. Ananth and D. Venkatesan, The Low-Frequency Spectral Behavior of Cosmic Ray Intensity, *J. Geophys. Res.*, 96, A9, 15,871-15,875, doi:10.1029/91JA01166, 1991.

Kudela, K., J. Rybák, A. Antalová, and M. Storini, Time evolution of low-frequency periodicities in cosmic ray intensity. *Sol. Phys.*, 35, pp. 165-175, 2002.

Kudela, K. and R. Brenkus, Cosmic ray decreases and geomagnetic activity: list of events 1982-2002, *J. Atmos. and Solar-Terrestrial Physics*, 66, 13-14, 1121-1126, 2004.

Kudela, K., D. Venkatesan and R. Langer, Variability of Cosmic Ray Power Spectra, *J. Geomagnet. Geoelectr.*, 48, 1017-1024, 1996.

Kudela, K., M. Storini, M.Y. Hofer and A. Belov: Cosmic rays in relation to space weather, *Space Sciences Series of ISSI*, Vol. 10, 153-174, pp. 424, 2000.

Kudela, K. and M. Storini, Cosmic ray variability and geomagnetic activity: a statistical study, Adv. Space Res., *J. Atmos. Sol. Terr. Phys.*, 67, 10, 907–912, 2005.

Kudela K., Bucik R., Bobik P., On transmissivity of low energy cosmic rays in disturbed magnetosphere, *Adv. Space Res.*, 42, 7, 1300-1306, 2008.

Kudela, K., Cosmic Ryas and space waether: direct and indirect relations, in D'Olivo J.C., G. Medina-Tanco, J.F. Valdes-Galicia, eds. *Proc. 30th ICRC*, Merida, Mexico, 6, 195-208, 2009.

Kudela, K., Space weather near Earth and energetic particles: selected results, *J. Phys.: Conf. Ser.* 409 012017, 2013.

Kurt, V.G., Yushkov B.Yu., Belov, A. V., On the Ground Level Enhancement Beginning, *Astron. Lett.,* Vol. 36, No. 7, pp. 520–530, 2010.

Kurt, V.G., Yushkov, B. Yu., Belov, A., Chertok, I., Grechnev, V. A Relation between Solar Flare Manifestations and the GLE Onset, *Proc. 32nd ICRC,* paper 441, 2011.

Kuwabara, T., J.W. Bieber, J. Clem et al., Development of a ground level enhancement alarm system based upon neutron monitors, *Space Weather,* 4, S10001, doi:10.1029/2006SW000223, 2006.

Laha, I., Banerjee, A. Bajaj, P.K. Barhai, A.K. Das, V.N. Bhoraskar and S.K. Mahapatra, Irradiation effects of 6 MeV electron on electrical properties of Al/ Al2O3/n-Si MOS capacitors, *Radiation Physics and Chemistry,* http://dx.doi.org/ 10.1016/j.radphyschem.2012.04.006, 2012.

Laken, B.A., E. Palle, J. Čalogovič and E.M. Dunne, A cosmic ray-climate link and cloud observations, *J. Space Weather Space Clim.* 2, A18. DOI: 10.1051/swsc/2012018, 2012.

Laurenza, M., E.W. Cliver, J. Hewitt et al., A technique for short-term warning of solar energetic particle events based on flare location, flare size, and evidence of particle escape, *Space Weather,* 7, S04008, doi:10.1029/2007SW000379, 2009.

Laurenza, M., A. Vecchio, M. Storini and V. Carbone, Quasi-biennal modulation of galactic cosmic rays, Astrophys. J., 749, 167, doi:10.1088/0004-637X/749/2/167, 2012.

Lazutin, L., M. Panasyuk and N. Hasebe, Acceleration and losses of energetic protons and electrons during magnetic storm on August 30-31, 2004, *Cosmic Res.*, 49, 1, 35, 2011a.

Lazutin, L., E.A. Muravieva, K. Kudela, M. Slivka, Verification of Magnetic Field Models Based on Measurements of Solar Cosmic Ray Protons in the Magnetosphere, *Geomagn. Aeron.*, 51, 2, 198–209, 2011b.

Lazutin, L., On radiation belt dynamics during magnetic storms, *Adv. Space Res.*, 49, 302-315, 2012.

Leerungnavarat, K., D. Ruffolo and J.W. Bieber, Loss Cone Precursors to Forbush Decreases and Advance Warning of Space Weather Effects, *Astrophys. J.,* 593: 587-596, August 10, 2003.

Maghrabi, A.H. H. Al Harbi, Z.A. Al-Mostafa et al., The KACST muon detector and its application to cosmic-ray variations studies, *J. Adv. Space Res.,* doi:10.1016/j.asr.2011.10.011, in press, 2011.

Mahapatra, S.K., S.D. Dhole, V.N. Bhoraskar and G.G. Raju, Dependence of charge buildup in the polyimide on the incident electron energy, *Journal of Applied Physics*, Volume 100, Issue 3, pp. 034913-034913-7, 2006.

Matthiä, D. , T. Berger, A.I. Mrigakshi and G. Reitz, A ready-to-use galactic cosmic ray model, *Adv. Space Res.,* Vol. 51, 329-338, 2013.

Mavromichalaki, H., P. Preka-Papadema, B. Petropoulos et al., Low- and high-frequency spectral behavior of cosmic ray intensity for the period 1953-1966, *Ann. Geophys.*, 21, 1681-1689, 2003.

Mavromichalaki, H., B. Petropoulos, C. Plainaki et al., Coronal index as a solar activity index applied to space weather, *Adv. Space Res.*, 35, 410–415, 2005.

Mavromichalaki, H., G. Souvatzoglou, C. Sarlanis et al., Space weather prediction by cosmic rays, *Adv. Space Res.* 37, 1141–1147, 2006.

Mavromichalaki, H., G. Souvatzoglou, Ch. Sarlanis et al., Using the real-time Neutron Monitor Database to establish an Alert signal, *Proc. 31st ICRC,* Lodz, paper icrc 1381, 2009.

Mavromichalaki, H., A. Papaioannou, C. Plainaki et al., Applications and usage of the real-time Neutron Monitor Database, *Adv. Space Res.,* 47, 12, 2210-2222, 2011.

Mavromichalaki, H et al., Solar Radiation Storm issued in real-time by the Athens Neutron Monitor Alert Code operated via NMDB, May 18, 2012.

McKenna-Lawlor S, Gonçalves P, Keating A, Reitz G, Matthiä D. Overview of energetic particle hazards during prospective manned missions to Mars. Planetary and Space Science 63−64:123−132; 2012. 2012.

Medina, J., J.J. Blanco, O. García et al., Castilla-La Mancha neutron monitor (CaLMa): status at July 2012, paper presented at ECRS 2012, Moscow, July 2-7, 2012.

Mendoza, B., V.M. Velasco and J.F. Valdes-Galicia, Mid-Term Periodicities in the Solar Magnetic Flux, *Sol. Phys.*, 233, 2, 319-330, DOI: 10.1007/s11207-006-4122-2, 2006.

Mendoza, B. and Velasco-Herrera, V.M., On Mid-Term Periodicities in Sunspot Groups and Flare Index, *Sol. Phys.*, 271, 169–182, DOI 10.1007/s11207-011-9802-x, 2011.

Miroshnichenko, L.I., Radiation hazard in space, *Astrophysics and Space Science Library*, vol. 297, Kluwer, pp. 238, 2003.

Mishev, A., Stamenov, J., Recent status and further possibilities for space weather studies at BEO Moussala. *Journal of Atmospheric and Solar-Terrestrial Physics* 70 (2e4), 680e685, 2008.

Mishev, A.L., P.I.Y. Velinov, L. Mateev and Y. Tassev, Ionisation effect of nuclei with solar and galactic origin in the Earth atmosphere during GLE 69 on January 20, 2005, *J. Atmos. And Solar-Terrestrial Phys.*, vol. 89, 1-7, November 2012

Mishev, A., Short- and Medium-Term Induced Ionization in the Earth Atmosphere by Galactic and Solar Cosmic Rays, *International Journal of Atmospheric Sciences Volume 2013* (2013), Article ID 184508, 9 pages, 2013.

Modzelewska, R. and M.V. Alania, Dependence of the 27-day variation of cosmic rays on the global magnetic field of the Sun, *Adv. Space Res.*, doi:10.1016/j.asr.2011.07.022, 2011.

Mrigakshi, A.I., D. Matthiä, T. Berger, G. Reitz and R.F. Wimmer-Schweingruber, Assessment of Galactic Cosmic Ray models, *J. Geophys. Res.*, in press, 2012.

Mulligan, T., J. B. Blake, D. Shaul et al., Short-period variability in the galactic cosmic ray intensity: High statistical resolution observations and interpretation around the time of a

Forbush decrease in August 2006, *J. Geophys. Res.*, 114, A07105, doi:10.1029/2008JA013783, 2009.

Mursula, K., Zieger, B. The 1.3-year variation in solar wind speed and geomagnetic activity. *Adv. Space Res.*, 25, 9, 1939–1942, 2000.

Mustajab, F. and Badruddin, Geoeffectiveness of the interplanetary manifestations of coronal mass ejections and solar-wind stream–stream interactions, *Astrophys. And Space Sci.*, 331, 1, 91-104, 2011.

Nuňez, M. Predicting solar energetic proton events (E > 10 MeV), *Space Weather*, 9, S07003, doi:10.1029/2010SW000640, 2011.

Nymmik R.A. Extremely large solar high-energy particle events: occurrence probability and characteristics, *Adv. Space Res.* 40, 326-330, 2007.

Obridko, V.N. and B.D. Shelting, Occurrence of the 1.3-year periodicity in the large-scale solar magnetic field for 8 solar cycles, *Advances in Space Research* 40, 1006–1014, 2007.

Oh, S.Y., J.W. Bieber, J. Clem et al., Neutron Monitor Forecasting of Radiation Storm Intensity, *Proc. 31st ICRC*, Lodz, p. icrc0602, 2009.

Oh, S.Y., Y. Yi, J. W. Bieber, P. Evenson and Y. K. Kim, Characteristics of solar proton events associated with GLEs, *J. Geophys. Res.*, 115, A10107, doi:10.1029/2009JA015171, 2010.

Okhlopkov, V.P., Distinctive properties of the frequency spectra of cosmic ray variations and parameters of solar activity and the interplanetary medium in solar cycles 20–23, Moscow University *Phys. Bull.*, 66, 1, 99-103, DOI: 10.3103/S0027134911101018, 2011.

Panasyuk, M.I., Cosmic Rays and Radiation Hazards for Space Missions, in Space Storms and Space Weather Hazards, ed. By I.A. Daglis, NATO *Science Ser.* 251-284, 2001.

Panasyuk, M.I., The Ion Radiation Belts: Experiments and Models, in Effects of Space Weather on Technology Infrastructure edited by Ioannis A. Daglis, Kluwer, 65-90, 2004.

Papailiou, M., H. Mavromichalaki, A. Belov et al., Precursor Effects in Different Cases of Forbush Decreases, *Solar Phys* ., 276:337–350, DOI 10.1007/s11207-011-9888-1, 2012a.

Papailiou, M., H. Mavromichalaki, A. Belov et al., The Asymptotic Longitudinal Cosmic Ray Intensity Distribution as a Precursor of Forbush Decreases, *Solar Phys.*, DOI 10.1007/s11207-012-9945-4, 2012b.

Papailiou, M., H. Mavromichalaki, K.Kudela et al., Cosmic radiation influence on the physiological state of aviators, *Natural Hazards*, 61, 719-727, 2012c.

Papailiou, M., H. Mavromichalaki, M. Abunina et al., Forbush Decreases Associated with Western Solar Sources and Geomagnetic Storms: A Study on Precursors, *Sol. Phys.*, 283, 2, 557-563, 2013.

Pavlov, N.N., L.V. Tverskaya, B.A. Tverskoy and E.A. Chuchkov, Variations in Radiation Belt Energetic Particles during a Strong Magnetic Storm on March 24-26, 1991, *Geomagn. Aeron.*, 33, 6, 41-45, 1993.

Pérez-Peraza, J., V. Velasco and S. Kavlakov, Wavelet coherence analysis of Atlantic hurricanes and cosmic rays, *Geofísica Internacional* 47, 3, 231-244, 2008.

Pérez-Peraza, J.A., V. M. Velasco-Herrera, J. Zapotitla, L. I. Miroshnichenko, and E. V. Vashenyuk, Search for Periodicities in Galactic Cosmic Rays, Sunspots and Coronal Index Before Arrival of Relativistic Protons from the Sun, Bulletin of the Russian Academy of Sciences. *Physics*, 2011, Vol. 75, No. 6, pp. 767–769, 2011.

Pérez-Peraza, J.A., V. Velasco, I. Ya. Libin and K.F. Yudakhin, Thirty-year periodicity in cosmic rays, *Advances in Astronomy*, Vol. 2012, article ID 691408, 11 p., 2012.

Posner, A. Up to 1-hour forecasting of radiation hazards from solar energetic ion events with relativistic electrons, *Space Weather*, 5, S05001, 28 PP., doi:10.1029/2006SW000268, 2007.

Reeves, G. D., S.K. Morley, R.H.W. Friedel et al., On the relationship between relativistic electron flux and solar wind velocity: Paulikas and Blake revisited, *J. Geophys. Res.*, 116, A02213, doi:10.1029/2010JA015735, 2011.

Riazantseva, M.O., I.N. Myagkova, M.V. Karavaev et al., Enhanced energetic electron fluxes at the region of the auroral oval during quiet geomagnetic conditions November 2009, *Adv. Space Res.*, doi: http://dx.doi.org/ 10.1016/j.asr.2012.05.015, 2012.

Rieger, E., G. H. Share, D. J. Forrest et al., A 154-day periodicity in the occurrence of hard solar flares, *Nature*, 312, 623, 1984.

Richardson, I.G., Energetic Paricles and Corotating Interaction Regions in the Solar Wind, *Space Sci. Rev.*, 111, 267-376, 2004.

Rockenbach, M., A. Dal Lago, W.D. Gonzalez et al., Geomagnetic storm's precursors observed from 2001 to 2007 with the Global Muon Detector Network (GMDN), *Geophys. Res. Let.* 38, L16108, 4 PP., doi:10.1029/2011GL048556, 2011.

Rouillard, A. and M. Lockwood, Oscillations in the open solar magnetic flux with a period of 1.68 years: imprint on galactic cosmic rays and implications for heliospheric shielding, *Ann. Geophys.*, 22, 4381-4395, 2004.

Ruffolo, D., Transport and Acceleration of Energetic Charged Particles near an Oblique Shock, *Astrophys. J.*, 515, 2, 787, doi:10.1086/307062, 1999.

Rybanský, M. Coronal index of solar activity. I - Line 5303 A, Bull. Astron. Inst. Czech. 26, 367–377, 1975.

Rybanský, M., V. Rušin and M. Minarovjech, Coronal Index of Solar Activity, *Space Sci. Rev.* 95, 227-234, 2001.

Sabbah, I., *personnal communication*, 2012.

Sabbah, I. and K. Kudela, On ~27 day periodicity in cosmic rays, *under preparation, manuscript,* 2012.

Sdobnov, V. Ye., Analysis of the Forbush effect in May 2005 using the spectrographic global survey method, *Bull. Russian Acad. Sci.: Physics,* 75, 6, 805-807, 2011.

Scherer, K., H. Fichtner, B. Heber, U. Mall (eds), Space Weather, The Physics Behind a Slogan, Springer, 2005.

Shea, M.A. and D.F. Smart, Space Weather and the Ground-Level Solar Proton Events of the 23rd Solar Cycle, *Space Sci. Rev.,* vol. 171, Issue 1-4, pp. 161-188, 2012.

Singh, A.K., Devendraa Siingh and R.P. Singh, Space Weather: Physics, Effects and Predictability, *Surv. Geophys.*, 31, 581-638, DOI: 10.1007/s10712-010-9103-1, 2010.

Singh, A.K., Devendraa Siingh and R.P. Singh, Impact of galactic cosmic rays on Earth's atmosphere and human health, *Atmospheric Environment* 45 (23) , pp. 3806-3818, 2011.

Sloan, T., Cosmic Rays, Solar Activity and the Climate, *J. Phys.: Conf. Ser.* 409 012020, 2013

Smart, D.F. and M.A. Shea, Fifty years of progress in geomagnetic cutoff rigidity determination, *Adv. Space Res.*, 44, 10, 1107-1123, 2009.

Starodubtsev, S.A., I.G. Usoskin and K. Mursula, Rapid cosmic ray fluctuations: Evidence for cyclic behaviour, *Sol. Phys.*, 224, 335/343, 2004.

Starodubtsev, S.A., I. G. Usoskin, A. V. Grigoryev, and K. Mursula, Long-term modulation if the cosmic ray fluctuation spectrum, *Ann. Geophys.*, 24, 779-783, 2006.

Stozhkov, Y. I., N.S. Svirzhevsky, G.A. Bazilevskaya et al., Cosmic rays in the stratosphere in 2008-2010, *Astrophys. Space Sci. Trans.*, 7, 379-382,doi:10.5194/astra-7-379-2011, 2011.

Stozhkov, Y.I., N.S. Svirzhevsky, G.A. Bazilevskaya et al, Fluxes of Cosmic Rays in the maximum of absorption curve in the atmosphere and at the atmosphere boundary (1957-2007), preprint FIAN, 14, 2007.

Storini, M., The Relevance of Cosmic Rays to Space and Earth Weather, ECRS 2006, Lisbon.

Storini, M., Energetic particles near Earth: relations to Space Weather studies, ECRS 2010 Turku, http://ecrs2010.utu.fi/done/presentations/EDU1/2A_ PA1_Wednesday/2_Storini.pdf

Suvorova, A., A. Dmitriev and Lung-C Tsai, Evidence for Near-Equatorial Deposition by Energetic Electrons in the Ionospheric F-layer, ArXiv 1301.241, accepted in journal, 2013a.

Suvorova, A.V., L.-C. Tsai and A,V. Dmitriev, TEC enhancement due to energetic electrons above Taiwan and the West Pacific, in press, Terrestrial, *Atmospheric and Oceanic Sciences* (TAO), Dec. 2012 A special issue on "Connection of solar and heliospheric activities with near-Earth space weather: Sun-Earth connection", 2013b.

Tsyganenko, N.A. and M.I. Sitnov, Modelling the dynamics of the inner magnetosphere during strong geomagnetic storms, *J. Geophys. Res.*, 110, A3, A03208, 2005.

Tverskaya, L.V., Diagnostics of the Magnetosphere Based on the Outer Belt Relativistic Electrons and Penetration of Solar Protons: *A Review, Geomagnetism and Aeronomy*, 51, 6-22, 2011.

Tyasto, M.I., O. A. Danilova, V. M. Dvornikov and V. E. Sdobnov, Strong Cosmic Ray Cutoff Rigidity Decreases during Great Magnetospheric Disturbances, *Bull. Russian Acad. Sci., Physics*, 73, 3, 367–369, 2009.

Tyasto, M.I., O. A. Danilova, and V. E. Sdobnov, Variations in the Geomagnetic Cutoff Rigidity of CR in the Period of Magnetospheric Disturbances of May 2005: Their Correlation with Interplanetary Parameters, *Bull. Russian Acad. Sci., Physics*, 75, 6, 808–811, 2011.

Tyasto, M.I., O.A. Danilova, N.G. Ptitsyna and V.E. Sdobnov, Variations in cosmic ray cutoff rigidities during the great geomagnetic storm of November 2004, *Adv. Space Res.*, 51, 7, 1230-1237, April 2013.

Tylka, A. J., Dietrich, W. F. and Boberg, P. R., Probability distributions of high-energy solar-heavy-ion fluxes from IMP-8: 1973-1996, *IEEE Trans. Nucl. Sci.*, vol. 44, no. 6, pp. 2140–2149, Dec. 1997.

Usoskin, I.G., G.A. Bazilevskaya, G.A. Kovaltsov, Solar modulation parameter for cosmic rays since 1936 reconstructed from ground-based neutron monitors and ionization chambers, *JGR*, 116, A02104, 9 PP., doi:10.1029/2010JA016105, 2011.

Usoskin, I.G., G.A. Kovaltsov, I.A. Mironova, A.J. Tylka and W.F. Dietrich, Ionisation effect of solar particle GLE events in low and middle atmosphere, *Atm. Chemistry and Physics*, 11, 5, 1979-1988, 2011.

Valach F., M. Revallo, P. Hejda, J. Bochníček, Predictions of SEP events by means of a linear filter and layer-recurrent neural network, *Acta Astronautica*, 69, 758–766, 2011.

Valach, F., M. Revallo, J. Bochníček, and P. Hejda, Solar energetic particle flux enhancement as a predictor of geomagnetic activity in a neural network-based model, *Space Weather*, 7, S04004, doi:10.1029/ 2008SW000421, 2009.

Valdés-Galicia, J. F., R. Pérez-Enríquez and J.A. Otaola, The cosmic ray 1.68-year variation: a clue to understand the nature of solar cycle?, *Sol. Phys.*, 167, 409-417, 1996.

Vashenyuk, E.V., Yu. V. Balabin, and B. B. Gvozdevsky, Features of relativistic solar proton spectra derived from ground level enhancement events (GLE) modeling, *Astrophys. Space Sci. Trans.*, 7, 459–463, www.astrophys-space-sci-trans.net/7/459/2011/, doi:10.5194/astra-7-459-2011, 2011.

Vecchio, A., M. Laurenza, D. Meduri, V. Carbone and M. Storini, The dynamics of the solar magnetic field: polarity reversals, butterfly diagram, and quasi-biennal oscillations, *The Astrophys. J.*, 749:27 (10 pp), April 10, 2012.

Velinov, P.I. Y., S. Asenovski, and L. Mateev, "Simulation of cosmic ray ionization profiles in the middle atmosphere and lower ionosphere on account of characteristic energy intervals," *Comptes rendus de l'Academie bulgare des Sciences*, vol. 64, no. 9, pp. 1303–1310, 2011.

Velinov, P.I.Y., S. Asenovski, L. Mateev, Ionization of anomalous cosmic rays in ionosphere and middle atmosphere simulated by CORIMIA code, *Comptes Rendus de L'Academie Bulgare des Sciences*, 65, 9, 1261-1268, 2012a

Velinov, P.I.Y., S. Asenovski, L. Mateev, Improved cosmic ray ionization model for the ionosphere (CORIMIA) with account of 6 characteristic intervals, Comptes Rendus de L. *Academie Bulgare des Sciences*, 65, 8, 1135-1144, 2012b.

Velinov, P.I.Y., S.N. Asenovski and L.N. Mateev, Numerical calculation of cosmic ray ionization rate profiles in the middle atmosphere and lower ionosphere with relation to characteristic energy intervals, *Acta Geophysica*, 61, 2, 494-509, April 2013.

Venkatesan, D. and Badruddin, Cosmic Ray Intensity Variation in the 3D Heliosphere, *Space Sci. Rev.* 52, 121-194, 1990.

Veselovsky, I.S. and O. S. Yakovchuk, On Forecasting Solar Proton Events Using a Ground Based Neutron Monitor, *Solar System Res.*, 45, 4, 354–364, 2011.

Vieira, L.R., A. Dal Lago, N. R. Rigozo et al., Near 13.5-day periodicity in Muon Detector data during late 2001 and early 2002, *Adv. Space Res.*, 49, 11, 1615-1622, 2012.

Watanabe, T. et al., Cosmic-Ray Data in 1953-2011, CAWSESDB-J-OB0062, Solar Terrestrial Environment Laboratory, Nagoya University, June 1, 2012.

Yashin, I.I., N. V. Ampilogov, I. I. Astapov et al., *Present status of muon diagnostics*, paper SH 580, ECRS 2012, Moscow, 4 p., 2012.

Chapter 7

THE JEM-EUSO SPACE MISSION: FRONTIER ASTROPARTICLE PHYSICS AT ZEV RANGE FROM SPACE

M. D. Rodríguez Frías[1], L. del Peral[1], G. Sáez Cano[1],
J. A. Morales de los Ríos[1,2,5], H. Prieto[1,2], J. H. Carretero[1],
K. Shinozaki[2], M. D. Sabau[3], T. Belenguer[3], C. González Alvarado[3],
S. Briz[4], A. J. de Castro[4], I. Fernández[4], F. Cortés[4], F. López[4],
J. Licandro[5], E. Joven[5], M. Reyes[5], O. Catalano[6], A. Anzalone[6],
M. Casolino[7], K. Tsuno[2] and S. Wada[2]

[1]Space & Astroparticle Group. UAH. Madrid, Spain
[2]RIKEN, 2-1 Hirosawa, Wako, Tokyo, Japan
[3]LINES laboratory, Instituto Nacional de Técnica Aeroespacial (INTA), Madrid, Spain
[4]LIR laboratory, University Carlos III of Madrid (UC3M), Spain
[5]Instituto de Astrofísica de Canarias (IAC), Tenerife, Spain
[6]INAF/IASF Istituto di Astrofisica Sapziale e Fisica Cosmica di Palermo, Italy
[7]University of Rome Tor Vergata, Rome, Italy

ABSTRACT

Cosmic Ray physics is one of the fundamental key issues and an essential tool of Astroparticle Physics that aims, in a unique way, to address many fundamental questions of the non-thermal Universe in the Astroparticle Physics domain. The huge physics potential of this field can be achieved by an upgrade of the performance of current ground-based instruments and new space-based missions, such as the JEM-EUSO space telescope. The JEM-EUSO space mission is the Extreme-Universe Space Observatory (EUSO) which will be located at the Exposure Facility of the Japanese Experiment Module (JEM/EF) on the International Space Station (ISS) and thus, this downward looking telescope will allow full-sky monitoring capability to watch for UHECR and Extremely High Energy Cosmic Rays (EHECR). An Atmospheric Monitoring System (AMS) is mandatory and a key element of a Space-based mission which aims to detect

Ultra-High Energy Cosmic Rays (UHECR). JEM-EUSO has a dedicated atmospheric monitoring system that plays a fundamental role in our understanding of the atmospheric conditions in the FoV of the main telescope. The JEM-EUSO AMS consists of an infrared camera and a LIDAR device that are being fully designed with space qualifications to fulfill the scientific requirements of this space mission.

INTRODUCTION

The Extreme Universe Space Observatory on the Japanese Experiment Module (JEM-EUSO) [1, 2, 3] of the International Space Station (ISS) is the first space-based mission worldwide in the field of Ultra High-Energy Cosmic Rays (UHECR) and will provide a real breakthrough toward the understanding of the Extreme Universe at the highest energies never detected from Space so far. JEM-EUSO will pioneer from Space the observation of the extensive air showers (EAS) produced by UHECR. The spectrum of scientific goals of the JEM-EUSO space mission includes as exploratory objectives the detection of high-energy gamma rays and cosmic rays, high-energy neutrinos, the study of galactic and extragalactic magnetic fields, and tests of relativity and the quantum gravity effects at these extreme energies. In parallel, along the life of the mission, JEM-EUSO will systematically survey atmospheric phenomena over the Earth surface. The JEM-EUSO collaboration is presently a joint effort of 13 countries and 270 researchers, where Europe has a consolidated position with 8 countries involved in this space mission led by Japan.

One of the major advantages of JEM-EUSO is its large aperture. It is expected that JEM-EUSO will reach, at the end of the decade, an exposure of 1 Mega Linsley (Figure 1). Since the structure of the GZK feature is highly dependent on distance, the combination of anisotropy and spectral information can help to pinpoint the type of astrophysical objects responsible for the origin of ultra high energy cosmic particles and even, possibly, to identify individual sources. The next figure shows the exposure expected by JEM-EUSO in both Nadir and Tilted mode, reaching 1 Mega Linsley by 2020.

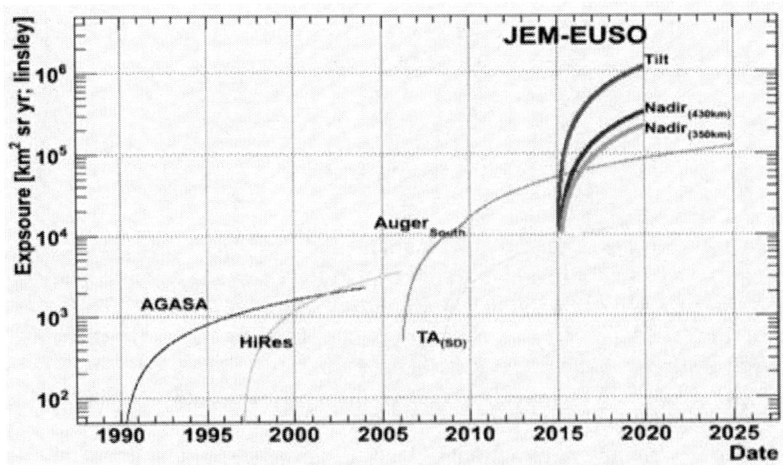

Figure 1. Exposure of current and proposed cosmic rays experiments. (Purple Book, internal JEM-EUSO publication).

IMPACT OF ATMOSPHERIC CLOUDS ON THE EXTENSIVE AIR SHOWER DETECTION

Since JEM-EUSO will use the atmosphere as a calorimetric detector, atmospheric conditions must be very accurately monitored. Some simulations using ESAF [4,5], the official JEM-EUSO software of the collaboration, has been done to study how different types of atmospheric clouds can affect the Extensive Air Showers (EAS) signal detected by the JEM-EUSO telescope.

Figure 2 shows a sample of sixteen different cloudy cases for the same EAS. All of them are compared with no cloud case (blue thin line). In case of a low and optically thick cloud (for example left, top part of the graph, with an optical depth of 5 and a cloud altitude of 2.5 km), most of the EAS will be developed above the cloud. When the EAS arrives to the cloud, a very high reflected Cherenkov peak will be produced on the top of the cloud. If this cloud is not detected by an atmospheric monitoring system, the arrival direction and main physical parameters of the EAS might be wrongly inferred. On the contrary, if the cloud is high but optically intermediate (for example, low, right plots of the graph, with an optical depth of 0.5 and a cloud altitude of 10 km), EAS signals will suffer some absorption. Therefore, if this cloud is not detected, this signal can be misunderstood with the signal of a lower energy cosmic ray EAS. For high and optically thick clouds (for example, top, right plots of the graph, with an optical depth of 5 and a cloud altitude of 10 km), the cloud will block the signal after the cloud. Again, an atmospheric monitoring system is crucial to discriminate these events.

The impact of atmospheric clouds depends on the arrival direction of the cosmic rays as well. Optically thick and lower clouds will more strongly affect vertical EAS, while more horizontal showers develop higher in the atmosphere. Therefore, they will be strongly affected only by very high and optically thick clouds.

THE INFRARED CAMERA OF JEM-EUSO

The IR Camera of JEM-EUSO [6,7,8] is an infrared imaging system aimed at detecting the presence of clouds and to obtain the cloud coverage and cloud top altitude during the observation period of the JEM-EUSO main instrument. Its full design, prototyping, space qualified construction, assembly, verification and integration is the responsibility of the Spanish Consortium involved in JEM-EUSO. Moreover, since measurements shall be performed at night, it shall be based on cloud IR emission. The observed radiation is basically related to the target temperature and emissivity and, in this particular case, it can be used to get an estimate of how high clouds are, since their temperatures decrease linearly with height at 6 K/km in the Troposphere.

The IR camera onboard JEM-EUSO will consist of refractive optics made of germanium and zinc selenide and an uncooled microbolometer array detector. Interferometer filters will limit the wavelength band to 10-12 µm. In the current configuration, two filters will be used centered at the wavelengths of 10.8 and 12 µm to increase the precision of the radiative temperature measurements. The IR camera is a very wide 60° FoV camera, totally matching the FoV of the main JEM-EUSO telescope. The angular resolution, which corresponds to one pixel, is about 0.1°. A temperature-controlled shutter in the camera and mirrors are used to calibrate background noise and gains of the detector to achieve an absolute temperature accuracy of 3 K. Though the IR camera takes images continuously at a video frame rate (equal to 1/30 s), the transfer of the images takes place every 30 s, in which the ISS moves half of the FoV of the JEM-EUSO telescope. Figure 3 shows the present configuration of the IR camera of JEM-EUSO.

Although there are no formal requirements for data retrieval, it has been assumed that the IR camera retrieval of the cloud top altitude could be performed on-ground by using stereo vision techniques or radiometric algorithms based on the radiance measured in one or several spectral channels (i.e. split-window techniques). Therefore, the IR camera preliminary design should be compliant with both types of data processing.

End to End simulation of the infrared camera will give us simulated cloud images of those we expect to obtain with the instrument. Therefore this simulation together with the data analysis will be included in the AMS detector simulation module of the JEM-EUSO analysis software. First the simulated radiation produced by the Earth's surface and atmosphere have been considered, taking into account the effect of the optics and the detector. In a later phase, the electronics and image compression algorithm will be implemented. A data analysis module is foreseen to take the data from the simulator, and real data from the IR-Camera to perform the analysis tasks with the algorithms for data retrieval. The output from this analysis module will be used as an input in the official codes for the event reconstruction of the main telescope.

The IR simulations, emitted by the ground and the atmosphere, are simulated using a modified version of the SDSU software developed in the Hydrospheric Atmospheric Research Center, Nagoya University [9], to simulate the wavelength of our detector, and thanks to the capabilities of this code, we are simulating the UV (Ultra-violet) range of our main instrument (250-500 μm) to get an approach for the slow data of JEM-EUSO. Moreover, the optics simulation has been performed using a MTF (Module transfer Function) and the diffraction simulation and the distortion of the FOV have been taken into account as well. The detector simulation has been implemented based on the data taken from the test of an INO IRXCAM-640 with the same ULIS microbolometer being planned to implement in the IR-Camera. At the end, the output from this end-to-end simulation of the infrared camera will be cloud images similar to what we expect from our infrared camera, which would allow us to test the data retrieval algorithms and calculate correction factors for the IR-Camera. A preliminary example of the IR simulations, done by the SDSU software, is shown in figure 4.

Retrieval of Cloud Temperature from the Brightness Temperature of the Infrared Camera

The main objective of the IR camera is to measure the cloud temperature from the radiance received by the detector. However, the radiance that impinges the detector is not the radiance emitted by the cloud because the atmosphere absorbs part of this energy. Therefore the temperature retrieved from the radiance measured by the detector (brightness temperature, T_b) is not exactly the cloud temperature.

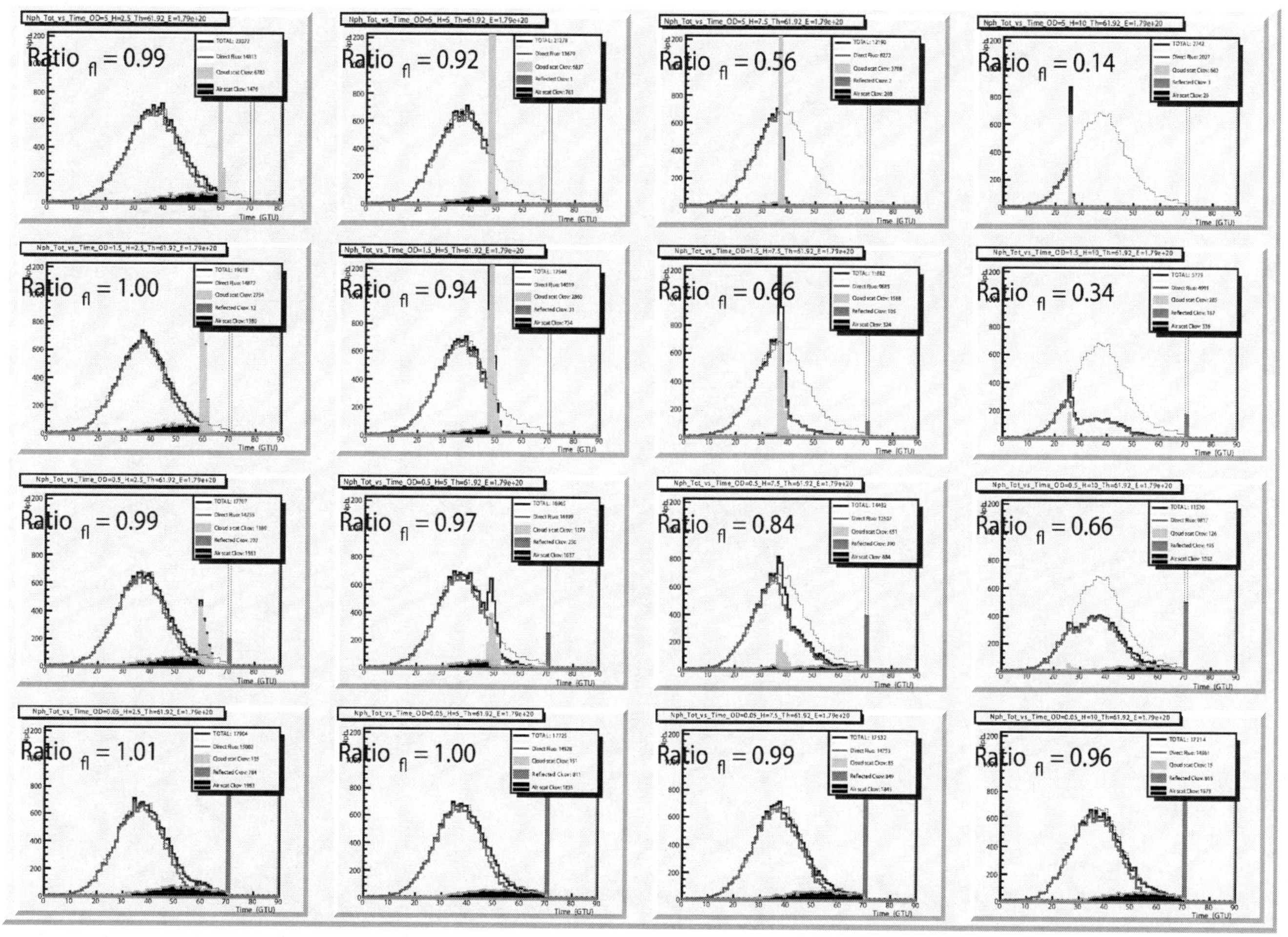

Figure 2. Detected signal by JEM-EUSO of a standard EAS with 1.8 EeV and 62°, under sixteen different cloudy cases. The sixteen cloudy cases consist of four different altitudes (H=2.5km, 5km, 7.5 km, 10 km) from left to right, and four different optical depths (OD= 0.05, 0.5, 1.5, 5 from bottom to top). The comparison of every case with clear sky (thin blue line) is presented as well.

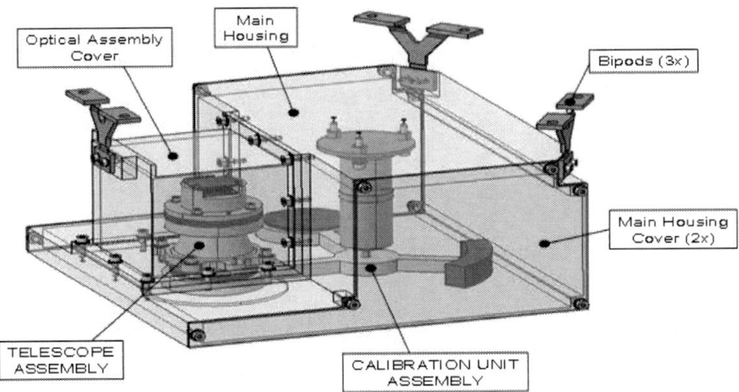

Figure 3. Baseline of the IR camera of the JEM-EUSO space mission.

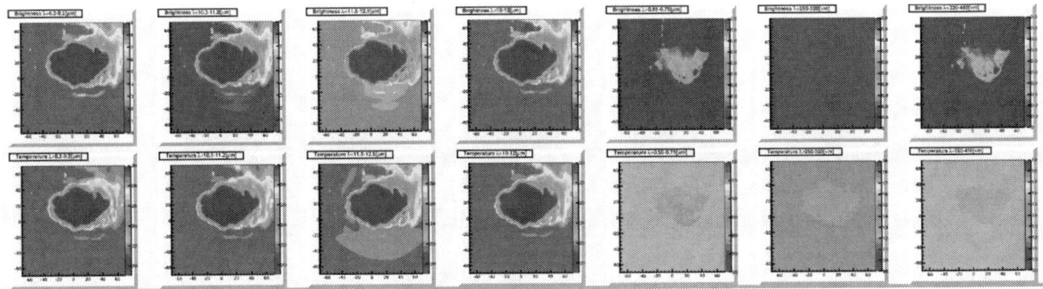

Figure 4. Image of an scenario simulated with the SDSU code, where the radiated brightness (W/(m² str μm)) and brightness temperature (K) in several bands (IR, visible and UV).

In order to retrieve cloud top temperature (T_c) from brightness temperature (T_b), provided by the IR camera, some algorithms to correct the atmospheric effect must be applied to the T_b data. Since the atmospheric attenuation is not the same along the spectrum, different spectral bands will be affected by the atmosphere in a different way. Therefore the radiance measured by the IR camera in two different spectral bands will be different. This difference in brightness temperature can be used to correct the atmospheric effects and this is the basis of the Split Window algorithm developed by McMillin and Crosby to obtain the surface temperature from the radiance measured in two channels in thermal infrared [10].

Since then, many authors have developed algorithms to retrieve the surface temperature from satellite measurements [11]. Nevertheless, all these algorithms are based on linearization of Planck's law and on the Radiative Transfer Equation (RTE) that consist of linear or quadratic functions of the temperatures retrieved in two bands and coefficients depending on the transmittance of the atmosphere. However, in real cases, the transmittance is not always known and for this reason different algorithms have been developed by some authors. The difference between algorithms lies in these coefficients and different authors propose different coefficients to retrieve the surface temperatures in different conditions.

The same methodology can be applied to measure clouds temperatures, especially to low-level and thick clouds. In fact, there are also some Split Window (SW) algorithms devoted to retrieve clouds temperatures from satellites such as AVHRR, MODIS, etc. [12, 13]. These algorithms are able to retrieve top-cloud temperature [14], cloud emissivity and type cloud [15] and cloud microphysics [16].

In this work this procedure is applied to retrieve the cloud top temperature from brightness temperatures measured in bands B1 (10.3 μm -11.3 μm) and B2 (11.5 μm -12.5 μm). In order to calculate the atmospheric transmittances required to compute the algorithm coefficients, standard atmospheric profiles have been used. Only profiles above 3 km are considered in the transmittance calculi since clouds below that level are not of interest for the JEM-EUSO project. In the following equation the proposed algorithm is shown.

$$T_C = -1.12515 + 1.639429(T_1) - 0.634044 6(T_2)$$

In order to validate this algorithm, the spectral radiance of clouds at different heights has been simulated in different atmospheric conditions. From these spectra, brightness temperatures in bands B1 and B2 have been retrieved and the algorithm has been applied to them to calculate the cloud temperature. The difference between the Tc retrieved by the algorithm and the cloud temperature used in the simulation is represented in Figure 5 (red squares) as a function of the cloud height. Different square values at the same height are related to the same cloud in different atmospheric conditions. Also the brightness temperature retrieved in a two micron band centered in 11 μm (Tb$_{TIR}$) is also represented in Figure 5 (blue squares). As can be seen in this figure, the algorithm proposed is able to significantly reduce the error in the cloud temperature retrieval. Moreover, the residual error is not dependent on the cloud height.

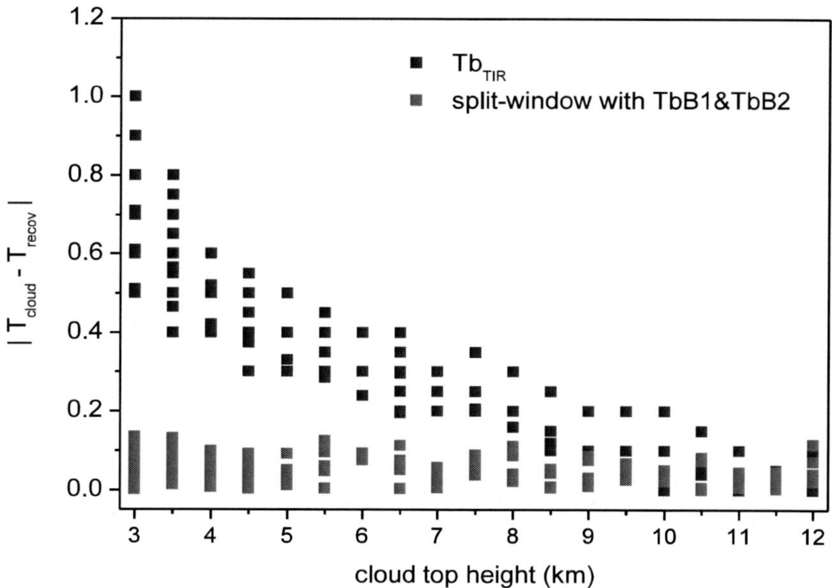

Figure 5. Retrieval errors after applying SW algorithms to the brightness temperatures of B1 and B2 bands (red squares) and errors associated to brightness temperature in TBtir band without applying SW algorithms (blue squares).

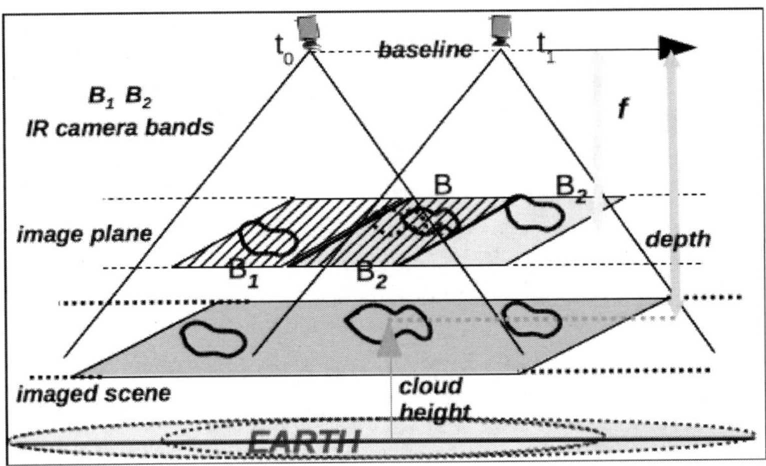

Figure 6. Schematic drawing of the 'Stereo System' and the bi-band acquisition of the stereo pair of a cloudy scene.

Retrieval of Top Cloud Height from Stereo Vision with the Infrared Camera

The Cloud Top Height (CTH) estimation in the JEM-EUSO atmospheric monitoring system will be approached by different methods. The one proposed here is based on the stereo vision technique [17] that mainly exploits the parallax effect of a binocular vision. The same effect can be obtained even if the observed object is static and the observer moves towards a selected direction. In our case the ISS displacement provides the movement of the IR camera along track, while the underlying scene is imaged with a suitable time interval between consecutive acquisitions, allowing overlap in their field of views. In this way a scene results, imaged by two different view angles, and the common part of the two corresponding images are analyzed by the stereo technique to retrieve the CTH. According to the IR camera design, bi-band stereo views are acquired first by observing a scene in one band, then acquiring the intersection part of the next observed scene, in the other band. In Figure 6 a schematic drawing of the system is shown.

The cloud height reconstruction strongly depends on the disparity evaluation i.e. on the apparent displacement of the pixels in the two images, due to the parallax effect (Figure 7). Corresponding points are searched in the stereo pair by an image matching process that could be affected by the differences in Brightness Temperature (BT) values present in the bi-band stereo case, because of the differences in emissivity of the scene elements at different wavelengths. Therefore, the methods normally used for mono-band are not straightforward applicable to a multi-band stereo. In Figure 8 an example of disparity map is shown, obtained using a method mainly based on [18, 19, 20] and applied to a MSG8, MSG9 stereo pair from the Meteosat Second Generation geostationary satellites. The darkest disparities correspond to sea and ground cloud free areas, whereas the brightest ones correspond to the cloudy areas, the higher the cloud, the darker the gray level.

Different methods need to be compared and optimized to meet the mission requirements. Studies are in progress on how the set of different approaches for CTH estimation can be combined to complement each other.

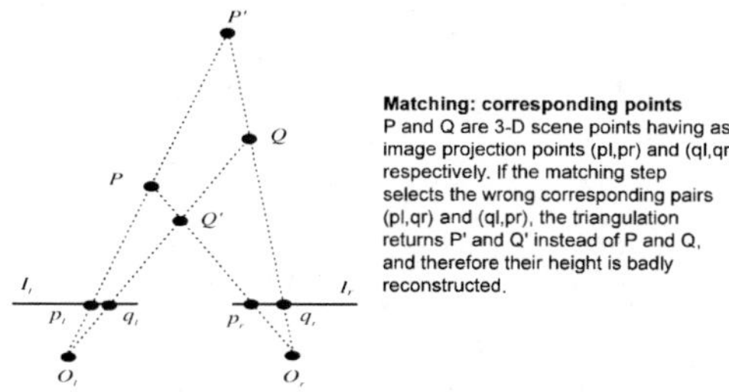

Figure 7. The matching crucial role in the process of reconstruction is shown.

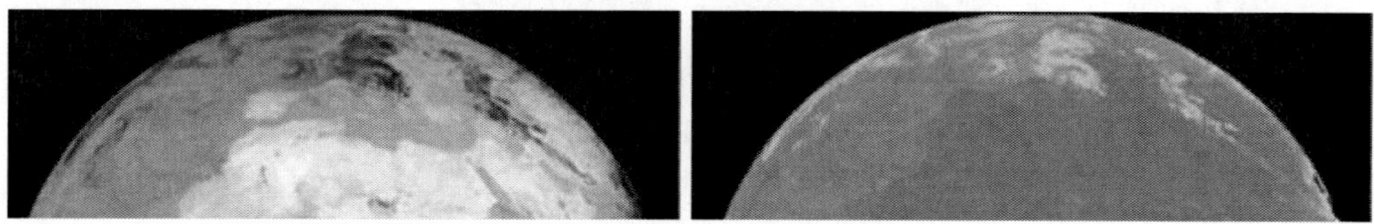

Figure 8. Left: one of the two MSG images of a stereo pair. Brightest (darkest) BT values, represent non-cloudy (cloudy) areas. Right: retrieved disparity map where the brightest (darkest) gray level values, represent the biggest (smallest) disparity values.

JEM-EUSO Computing Model

For High Energy Physics (HEP) experiments the huge amount of data and complex analysis algorithms require the use of advanced GRID computational resources. Therefore a complete infrastructure is needed for the simulation and data analysis, in the frame of the GRID architecture, and computing infrastructure, intended for the JEM-EUSO space mission software. Moreover solutions to account for a complete installed cluster system, with the software and data repositories as well as a many other details, should be studied and pointed out.

The JEM-EUSO mission needs to evaluate the complete physical process of the detection of an ultra-high cosmic ray reaching the Earth. In order to take into account all the processes involved in the detection of the EAS, several kinds of software are involved and have to be carefully evaluated. The injection of the primary cosmic ray into the Earth's atmosphere and the generation of the EAS; simulation of the detector response; the reconstruction of the physical parameters of the primary cosmic ray from the JEM-EUSO detector data; and finally backtracking the particles from the Earth surface through space to find out the arrival direction of the primary particle.

Once we have taken into account all the software listed previously, we proceeded to make a fully computational requirements study, first defining a set of Use-Cases Scenarios and then evaluating the needs for each set of tasks, like; Monte-Carlo Simulation, RAW data and Analysis, Calibration tasks and resources for Unscheduled tasks made by the scientists working on the collaboration. In Figure 9, we estimated the needs for CPU requirements [21, 22, 23].

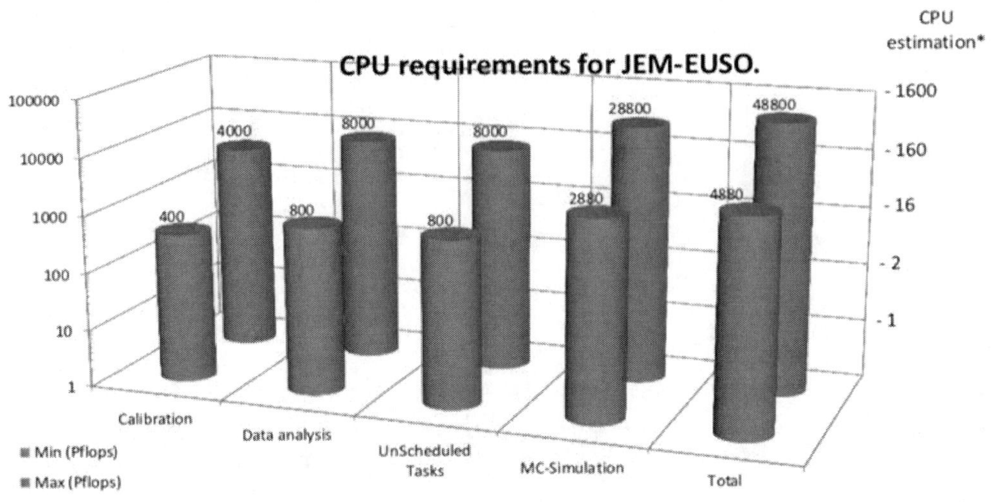

Figure 9. Estimation of the CPU requirements for the JEM-EUSO Space Mission.

REFERENCES

[1]   The JEM-EUSO Collaboration website, http://jemeuso.riken.jp/en/ index.html

[2]   The JEM-EUSO Collaboration, *Phys. B (Proc. Suppl.)*, 237, 175-176 2008.

[3]   The JEM-EUSO Collaboration. The JEM-EUSO Mission, *New J. Phy.*, 2009, 11, 065009.

[4]   Sáez-Cano, G; et al. *"Observation of Ultra High Energy Cosmic Rays in cloudy conditions by the JEM-EUSO Space Observatory"*. Proc 32st International Cosmic Ray Conference, 2011.

[5]   Sáez-Cano, G; et al. *Journal of Physics: Conference Series*, ISBN: 1742-6588, In press, 2012.

[6]   Morales de los Ríos, JA; et al. *"The IR-Camera of the JEM-EUSO (JAXA) Space Observatory"*. Proc 32st International Cosmic Ray Conference, Beijing., 2011.

[7]   Rodríguez MD. frías for the JEM-EUSO Collaboration. *"The JEM-EUSO Space Mission @ forefront of the highest energies never observed from Space"*. Proc. Frascatti Workshop, In Press, 2012.

[8]   Rodríguez Frías MD. for the JEM-EUSO Collaboration. "The Atmospheric Monitoring System" of the JEM-EUSO Space Mission. *Ultra-High Energy Cosmic Ray*, 2012 conference, CERN.

[9]   Masunaga, H; et al. Satellite Data Simulator Unit, A multisensor, multispectral Satellite Simulator Package, *American Meteorological Society*, 2010.

[10] McMillin, L; Crosby, D. Theory and validation of the multiple window sea surface temperatures technique. *J. Geophys. Res.*, 1984, 89, 3655-3661.

[11] King, MD; Tsay, SC; Platnick, SE; Wang, M; Liou, KN; 'Cloud Retrieval Algorithms for MODIS: Optical Thickness, Effective Particle Radius, and Thermodynamic Phase', MODIS Algorithm Theoretical Basis Document No ATBD-MOD-05, 1997.

[12] Hayasaka, 'Recent Studies on Satellite Remote Sensing of Clouds in *Japan' Adv. Space Res.*, 1996, Vol. 18, No. 7, pp (7) 29-pp 36.

[13] Heidiger, AK; Pavolonis, MJ. 'Gazing at Cirrus Clouds for 25 Years through a Split Window. Part I: Methodology'. *Journal of Applied Meterology and Climatology*, Vol. 48, 1100-1116.

[14] Nieman, SJ; Schmetz, J; Menzel, WP. 'A comparison of several techniques to assign heights to cloud tracers', *J. Appl. Meteor.*, 1993, 32, 1559–1568.

[15] Inoue, T. 'A cloud type classification with NOAA 7 split-window measurements? *J. Geophys. Res.*, 1987, 92, 3991–4000.

[16] Inoue, T. 'On the temperature and effective emissivity determination of semi-transparent cirrus clouds by bispectral measurements in the 10 micron window region'. *J. Meteor. Soc. Japan*, 1985, 63, 88–99.

[17] Faugeras, O. Three-Dimensional computer vision: a geometric viewpoint. MIT Press, 1993.

[18] Manizade, KF; Spinhirne, JD. Stereo Cloud heights from multispectral IR imagery via Region-of-Interest Segmentation. *IEEE Transactions on Geoscience and Remote Sensing*, 2006, 44(9), 2481–2491.

[19] Anzalone, A; et al., Combining Fuzzy C-Mean and Normalized Convolution for Cloud detection in IR images, WILF'09, June (2009)

[20] Anzalone, et al.: A study on recovering the cloud top-height from infra-red video sequences. In MDIC04 Proc, 2004.

[21] Morales de los Rios, JA; et al, The SPAS-UAH Cluster for the JEM-EUSO Computing Model, ISPA 2012, IEEE CPS, Madrid, Spain, In press. 2012.

[22] Morales de los Rios, JA; et al., "A computing model for the Science Data Center of high energy Physics space-based experiments". Contribution to the "*I Encuentro de Investigadores CI3*. 2011.

[23] Morales de los Rios, JA; et al. "Proposal of a Computing Model and GRID resources for the JEM-EUSO space mission". Communications in Computer and Information Science. *SPRINGER-VERLAG GMBH*. (1865-0929), In Press, 2012.

In: Homage to the Discover of Cosmic Rays ...
Editor: Jorge A. Perez-Peraza

ISBN: 978-1-63483-084-3
© 2015 Nova Science Publishers, Inc.

Chapter 8

PARTICLES IN THE HELIOSPHERE

Luis Del Peral[1,*]***, M.D. Rodríguez-Frías***[1] ***and Raúl Gómez-Herrero***[2]
[1]Space & Astroparticle Physics Group. UAH. Madrid. Spain
[2]Institut fr Experimentelle und Angewandte Physik,
Christian-Albrechts-Universität Kiel, Kiel Germany

Abstract

The Heliosphere is the region of the space influenced by the Sun and its expanding Corona, the Solar Wind. The Heliosphere ends at the termination bow shock or bow wave [1] predicted at arround 150 AU from the Sun. Charged Energetic particles below the rigidity cutoff from outside the Heliosphere are prevented from entering by the solar magnetic field. This assures us that energetic particles below the heliospheric cutoff are of heliospheric origin. Particle sources in the Heliosphere are different and complex. There are particles energized directly at the Sun but also in the interplanetary medium and at planetary magnetospheres.

Keywords: Heliosphere, solar energetic particle, SOHO, interplanetary medium

1. Sources of Heliospheric Energetic Particles

Energetic particles observed in the Heliosphere have different origins:

- Galactic and Extragalactic Cosmic Rays (GCR) from outside the solar system. GCR of energies above hundred of MeV can penetrate the solar magnetic field shielding and entering into the Heliosphere. In the energy range from hundred of MeV to few GeV their intensity exhibits an inverse correlation with the solar activity cycle of 11 years of periodicity. At higher energies GCR are not affected by heliospheric modulation in a significant way. In this review we will focus in Energetic Particles from heliospheric origin.

- Solar Energetic Particles (SEP) generated by the Sun. The solar magnetic activity explosively releases huge amounts of energy in the solar atmosphere that produces

---

*E-mail address: luis.delperal@gmail.com

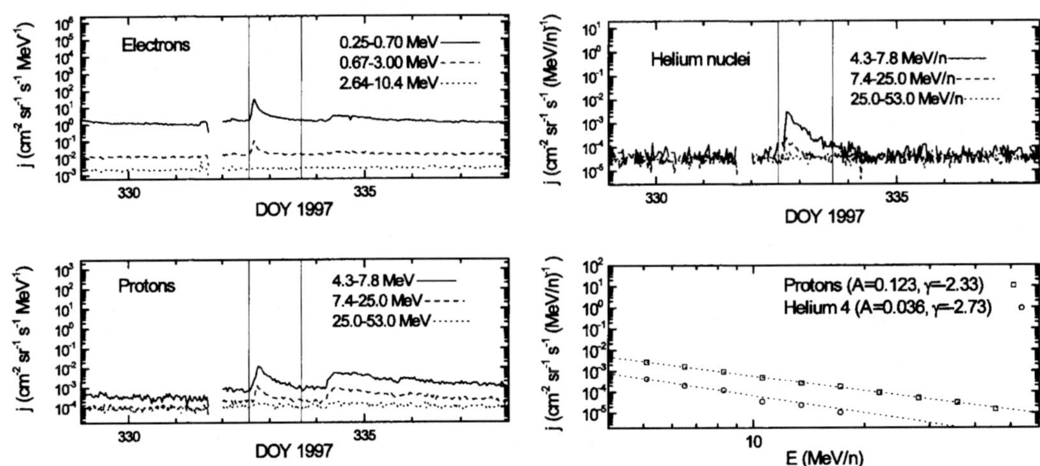

Figure 1. Abundance ratios, differential fluxes and spectra of November 26, 1997 ^{3}He-rich ISEP detected by EPHIN/SOHO. Lower intensity event with no appreciable acceleration of p and ^{4}He beyong 25 MeV. Possibly associated with a 2B/X2.6 flare at 331:12:59 UT, N17E63 (NOAA 8113).

SEP called Solar Cosmic Rays. SEP with energies ranging between few keV to several GeV mainly protons and electrons are ejected to the interplanetary medium traveling guided by the solar magnetic field lines interacting with the solar wind, planetary magnetospheres, and shock waves that can reaccelerate them.

- Energetic Particles accelerate in the Interplanetary Medium by different processes as shock waves and ion pickup. This population of energetic particles are of interplanetary origin. They are accelerated in the Interplanetary Medium by transient events such as shock waves, Corrotating Interaction Regions (CIRs), stochastic acceleration, plasma waves and turbulences, and ion pickup.

- Particles accelerated in planetary magnetospheres. Planetary magnetospheres, mainly Jovian and Earth magnetospheres become a source of energetic particles, mainly electrons and protons accelerated in magnetic reconnection regions. During solar quiet time periods they may be observed in the Earth vicinities.

- Interestelar pickup ions and anomalous cosmic rays. Finally, neutral atoms from the Very Local Interestellar Medium (VLISM), penetrate the Heliosphere where may be ionized by photoionization, a charge exchange or electron impact so called Anomalous Cosmic Rays that can be detected during quiet time periods with very low energy (1-30 MeV/nucleon) and low ionization levels.

2. Solar Energetic Particles

The Sun is continuously emitting a flow of ionized plasma, known as Solar Wind (SW), that pervades interplanetary space guided by the solar magnetic field and forced by the dif-

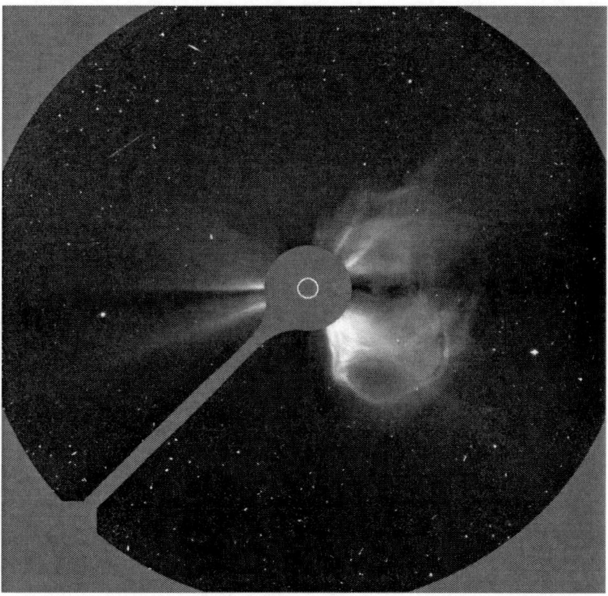

Figure 2. LASCO C3 image of a large coronal mass ejection of 20 April 1998. The dark disk blocks the Sun so that the LASCO instrument can observe the structures of the corona in visible light. The white circle represents the size and position of the Sun. Courtesy of SOHO/LASCO consortium. SOHO is a project of international cooperation between ESA and NASA.

ference in gas pressure between the solar corona and interstellar space. The solar wind is strongly influenced by the solar activity. Solar Wind becomes a complex plasma flowing in an interplanetary spiral with fast and slow flows that generate compression zones between fast and slow winds and rarefaction between slow and fast. These structures co-rotate with the Sun and generate shock-bounded compressed shells beyond 3 or 4 AU known as CIRs. Solar Flares generate energetic particles (100 keV-100 MeV) that liberate into the interplanetary medium in an impulsive way. They are observed at the Earth as ISEP (Impulsive Solar Energetic Particle) events with the following observational features [2] [3]:

- They are low intensity events with non-relativistic energy particles.

- Length lower than a day.

- Associated to Impulsive Solar Flares with absence of Coronal Mass Ejections (CME) or Interplanetary Shock Waves (ISW).

- Sometimes chained ISEP has been observed after a CME occurrence.

- Charge states of ISEP corresponding to those of the plasma temperature of the flaring region ($T \geq 5 \cdot 10^6$ K)

- Their frequency is as high as 1000 by year in solar maximum periods but only detectable with good magnetic connection.

- Particles are ejected in a 30° cone from the SF.

- Present high e^-/p ratio, high content of ^{3}He and heavy elements (^{3}He rich ISEP events) and high velocity dispersion.

- Energy spectra in good agreement with stochastic acceleration with spectral index between 1.8 and 3.4.

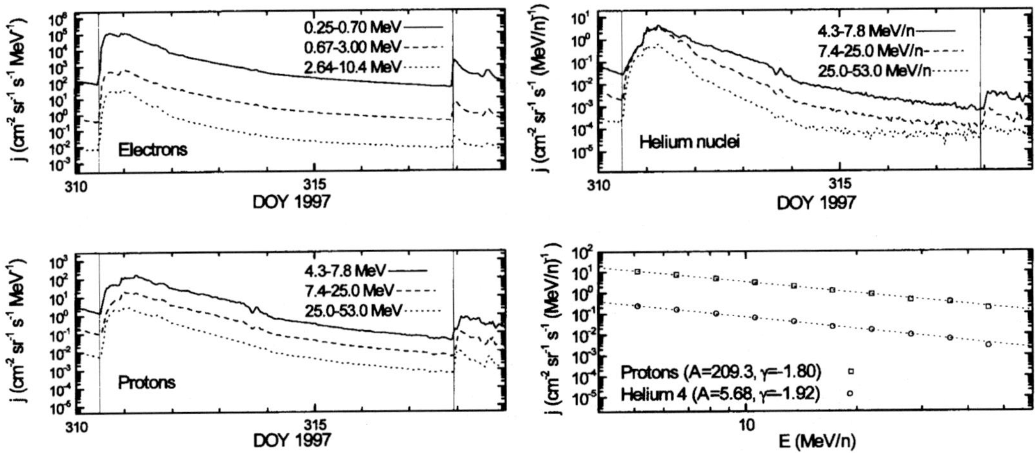

Figure 3. Abundance ratios, differential fluxes and spectra of November 6, 1997 GSEP detected by EPHIN/SOHO. Very energetic GSEP event, with strong acceleration of p and He beyond 50 MeV/nucleon and electron acceleration beyond 10 MeV. It is associated with a 2B/X9.4 flares at 310:11:49 UT , S18W63 (NOAA 8100) followed by a CME ejected at 310:12:10 UT. A magnetic cloud ejected Nov 4 passed by spacecraft between days 311.2-312.5 driven by a strong shock detected at 310.9 followed by other shocks at 313.4 and 313.9. [2].

Figure 1 shows a typical ^{3}He-rich ISEP event of Nov 26th, 1997 detected by EPHIN/SOHO. The ISEP was of short duration. There was not appreciable acceleration of p and He beyond 25 MeV/n. There was not a clear candidate for this event, possibly it was associated with a 2B/X2.6 flare at 331:12:59 UT N17E63 (NOAA 8113). Other Solar Energetic particle sources are CME. A CME (Figure 2) is a massive burst of coronal material ejected into the interplanetary space. CMEs are originated in active regions with closed magnetic field lines able to contain the plasma. Recombination or disruption of magnetic field lines causes the ejection of the coronal material ($1.6 \cdot 10^{12}$ kg in average) at velocities ranging from 20 to 3200 km/s. As results, a magnetic cloud of coronal plasma is liberated to the Interplanetary Space. The CME passage through the Earth, generate geomagnetic effects such as auroras, increase of the ionization in the Ionosphere.

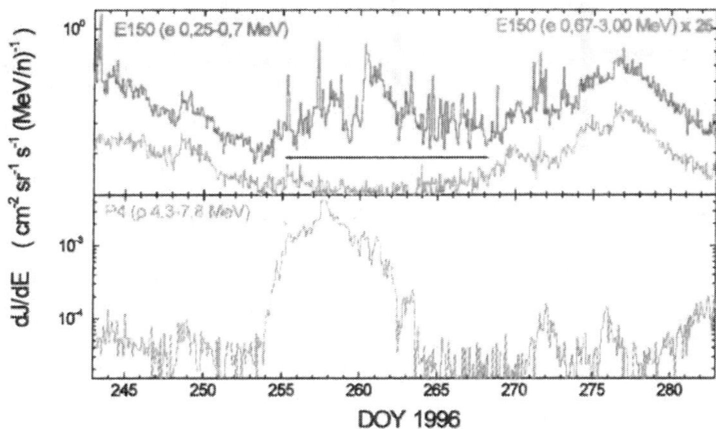

Figure 4. Electron and proton flux during CIR passage detected by EPHIN on board SOHO satellite. Increases observed in the electron flux over the jovian electron background. Proton flux denotes clearly a CIR event.

3. Interplanetary Energetic Particles

Figure 3 shows a typical GSEP. When a CME is enough fast (> 1000 km/s), a shock wave is generated driving the CME passage through the Heliosphere . This shock wave accelerate particle generating a Gradual Solar Energetic Particle Event (GSEP) with following features [2] [3] [4]:

- Are intense and energetic events with relativistic particles (up to several GeV).

- Are long duration events, usually several days.

- Usually associated to SF and Shock Waves driven fast CMEs. With detection of radio-bursts type II and IV.

- Have low frequency (around 10 by year in solar maximum activity). But they are easier to detect because of their high fluxes.

- e-/p ratio is low in contrast with ISEP. Most of the energy are carried by protons.

- 3He/4He ratio and heavy element take values from the Corona and SW.

- Particles are generated in a $180°$ cone.

- Charge states in agreement with the source region in the corona ($T = 1 - 2 \cdot 10^6$ K)

- Velocity dispersion is appreciable only before de shock passage.

- Energy spectra fit a stable power law in good agreement with shock acceleration.

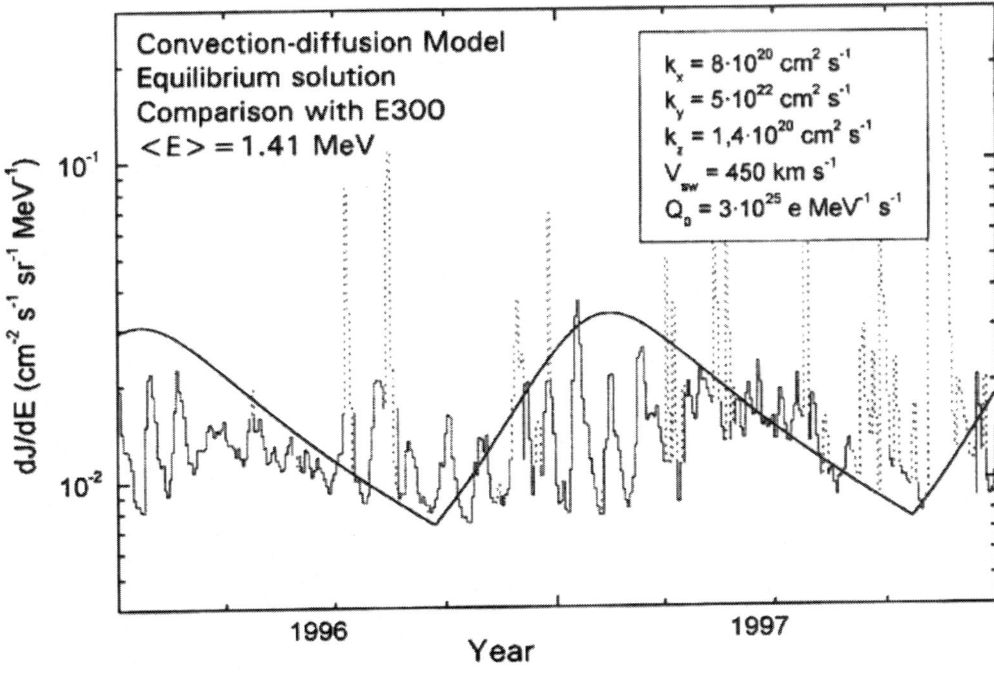

Figure 5. Differential electron flux values are plotted in comparison with propagation model predictions (solid line) with the propagation parameters given in the figure. Dotted lines correspond to SEP events [6].

4.   Energetic Particles from Planetary Magnetospheres

CIRs usually accelerate particles in the Interplanetary Medium. But they become observable only during quiet time because of its low intensity. Among their features are soft spectral index between 5.3 and 3.4 denoting a soft Interplanetary Shock as responsible of the acceleration. The temporal behavior of the spectral index (Figure 4) is a progressive softening of the spectrum during the ascending phase (attributable to the forward shock), reaching minimum values -6 two days before the central maximum flow, hardening slowly after reaching the final phase. This late hardening seems to occurs because the observer is connected to regions increasingly distant of the reverse shock, where the acceleration is more intense.

The SW interaction with planetary magnetospheres generates bow shock and magnetic reconnection in magnetospheric tails where intense electric fields are created and accelerate particles, mainly electron but also protons reaching the Earth and become detectable during quiet time of solar activity. During solar quiet time periods the dominant electron population in the energy range from 0.2 to 25 MeV is of jovian origin [6] [7] [8]. The jovian electron abundance is 50 times higher than galactic electrons at these energies. Jovian electrons follow the interplanetary magnetic field lines, therefore, their intensities, observed near Earth, fluctuate depending on magnetic connection between Earth and Jupiter, with a

13 months periodicity (synodic period of Jupiter). While proton spectra are dominated by solar cycle variations, electron intensities do not show a significant variation under solar modulation. Differences in the intensity variation of both particle species are due to different heliospheric regions crossed. Protons come from interstellar space and go across the complete heliosphere to reach 1 AU. Meanwhile electrons are generated in the inner solar system, mainly in the jovian magnetosphere. CIRs inhibit perpendicular diffusion of electrons. Therefore, when a CIR is interposed between Jupiter and Earth the propagation of jovian electrons is inhibited. The jovian electron flux observed at Earth should remain minimum since the CIR evolves from Earth to Jupiter. Thereafter, the jovian electron flux at Earth should begin to increase. This behavior is found with a periodicity of approximately 27 days during 5 or 6 months that recurred after each 13 month period. Figure 5 show electron flux detected by SOHO during 1996 solar minimum.

References

[1] Mc Comas et al., *Science* 1221054 (2012).

[2] Gómez-Herrero, R. Rodríguez-Frías, M. D. y Del Peral, L., *Análes de Física*, 95, 190-196 (2000)

[3] Gómez-Herrero, R., Del Peral, L., Rodrguez-Fras, M. D. *Advances in Space Research* 35, 617-624 (2005)

[4] Rodríguez-Frías, M.D., Gómez-Herrero, R. Del Peral, L., *Lectures Notes and Essays in Astrophysics* I, (2004)

[5] GLEs Gómez-Herrero, R, Rodríguez-Frías, M.D., Del Peral, L. *Astroparticle Physics* 17, 1-12 (2002)

[6] Del Peral, L. Gómez-Herrero, R., Rodríguez-Frías, M.D., *Astroparticle Physics* 20, 235-245 (2003).

[7] Gómez-Herrero, R., Rodríguez-Frías, M.D., Del Peral, L. *Solar Physics* 194, 405-414, (2000)

[8] Gómez-Herrero et al., *Lectures Notes and Essays in Astrophysics* III, 213-222 (2008)

In: Homage to the Discover of Cosmic Rays ...
Editor: Jorge A. Perez-Peraza

ISBN: 978-1-63483-084-3
© 2015 Nova Science Publishers, Inc.

Chapter 9

ON THE OBSERVATION OF THE COSMIC RAY ANISOTROPY BELOW 10^{15} EV

G. Di Sciascio[1] *and R. Iuppa*[1,2]
[1]INFN, sez. di Roma Tor Vergata
[2]Dipartimento di Fisica dell'Universitá degli studi di Roma Tor Vergata

Abstract

The measurement of the anisotropy in the arrival direction of cosmic rays is complementary to the study of their energy spectrum and chemical composition to understand their origin and propagation. It is also a tool to probe the structure of the magnetic fields through which cosmic rays travel.

As cosmic rays are mostly charged nuclei, their trajectories are deflected by the action of galactic magnetic field they propagate through before reaching the Earth atmosphere, so that their detection carries directional information only up to distances as large as their gyro-radius. If cosmic rays below 10^{15} eV are considered and the local galactic magnetic field ($\sim 3\,\mu G$) is accounted for, gyro-radii are so short that isotropy is expected. At most, a weak di-polar distribution may exist, reflecting the contribution of the closest CR sources.

However, a number of experiments observed an energy-dependent *"large scale"* anisotropy in the sidereal time frame with an amplitude of about 10^{-4} - 10^{-3}, revealing the existence of two distinct broad regions: an excess distributed around 40° to 90° in Right Ascension (commonly referred to as "tail.in" excess) and a deficit (the "loss cone") around 150° to 240° in Right Ascension. In recent years the Milagro and ARGO-YBJ collaborations reported the of a "medium" scale anisotropy inside the tail-in region. The observation of such small features has been recently claimed by the Icecube experiment also in the Southern hemisphere.

So far, no theory of cosmic rays in the Galaxy exists which is able to explain the origin of these different anisotropies leaving the standard model of cosmic rays and that of the galactic magnetic field unchanged at the same time.

Although observations of the CR anisotropy were reported up to 10^{20} eV, this work is focused on the energy range below 10^{15} eV, where the counting rates of the detectors are high enough and the evidence for the anisotropy well established.

1. Introduction

Cosmic rays (CRs) are the most outstanding example of accelerated particles. Understanding their origin and propagation through the interstellar medium is a fundamental problem which has a major impact on models of the structure and nature of the Universe.

The observed primary CR energy spectrum exceeds 10^{20} eV showing a few basic characteristics: (a) a power-law behaviour $\sim E^{-2.7}$ until the so-called "knee", a small downwards bend around few PeV; (b) a power-law behaviour $\sim E^{-3.1}$ beyond the knee, with a slight dip near 10^{17} eV, sometimes referred to as the "second knee"; (c) a transition back to a power-law $\sim E^{-2.7}$ (the "ankle") around 10^{18} eV; (d) a cutoff due to extra-galactic CR interactions with the Cosmic Microwave Background (CMB) around 10^{20} eV (the Greisen-Zatsepin-Kuzmin effect). All these features are believed to carry fundamental information to shed light on the key question of the CR origin.

CRs below 10^{17} eV are expected to be mainly galactic, produced and accelerated in SuperNova (SN) blast waves, which can provide the necessary energy input and naturally produce particles with a power law energy spectrum and spectral index close to the value inferred from the observations. The SN acceleration mechanisms are limited by the rate at which the particles can gain energy and by the lifetime of the shock. As an example, for a typical SN, Lagage and Cesarsky [1] calculated that this limit corresponds, for a particle of charge Z (in natural units), to a total energy of $\sim Z \times 10^{14}$ eV.

As CRs are mostly charged nuclei, their paths are deflected and highly isotropized by the action of galactic magnetic field (GMF) they propagate through before reaching the Earth atmosphere. The GMF is the superposition of regular field lines and chaotic contributions. Although the strength of the non-regular component is still under debate, the local total intensity is supposed to be $B = 2 \div 4$ μG [2]. In such a field, the gyro-radius of CRs is given by $r_{a.u.} \approx 100_{TV}$, where $r_{a.u.}$ is in astronomic units and R_{TV} is the rigidity in TeraVolt. Clearly, there is very little chance of observing a point-like signal from any radiation source below 10^{17}eV, as they are known to be at least several hundreds parsecs away.

If it is true that magnetic fields are the most important "isotropizing" factor when they randomly vary on short distances, it is clear as much that some particular features of the magnetic field at the boundary of the solar system or farther might focus CRs along certain lines and the observed arrival direction distribution turns out to be consequently anisotropic.

As it will be discussed in this paper, different experiments observed an energy-dependent *"large scale"* anisotropy in the sidereal time frame with amplitude spanning 10^{-4} to 10^{-3}, from tens GeV to hundreds TeV, suggesting the existence of two distinct broad regions, one showing an excess of CRs (named *"tail-in"*), distributed around $40°$ to $90°$ in Right Ascension (R.A.). The other a deficit (the so-called *"loss cone"*), distributed around $150°$ to $240°$ in R.A..

The origin of the galactic CR anisotropy is still unknown, but the study of its evolution over the energy spectrum has an important valence to understand the propagation mechanisms and the structure of the magnetic fields through which CRs have traveled.

The propagation in the Galaxy of CRs trapped by the magnetic field is usually described in terms of diffusion at least up to 10^{16-17} eV [3]. The gyro-radius of particles is smaller than the largest scale (100 - 300 pc) on which the turbulent component of the GMF is believed to vary. Therefore, the CR propagation resembles a random walk. The diffusion produces density gradients and thus an anisotropy as intense as δ takes place: $\delta = \frac{3D}{v}\frac{\nabla N}{N}$, where v is the particle velocity, N the particle density and D the diffusion coefficient expected to increase with the rigidity R. The models proposed to describe the CR propagation adopt different dependence of D with energy: $D \propto R^{0.6}$ cm^2 s^{-1} in the "leaky box" approximation with a regular magnetic field [4]; $D \propto R^{0.3}$ cm^2 s^{-1}, for models with re-acceleration and an additional random turbulent component of the GMF; $D \propto R^{0.15-0.20}$ cm^2 s^{-1} for models with a Hall diffusion [5]. A measurement of the evolution of anisotropy with the energy is therefore important to disentangle between different models of CR propagation in the Galaxy.

Moreover, the study of the anisotropy can clarify the key problem of galactic CRs, i.e. the origin of the knee. Indeed, if the knee is due to an increasing inefficiency in CR containment in the Galaxy a change in δ is expected. If, on the contrary, the knee is related to the inefficiency of the acceleration mechanism, we do not expect a change of anisotropy when nearing the knee.

Around 10^{18} eV the gyro-radius of CRs becomes comparable to the thickness of the galactic disk, therefore their confinement in the Galaxy is less effective and the diffusive motion loses importance. As a consequence, if the CR sources are distributed mainly in the galactic plane, an increase of the anisotropy toward the galactic disk is expected, due to the proton component.

Above 10^{18} eV, the CRs are believed to be extra-galactic. With increasing energy the *"charged-particle astronomy"* becomes possible in principle, because the magnetic field should not affect significantly the CR propagation. Nevertheless, the uncertainty about the extra-galactic magnetic field structures and about the CR chemical composition at that energy makes challenging to achieve significant results. Although observations of anisotropy were reported up to energy around 10^{20} eV, in this paper we review measurements and models concerning CRs below 10^{15} eV, where the counting rate of detectors is high enough to give solid arguments on which to ground the discussion.

The paper is organized as follows. In the section 2 the detection methods used so far are summarized. The existing data on the CR anisotropy below the knee are reviewed in the section 3. In the section 4 the different models proposed to interpret data are reported. Some conclusions follow.

2. Methods of Detection

The amplitude of the anisotropy is so small ($\approx 10^{-3}$ or less) that the experimental approach to detect it has to be sensitive on necessity. In order to measure such a tiny effect, large exposures are needed, i.e. large instrumented areas and long-lasting data acquisition campaigns. Until now, only ground-based detectors demonstrated to have the required sensitivity, making use of the Earth's rotation to order data in solar and sidereal time.

As the Earth rotates, the field of view of ground-based detectors points towards different directions at different times, sweeping out a cone of constant declination.

The CR arrival direction is given after the definition of a reference frame. If Sun-related effects are looked for, a system of coordinates co-moving with the Earth has to be used.

Usually, it is a spherical system centered on the Earth. Instead, if phenomena originated outside the heliosphere are considered, a reference frame co-moving with the Sun is set, usually a spherical system centered there[1]. Whatever the spherical system is, because of the Earth rotation, the direction ϕ is periodically observed from ground-based detectors, that is why the ϕ axis is regarded as a time axis. The "solar" or "universal" time is used when the Sun is wanted to be at rest in the reference frame, whereas the "sidereal" time is used when the Galactic center is wanted to be at rest.

The *sidereal anisotropy* is spoken about when the sidereal time is used, whereas the *solar anisotropy* is measured by using the solar time. The *anti-sidereal* time analysis is often performed too, as no signal is expected in this non-physical reference frame and any positive result there can be used as estimation of the systematic uncertainty on the sidereal time detection.

Once the reference frame is defined, data are collected and ordered accordingly. There are a number of ways in which the degree of anisotropy of a given distribution can be defined. Perhaps the most traditional one in CR physics is:

$$\delta = \frac{I_{max} - I_{min}}{I_{max} + I_{min}} \tag{1}$$

where I_{max} and I_{min} are the maximum and the minimum observed intensity. This definition is the most general to be given, as no hypothesis on the form of the anisotropy is needed: either it is a peak in a smooth distribution or a di-polar modulation, the definition (1) can be applied.

In 1975 Linsley proposed to perform analyses in right ascension only, through the so-called "harmonic analysis" of the counting rate within a defined declination band, given by the field of view of the detector [8]. In general, one calculate the first and second harmonic from data by measuring the counting rate as a function of the sidereal time (or right ascension), and fitting the result to a sine wave. The Rayleigh formalism allows to evaluate the amplitude of the different harmonics, the corresponding phase (the hour angle of the maximum intensity) and the probability for detecting a spurious amplitude due to fluctuations of a uniform distribution.

Let $\alpha_1, \alpha_2, \ldots \alpha_n$ be the right-ascension of the n collected events. From this data series, the experimenter has to determine the components of the two-dimensional vector amplitude (or, equivalently, the scalar amplitude r and phase ϕ).

The R.A. distribution $f(\alpha)$ can be represented with a Fourier series:

$$f(\alpha) = \frac{a_0}{2} + \sum_{k=1}^{\infty} (a_k \cos k\alpha + b_k \sin k\alpha) = \frac{a_0}{2} + \sum_{k=1}^{\infty} r_k \sin(k\alpha + \phi_k) \tag{2}$$

where $a_k = r_k \sin \phi_k$ and $b_k = r_k \cos \phi_k$ are the Fourier coefficients and

$$r_k = \sqrt{a_k^2 + b_k^2} \tag{3}$$

$$\phi_k = \tan \frac{a_k}{b_k} \tag{4}$$

---

[1] Or on the Earth, that is the same because of the distances under consideration.

are the amplitude and the phase of the kth harmonic, respectively. It is known that the Fourier coefficients can be computed from the α_m data series by means of the equations:

$$a_k = \frac{2}{n} \sum_{i=1}^{n} \cos k\alpha_i \tag{5}$$

$$b_k = \frac{2}{n} \sum_{i=1}^{n} \sin k\alpha_i \tag{6}$$

where n is the number of data points. This formalism can be applied up to whichever order k, but historically experiments never had the sensitivity to go beyond $k = 3$ and anthologies are usually compiled about $k = 1$ only. If a non-zero first harmonic amplitude is found, it is important to estimate the chance probability P that it is a fluctuation of an isotropic distribution. If the sample α_1, α_2, ..., α_n is randomly distributed between 0 and 2π, then, as $n \to \infty$, the probability P of obtaining an amplitude greater than or equal to r is given by the well-known Rayleigh formula

$$P(\geq r) = \exp(-k_0) \tag{7}$$

where $k_0 = (nr^2)/4$.

In general, harmonic analysis is effective for revealing broad directional features in large samples of events measured with poor angular resolution but good stability.

The technique is rather simple: the greatest difficulty lies in the treatment of the data, that is, of the counting rate themselves. In fact, the expected amplitudes are very small with related statistical problems: long term observations and large collecting areas are required. Spurious effects must be kept as low as possible: uniform detector performance both over instrumented area and over time are necessary as well as operational stability. In addition, CR experiments typically suffer from large variations of atmospheric parameters as temperature and pressure, which translate into changes of the effective atmospheric depth, affecting the CR arrival rate.

Therefore, the measured rate must be corrected precisely for instrumental and atmospheric effects to prevent any misinterpretation of CR flux variations.

The *East-West method* was designed to avoid performing such corrections, preventing potential subsequent systematic errors introduced by data analysis. The original idea was proposed in the early 1940s to study asymmetries in the flux of solar CRs [9, 10, 11] and was later applied by Nagashima to extensive air showers (EASs) at higher energy [12]. It is a differential method, as it is based on the analysis of the difference of the counting rates in the East and West directions [13]. By considering this, the method is able to eliminate any atmospheric effect or detector bias producing a common variation in both data groups.

For primary CRs with energy lower than 10^{12} - 10^{13} eV, underground muon detectors have been widely used. The muon rate deep underground can be affected by a spurious periodic modulation as the result of the competition between pion decay and interaction in the upper atmosphere. As the atmosphere cools at night (or during the winter), the density increases and the pions produced in CR collisions with the atmosphere are fractionally more likely to interact than decay when compared with the day (or summer) when the climate it is warmer. The daily modulation introduced by solar heating and cooling is seen when muons are binned in solar-diurnal time; yearly or seasonal modulations are seen when events are binned with a period of a (tropical) year [6, 7]. With increasing energy an increasing fraction of high energy secondary pions interact in the higher atmosphere instead of decaying to muons and the muon intensities decrease too rapidly to yield sufficient statistics for high-energy primary CRs.

The detection of EASs by employing ground detector arrays with large extensions (from 10^4 to 10^9 m^2) and long operation time (up to 20 years), extended the anisotropy studies up the highest energy region. The information about the individual shower was very limited but the very high counting rate, in particular for arrays located at high altitude, made this energy range attractive for anisotropy studies.

The observation of anisotropy effects at a level of 10^{-4} with an EAS array is a difficult job, because of how difficult is to control this kind of devices. A wrong estimation of the exposure may affect the CR arrival rate distribution, even creating artifacts (i.e. fake excesses or deficit regions). In fact, drifts in detector response and atmospheric effects on air shower development are quite hard to be modeled to sufficient accuracy[2]. The envisaged solution is to use the data to estimate the detector exposure, but data contain either signal and background events, so that some distortions could be present in the results. The shape and the size of possible artifacts depend on the characteristic angle and time scale over which all the aspects of the data acquisition vary more than the effect to be observed. Therefore, if an anisotropy of the order 10^{-4} is looked for, operating conditions must be kept (or made up) stable down to this level, all across the field of view and during all the acquisition time. In addition, as it is well known, the partial sky coverage carried out by a given experiment will bias the observations of structure on largest scales. Finally, the observation of a possible small angular scale anisotropy region contained inside a larger one relies on the capability for suppressing the anisotropic structures at larger scales without, simultaneously, introducing effects of the analysis on smaller scales.

Both multi-directional and uni-directional detectors have been widely used to investigate different declination intervals defined by the center direction of viewing of the telescopes.

Former experiments were not able to reconstruct the arrival direction of the primary, but the anisotropy induces periodicity in the response from a CR detector, having a fundamental frequency corresponding to the duration of a sidereal day. Physicists exploited the rotation of the Earth to build 1D distributions of the CR relative intensity. Due to the poor statistics, any attempt to reconstruct the all-sky structure of the anisotropy was quite unrealistic and the only chance was to check if any strong statistical indication of a genuine CR intensity variation was there or not.

In the last decade, some experiments have been able to reconstruct the primary arrival direction with good precision on large amount of data, thus allowing to build two-dimensional (2D) maps of the anisotropy. When 2D sky maps are realized, excesses and deficits observed in one dimension appear to be localized and to have their own extension and morphology. Any hypothesis about the correlation of the observed features with astrophysical sources can be confirmed or denied by the exact information on the direction. As 2D sky-maps for the CR anisotropy appeared only in the last few years, there is not any

---

[2]In fact, the temperature can modify the lateral extension of a shower trough the variation of the Moliere radius, and the pressure can influence the absorption of the electromagnetic component.

standard way to present data yet. Data are usually compared on the basis of the position and the extension of the observed features, as well as on their relative intensity. Recently, the Icecube experiments introduced a multi-scale analysis in spherical harmonics [91].

3. Experimental Results

In the figure 1 the amplitude and the phase of the first harmonic measured by different experiments starting since 1973 are shown as a function of the primary CR energy.

A history of the experimental results is outlined in the next subsections.

3.1. Observations pre-1975

The earliest evidence of periodicity in the CR intensity was reported in 1932 by Hess and Steinmaurer [43], see the figure 2. Indeed, the question of a possible variation of the cosmic radiation flux with time arose soon after its discovery. This observation, interpreted by Compton and Getting (CG) as an effect of the Earth motion relative to the sources of CRs [44], stimulated many measurements to verify the CG model.

Although different ground-based experiments reported evidence of a diurnal variation of galactic CRs long before the 50s, solar modulation, the geomagnetic field and some atmospheric effects were considered a considerable obstacle for primary CRs of energy less than 10^{12} eV. Indeed, this energy range, mainly investigated by underground muon telescopes, turned out to be very critical to get clear evidence on the anisotropy of CRs before they enter the solar cavity.

Variations due to the geomagnetic field or to the solar activity become negligible with increasing energy but, for a long time, the existence of the CR anisotropy in the multi-TeV energy range remained uncertain. In fact, the amplitude of the temporal variation of the EAS trigger rate (a few percent), mostly due to the atmospheric pressure and temperature effects, is much larger than the expected amplitude of the galactic CR anisotropy. The additional problem of the low event rate with increasing energy led to a number of contradictory results (see e.g., [45, 46, 47, 36]).

Only during 50s, when large detectors were operated underground or at the surface monitoring of the atmospheric variations, a clear evidence of the existence of the galactic anisotropy was obtained.

In 1954 Farley and Storey [48] published the results of the analysis of more than 10^5 EAS events recorded in Auckland, in the Southern hemisphere, with 3 counter trays at the corners of a 4 m-side triangle, in the period February 1951 - February 1952. They reported a significant observation of solar and sidereal diurnal variations of the CR intensity. By introducing the concept of anti-sidereal time an allowance were made, for the first time, for the seasonal amplitude modulation of the solar diurnal variation. The comparison with two other experiments carried out in the northern hemisphere showed that the variations were in phase and the maxima coincident in local sidereal time. This gave important grounds for believing that the observed variation was a genuine sidereal effect and not a residual due to seasonal changes in the solar diurnal effect.

In 1957 two large wide angle muon telescopes were installed under 18 m of sandstone near Hobart [49] in the Southern hemisphere. The median primary energy was 1.8×10^{11} eV

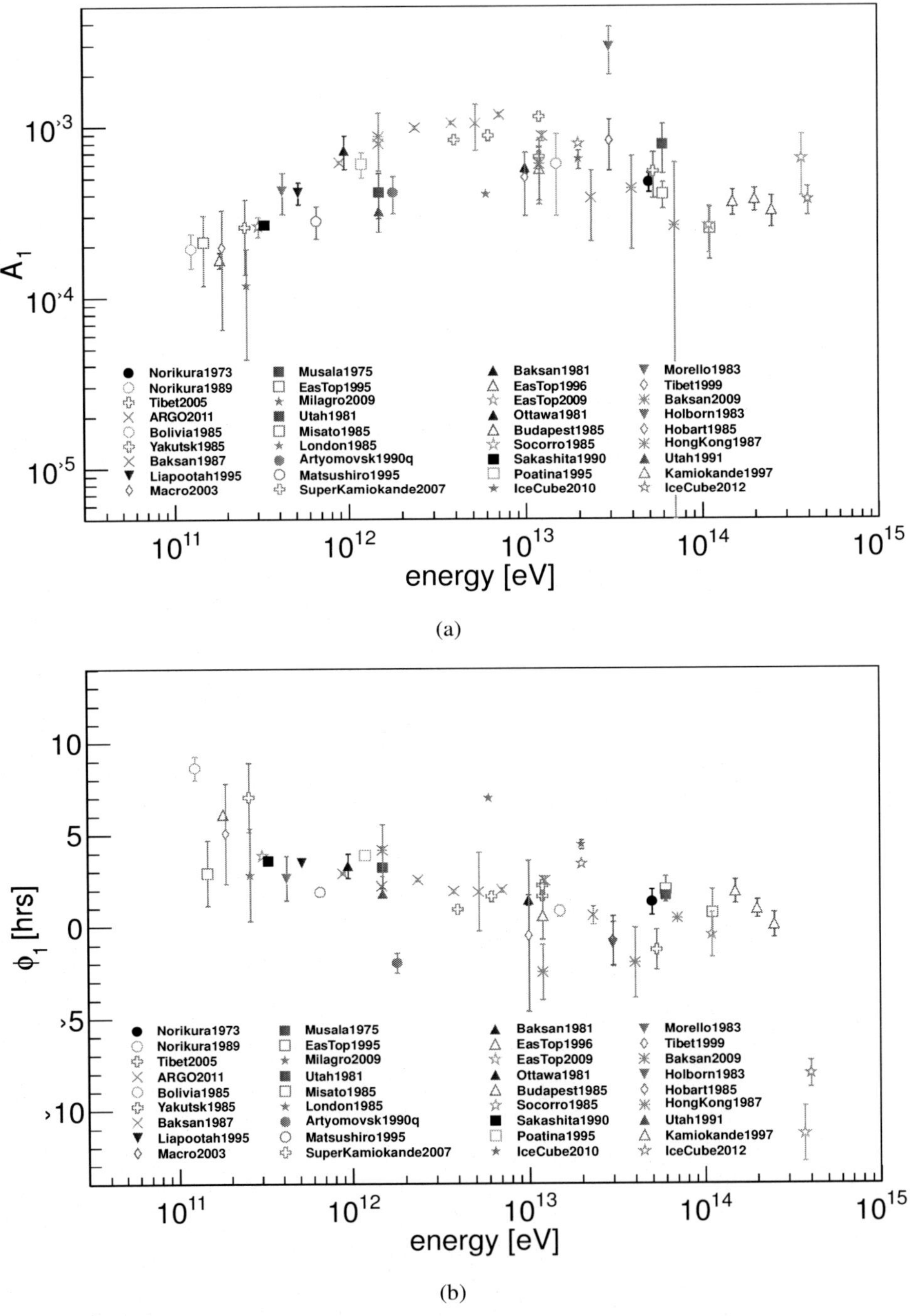

(a)

(b)

Figure 1. First harmonics of the sidereal daily variations measured by underground muon detectors ([14, 15, 16, 17, 18, 19, 20, 21, 22, 23, 24, 25, 26, 27, 28, 29, 30, 31]) and EAS arrays ([32, 33, 34, 35, 12, 36, 37, 38, 39, 40, 41, 42]). The amplitude (a) and the phase (b) are plotted as a function of the primary CR energy. For clarity, the points are not labeled with citation information.

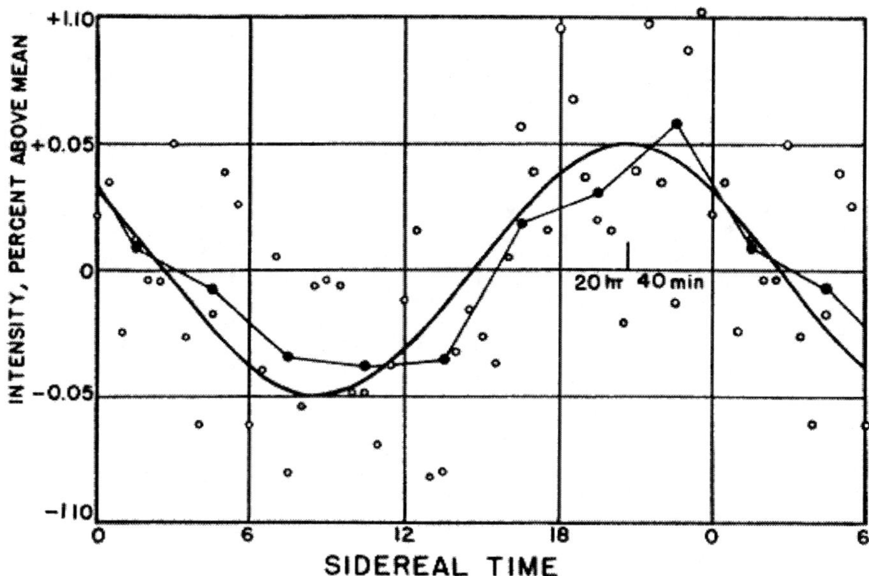

Figure 2. Percentage variation in intensity of the CRs with sidereal time. Data were collected by Hess and Steinmaurer in 1932. The open circles are half-our means; the solid circles are 3 hr means. The solid line represents the predicted Compton-Getting effect.

and the counting rate 7×10^4 particles/hour. Early in 1958 in the Northern hemisphere two almost similar muon telescopes were operated at a similar depth underground in Budapest [50].

The combined analysis of the Hobart-Budapest data was very important to get evidence that the galactic anisotropy has higher order features besides the fundamental harmonic. Observations of the sidereal daily variation revealed a latitude-dependent pattern in the first and second harmonics, as shown in the figure 3. A 12-hour difference between the phase of maxima in the two hemispheres was observed, as well as larger semi-diurnal maxima at the low latitudes.

These results suggested the existence of a global galactic anisotropy consisting of a collinear types of bidirectional anisotropy and a smaller unidirectional anisotropy in line with it. The particular bidirectional type was indicated by the diurnal maxima in phase with one of the semi-diurnal maxima and the remarkable dependence on the latitude of the intensity of the semi-diurnal modulation. The fact that the diurnal amplitude in Hobart was considerably larger than in Budapest suggested the presence of a unidirectional diurnal variation in phase with the bidirectional component in Hobart and therefore 12 hours out of phase with that in Budapest [51, 52].

However, the 12-hour phase difference between the two hemispheres naturally suggests the possibility that the result is not due to a genuine sidereal anisotropy but to a spurious sidereal effect arising from the seasonal modulation of the solar daily variation of CR intensity. To test the difference between the two experiments a narrow angle telescope was

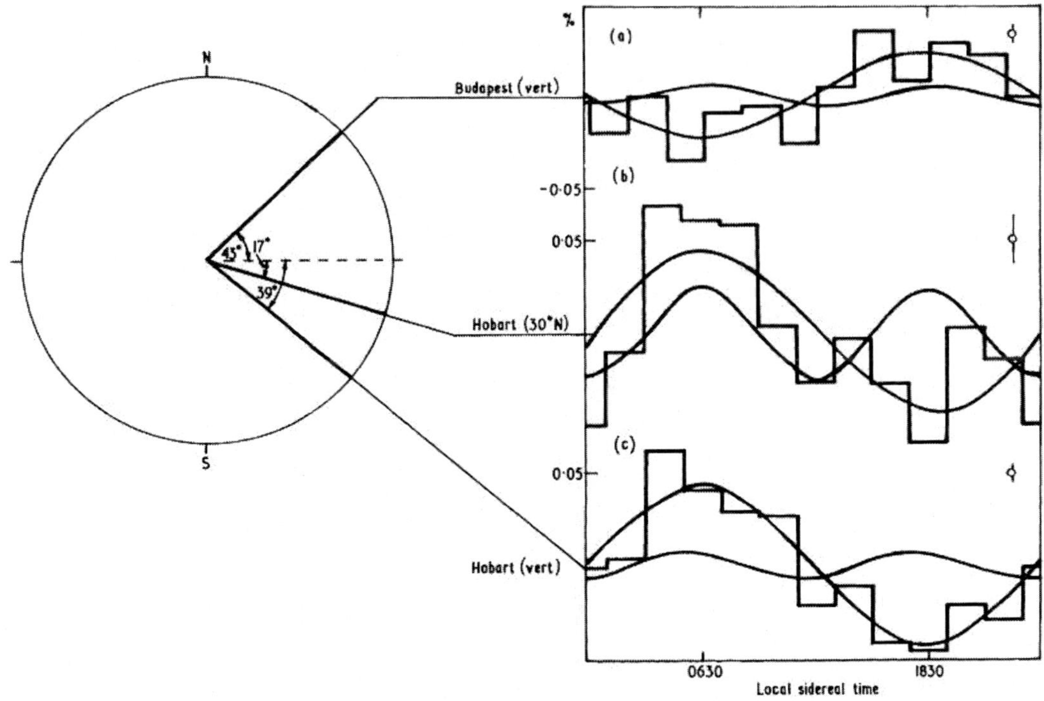

Figure 3. The sidereal daily variation of muon intensity observed underground between 1959 and 1962 in both hemispheres: (a) with vertical telescopes in Budapest, (b) with a telescope inclined 30° N in Hobart, (c) with vertical telescopes in Hobart. The first and second harmonics of best fit are shown. The latitude angles are those of the asymptotic directions of viewing. The error tails shown are the standard errors of the fitted amplitudes.

installed in Hobart pointing north at a sufficiently high angle (70°) to the zenith so as to view into the northern hemisphere. The telescope observed a sidereal daily variation not in phase with the vertical telescope in Hobart, but with the Budapest phase, see the figure 4, thus excluding the presence of a spurious component [51].

A very large narrow angle air Cerenkov telescopes operated in Nagoya confirmed these results recording the variation of CR intensity at a median energy of about 2×10^{11} eV in the period 1968-1971 [53]. Further evidences came from a couple of narrow angle telescopes operated in Mawson in 1968 [54, 55].

From 1965 onwards a great increase took place in the number and size of underground experiments. A highly significant development in the early 1970s was the start of observations of small air showers at a median energy around 10^{13} eV at Mt. Norikura in Japan [56].

In fact, the observations made in the energy range 10^{13} - 10^{15} eV until that moment did not give concrete evidence for the existence of the sidereal anisotropy due to several reasons. First, the counting rates attained was not sufficient to obtain the necessary accuracy of the order of 10^{-4}. Moreover, the experiments were carried out with no consideration of the stability of the apparatus, thus with dispersion of actual data far greater than the statistical

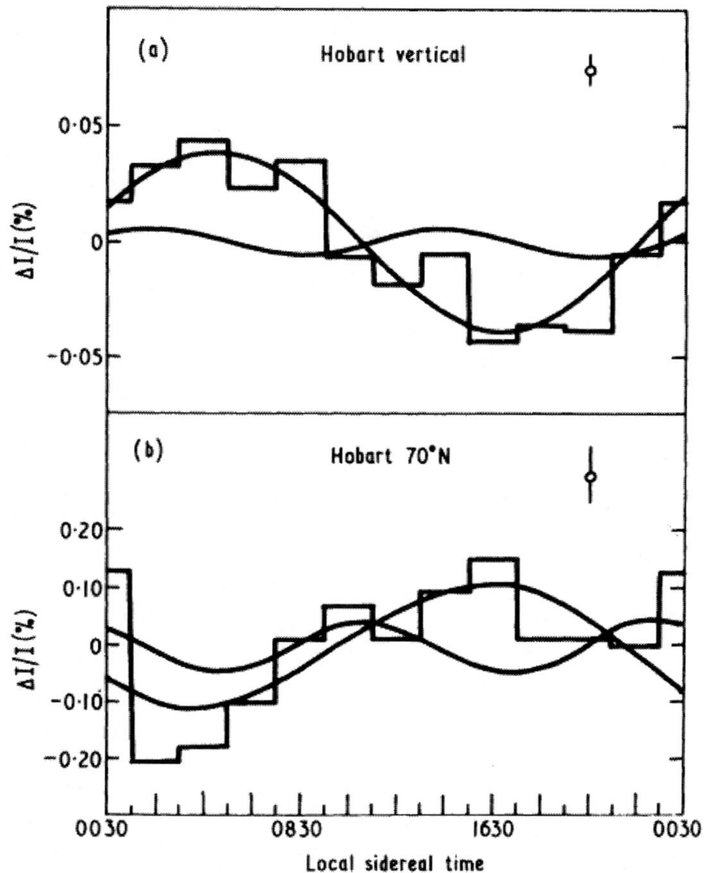

Figure 4. The sidereal daily variation of muon intensity observed underground in Hobart (a) from vertical telescopes scanning an average southern latitude $\approx 39°$, (b) from a narrow angle telescope inclined 70° N of zenith, scanning an average northern latitude $\approx 20°$. The first and second harmonics of best fit are shown. The error tails shown are the standard errors of fitted amplitudes.

one.

In the Mt. Norikura experiment these problems were worked around because the apparatus was a multi-directional telescope housed in an air-conditioned laboratory with daily temperature variation less than $\sim 0.1°$C. In this way, th effect of the temperature on the detector was controlled to be less than $\sim 0.01\%$. The high counting rate ($\sim 45,000$/hour) was attained due to the large detector area and the high altitude location (2770 m a.s.l.). In addition, to reduce the atmospheric effects, especially that of external temperature variations, the observation of directional air showers (East and West component) was carried out utilizing the absorption effect of air showers by a 10 cm lead wall [57]. The omni-directional Norikura array measured a sidereal anisotropy consistent with the Hobart-Budapest results, thus making the scientific community confident that no significant spurious component had affected the experimental results. The importance of this measurement was due to its being the first directional-measurement of anisotropy ever.

The Mt. Norikura observations were confirmed by other EAS arrays from Peak Musala in Bulgaria and from Baksan in Georgia, in the energy range 10^{13} - 10^{14} eV [58]. In 1970s another important experiment, made of three vertical telescopes for muon detection, started operations in Poatina, Tasmania, at a median energy of about 5×10^{11} eV. Other important results came from a group of large muon-directional plastic scintillator telescopes located at ground in Nagoya and underground in Misato and Sakashita [60, 61].

Nevertheless, in spite of the progress in detection techniques, the reported amplitudes were so small in relation to factors affecting observations (geomagnetic and helio-magnetic fields, solar modulations, temperature and pressure variations, low counting rates) that their accuracy as a measure of the CR anisotropy remained in doubt for many years and the results not conclusive yet.

3.2. Observations post-1975

The Munich International Cosmic Ray Conference (ICRC) in 1975 marked a change in viewpoint regarding the CR arrival direction anisotropy [62, 63]. In fact, in three papers by Gombosi et al. [64], Nagashima et al. [65] and Fenton and Fenton [66], the observation of a sidereal anisotropy seen with approximately the same amplitude ($\sim 0.07\%$) and phase (~ 00 hr RA) at approximately the same latitude in both hemispheres was reported in the energy interval 10^{12} - 10^{14} eV/nucleon. In addition, in this conference there was a clear experimental evidence that solar effects overwhelm the results of underground telescopes detecting muons produced by primary CRs with energy less than $\leq 10^{12}$ eV/nucleon. Such effects prevented any clear determination of the anisotropy from being made in this energy range [62].

In 1977 Linsley and Watson reviewed the data on CR anisotropy pre-1965 in the energy range 10^{14} - $3\cdot10^{17}$ eV concluding that the combined results cannot be explained by random fluctuations or selection effects [67].

In the following years other measurements carried out at Musala peak [68], Mt Norikura [69, 70] and in Baksan [34] provided additional confirmation of the existence of a sidereal CR anisotropy. In addition these results from EAS-arrays in the energy range 10^{13} - 10^{14} eV were in good agreement with observations from experiments with underground muon telescopes around 10^{12} eV [49, 71, 72, 73].

In 1983 Linsley concluded that the results obtained by a harmonic analysis in sidereal time "should be accepted without any reservations about spurious effects of a statistical, instrumental or atmospheric nature. (...) The success of the small air shower experiments shows that the residual effect of instrumental instabilities can be held to 1 part in 10^{4}" [63].

In the mid 80s, new analysis techniques for underground muon observations were developed, to eliminate the spurious sidereal variation from the apparent variation. Those methods were based on the observed anti-sidereal diurnal variation [15], which was demonstrated to be capable to eliminate effects on the low energy CRs due to the reversal of the Sun's polar magnetic field [74].

The same methods were soon applied to the observation of EASs conducted since 1970 with an air shower detector at Mount Norikura in Japan (median energy $\approx 1.5\times10^{13}$ eV). A significant sidereal diurnal variation at higher energy was observed, together with the strong evidence of semi-diurnal and tri-diurnal variations [12]. As semi-diurnal and tri-diurnal modulations correspond to 180° and 120° periodicity respectively, excesses as wide as 60° were found, showing some hints of sub-structures and medium scale features already at that time.

A continuous observation of the 24 hr profile of the CR rate in sidereal time over 12 years revealed that the sidereal diurnal variation of 10 TeV CR intensity exhibits a deficit with a minimum around 12 hr [12]. A similar anisotropy was reported in those years by other experiments in the same energy region [75, 25], and also down to 6×10^{10} eV, by underground muon detectors, with the same form and almost the same phase.

Until this moment, the observation of the galactic anisotropy was concerned with two types, unidirectional and bidirectional [52]. They may occur independently of each other or as a component of a more complex global anisotropy. The unidirectional anisotropy can be produced by the motion of the solar system relative to a frame of reference in which the CR system can be regarded as isotropic (the CG effect). The diurnal variation is symmetric, in the sense that it is the same at the same latitude in both hemispheres. On the contrary, a bidirectional diurnal variation is asymmetric with respect to the hemisphere, since times of maximum will differ by 12 hr between northern and southern latitudes of observation. Consequently, if the total observed diurnal variation consists of superimposed unidirectional and bidirectional components they can be separately identified by taking the sum and difference of observations, made under the same detecting conditions, at the equal latitudes in both hemispheres. The unidirectional component is separated out in the sum of observations and the bidirectional one in the difference.

In 1998 Nagashima, Fujimoto, and Jacklyn (NFJ hereafter) reported the first comprehensive observation of a large angular scale anisotropy in the sub-TeV CRs arrival direction by combining data from different experiments in the northern and southern hemispheres [76]. They analyzed data of the muon telescopes underground in Sakashita and Hobart (median energies $\approx$ 331-387 GeV and $\approx$ 184 GeV, respectively), and at the Nagoya ground station ($\approx$ 60-66 GeV). Data recorded with the small air shower array at Mt. Norikura were also used.

The results obtained by the NFJ analysis are shown in the figure 5. In addition to a *dipolar anisotropy* (the CR deficit around $\alpha = 0$ hr $\delta = -20°$, with intensity independent of the energy), which they named "galactic" anisotropy, NFJ found a *new anisotropy component* causing an excess of intensity with a maximum around 6 hr in the sub-TeV region.

This directional excess flux, confined in a narrow cone with a half opening angle of $\sim 68°$, seemed to coincide with the expected helio-magneto-tail direction ($\alpha = 6.0$ hr $\delta = -29.2°$), opposite to the proper motion of the solar system and it was supposed to be of solar origin and was named *"tail-in"* by NFJ.

The amplitude of these anisotropy components increases with energy up to ≈ 1 TeV. After this energy, the new contribution was observed to be less and less important, even if disappearing at ~ 10 TeV (Mt Norikura).

NFJ suggested that such a decrease is consistent with a signal related to the acceleration of particles arriving from the helio-tail direction, i.e., ~ 6 hr in sidereal time.

It was an indirect confirmation that higher energy particles are an ideal tool to investigate the galactic anisotropy, being insensitive to Sun-related effects.

In the earliest "map" of the large scale anisotropy, figure 6, referred to as the *NFJ*

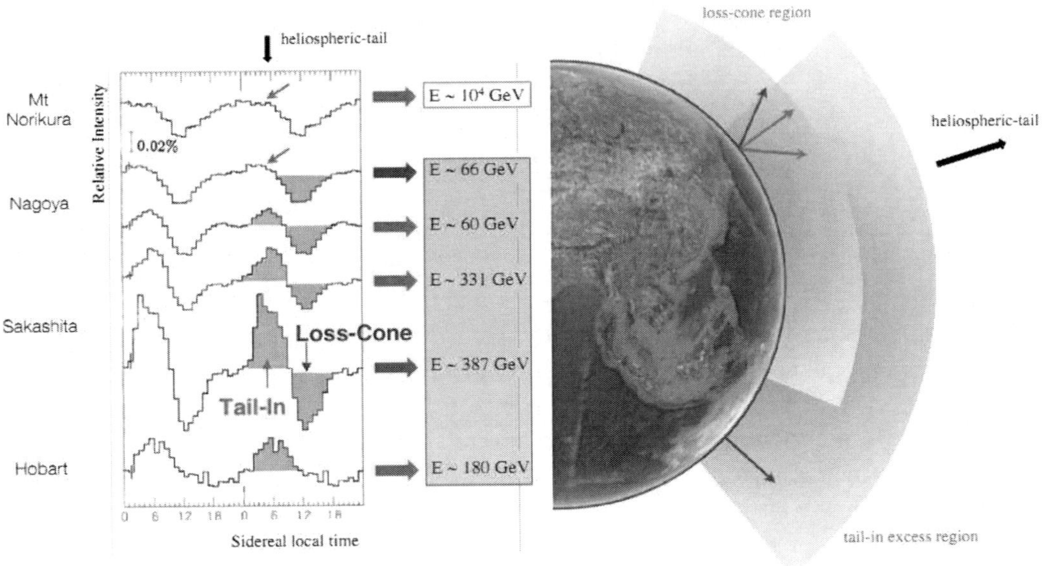

Figure 5. CR sidereal daily variation from different muon telescopes or air shower detectors: from top to bottom, Mount Norikura air shower array (shown for reference at high energy and with the amplitude multiplied by 1/4 to account for the larger anisotropy amplitude at high energy), Nagoya station (looking 30° northward, and looking vertically upward), Sakashita station (looking vertically upward and 41° southward), and Hobart station (looking vertically upward). The CR median energy of different detectors is shown. In order to clearly show the peaks and valleys, the 24 hr variation is repeatedly shown in a 2-day time interval. The error expresses the dispersion of the hourly relative intensity (courtesy of P. Desiati).

model, the excess and deficit cones were obtained by interpolating between one dimensional anisotropy measurements made in several different declination strips.

the result was qualitative in nature because data were from different detector types (shallow underground muon telescopes and surface EAS array) with large spread in energy sensitivity and different systematic uncertainties.

The tail-in excess was recognized to be not symmetric along the R.A. direction and this result confirmed the discovery of the semi- and tri-diurnal variations ten years before. Some long-period measurements from neutron's monitors did not observe any decrease of the tail-in excess amplitude during the reversal of the solar magnetic field from the negative to the positive polarity state (1989). This result strengthened the hypothesis of the "helio-spheric" origin of the phenomenon, against the "galactic" one [77].

The positive pole of the di-polar anisotropy slightly overlaps with the tail-in zone, what makes the deficit di-polar region more evident alone. This region is often referred as *"loss-cone"*.

The discovery of the tail-in gave an important piece for the solution of the problem of the observed phase shift in the sidereal diurnal variation, from 6 to 0 hours with the increase of the energy. In fact, NFJ explained the phenomenology by distinct contributions

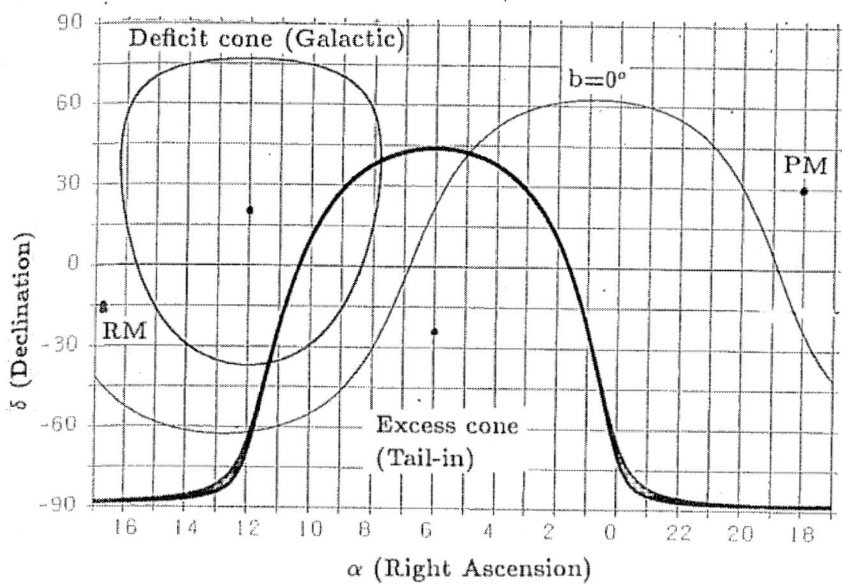

Figure 6. The distribution of the tail-in and loss-cone (galactic) anisotropies on the equatorial coordinate grid. The excess cone of the tail-in anisotropy, within which all the directional excess flux is confined, is shown bounded by the thick line including the south pole. A thinner line bounds the deficit cone of the galactic anisotropy, in which all the directional flux is confined. PM shows the direction of the proper motion of the solar system. RM shows the direction of the relative motion of the system to the neutral gas. The galactic equator is represented by the thin line labelled by $b = 0°$.

from two anisotropies. The amplitude of the tail-in anisotropy, with a maximum at ~ 6 hr in sidereal time, decreases with increasing primary energy above ~ 1 TeV, while the Galactic anisotropy, with a minimum at ~ 12 hr, remains constant. Furthermore, NFJ suggested that the phase of the composite first harmonic vector turns counterclockwise from ~ 3 to ~ 0 hr with increasing energy (the composite amplitude is also expected to decrease by $\sim 30\%$ when the amplitudes of two anisotropy components are equal at ≤ 1 TeV).

Finally, it was clearly demonstrated that the observed sidereal variation was not compatible with the CG effect expected from the motion of the solar system in the interstellar space. The authors drew the conclusion that the solar system drags with it in its motion the surrounding interstellar magnetic field within which the CRs with low energy are isotropically confined.

The work [76] was so important to make the "tail-in" and "loss-cone" names used by the scientific community, although there is no agreement yet about their origin.

A detailed analysis of the sidereal diurnal modulations observed in a total of 48 directional channels of the underground muon detectors monitoring both the northern and southern sky indicated that the maximum phase of the new anisotropy component, which is ~ 6 hr in the northern hemisphere, shifts toward earlier times as the declination of the incident CRs moves southward to the equator [78, 79].

In 1996 the EAS-TOP collaboration reported a detailed analysis of high energy CR

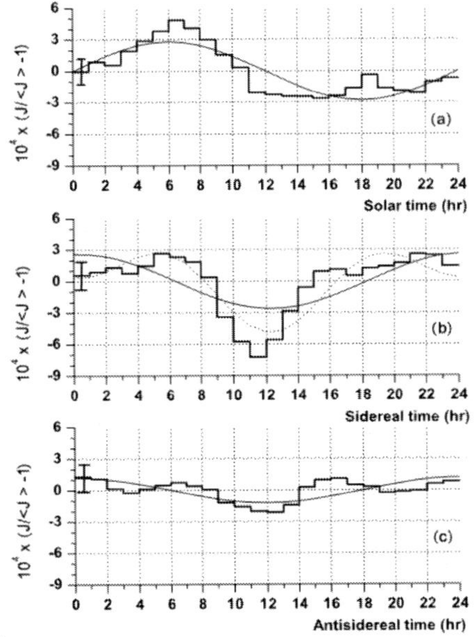

(a) Thick black lines: counting rate curves in solar (a), sidereal (b), and anti-sidereal (c) time at 1.1×10^{14} eV. The statistical uncertainty for each bin is given in the first one. The curves resulting from the first harmonic analysis are also shown (light black lines); for the sidereal time curve, the combination of the first and second harmonics (dotted black line) is additionally superimposed.

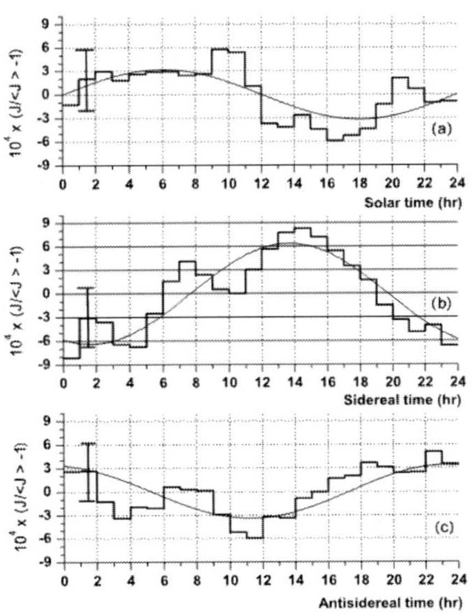

(b) Thick black lines: counting rate curves in solar (a), sidereal (b), and anti-sidereal (c) time at 3.7×10^{14} eV. The curves resulting from the first harmonic analysis are also shown (light black lines).

anisotropy at $\approx 10^{14}$ eV. Data were analyzed in solar, sidereal and anti-sidereal time using the East-West method, see the figure 7(a). A signal compatible with the CG effect is clearly observed in solar time, for the first time at energy $> 10^{14}$ eV, whereas no significant structure was observed in anti-sidereal time [41]. In 2009 the EAS-TOP collaboration reported the first evidence ever of anisotropy around 400 TeV [42]. Hints of increasing amplitude and change of phase above 100 TeV resulted from the analysis of the full data set, as shown in the figure 7(b).

In the same period the underground MACRO experiment, measuring >10 TeV muons, found evidence for modulations $(< 10^{-3})$ in solar and sidereal diurnal periods, at the limit of the detector statistics. The solar diurnal modulation was attributed to the daily atmospheric temperature variations at 20 km, the altitude of primary CR interaction with the atmosphere [29].

From the measurements done by "old" experiments (see [80, 81] for a review) it was claimed that the amplitude of the first harmonic increases with increasing energy, starting from 10^{15} eV. The phase too seemed to be energy dependent, with changes correlated with the Galactic structure. However, the statistical uncertainty was too large to drew any conclusions. In fact, the anisotropy apparent amplitude naturally increases as the number of events decreases (i.e., as the energy increases), as a simple consequence of (lack of) statistics: $\delta =$

$(I_{max} - I_{min})/(I_{max} + I_{min}) \propto N^{0.5}/N = 1/N^{0.5}$ [82].

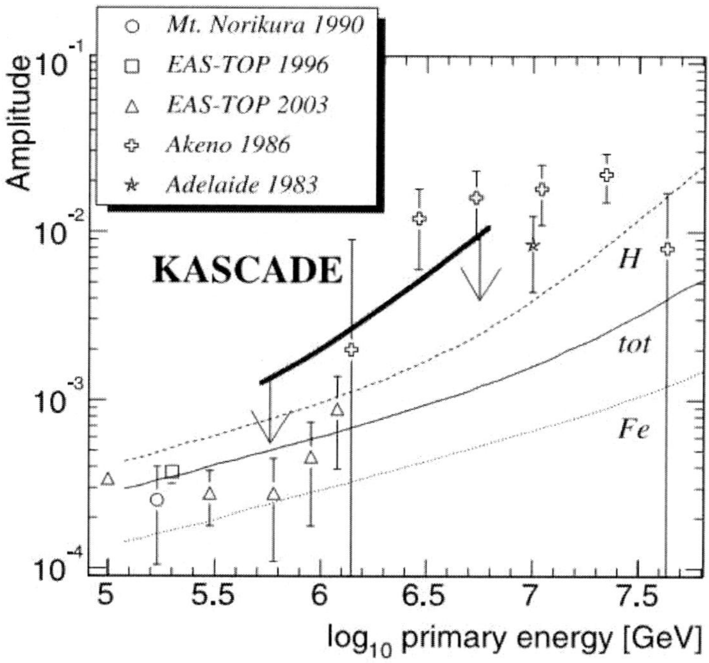

Figure 7. KASCADE upper limits (95%) of Rayleigh amplitudes vs. primary energy (thick line) compared to other results [84, 41, 85, 86, 87]. Model predictions from [88] for the total anisotropy and for the anisotropies of the proton and iron components are also shown (thin lines).

In 2004 the KASCADE experiment presented an analysis of the CR anisotropy in the PeV energy range. No hints of anisotropy in the R.A. distributions were reported and upper limits for the Rayleigh amplitude between 10^{-3} at a primary energy of 0.7 PeV and 10^{-2} at 6 PeV were set, see the figure 7 [83].

The EAS-TOP collaboration in 2003 published a measurement of the first harmonic parameters up to 1.2 PeV [85]. The amplitudes are shown in the figure 7. Since the signal above 300 TeV was not statistically significant, they set upper limits ranging between $4 \cdot 10^{-4}$ at 0.3 PeV and $2 \cdot 10^{-3}$ at 1.2 PeV. We recall what already mentioned in this section, i.e. in 2009 the EAS-TOP collaboration published a new analysis providing the amplitude and phase of the first harmonic at 370 TeV.

The EAS-TOP and KASCADE experiments did not confirm the increase of anisotropy with energy (from 10^{15} eV upward) claimed by old EAS arrays, though limited statistics still prevents to draw any firm conclusion.

3.3. Recent Observations: 2D Sky Maps

Based on a harmonic analysis, these observations are consistent with the large-scale diffusive propagation of CRs in the Galaxy [89], but there was no consensus for a production mechanism in the local interstellar region surrounding the heliosphere that would reproduce the sidereal diurnal variation.

In fact, although a general view of the situation could be obtained by merging data from different experiments (e.g., see the figure 6) the lack of homogeneous (i.e., one-experiment) information on the dependence of the anisotropy on declination impeded a complete experimental formulation of the problem [37].

On the other hand, the 24 hr profile and its declination dependence can be measured *together* only with direction-sensitive devices.

In the last decade large area detectors with correspondingly large statistics and good pointing accuracy came on line, each one able to make a two-dimensional (2D) representation of the CR arrival direction distribution, thus allowing detailed morphological studies of the anisotropy either along R.A. and declination directions.

In 2006 the Tibet ASγ experiment, located at Yangbajing (4300 m a.s.l.), published the first 2D high-precision measurement of the CR anisotropy in the Northern hemisphere in the energy range from few to several hundred TeV [90]. in the figure 8 the CR intensity maps observed by Tibet ASγ are shown. The 1D projection of the 2D maps in R.A. is also shown. This measurement revealed finer details of the known anisotropy and established a new component of the galactic anisotropy around the Cygnus region (region III of panel D).

In solar time the observed di-polar anisotropy was in fair agreement with the expected CG effect due to the Earth's orbital motion around the Sun.

For CR energy higher than a few hundred TeV, all the components of the anisotropy fade out, showing a co-rotation of galactic CRs with the local Galactic magnetic environment.

The Tibet ASγ collaboration carried out the first measurement of the energy and declination dependences of the R.A. profiles in the multi-TeV region with a single EAS array [37]. They found that the first harmonic amplitude is remarkably energy-independent in the range 4 - 53 TeV, contrary to the suggestions of the NFJ model.

The panels (a) and (b) of figure 9 show the full 24 hr profile of the sidereal diurnal variation averaged over all declinations and primary energies, together with the results of the Mt Norikura EAS array, which was not capable of resolving the declination of the incident direction. The Tibet ASγ results are in good agreement with the Norikura profile, if the average over all declinations is considered.

The importance of this new result became even clearer when looking at the panel (c), where the sidereal profiles in each declination band separately are shown. The phase of maximum changes from ~ 7 to ~ 4 hr in sidereal time, and the amplitude of the excess increases as the viewing direction moves from the northern hemisphere to the equatorial region. This is consistent with the anisotropy component firstly found in the sub-TeV region by underground muon detectors and referred to as tail-in anisotropy. The Tibet ASγ experiment clearly showed that the tail-in anisotropy continues to exist in the multi-TeV region, what have deep implications for all models of CR acceleration and propagation providing predictions on the anisotropy.

Though with less sensitivity, a 2D map of the CR anisotropy was reported also by SuperKamiokande-I (SK-I) in 2007 [28]. The experiment, realized with an underground imaging water Cerenkov detector, published the map shown in the figure 10, representing high energy muons (> 10 TeV) coming from different sky regions. The observation con-

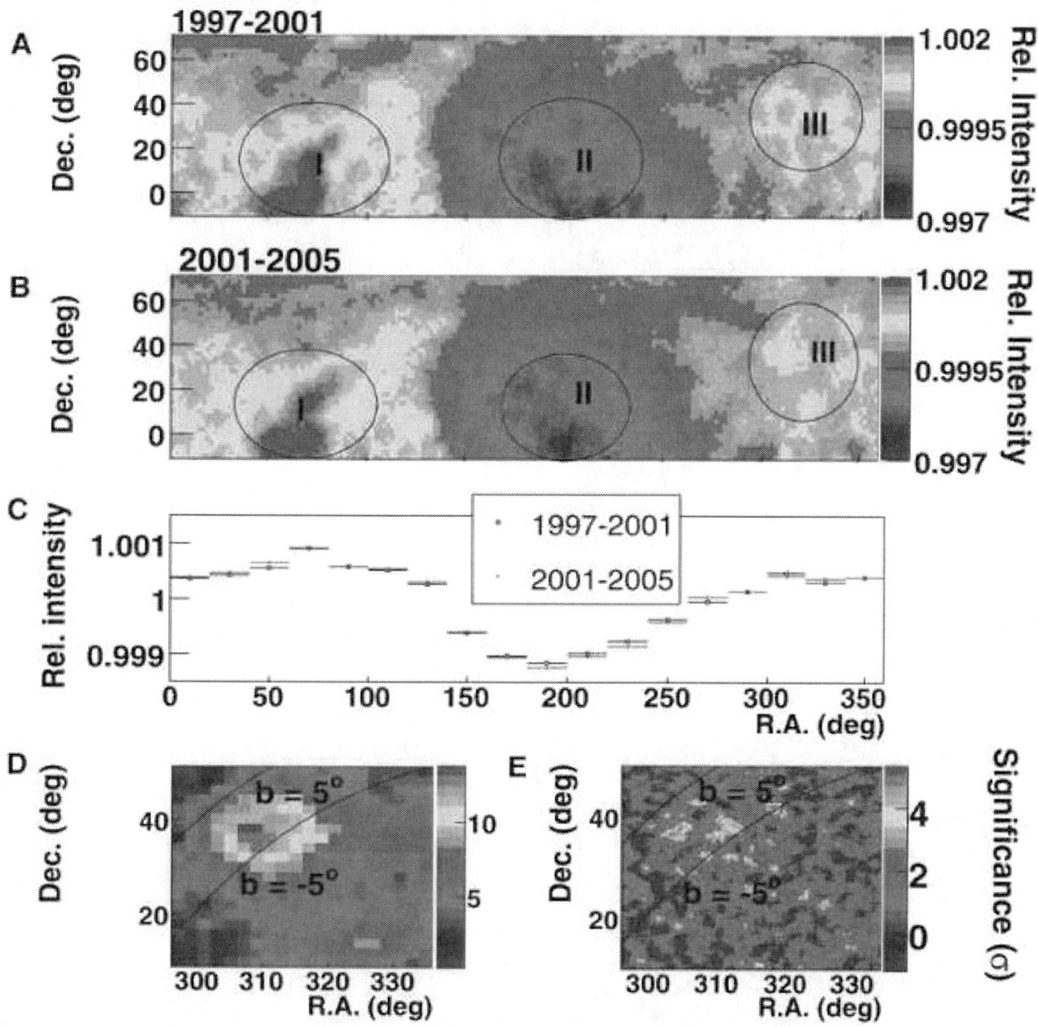

Figure 8. Equatorial CR intensity map for Tibet ASγ data taken from (A) 1997 to 2001 and (B) 2001 to 2005. The vertical color bin width is 2.5×10^{-4} for the relative intensity in both (A) and (B). The circled regions labeled by I, II, and III are the tail-in component, the loss-cone component, and the newly found anisotropy component around the Cygnus region, respectively. (C) The 1D projection of the 2D maps in R.A. for comparison. (D) and (E) show significance maps of the Cygnus region for data from 1997 to 2005. Two thin curves in (D) and (E) stand for the Galactic parallel b$\pm$5°. Small-scale anisotropies (E) superposed onto the large-scale anisotropy hint at the extended gamma-ray emission.

firmed the existence of two large scale anisotropy regions, an excess and a deficit region, in agreement with the NFJ model and with the Tibet ASγ measurement.

The Milagro collaboration published in 2009 a 2D display of the sidereal anisotropy

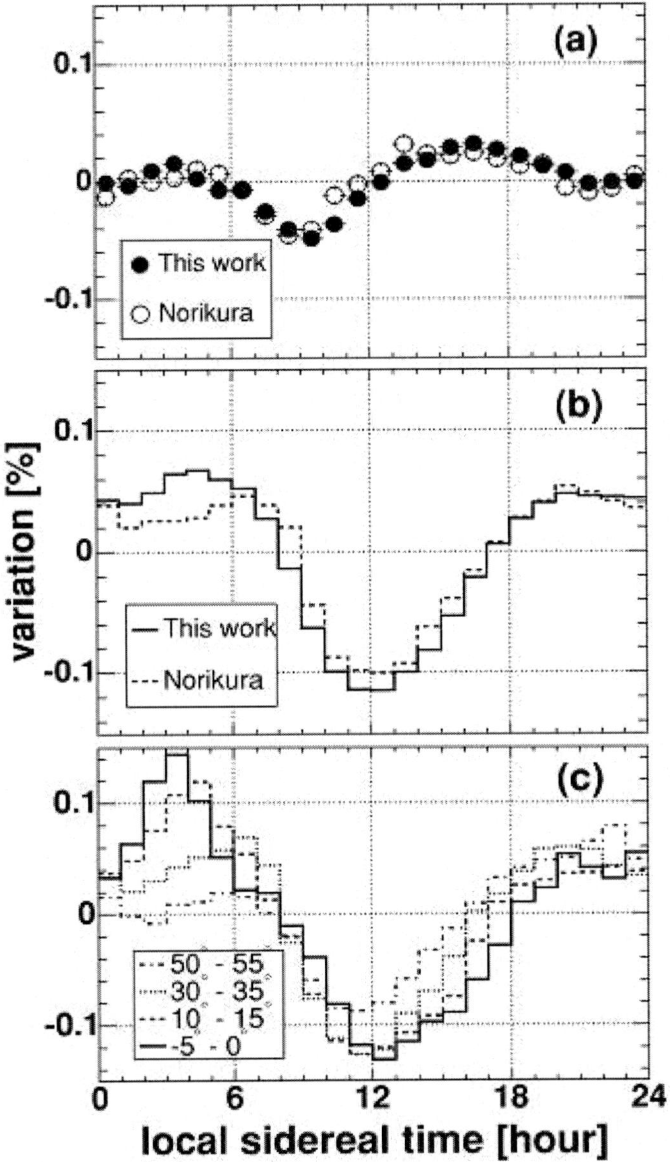

Figure 9. Sidereal daily variations averaged over all primary energies. Panels (a) and (b) show, respectively, the differential variation and the physical variation averaged over all declinations, while panel (c) shows the physical variation in four declination bands. The open circles in (a) and the dashed histograms in (b) display the variations reported from the Norikura experiment (for details see [12]).

projections in R.A. at a primary CR energy of about 6 TeV [38]. The result is shown in the figure 11. The map is generated by combining 18 separate profiles of the anisotropy projection in R.A. of width 5° in declination. The measurement is in the R.A. direction only and no information in given about the R.A.-averaged CR rate difference from one

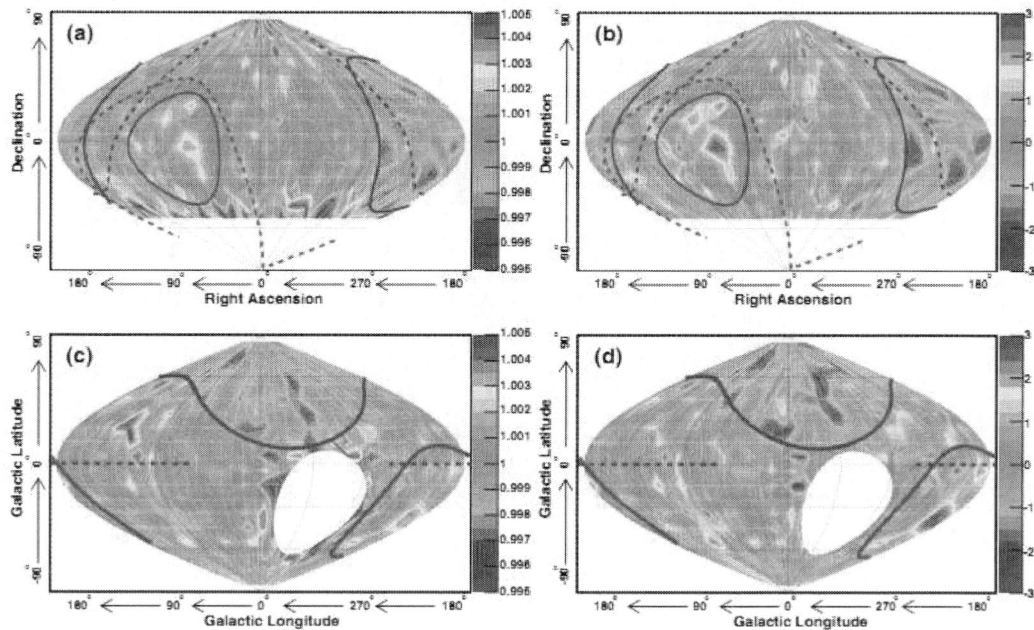

Figure 10. Sky map of the anisotropy in equatorial coordinates. The sky is divided into $10° \times 10°$ cells. Declinations less than -53.58° (white region) always lie below the horizon and are thus invisible to the detector. In (a), each cell shows the fractional variation from the isotropic flux, while in (b) it shows the standard deviation of this variation. The solid red and blue curves show the excess and deficit cones obtained using a clustering algorithm applied to the data. The dashed curves in (a,b) show excess and deficit cones from the NFJ model. (c) and (d) show the maps in (a) and (b) transformed to the galactic coordinates. The solid red and blue curves are the same cones as described above. The dashed red horizontal line indicates the direction of the Orion arm. The white patch indicates the below-horizon region.

declination band to another, i.e. on the anisotropy as a function of the declination band. In this sense, the figure 11 does not purport to be a picture of the anisotropy.

They observed a steady increase in the magnitude of the signal over seven years, in disagreement with the Tibet ASγ results [92]. It is worth noting that the energy at which the Tibet ASγ and Milagro results were obtained (~ 10 TeV) is too high for Sun effects play an important role.

The ARGO-YBJ experiment, located in Tibet (Yangbajing, 4300 m a.s.l.), reported the observation of 2D sky maps of large scale CR anisotropy in 2009 and 2011 for different primary energies [32, 33]. This is the first measurement of the CR anisotropy with an EAS array below the TeV energy region so far investigated only by underground muon detectors. The results, shown in the figure 12, are in agreement with the findings of the Tibet ASγ experiment, suggesting that the large anisotropy structures fade out to smaller spots when the energy increases.

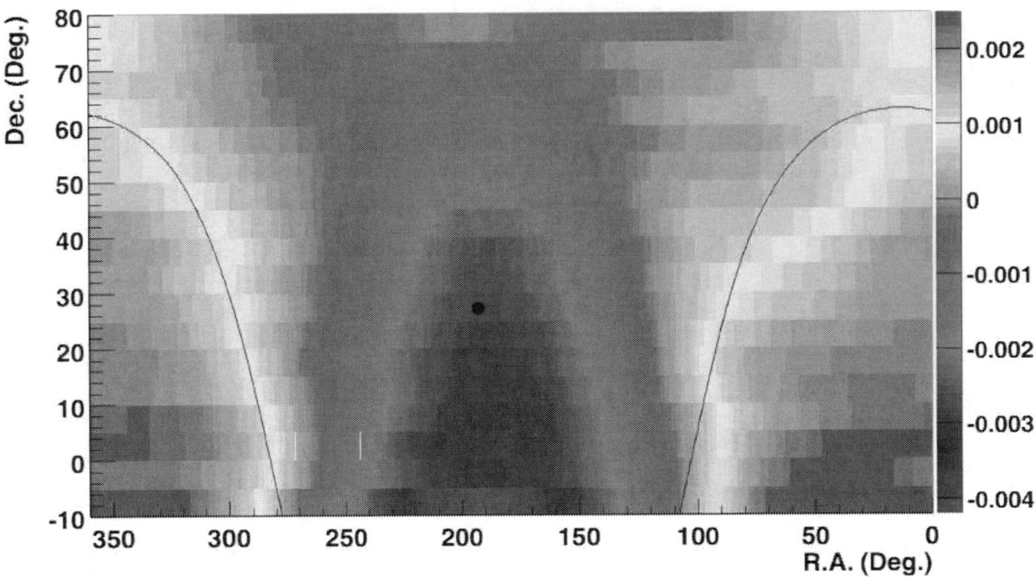

Figure 11. Result of a harmonic fit to the fractional difference of the CR rates from isotropic in equatorial coordinates as viewed by Milagro for the years 2000 - 2007. The color bin width is 1.0×10^{-4} reflecting the average statistical error. The two black lines show the position of the Galactic equator and the solid circle shows the position of the Galactic north pole. This map is constructed by combining 18 individual profiles of the anisotropy projection in R.A. of width $5°$ in decl. It is not a 2D map of the sky. The median energy of the events in this map is 6 TeV.

In 2010 the Icecube neutrino detector located at the South Pole reported the first 2D observation ever of the CR anisotropy in the southern hemisphere. This measurement was performed using muons generated in air showers by CRs with a median energy of 20 TeV [30]. Applying a spherical harmonic analysis to the relative intensity map of the CR flux, they observed significant anisotropy structures on several angular scales. Besides the large scale structure (see the figure 13), in form of intense dipole and quadrupole components, data showed several localized smaller features on scales between $15°$ and $30°$. The intensity of these spots is about a factor of five weaker than $\ell \leq 2$ ones [91].

Another important result from the Icecube collaboration was the confirmation of the EAS-TOP finding, in the northern hemisphere, that a phase shift occurs around 400 TeV [31].

3.4. Recent Observations: Medium Scale Anisotropy

In 2007, modeling the large scale anisotropy of 5 TeV CR, the Tibet-ASγ collaboration ran into a "skewed" feature over-imposed to the broad structure of the so-called tail-in region [100, 101]. They modeled it with a couple of intensity excesses in the hydrogen deflection plane [93, 94], each of them $10°$-$30°$ wide. A residual excess remained in coincidence with

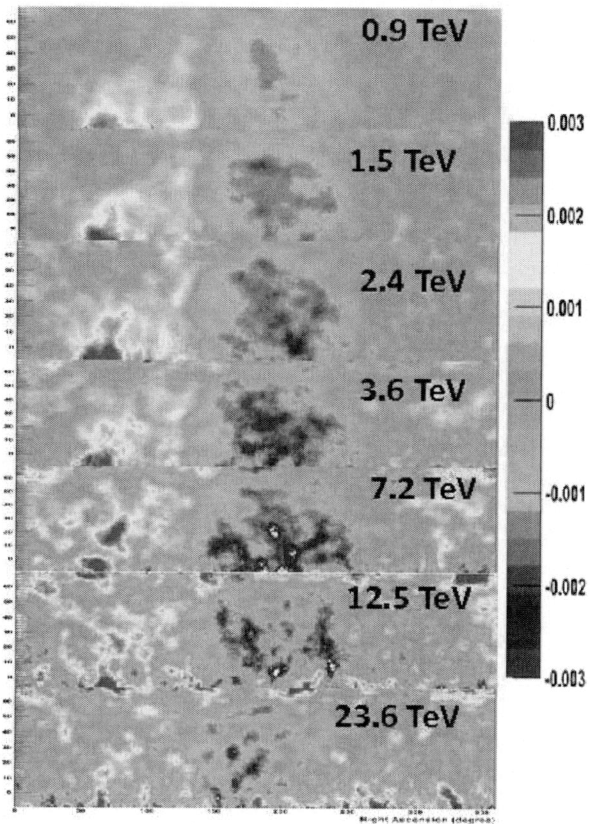

Figure 12. Large scale CR anisotropy observed by ARGO-YBJ as a function of the energy. The color scale gives the relative CR intensity.

the helio-tail. See the figure 14 (d) and its caption for more details.

Afterwards the Milagro collaboration claimed the discovery of two localized regions of excess 10 TeV CRs on angular scales of $10°$ with greater than 12 σ significance [102]. The figure 15 reports the pre-trial significance map of the observation. Regions "A" and "B", as they were named, are positionally consistent with the "skewed feature" observed by Tibet-ASγ.

The strongest and most localized of them (with an angular size of about $10°$) coincides with the direction of the helio-tail. The fractional excess of region A is $\sim 6 \times 10^{-4}$, while for region B it is $\sim 4 \times 10^{-4}$. The deep deficits bordering the excesses are due to a bias in the reference flux calculation. This effect slightly underestimates the significance of the detection. The Milagro collaboration excluded the hypothesis of gamma-ray induced excesses. In addition, they showed the excess over the large scale feature without any data handling (see the figure 2 of [102]).

The excesses in both regions are harder than the spectrum of the isotropic part of CRs.

Easy to understand, more beamed the anisotropies and lower their energy, more difficult to fit the standard model of CRs and galactic magnetic field to experimental results. That

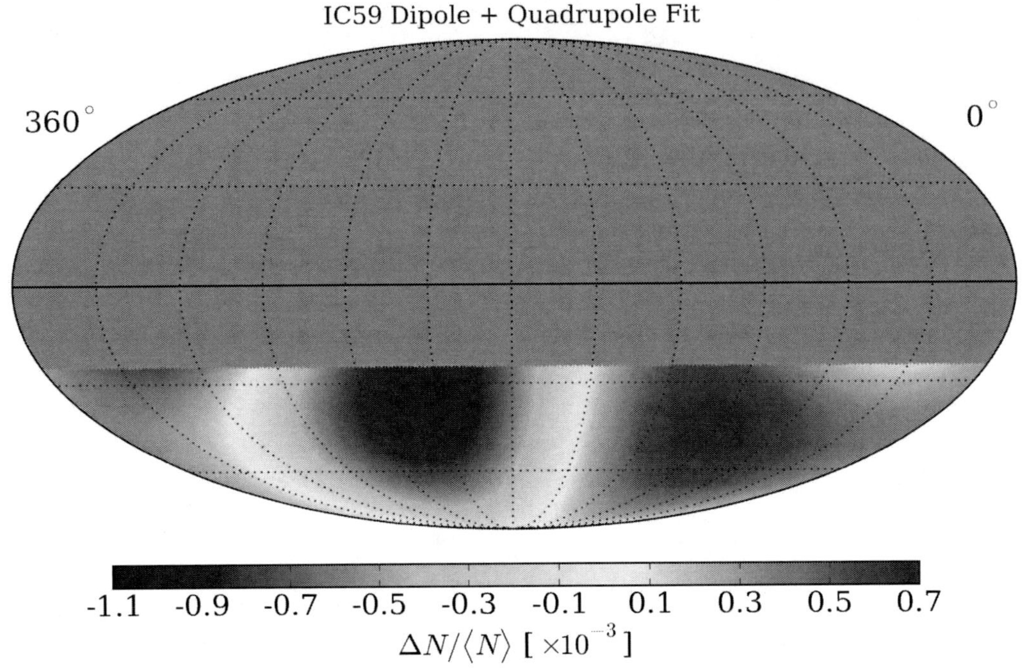

Figure 13. Sky distribution obtained by the Icecube experiment (IC59 configuration) by using the best-fit dipole and quadrupole coefficients.

is why these observations were rather surprising. In addition, the observation of a possible small angular scale anisotropy region contained inside a larger one rely on the capability for suppressing the smooth global CR anisotropy at larger scales without, at the same time, introducing effects of the analysis on smaller scales.

Nonetheless, this observation has been confirmed by the ARGO-YBJ experiment in 2009 [103]. The figure 16 shows the ARGO-YBJ sky map realized with $\approx 2 \cdot 10^{11}$ events in the declination region $\delta \sim -20° \div 80°$. The median energy of the isotropic CR proton flux is $E_p^{50} \approx 1.8$ TeV (mode energy ≈ 0.7 TeV) [104].

The most evident features are observed by ARGO-YBJ around the positions $\alpha \sim 120°$, $\delta \sim 40°$ and $\alpha \sim 60°$, $\delta \sim -5°$, positionally consistent with the Milagro A and B regions. The "region 1" and "region 2", as they there named after some differences with respect to the Milagro observation, were detected with a statistical significance of about 14 s.d. .

The multiplicity spectrum of these regions is shown in the figure 17. The spectrum of region 1 is harder than that of isotropic CRs and a cutoff around 600 shower particles (proton median energy $E_p^{50} = 8$ TeV) seems to be there. On the other hand, the excess hosted in region 2 is less intense, with a spectrum closer to that of isotropic cosmic rays. The steepening from 100 shower particles on ($E_p^{50} = 2$ TeV) is likely related to efficiency effects [105].

On the left side of the sky map, several new extended features are visible, though less

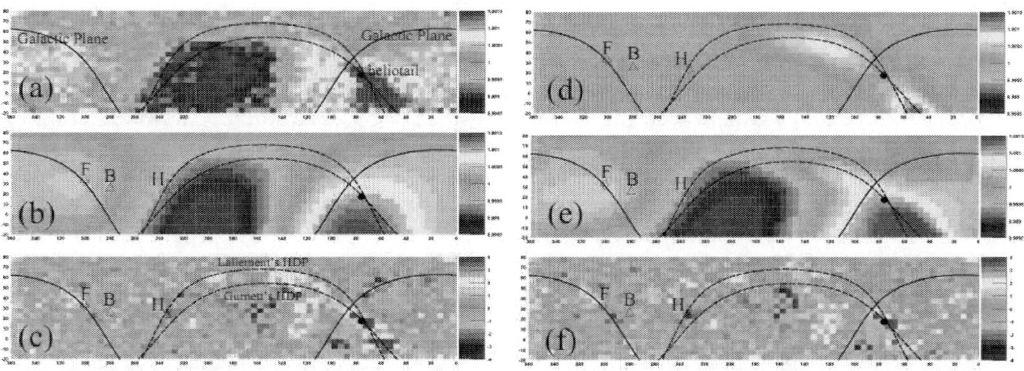

Figure 14. 2D anisotropy maps of galactic CRs observed and reproduced at the modal energy of 7 TeV by the Tibet-ASγ experiment [92]. (a): the observed CR intensity; (b): the best-fit large scale component; (c): the significance map of the residual anisotropy after subtracting the large scale component; (d): the best-fit medium scale component; (e): the best-fit large+medium scale components; (f): the significance map of the residual anisotropy after subtracting the large and the medium scale component. The solid black curves represent the galactic plane. The dashed black curves represent the Hydrogen Deflection Plane reported by [93] and [94]. The helio-tail direction $(\alpha, \delta) = (75.9°, 17.4°)$ is indicated by the black filled circle. The open cross and the inverted star with the attached characters "F" and "H" represent the orientation of the local interstellar magnetic field by [95] and [96], respectively. The open triangle with "B" indicates the orientation of the best-fit bi-directional cosmic-ray flow obtained in the reference [92].

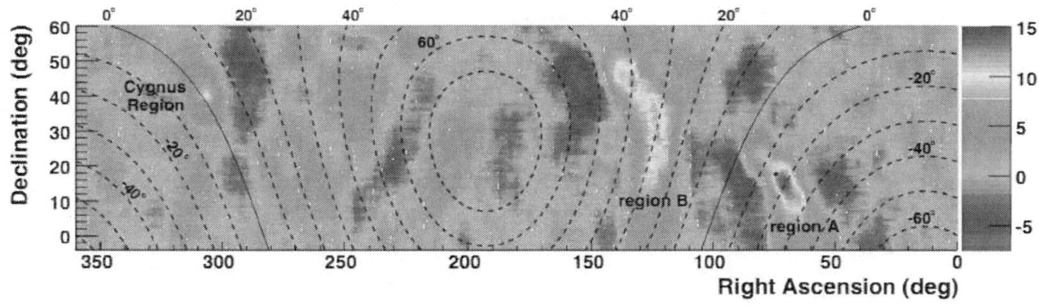

Figure 15. Significance map for the Milagro data set without any cuts to remove the hadronic CR background. A 10° bin was used to smooth the data, and the color scale gives the statistical significance. The solid line marks the Galactic plane, and every 10° in Galactic latitude are shown by the dashed lines. The black dot marks the direction of the helio-tail, which is the direction opposite the motion of the solar system with respect to the local interstellar matter.

intense than those aforementioned. The area $195° \leq R.A. \leq 315°$ seems to be full of few-degree excesses not compatible with random fluctuations (the statistical significance is more than 6 s.d.). The observation of these structures was reported by the ARGO-YBJ collaboration for the first time and together with that of regions 1 and 2 it may open the way to an interesting study of the TeV CR sky.

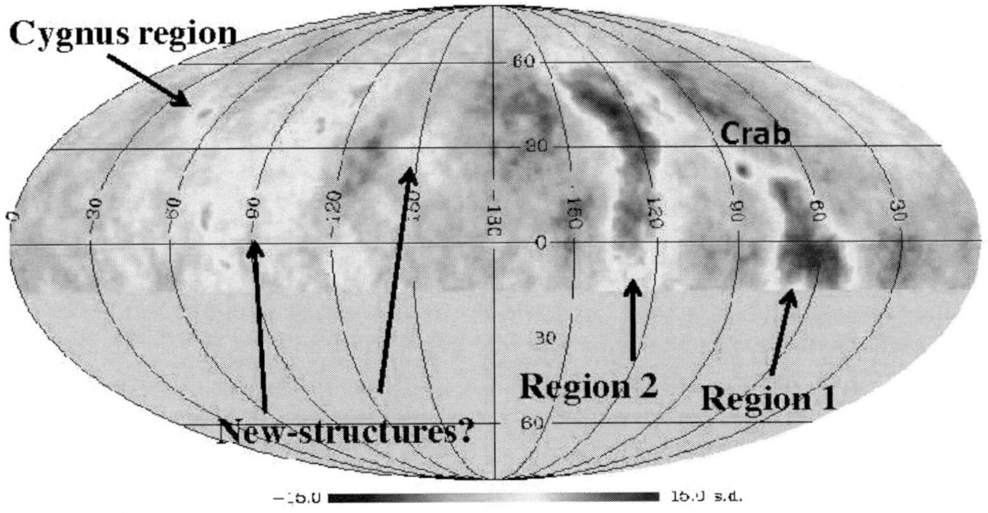

Figure 16. Intermediate scale CR anisotropy observed by ARGO-YBJ. The color scale gives the statistical significance of the observation in standard deviations.

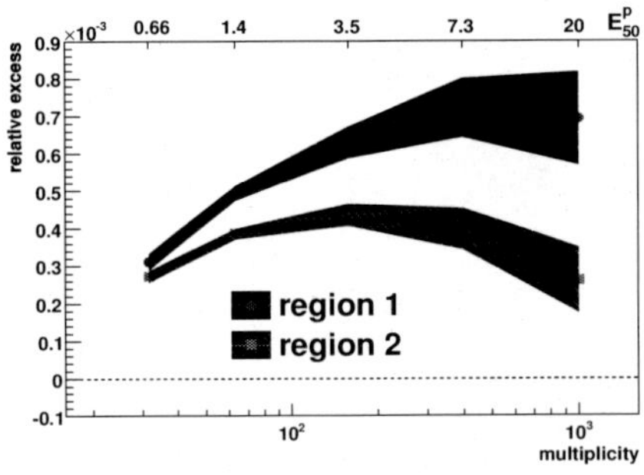

Figure 17. Size spectrum of the regions 1 and 2. The vertical axis represents the relative excess (Ev-Bg)/Bg. The upper scale shows the corresponding proton median energy.

Anticipated in the previous section, the Icecube collaboration found small scale features also in the Southern hemisphere [91]. In the figure 18 a pictorial view of the CR arrival distribution small scale structures all over the sky is reported. It has been realized for this paper by merging maps published by ARGO-YBJ and Icecube.

It is worth recalling that the Icecube experiment measures muons, making us confident

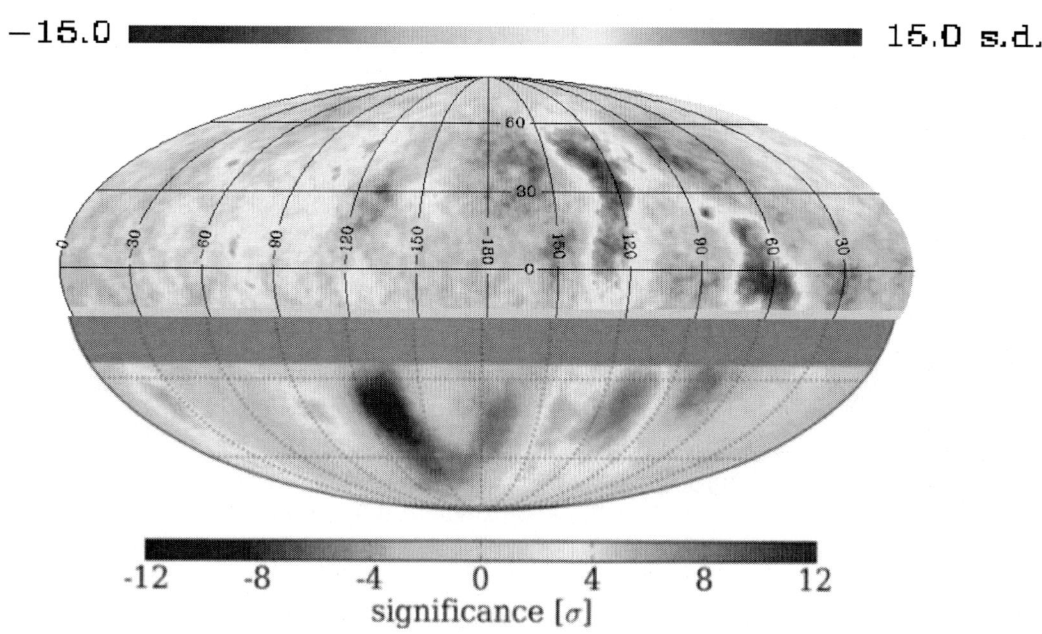

Figure 18. Pictorial view of the CR TeV sky obtained by merging data from the ARGO.YBJ and the Icecube experiments. ARGO-YBJ covered the declination range $-20° - 80°$, whereas the Icecube experiment observed the sky below $\delta = -65°$. The image has illustrative purposes, as the median energy and the angular scale for which data were optimized are different for the experiments.

that charged CRs of energy above 10 TeV are observed.

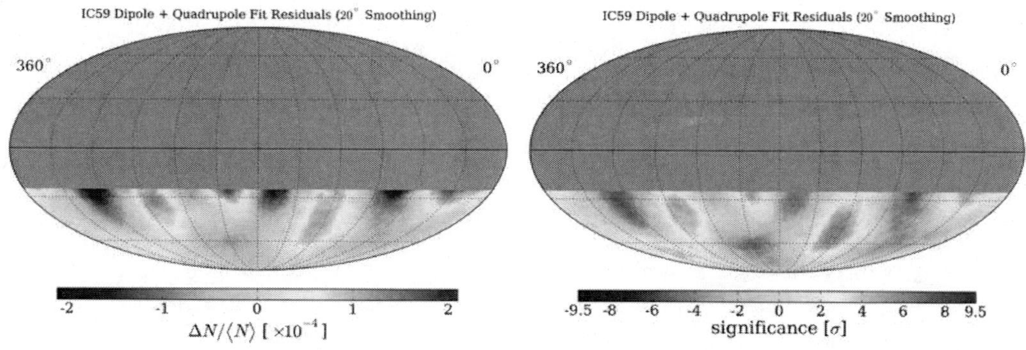

Figure 19. Left: residual intensity map plotted with 20° smoothing. Right: significances of the residual map (pre-trials), plotted with 20° smoothing.

4.   Models and Interpretations

Dealing with models of the results about the CR anisotropy, it should be noticed that people spent less effort for interpreting these observations than other traditional CR measurements, like the energy spectrum and the chemical composition. Perhaps, this is related to the few results available up to 1980s, as well as to the particular aspect of the anisotropy, which can be interpreted only if data are enough to cover wide portions of the sky. Surely, the problem of the anisotropy made a comeback with the last decade results and a number of theoretical models are expected to be formulated in the next future.

As discussed in the previous sections, on the basis of the sidereal diurnal modulations observed up to 100 TeV, most of the experiments reported a 24 hrs modulation as intense as 10^{-4} - 10^{-3} with phase of the maximum somewhere between 23 and 3 hr.

The origin of this anisotropy of galactic CRs is still unknown.

In principle any anisotropy reflects a motion: (a) the motion of the Earth/Solar System with respect to the isotropic CRs rest frame, as suggested by Compton and Getting; (b) the motion of CRs from galactic sources to the extra-galactic space and/or from extra-galactic sources to the Earth.

An observer moving with velocity v relative to the rest frame of a CR plasma will detect a deviation due to this motion from the average CR intensity, an effect first described by Compton and Getting [44]. If the CRs have a differential power-law energy spectrum, $E^{-\gamma}$, then the fractional intensity enhancement due to the CG anisotropy is expressed as

$$\frac{\Delta I}{\langle I \rangle} = (\gamma + 2) \, \frac{v}{c} \, \cos\theta \tag{8}$$

with I denoting the CR intensity, v/c the ratio of the detector velocity in the CR plasma rest frame and θ the angle between the observed CR and the moving direction of the detector [106]. The eq. 8 reads as a di-polar anisotropy with the maximum in the direction of the motion and a deficit in the opposite direction. We note that, for relativistic particles, the CG anisotropy does not depend on CR particle energy by equation 8.

As it proceeds from a general principle, the CG effect is expected to generate observable anisotropy whenever the reference frame is suitably set up.

- A CG effect due to the Earth's motion around the Sun (Solar CG, SCG) is expected with an amplitude of about 0.05% or less, depending on the geographic latitude of the detector, and with a maximum at 6:00 hr in the local solar time, as observed by different experiments;

- if CRs did not co-rotate with the Galaxy, an observer on Earth would see an apparent excess of CR intensity towards the direction of galactic rotation and a deficit in the opposite one (Galactic CG, GCG). The expected amplitude is of order 10^{-3} with an apparent relative excess around 21 hr in sidereal time.

The CG effect has been always considered as a benchmark for the reliability of the detector and the analysis method. In fact, at least as far as the SCG is concerned, all the features (period, amplitude and phase) of the signal are predictable without uncertainty, due to the exquisitely kinetic nature of the effect. Actually, there may be some noise at low energy, due to the solar modulation and other possible local magnetic features not fully understood yet.

Nonetheless, above few TeV, where the solar modulation becomes negligible, several experiments measured the SCG effect in solar time and found it in agreement with the expectation.

In 1986 using underground observations, Cutler and Groom reported the first clear signature of the SCG anisotropy for multi-TeV CRs observing a small diurnal amplitude modulation of the CR muon intensity [107]. The parent particles are sufficiently rigid (~ 1.5 TeV/c) that solar and geomagnetic effects should be negligible. The expected SCG at that latitude is 3.40×10^{-4}, with the maximum at 06:00 local solar time. Analysis of the arrival times of 5×10^{8} muons during a period of 5.4 yr yields a fractional amplitude variation of $2.5^{+0.7}_{-0.6} \times 10^{-4}$, with a maximum at $08:18 \pm 1.0$ h local mean solar time, deviating from 6:00 h by +2h at 2 σ significance. They attributed the deviation to the meteorological effects on the underground muon intensity.

The first clear observation of the SCG effect with an EAS array was reported by the EAS-TOP collaboration in 1996 with a statistical significance of 7.3 σ at an energy of about 10^{14} eV [41]. A detailed study of the SCG effect has been carried out by the Tibet ASγ experiment. In the figure 20 the solar time anisotropy observed in two different periods is shown. In the figure 21 the anisotropy is reported for two different primary CR energies. The modulations show that the observations follow the SCG model at 6.7 TeV (right panel), but some deviations are there at lower energy, 3.8 TeV (left panel). This finding suggested that some related effects may influence the CR arrival direction distribution also in the few TeV energy region [108].

In spite of these successful results about the SCG, the arrival distribution in sidereal time was never found to be purely di-polar, neither any signature of the GCG effect was ever observed in sidereal time.

Consequently, the CR plasma is supposed to co-move with the solar system and the origin of the observed anisotropy is thought to be related to "harder" effects, to be searched for in unknown features of the local interstellar medium (LISM), either for the magnetic field and the closest CR sources.

Some authors suggested that the large scale anisotropy can be explained within the diffusion approximation taking into account the role of the few most nearby and recent sources [110, 111, 112]. Other studies suggest that a non-di-polar anisotropy could be due to a combined effect of the regular and turbulent GMF [113], or to local uni- and bi-dimensional inflows [92]. In particular the authors modeled the observed anisotropy by a superposition of a large, global anisotropy and a midscale one. The first one is proposed to be generated by galactic CRs interacting with the magnetic field in the local interstellar space surrounding the heliosphere (scale $\sim$2 pc). The midscale anisotropy is possibly caused by a modulation of galactic CRs in the helio-tail [114].

The highest energy observations by EAS-TOP and ICECUBE [42, 31] of the existence of a new anisotropy around 400 TeV suggest that the global anisotropy is the superposition of different contributions due to phenomenologies at different distances from the Earth [92, 97]. A possible underlying anisotropy due to the CR sources distribution seems to manifest itself at energies above 100 TeV, being at lower energies the proton gyro-radius of the same order of the helio-tail [98, 99]. Therefore, probably in the one hundred TeV

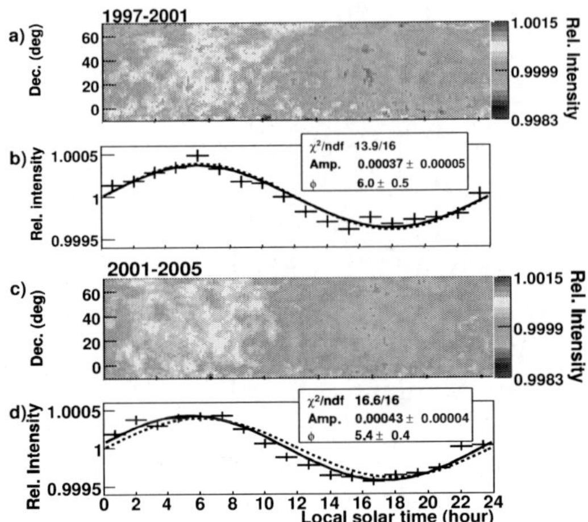

Figure 20. Solar time anisotropy observed by the Tibet ASγ experiment in 2D at modal energy of 10 TeV [90]. Figures a) and b) are the 2D map and the 1D projection of data collected in 1997-2001. Figures c) and d) refer to the period 2001-2005. The black solid line is the best-fit sine function, whereas the dashed line is the expected CG effect.

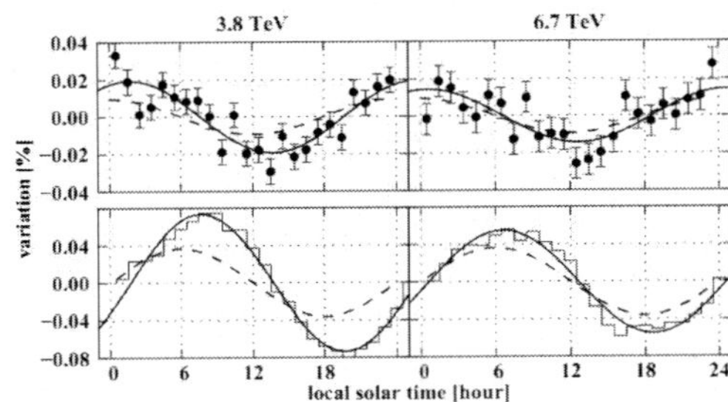

Figure 21. Observed solar daily variations divided into two energy regions compared with the expected anisotropy due to the CG effect (the broken lines). The plot in the upper panel is the differential form of the solar daily variations, and histogram in the lower panel is physical daily variations. Solid curves are the two-parameter χ^2-fitted sine curves.

energy region we are observing a transition in the cause of the anisotropy, being mainly due to local solar effects the lower energy phenomenology.

About the medium scale anisotropy, no theory of CRs in the Galaxy exists yet which is able to explain few degrees anisotropies in the rigidity region 1-10 TV leaving the standard model of CRs and that of the local galactic magnetic field unchanged at the same time. First interpretations based on observing the excess are inside the "tail-in" zone, that induced authors to drag in interactions of CRs with the heliosphere[102]. Several authors, noticing that the TeV region is usually free of heliosphere-induced effects, proposed a model where the excesses are produced in the Geminga supernova explosion [115]. In the first variant of the model CRs simply diffuse (Bohm regime) up to the solar system, while the second version limits the diffusion to the very first phase of the process and appeals to non-standard diverging magnetic field structure to bring them to the Earth. Other people [116] proposed similar schemes involving local sources and magnetic traps guiding CRs to the Earth. It must be noticed that sources are always intended to be near-by, at less than 100 - 200 pc. Moreover the position of the excesses in galactic coordinates, symmetrical with respect to the galactic plane, played an important role in inspiring such models.

Grounding on the observation that all nearby sources and new magnetic structures brought in to explain the medium scale anisotropy should imply other experimental signatures that would have been observed in the past, some other models were proposed. In [92, 114] the authors suggest that the magnetic field in the helio-tail (that is, within $\sim$70 AU to $\sim$340 AU from the Sun) is responsible for the observed midscale anisotropy in the energy range of 1 - 30 TeV. In particular, this effect is expressed as two intensity enhancements placed along the hydrogen deflection plane, each symmetrically centered away from the helio-tail direction.

The hypothesis that the effect could be related to the interaction of isotropic CRs with the heliosphere has been re-proposed in [97]. Grounding on the coincidence of the most significant localized regions with the helio-spheric tail, magnetic reconnection in the magneto-tail has been shown to account for beaming particles up to TeV energies. On a similar line, in [117] the authors suggest that even the large scale anisotropy below 100 TeV is mostly due to particle interactions with the turbulent ripples generated by the interaction of the helio-spheric and interstellar magnetic field. This interaction could be the dominant factor that re-distribute the large scale anisotropy from an underlying existent anisotropy. At the same time, such scattering can produce small angular scale features along line of sights that are almost perpendicular to the local interstellar magnetic fields.

Recently, the role of interactions of TeV - PeV CRs with a turbulent magnetic field across they propagate through has been emphasized to explain the small scale anisotropy [112]. The authors show that energy-dependent medium and small scale anisotropies necessarily appear, provided that there exists a large scale di-polar anisotropy, for instance from the inhomogeneous source distribution. The small scale anisotropies naturally arise from the structure of the local turbulent GMF, typically within the CR scattering length.

There has been also the suggestion that CRs might be scattered by strongly anisotropic Alfven waves originating from turbulence across the local field direction [118].

Besides all these "ad hoc" interpretations, several attempts occurred in trying to insert the CR excesses in the framework of recent discoveries from satellite-borne experiments, mostly as far as leptons are concerned. In principle there is no objection in stating that few-degree CR anisotropies are related to the positron excess observed by Pamela [119] and to the electrons excess observed by Fermi [120]. All observations can be looked at as different signatures of common underlying physical phenomena.

5. Conclusions

The cosmic ray arrival direction distribution and its anisotropy has been a long-standing problem ever since the 1930s. In fact, the study of the anisotropy is a powerful tool to investigate the acceleration and propagation mechanism determining the CR world as we know it. Nonetheless, with respect to other classics of CR physics, fewer experiments managed to get significant results, due to the small intensity of the effect.

From the theoretical viewpoint, apart from some effects of kinetic nature (SCG and GCG effect), no signal is expected either in the sidereal and the solar time reference frame. If the energy is high enough, i.e. solar effects are negligible, the standard model of production, acceleration and propagation of CRs does not foresee any deviation from the isotropy. In this picture of the outer space, a major role is played by the magnetic fields the CRs pass through before touching the Earth atmosphere. When the energy increases enough, a weak dipole may arise, as a mild signature of the closest and most recent CR sources.

So far, the anisotropy in the CR arrival direction distribution have been observed by different experiments with increasing sensitivity and details at different angular scales.

Current experimental results show that the main features of the anisotropy are uniform in the energy range (10^{11} - 10^{14} eV), both with respect to amplitude (10^{-4} - 10^{-3}) and phase ((0 - 4) hr). The existence of two distinct broad regions, one showing an excess of CRs (called "tail-in"), distributed around $40°$ to $90°$ in R.A., the other a deficit (the "loss cone"), distributed around $150°$ to $240°$ in R.A., has been clearly observed.

The framework became clearer in the last decade, when more and more experiments demonstrated to have the capability of drawing 2D sky-maps, thus giving to the scientific community the evidence of localized regions whose physical nature is still unknown.

In the last years the Milagro and ARGO-YBJ Collaborations reported evidence of the existence of a medium angular scale anisotropy contained in the tail-in region. The observation of similar small scale anisotropies has been recently claimed also by the Icecube experiment in the southern hemisphere.

So far, no theory of CRs in the Galaxy exists which is able to explain both large scale and few degrees anisotropies leaving the standard model of CRs and that of the local galactic magnetic field unchanged at the same time.

A joint analysis of concurrent data recorded by different experiments in both hemispheres, as well as a correlation with other observables like the interstellar energetic neutral atoms distribution [121, 122], should be a high priority to clarify the observations.

References

[1] Lagage P.O. and Cesarsky C.J., *Astron. Astrophys.*, 125, 249 (1983).

[2] Beck R., *Space Sci. Rev.*, 99, 243 (2001)

[3] Berezinsky V.S. et al., *Astrophysics of cosmic rays*, V.S. Ginzburg editor, North-Holland (1990).

[4] Swordy S.P., *24th ICRC Proc.*, Rome, IT; 2, 697 (1995).

[5] Ptuskin V.S. et al., *Astron. Astrophys.*, 268, 726 (1993).

[6] Ambrosio M. et al., *Astropart. Phys.*, 7, 109 (1997).

[7] Ambrosio M. et al., *Nucl. Instrum. Methods Phys. Res. A*, 486, 663 (2002).

[8] Linsley J., *Phys. Rev. Lett.*, 34, 1530 (1975).

[9] Kolhorster W., *Phys. Z*, 42, 55 (1941).

[10] Alfven H. et al., *Arkiv. Mat. Astron. Fysik,* 29A, 24 (1943).

[11] Elliot H. et al., *J. Atmos. Terr. Phys.,* 1, 205 (1951).

[12] Nagashima K. et al., *Nuovo Cimento C,* 12, 695 (1989).

[13] Bonino R. et al., *ApJ,* 738, 67 (2011).

[14] Swinson D.B. and Nagashima K., *Planet Space Sci.,* 33, 1069 (1985)

[15] Nagashima K. et al., *Planetary Space Science,* 33, 395 (1985a).

[16] Ueno H. et al., *Proc. 21th ICRC,* Adelaide, AUS; 6, 361 (1990).

[17] Thambyaphillai T., *Proc. 18th ICRC,* Bangalore, IN; 3, 383 (1983).

[18] Munakata K. et al., *Proc. 24th ICRC,* Rome, IT; 4, 639 (1995).

[19] Mori S. et al., *Proc. 24th ICRC,* Rome, IT; 4, 648 (1995).

[20] Bercovich M. and Agrawal S.P., *Proc. 17th ICRC,* Paris, FR; 10, 246 (1981).

[21] Fenton K.B. et al., *Proc. 24th ICRC,* Rome, IT; 4, 635 (1995).

[22] Cutler D.J. et al., *ApJ* 248, 1166 (1981).

[23] Cutler D.J. et al., *ApJ* 376, 322 (1991).

[24] Lee Y.W. and Ng L.K., *Proc. 20th ICRC,* Moscow, URSS; 2, 18 (1987).

[25] Bergamasco L. et al., *Proc. 21th ICRC,* Adelaide, AUS; 6, 372 (1990).

[26] Andreyev Y.M. et al., *Proc. 20th ICRC,* Moscow, URSS; 2, 22 (1987).

[27] Munakata K. et al., *Phys. Rev. D* 56, 23 (1997).

[28] Guillian G. et al., *Phys. Rev. D,* 75, 062003 (2007).

[29] Ambrosio M. et al., *Phys. Rev. D,* 67, 042002 (2003).

[30] Abbasi R. et al., *ApJL,* 718, L194 (2010).

[31] Abbasi R. et al., *ApJ,* 746, 33 (2012).

[32] Cui S.W. et al., *Proc. of the 31th ICRC,* Lodz, PL; ID 0814 (2009).

[33] Cui S.W. et al., *Proc. of the 32th ICRC,* Beijing, CN; 041 (2011).

[34] Alexeenko V.V. et al., *Proc. 17th ICRC,* Paris, FR; 2, 146 (1981).

[35] Munakata K. et al., *Proc. of the 26th ICRC,* Salt Lake City, USA; 7, 293 (1999).

[36] Gombosi T. et al., *Nature,* 255, 687 (1975a).

[37] Amenomori M. et al., *ApJL,* 626, L32 (2005).

[38] Abdo A.A. et al., *ApJ*, 698, 2121 (2009).

[39] Morello C. et al., *Proc. of the 18th ICRC*, Bangalore, IN; 2, 137 (1983).

[40] Aglietta M. et al., *Proc. of the 24th ICRC*, Rome, IT; 2, 800 (1995).

[41] Aglietta M. et al., *ApJ* 470, 501 (1996).

[42] Aglietta M. et al., *ApJL*, 692, L130 (2009).

[43] Wollan E.O., Rev. Mod. Phys., 11, 160 (1939).

[44] Compton A.H. and Getting I.A., *Phys. Rev.,* 47, 817 (1935).

[45] Daudin A. and Daudin J., *J. Atm. Terr. Phys.,* 3, 245 (1953).

[46] Escobar I. et al., *Planet Space Sci.,* 1, 155 (1959).

[47] Cachon A., *Proc. Fith Inter-American Seminar on Cosmic Rays*, 2 (1962).

[48] Farley F.J.M. and Storey J.M., *Proc. Phys. Soc.,* A67, 996 (1954).

[49] Fenton A.G., *Proc. Int. Symp. on High Energy Cosmic Ray Modulation*, Tokio, JP; 92 (1976).

[50] Somogyi A.J., *Proc. Int. Symp. on High Energy Cosmic Ray Modulation*, Tokio, JP; 85 (1976).

[51] Jacklyn R.M., *Proc. 9th ICRC*, London, UK; 1, 141 (1965).

[52] Jacklyn R.M., *Proc. ASA*, 6, 425 (1986).

[53] Sekido Y et al., *Proc. 12th ICRC*, Hobart, AUS; 1, 302 (1971).

[54] Jacklyn R.M. and Vrana A., *Proc. ASA*, 1, 278 (1969).

[55] Jacklyn R.M. and Cooke D.J., *Proc. 12th ICRC*, Hobart, AUS; 1, 290 (1971).

[56] Sakakibara S. et al., *Proc. 13th ICRC*, Denver, USA; 2, 1058 (1973).

[57] Sakakibara S. et al., *Proc. 14th ICRC*, Munich, GE; 4, 1503 (1975).

[58] Sakakibara S. et al., *Proc. Int. Symp. on Cosmic Ray Modulation in the Heliosphere*, Morioka, JP; 314 (1984).

[59] Fenton A.G. and Fenton K.B., *Astron. Soc. Aust.*; 2, 139 (1972).

[60] Ueno H. et al., *Proc. Int. Symp. on Cosmic Ray Modulation in the Heliosphere*, Morioka, JP; 349 (1984).

[61] Ueno H. et al., *Proc. 19th ICRC*, La Jolla, USA; 5, 35 (1985).

[62] Fenton A.G., *Proc. 14th ICRC*, Munich, GE; 11, 3907 (1975).

[63] Linsley J., *Proc. 18th ICRC*, Bangalore, IN; 12, 135 (1983).

[64] Gombosi T. et al., *Proc. 14th ICRC Proc.*, Munich, GE; 2, 586 (1975b).

[65] Nagashima et al., *Proc. 14th ICRC*, Munich, GE; 4, 1503 (1975).

[66] Fenton A.G. and Fenton K.B., *Proc. 14th ICRC*, Munich, GE; 4, 1482 (1975).

[67] Linsley J. and Watson A.A., *Proc. 15th ICRC*, Plovdiv, BUL; 12, 203 (1977).

[68] Gombosi T. et al., *Proc. 15th ICRC*, Plovdiv, BUL; 11, 109 (1977).

[69] Sakakibara S., *Proc. Int. Symp. on High Energy Cosmic Ray Modulation*, Tokio, JP; 316 (1976).

[70] Sakakibara S. et al., *Proc. 16th ICRC*, Kyoto, JP; 4, 216 (1979).

[71] Davies E. et al., *Proc. 16th ICRC*, Kyoto, JP; 4, 210 (1979).

[72] Bergeson C. et al., *Proc. 16th ICRC*, Kyoto, JP; 4, 188 (1979).

[73] Bercovitch M. et al., *Proc. 17th ICRC*, Paris, FR; 10, 246 (1981).

[74] Nagashima K. et al., *Planetary Space Science*, 33, 1069 (1985b).

[75] Alexeenko V.V. and Navarra G., *Nuovo Cimento Lett,* 42, 321 (1985).

[76] Nagashima K. et al., *J. of Geoph. Res.,* 103, 17429 (1998).

[77] Karapetyan G.G., *Astrop. Phys.*, 33, 146 (2010).

[78] Hall D.L. et al., *J. of Geoph. Res.,* 103, 367 (1998).

[79] Hall D.L. et al., *J. of Geoph. Res.,* 104, 6737 (1999).

[80] Hillas A.M., *Ann. Rev. Astron. Astroph.*, 22, 425 (1984).

[81] Fitchel C.E. and Linsley J., *ApJ*, 300, 474 (1986).

[82] Ghia P., *Proc. Vulcano Workshop 1987, SIF Conf. Proc.* vol. 93, 475 (1987).

[83] Antoni T. et al., *ApJ*, 604, 687 (2004).

[84] Nagashima K. et al., *Proc. 21st ICRC*, Adelaide, AUS: 3, 180 (1990).

[85] Aglietta M. et al., *Proc. 28th ICRC*, Tsukuba, JP; 1, 183 (2003).

[86] Kifune, T. et al., *J. Phys. G,* 12, 129 (1986).

[87] Gerhardy, P. and Clay, R., *J. Phys. G*, 9, 1279 (1983).

[88] Candia J. et al., J. Cosmol. *Astropart. Phys.*, 5, 3 (2003).

[89] Shibata T. et al., *ApJ*, 612, 238 (2004).

[90] Amenomori M. et al., *Science*, 314, 439 (2006).

[91] Abbasi R. et al., *ApJ*, 740, 16 (2011).

[92] Amenomori M. et al., *Astrophys. Space Sci. Trans.*, 6, 49 (2010).

[93] Gurnett D.A. et al., *AIP Conf. Proc.*, 858, 129 (2006).

[94] Lallement R. et al., *Science*, 307, 1447 (2005).

[95] Frisch P.C., *Space Sci. Rev.,* 78, 213 (1996).

[96] Heerikhuisen J. et al., *ApJL* 708, L126 (2010).

[97] Lazarian A. and Desiati P., *ApJ*, 722, 188 (2010).

[98] Pogorelov N.V. et al., *Adv. Space Res.,* 44, 1337 (2009).

[99] Pogorelov N.V. et al., *ApJ*, 696, 1478 (2009).

[100] Amenomori M. et al., *AIP Conf. Proc.*, 932, 283 (2007).

[101] Amenomori M. et al., *Proc. of the 31th ICRC*, Lodz, PL; 296 (2009).

[102] Abdo A.A. et al., *Phys. Rev. Lett.,* 101, 221101 (2008).

[103] Vernetto S. et al., *Proc. of the 31th ICRC*, Lodz, PL; (2009) [arXiv:0907.4615].

[104] Di Sciascio G. and Iuppa R., *Proc. of the 32th ICRC*, Beijing, CH; 0507 (2011).

[105] Di Sciascio G. and Iuppa R., [arXiv:1112.0666]

[106] Gleeson L.I. and Axford W.L., *Ap. Science Sci. ,* 13, 115 (1965).

[107] Cutler D.J. and Groom D.E., *Nature*, 322, 434 (1986).

[108] Amenomori M. et al., *Proc. of the 28th ICRC*, Tsukuba, JP; 7, 3917 (2003).

[109] Amenomori M. et al., *Phys. Rev. Lett.* 93, 061101 (2004).

[110] Blasi P. and Amato E., arXiv:1105.4521 and arXiv:1105.4529 (2011).

[111] Erlykin A.D. and Wolfendale A.W., *Astropart. Phys.*, 25, 183 (2006).

[112] Giacinti G. and Sigl G., arXiv:1111.2536 (2011).

[113] Battaner E. et al. *ApJ* 703, L90 (2009).

[114] Amenomori M. et al. *Proc. of the 32th ICRC*, Beijing, CH; 361 (2011).

[115] Salvati M. and Sacco B., å, 485, 527 (2008).

[116] Drury L. and Aharonian F., *Astropart. Phys.*, 29, 420 (2008).

[117] Desiati P. and Lazarian A., arXiv:1111.3075 (2011).

[118] Malkov M.A. et al., *ApJ*, 721, 750 (2010)

[119] Adriani A. et al., *Nature*, 458 607 (2009).

[120] Abdo A.A. et al., *Phys. Rev. Lett.,* 102 181101 (2009).

[121] McComas D.J. et al., *Science*, 326 959 (2009).

[122] McComas D.J. et al., *Geoph. Res. Lett.*, 38 L18101 (2011).

In: Homage to the Discovery of Cosmic Rays …
Editor: Jorge A. Perez-Peraza

ISBN: 978-1-63483-084-3
© 2015 Nova Science Publishers, Inc.

Chapter 10

THE INFLUENCE OF COSMIC RAYS ON ANTARCTIC OZONE DEPLETION

Manuel Alvarez-Madrigal[*]
Tecnológico de Monterrey Campus Ciudad de México,
Col. Ejidos de Huipulco, Tlalpan, México D. F. México

ABSTRACT

Cosmic ray (CR) influence on the Earth's climate system has been documented in many cases, from ozone depletion during Ground Level Enhancements (GLEs) to cloudiness within the Earth's atmosphere as a function of CR intensity over time. In this chapter, CR influence on Antarctic ozone depletion over the short and long term is discussed. Such relationships have been studied for several years but have been difficult to investigate. Several indicators have shown the existence of relationships.However, over the long term the quantification of their strength has remained inconclusive. First, a mechanism for short term influence was suggested. More recent, the physical mechanism needed to explain the long term influence of CR on ozone depletion and, also, the contribution of the CR on long term ozone depletion in the Antarctic region has been estimated. A review of these findings is provided here.

INTRODUCTION

Cosmic radiation incident on the Earth has an impact on the atmosphere and climate. However, these connections have not been easy to prove. Since charged particles first interact with the Earth's electromagnetic fields, and since CR mainly consists of charged particles (protons, electrons, and ions) with very high energies (from 10^6eV to 10^{20}eV), CR interacts with the Earth in many ways. The shock wave at the border of the magnetosphere is the first obstacle that these particles pass with relative ease. As a result of high kinetic energy, protons with energies of approximately 1 MeV reach the mesopause; 10 MeV protons reach a height

---

[*] E-mail address: mmadrigal@itesm.mx

of approximately 65 km over the Earth surface; and protons with 100 MeV reach approximately 30-35 km [1]. Once inside the magnetosphere, these particles begin their interaction with our planet producing small electromagnetic disturbances in thegeomagnetic field. When these disturbances come together and extend throughout the magnetosphere, they produce alterations that are detected at ground level and that cause the precipitation of charged particles originally trapped around the Earth. All of these particle precipitations collide and interact with the upper layers of the atmosphere, producing effects on the entire Earth system. CR is present throughout the interplanetary medium where the Earth moves and is a type of background that continuously bombards the planet. However, this "rain" of fast particles presents variations in intensity caused by solar activity that produces periodic disturbances in the interplanetary medium that surrounds Earth and that forms better shielding tothe CR every 11 years when solar activity reaches a maximum. The Sun itself produces fast particles of very high energies that are classified as CR of solar origin [2].

CR charged particles that approach and precipitate within the lower layers of Earth's atmosphere potentially impact global weather conditions in various layers of the atmosphere. The particles can even impact global climate because the precipitation of cosmic radiation occurs everywhere on Earth. However, although these particles precipitate around the globe because they have an electric charge, this type of radiation has a greater ease of penetration in regions where magnetic field lines are perpendicular to the Earth's surface. As a result, the magnetic field does not oppose resistance to charged particles that move toward the surface, following a trajectory parallel to magnetic field lines. Thus, regions near Earth's magnetic poles have a higher precipitation of charged particles that reach lower layers and that interact with atmospheric gas molecules. Since magnetic field lines are parallel to the surface at the equator, and tend to prevent the passage of charged particles having a trajectory orthogonal to the lines [3], the area under the magnetic equator has a greater resistance in regards to spreading particles to the surface. Therefore, the polar regions (Arctic and Antarctic) are favorable for searching for evidence of the direct influence of CR over global climate events.

Since it is necessary to perform experimental observations, finding a relationship to the atmosphere has its difficulties, not only at the surface, but at different heights. The difficulty of measuring in polar regions results from its unfriendly climate for humans and from its distance from human populations. The establishment of permanent research stations has alleviated the difficulties of obtaining measurements at the surface. However, when measurements are needed at heights above the surface aircraft, rockets, and balloons are necessary. However, the amount of data that can be recorded at once is short, making long-term research difficult, since record keeping for various heights and long periods of time is, thus far, not possible [e.g. 4].

Recently, thanks to measurements made with lasers and with meteorological satellites and indirect measurements on board, progress has been made on the study of the relationship between cosmic radiation and the global phenomena that impact climate, although there is not yet enough data to investigate the types of phenomena that occur at various altitudes. In some cases, investments necessary to study phenomena over the long term have been made, helping us to achieve a nearly continuous database of some atmospheric features and to investigate possible relationships such as ozone depletion and CR flux in Antarctica. Due to the fact that ozone depletion and CR flux are phenomenon of general interest and have a global impact on ecosystems, investments have been made to collect data for both and to investigate their likely relationship.

Antarctic ozone has been monitored in detail since scientists predicted the possible existence of a hole in the stratospheric ozone layer due to anthropogenic pollution from chlorofluorocarbons (CFCs) [5]. The confirmation of the existence of such a recursive structure had a positive impact on the scientific and technological fields, allowing confirmation of the hypothesis that human activities cause environmental conditions to climatic change, over a time scale similar to the human lifespan. Thus, the possibility that one person can determine environmental damage and can observe its results exists. Such a realization led to agreements such as the Montreal Protocol, to investments in manufacturing and product designto reduce CFC useand to the adoption of actions to mitigate the effects of CFCs and to reverse the negative environmental impacts caused by these products. The success of agreements such as the Montreal Protocol for maintaining the control of the production of ozone-depleting substances [6] has raised the appearance of ways of thinking in order to protect the environment, such as recent programs in sustainable development that seek to harmonize human activities and the proper use of natural resources so as not to diminish resources for future generations, while taking care of social and economic impacts over the long term. Such an example indicates the close relationship between scientific activity, lifestyles, and the industrial activity of mankind as a whole. The stratospheric ozone layer has great relevance for the protection of life on the planet [e.g. 6]. Research regarding the causes and the origins of stratospheric ozone depletion have led to much progress in technological developments (such as the development of substitutes for CFCs) and theoretical discoveries in science (such as heterogeneous chemistry). The causes of stratospheric ozone depletion can be classified as natural, those generated by human activity, and those arising from the interaction of both.

Natural causes may include volcanic eruptions, the emission of ozone-depleting substances emitted from the oceans, increases in solar radiation, temperature variations in the Earth's atmosphere, and, of course, CR. Fast particles such as CR are very energetic and are precipitated mainly in polar regions toward the Earth's surface. In their path collide with the constituents of the atmosphere, causing warming, chemical reactions, cloud formation [7], and changes in the electrical conductivity of the atmosphere. As a coincidence, the height which has the most interactions is at the same height at which the ozone hole forms, a fact that not necessarily imply a connection between both phenomena but encouragesto searching for evidence of the physical relationships between both. The results of this search have shown a relationship between the destruction of stratospheric ozone and CR on different time scales [e.g. 8]. Over the short term, the influence of very energetic events perceived as large enhancements of counts in neutron monitors at ground level have been quantified, and predictions based on very specific simulations have been confirmed by field observations. The result was possible because these confirmations require monitoring only for short periods of time, at most in the range of months [e.g. 9]. However, to confirm assumptions and predictions over the long term requires monitoring periods in the range of years in order to determine evidence of action for the proposed mechanisms [e.g. 10, 11] which are more complicated. Below a review of the main findings regarding the influence of CR on stratospheric ozone depletion is provided.

INFLUENCE ON THE SHORT TERM

High-speed charged particles that impact the Earth from outer space and secondary charged particles created when they collide with atmospheric constituents influence the ozone creation-destruction process. The Earth's magnetic field and electric current systems within the magnetosphere surrounding Earth direct the total flow of particles into the atmosphere where most interact with gases and particles in the mesosphere and stratosphere. Sometimes the flux of charged particles produce variations in the atmospheric abundance of nitrogen and hydrogen compounds. The changes are associated also with a decrease in the level of atmospheric ozone in the corresponding layers. Frequently, the effects on local ozone are temporary, and a recovery is observed in a matter of days in cases of disturbances in the atmosphere at medium or high altitudes. At other times, when the disturbance associated with particles is in the stratosphere, the effects may last for years [11], as recently documented as a result of the availability of almost continuous satellite measurements for the abundance of ozone at different heights in the atmosphere. Precipitation events of charged particles in the atmosphere can be grouped into three types of disturbances: (1) solar particle events that mainly consist of fast protons entering the polar regions, often called solar proton events (SPEs); (2) energetic electrons that precipitate into aurora regions; and (3) galactic cosmic rays (GCRs) that continually produce odd nitrogen and hydrogenous compoundsin the low stratosphere and upper troposphere.

The SPEs and precipitating electrons influence ozone, mainly in polar regions. Such particles produce the ionization, excitation, dissociation, and dissociative ionization of atmospheric constituents within their paths. SPEs mainly deposit their energy in the mesosphere and stratosphere, while energetic electrons deposit their energy in the thermosphere and upper atmosphere (at greater heights). HO_x (odd hydrogen, atomic hydrogen (H), hydroperoxyl radical (HO_2), and hydroxyl radical (OH)) is created by a series of chemical reactions [e.g.12] resulting from the interaction of the media with these charged particles. Solar particles and associated secondary particles (a product of their collisions within the atmosphere) also create atomic nitrogen (N_2). N_2 allows the production of odd nitrogen, NO_y (reactive nitrogen that also includes atomic nitrogen (N), nitric oxide (NO, NO_2), nitrogen trioxide (NO_3), dinitrogen radical (N_2O_5), nitric acid (HNO_3), peroxynitric acid (HNO_4), $BrONO_2$, and $ClONO_2$, through various chemical reactions. HO_x increases in the polar mesosphere and upper atmosphere, and provokes an ozone decrease in the short-term due to the short lifetimes of the various compounds of HO_x. Such effects were first reported by Weeks et al. 1972 [13] and explained in Swider and Keneshea 1973 [14]. In contrast, due to the long lifetime for the family of NO_y compounds in polar region, an increase in NO_y produces changes in long and short-term polar stratospheric ozone. A physical mechanism for this was suggested by Crutzen et al. 1975 [15] and reported by Heath et al. 1977 [16]. Since that time, much work has been published documenting the changes in stratospheric ozone caused by SPEs within polar regions. A review of these processes can be found in Jackman and McPeters 2004 [17].

Although HO_x induces (mainly) short-lived ozone decreases, the contribution of HO_x is important for understanding the dynamics of the mesosphere and stratosphere. However, increases in NO_y are responsible for ozone decreases of a greater magnitude when they are associated with charged particles in stratospheric ozone. Currently there is evidence of substantial increases in the amount of NO_y in the mesosphere and stratosphere as a direct result of SPEs. For example, it has been reported that the large solar proton fluxes that occurred during July 2000 produced large increases of NO_x ($NO + NO_2$) in the Arctic [18]. Also reported is that large proton fluxes associated with SPEs, registered during October and November 2003, caused large amounts of NO_x [e.g. 19]. Additionally, there was an increase in other constituents such as HNO_3, NO_2O_5, and $ClONO_2$ as reported in [11]. All of these occurred as a direct result of SPEs that occurred in October/November 2003. Also, increases in NO_y over a longer duration that impact ozone depletion have been observed. For example, Randall et al. 2001 [20] found that during September 2000 an increase of NO_x occurred in the stratosphere, corresponding to the production associated with such event SPEs occurring during July 2000. An increase in NO_2 during December 2003 was also reported, likely caused by the SPEs that occurred during October-November 2003 [e.g. 21].

The production of these compounds leads to the destruction of stratospheric ozone. Once the processes that produce these additional compounds end ozone returns to its previous levels as a resultof the balance that occurs between the processes of creation and destruction for ozone in the stratosphere. Therefore, the effects of SPE events are not continuous because although the occurrence frequency follows the solar activity cycle SPEs occur sporadically. The magnitude of ozone decreases observed in stratospheric ozone, and associated with SPEs, are of a large scale. For example, during September 2000 a decrease in stratospheric ozone reached 45% (at 33 km) and was an SPE outcome during July 2000 [11]. Also, during November-December 2003 a decrease >30% in stratospheric ozone was recorded at the same height, and was attributed to the effects of SPEs during October-November 2003. However, although the impact of SPEs on stratospheric ozone is large, the measured impact on total columnar ozone is not. Registered decreases in total ozone have been on the order of 0.5-1% (between 1.5-3 DU) as compared to "normal ozone" of 300 DU [11]. A Dobson unit approximately corresponds to 0.01 mm of atmosphere above the Earth's surface. Under "normal" circumstances an average of 3 mm is expected for ozone over the planet's surface. These values limits the participation of SPEs as a major cause in the formation of the ozone hole, which requires that the total columnar ozone be below 220 DU, 80 DU inferior to the normal level. However, in this range SPEs do not contribute to ozone depletion.

LONG TERM EFFECTS

The precipitation of CR on Earth occurs continuously. Despite a constant supply of these particles, their intensity varies when special events occur on the Sun and/or in an interplanetary medium producing large-scale electromagnetic disturbances that partially isolate Earth from galactic CR, temporarily blocking and diverting the path of fast particles that come from the solar system or beyond. Such isolation is perceived at ground level as a decrease in the measured CR flux when such perturbations are present. Disturbances most often occur when the sun is near its peak in activity, approximately every 11 years. In CR a cycle of 11 years has also been observed, although in phase opposition (i.e. when the solar cycle is at its maximum intensity, the CR flux is at a minimum and vice versa). The CR cycle is not impacted much by SPE activity since, although SPE occurs more frequently near the solar cycle maxima, SPEs do not form a sustained perturbation in the long term because they are sporadic. Therefore, the interaction of the almost continuous CR flux within the Earth's atmosphere can be noticed in atmospheric variables, above climate influences, by looking at signals that are manifest mainly in 11-year cycles with an opposite phase to the solar cycle. The CR flux additionally possess additional periodicities, helping to highlight their effects within the complexity of various signals in atmospheric variables. According to [22], periodicities of 11, 7.7, 5.5, 2, and 1.7 years in the CR flux have been identified. Therefore, it is expected that the phenomena induced by CR may have similar frequencies depending on the physical mechanism that causes such disturbances in the atmosphere. By analyzing tropical stratospheric ozone, modulation signals of 11 years and 27 days have been determined and directly linked to solar activity [23]. However, the sign of the CR has been more elusive. Recently, Lu 2009 [24] found a relationship between CR flux and polar ozone depletion inthe Antarctic region. The results of satellite measurements indicated evidence of a relationship between CR and ozone depletion over Antarctica. Using this connection, the flux of CR can explain up to 27% of the observed variation of Antarctic ozone [25]. The findings imply that although anthropogenic pollution is primarily responsible for polar ozone depletion, CR flux has a contribution similar to traditional mechanisms that contribute to polar ozone depletion.

The findings have important implications for understanding and predicting the evolution of the Antarctic Ozone Hole, an expression of the severe impact that man has on his environment. The results help us to correctly assess expected variations due to natural causes and variations generated by human contamination with ODSs. The precise assessment of the impact of natural causes on ozone depletion may be useful quantifying damage due to anthropogenic pollution, serving to weigh the effectiveness of various actions taken to protect ozone, serving to calculate expected changes in the various indicators of ozone recovery, and serving to classify as successful (or not) the adoption of actions proposed in agreements and protocols signed for controlling pollution such as the Montreal Protocol (1987) and subsequent agreements or amendments. Although the major causes that produce the ozone hole are attributed to the contamination of anthropogenic ODSs, the influence of natural factors such as temperature variations, the dynamic effects of wind circulation, the polar vortex, planetary waves, volcanism, and the influence of fast particles that precipitate in the atmosphere, contribute to explaining the magnitude of the hole, but also complicate interpretations of observed trends and the success of policies to control the production of ODSs. Since ozone hole area is the basic measure of the severity of stratospheric ozone depletion over Antarctica, if environmental protection policies are successful a reduction in Antarctic ozone hole area should be detectable, although ozone hole area is not a direct measure of the abundance of stratospheric ozone. An ozone hole is considered to be present when the amount of total ozone column over a place is less than 220 Dobson units, corresponding to ozone levels observed prior to 1980, when the phenomenon began to occur each year during Austral spring [26]. In Figure 1, monthly data forAntarctic ozone hole area up to 2005 (values taken from [8]) are provided. Figure (1) provides the complexity of the evolution in ozone hole area. The solid line corresponds to the average for September, the dotted line provides data for October, and the dashed-dotted line provides values for November, during 1982-2005. Looking for the impact of CR in these signals is a complex task, as a result of a large variability in reported areas.

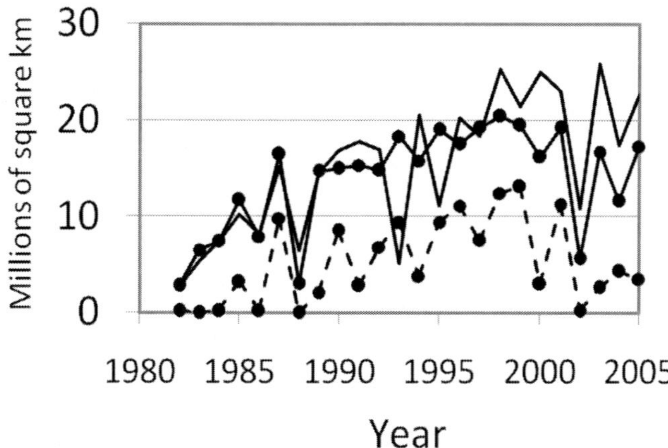

Figure 1. The monthly mean average of the Antarctic ozone hole area is shown from 1982 to 2005. The continuous line illustrates the September average, the dotted line shows October, and the dashed dotted line shows November data. Values taken from [8].

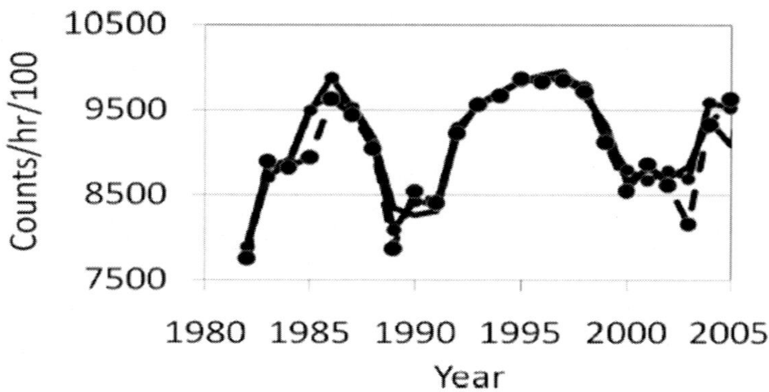

Figure 2. The monthly mean average of the counts from the neutron monitor detectors at the South Pole station is shown. The continuous line illustrates the September average, the dotted line shows October, and the dashed dotted line shows November data. Data taken from [8] for 1982-2005.

Although the effects of fast particles (SPEs, GCRs etc.) on the abundance of ozone have been reported previously, it is difficult to observe the CR signal in area data for the ozone hole, because it is an indirect measure of the amount of ozone in the zone. To contrast the data for the ozone hole, in Figure (2) counts for neutron monitors located at the South Pole station, as a measure of the CR flux that precipitates in this area [8], are provided. The figure illustrates the average monthly counts, in a similar manner to those in Figure (1), to the solid curve for the monthly mean for September. The dotted curve provides the monthly mean for the month of October.

The dashed and dotted curves indicate the average for the month of November. A different type of variability in the CR data, in comparison to the variability in ozone hole area, can be immediately identified.

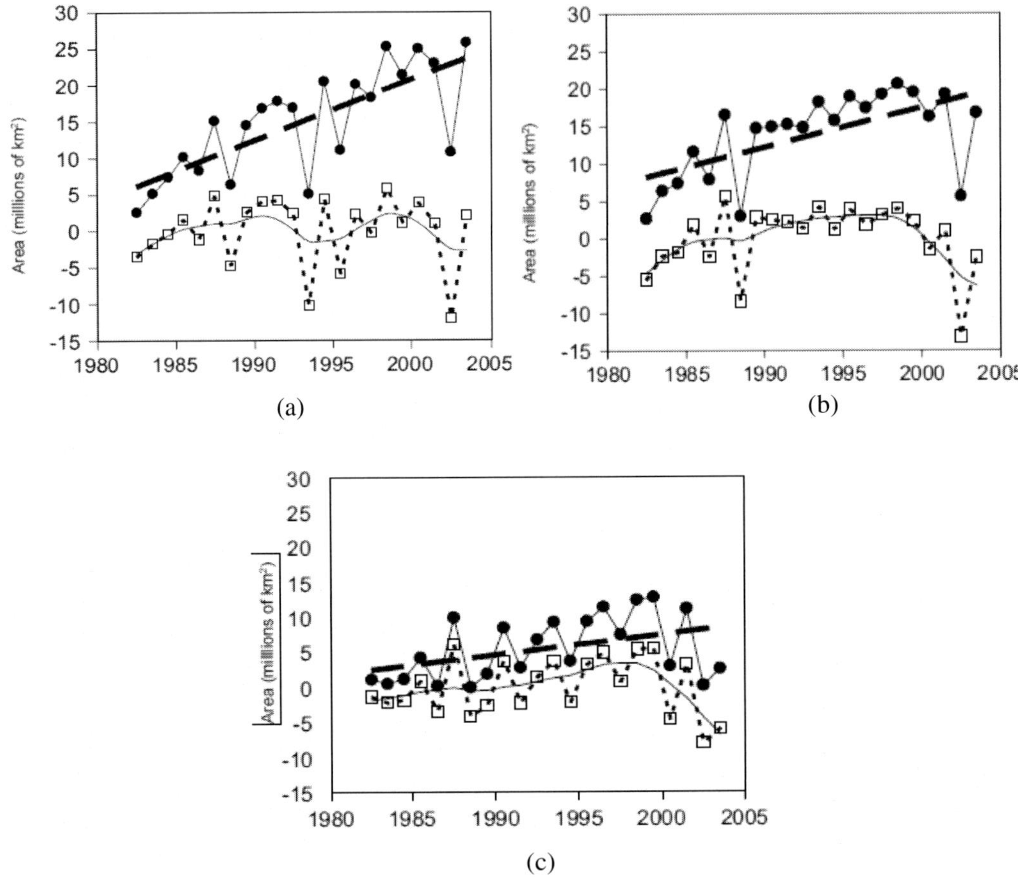

Figure 3. Trend analysis of the monthly mean ozone hole area during September (A), October (B) and November (C), for the period of 1982-2003. In each panel the dotted line represents the ozone hole area data. The squared dashed line shows the result of subtracting the linear trend (illustrated as a bold dashed line) from the original data and the continuous line illustrates a smoothing over the squared dashed line. Values taken from [27].

To determine a relationship between ozone hole area data and counts in the neutron monitor using a special technique to compare the data series is required. A direct comparison is not appropriate in this case, because while ozone hole area data evolves noticeably different for each month under review, the data for counts in the neutron monitor evolve similarly month to month, with only small variations between months. In this sense, the CR flux is more stable than the area of the ozone hole. Ozone hole areas that varied each month are not only large but the tendency of one of the month's analyzed was different. The amplitude of the variations were different and the behavior was also different, as reported by Alvarez-Madrigal and Pérez-Peraza 2007 [27]. In Figure (3), the behavior for ozone hole area during the month of September (Figure 3A), their tendency, and their variations were different in relation to those observed for the months of October (Figure 3B) and November (Figure 3C). According to [8] the result of the wavelet analysis indicated common periodicities at 1.3 and 3 years for the month of September. The 1.3 year periodicity was in phase while the 3 year phase was the inverse. For the month of October, a common frequency of 1.7 years that is clearly in phase during the complete period of the analysis and other common periodicities found for this month fluctuated with the kind of relationship depending on the lapse of the years analyzed. For the month of November, the common periodicities occurred for 1.3 years, for 5.5 years, and a band for 2-4 years. The relationship at 1.3 years was the reverse (when CR flux increases in the ozone hole were reduced) during November, unlike the previous months.

The common periodicities, some in a direct linear relationship (in phase) and others in an opposite relationship (anti-phase) indicated that at least two physical mechanisms linked CR to ozone hole area.The basic physical mechanisms that are known are the following:

1) The destruction of ozone through the dissociation of CFCs by means of the capture of electrons produced by the passage of CR. Number of such electrons are maximum at a height where there are polar stratospheric polar clouds, where the ice particles in clouds capture these electrons which result in the dissociation of a CFC within those clouds, increasing the ozone depletion [10],

2) The increase of ozone depletion by physical-chemical processes following the creation of NO_x and HO_x caused by the direct impact of CR [11]. The effects are not necessarily linear since an increase in the flow of CR does not always correspond to greater ozone depletion.

The combination of both mechanisms produces three possible scenarios that may explain some of the relationships obtained from a Wavelet analysis. The first scenario occurs where the first mechanism dominates.Therefore, the relationship observed is direct and corresponds to an in-phase relationship. The second scenario occurs where the second mechanism dominates.The third scenario occurs where no mechanism is dominant. When the scenarios the second and/or third occurs, relationship produced is not linear (or complex) but not inverse. The combination of mechanisms for the different scenarios could explain direct or non-linear relationships, but would not explain relationships in phase opposition.

The inverse relationship case implies that an increased flow of CR produces a decrease in the area of the ozone hole, and an ozone recovery. Currently, it is not clear how this could occur. However, the effect has been seen in wavelet coherence analyses when both series are compared. To explain the observed relationships, in order to clarify the mechanisms that take place and the changes that occur in the analyzed variables, more research is needed. Recently a new mechanism has been proposed in order to help to explain how the flux of CR affects the destructive processes of stratospheric ozone in the Antarctic; and increases in the size of the ozone hole, this mechanism takes place in the atmosphere producing increases in cloudiness when passing the CR.

ANTARCTIC OZONE DEPLETION, CLOUDINESS, AND THE CR FLUX

The physical mechanism which relates the CR flux and cloudiness is basically as described below.The ions created by the collision of CR within the atmosphere interact rapidly with local molecules, making them a cluster of ion complexes (aerosols) [28]. These clusters of ions can grow by means of ion-ion recombination or by ion-aerosol interactions and change the number of aerosols acting as cloud condensation nuclei (CCN) [29]. Other authors have suggested that the link between CR and cloudiness occurs through the global circuit of currents. As CR produces ionization in the atmosphere, these ions change the conductivity by increasing the electrical current in the global circuit, thereby modifying the conditions of aerosols formation on the edge of clouds and increasing cloud formation. A detailed review of these proposed mechanisms can be found in Carslaw et al. 2002 [30]. However, although the exact mechanism (or mechanisms) remains unknown, correlation studies support the mechanism's validity [7].

For example, a significant correlation between CR flux and low cloud cover (at approximately 3.2 km) during the period from 1983-1994 has been reported [31]. The basic result has been confirmed by additional independent studies [32]. Also determined is that during the Forbush decrease (involving a reduction in the CR flux to Earth) there was a reduction in the area covered by clouds, indicating that this relationship was also valid over the short term (in the range of days when the Forbush decrease takes place) [33]. All of the relationships found should be enough to consider the relationship as solid, but are not because solar activity has several indexes that are correlated mathematically to the CR flux and with climate parameters, but without a physical connection. The best known of these activity indices is the solar radiation index for various wavelengths, such as that for visible and ultraviolet wavelengths that produce appreciable changes in atmospheric dynamics and weather conditions on the planet, but that are not physically associated with the flow of CR although they have a similar periodicity. In order to find more conclusive evidence, studies with global coverage have recently been performed using the area covered by clouds and data from several neutron monitors around the world. Using these data a significant correlation was determined between the annual CR flux and low altitude cloud cover, over a period of 20 years. Although the corresponding checks remain to be validated with the theoretical model predictions, latitudinal variation has been determined to be similar between the amount of low-altitude clouds and the CR flux.The results support the idea that ionization produced by CR intensifies the formation of CCN which produces cloudiness [7]. Since the main destruction of the polar ozone takes place in the ice particles of Polar Stratospheric Clouds, thus the CR flux also impacts the abundance of stratospheric ozone via the modulation of the area covered by clouds. Although the relationship is not direct, it helps to explain the influence of CR on the quantity of ozone destroyed in Antarctica.

REFERENCES

[1]   G. C. Reid, Solar energetic particles and their effects on the terrestrial environment, in *Physics of the Sun*, vol. 3, edited by P. A.Sturrrock, pp. 251–278, chap. 12, Springer, New York (1986).

[2]   L. I. Miroshnichenko, *Solar Cosmic Rays*, Kluwer Academic publishers, 492, (2001).

[3]   M. A. Pomerantz, *Cosmic Physics*, Van Nostrand Reinhold Company, New York, (1971).

[4]   J. P. Burrows, A. Richter, A. Dehn. B. Deters, S. Himmelmann, S. Voigt and J. Orphal, *Journal of Quantitative Spectroscopy and Radiative Transfer*, 61, 509–517 (1999).

[5] M. Molina, S. Rowland, *Nature* 249, 810-812; doi:10.1038/249810a0 (1974)

[6] WMO (World Meteorological Organization), *Scientific Assessment of Ozone Depletion*: 2010 (Global Ozone Research and Monitoring Project Report No. 52, Geneva Switzerland, 2011).

[7] G. Usoskin, N. Marsh, G. A. Kovaltsov, K. Mursula, and O. G. Gladysheva, *Geophys. Res. Lett.*, 31, L16109, doi:10.1029/2004GL019507.(2004).

[8] M.Alvarez-Madrigal, J. Pérez-Peraza, and V. M. Velasco, CDTAW, 3(3),233-248 (2009).

[9] Rohen, G., et al., *J. Geophys. Res.*, 110, A09S39, doi:10.1029/2004JA010984 (2005).

[10] Q.-B. Lu and L. Sanche, *Phys. Rev. Lett.* 87, 078501 (2001).

[11] WMO (World Meteorological Organization), *Scientific Assessment of Ozone Depletion*: 2006 (Global Ozone Research and Monitoring Project Report No. 50, Geneva Switzerland, 2007).

[12] S. Solomon, D.W. Rusch, J.-C. Gérard, G.C. Reid, and P.J. Crutzen, *Planet. Space Sci.*, 29 (8), 885-893, (1981).

[13] L. H. Weeks, R.S. Cuikay, and J.R. Corbin, *J. Atmos. Sci.*, 29 (6), 1138-1142, (1972).

[14] W.Swider, and T.J. Keneshea, *Planet. Space Sci.*, 21 (11), 1969-1973, (1973).

[15] P. J. Crutzen, I.S.A. Isaksen, and G.C. *Reid, Science,* 189 (4201), 457-459, (1975).

[16] D. F. Heath, A.J. Krueger, and P.J. Crutzen, *Science,* 197 (4306), 886-889, (1977).

[17] C. H. Jackman, and R.D. McPeters, The effect of solar proton events on ozone and other constituents, in Solar Variability and its Effects on Climate, edited by J.M. Pap, P.A. Fox, and C. Frohlich, *AGU Monograph* 141, 305-319, Washington, D.C., (2004).

[18] C. H. Jackman, R.D. McPeters, G.J. Labow, E.L. Fleming, C.J. Praderas, and J.M. Russell, Geophys. *Res. Lett.,* 28 (15), 2883-2886, (2001).

[19] C. H. Jackman, M.T. DeLand, G.J. Labow, E.L. Fleming, D.K. Weisenstein, M.K.W. Ko, M. Sinnhuber, and J.M. Russell, *J. Geophys. Res.,* 110, A09S27, doi: 10.1029/2004-JA010888, (2005).

[20] C. E. Randall, D.E. Siskind, and R.M. Bevilacqua, *Geophys. Res. Lett.,* 28 (12), 2385-2388, (2001).

[21] M. López-Puertas, B. Funke, S. Gil-López, T. vonClarmann, G.P. Stiller, M. Höpfner, S. Kellman, H.Fischer, and C.H. Jackman, *J. Geophys. Res.,* 110, A09S43, doi: 10.1029/2005JA011050, (2005).

[22] H. Mavromichalaki, Preka-Papadema, P., Petropoulos, B., Tsagouri, I., Georgakopoulos, S., and *J. Polygiannakis, Ann. Geophys.,* 21, 1681-1689, doi: 10.5194/angeo-21-1681-2003, (2003).

[23] V. E. Fioletov, *J. Geophys. Res.,* 114, D02302, doi:10.1029/2008JD010499(2009).

[24] Q.-B. Lu, *Phys. Rev. Lett.,* 102, 118501 (2009).

[25] M. Alvarez-Madrigal, *Phys. Rev. Lett.* 105, 169801 (2010).

[26] P. A. Newman, S. R. Kawa, and E. R. Nash, *Geophys. Res. Lett.,* 31, L21104, doi: 10.1029/2004GL020596 (2004).

[27] M. Alvarez-Madrigal, and *J. Pérez-Peraza, Atmósfera,* 20(2), 215-221, (2007).

[28] W. J. Gringel, J. A. Rosen, and D. J. Hofmann, Electrical structure from 0 to 30 kilometers, in Studies in Geophysics: The Earth's Electrical Environment, pp. 166–182, *Natl. Acad.,* Washington D. C. (1986).

[29] A. Viggiano, and F. Arnold, Ion chemistry and composition of the Atmosphere, in
      Handbook of Atmospheric Electrodynamics, vol. 1,edited by H. Volland, pp. 1 –26,
      CRC Press, *Boca Raton,* Fl. (1995).
[30] K. S. Carslaw, R. G. Harrison, and J. Kirkby, *Science,* 298, 1732– 1737 (2002).
[31] H. Svensmark, and E. Friis-Christensen, *J. Atmos. Terr. Phys.,* 59, 1225– 1232 (1997).
[32] E. Pallé, and C. J. Butler (2000), *Astron. Geophys.,* 41(4), 18– 22 (2000).
[33] M. Pudovkin, and S. Veretenenko, *Adv. Space Res.,* 17(11), 161– 164 (1996).

In: Homage to the Discovery of Cosmic Rays …
Editor: Jorge A. Perez-Peraza

ISBN: 978-1-63483-084-3
© 2015 Nova Science Publishers, Inc.

Chapter 11

COSMIC RAYS COMPOSITION: PULSAR SOURCE

Neïla Zarrouk[*] and Raouf Bennaceur

Laboratoire de Physique de la Matière Condensée,
Faculté des Sciences de Tunis, Université Tunis El Manar, Tunisia

ABSTRACT

Radiation field produced by cosmic radiations in the earth's atmosphere is very complex and is significantly different from that found in the nuclear industry and other environments at ground level. High-energy ionising radiations (mostly protons) that enter the earth's atmosphere from outer space are known as primary cosmic rays. When they interact with atomic nuclei in the atmosphere (nitrogen, oxygen and other atoms), secondary particles and electromagnetic radiation are produced and they are called secondary cosmic rays. These secondary spectra depend on solar activity, longitude, latitude and altitude. In the atmosphere and at ground level, the flux of cosmic ray particles is mostly due to galactic protons incident on the atmosphere. Primary cosmic rays (85% protons, 12% alpha particles, 1% heavy nuclei ranging from carbon to iron, and 2% electrons and positrons) arrive at the heliosphere isotropically. Their sources are thought to include supernovae, pulsar acceleration, and explosion of galactic nuclei. These particles can have energies in excess of 1020 eV. One of the indirect observations we can make is the "composition" of GCRs. This can tell us a lot about the GCR sources and the cosmic rays' trip through the Galaxy. In a part of our work we have analyzed time dependence of the cosmic ray particles density as a function of a modelled pulsar source variability using Morlet wavelets. Semi-analytical solutions of cosmic rays transport equation are presented for two cases of sources: time dependant discrete source and a pulsar source. We chose RP J0737-3039 A pulsar as a model of source. We used the Morlet wavelets to describe the oscillations of pulsar source to analyse the CR particles density. The more realistic description is given, modeling the source oscillations by a sum of Morlet wavelets. The CR density response to this source has mainly changed in paces showing more realistic behaviour. Informations contained in Morlet decompositions and reconstructions show essentially periods of 250, 300 and 25 s.

Keywords: Cosmic rays composition, pulsar, wavelets

---

[*] Email address: neila.zarrouk@yahoo.fr; Phone number : 0021623066470.

INTRODUCTION

Most galactic cosmic rays are probably accelerated in the blast waves of supernova remnants. This doesn't mean that the supernova explosion itself gets the particles up to these speeds. The remnants of the explosions, expanding clouds of gas and magnetic field, can last for thousands of years, and this is where cosmic rays are accelerated. Bouncing back and forth in the magnetic field of the remnant randomly lets some of the particles gain energy, and become cosmic rays. Eventually they build up enough speed that the remnant can no longer contain them, and they escape into the Galaxy.

Because the cosmic rays eventually escape the supernova remnant, they can only be accelerated up to a certain maximum energy, which depends upon the size of the acceleration region and the magnetic field strength.

In a first part of our work we have presented a summary dealing with cosmic rays composition in early research then we have given the ideas and results of our work analyzing time dependence of the cosmic ray particles density as a function of a modelled pulsar source variability using Morlet wavelets.

Unlike the conventional spectral analysis the wavelet transform is a suitable tool for description of non-stationary processes containing multiscale features detection of singularities and analysis of transient phenomena. Methods such as wavelet should be very suitable; thus it can be applied to the problem under consideration. Using Morlet reconstructions [24-27], we have detected and predict new period details or singularities hidden behind the original spectra describing galactic cosmic ray density modulations as a response to a certain source [13].

The most important step is the wavelet choice; the wavelet type influences the time and frequency resolution of results. Indeed while the Derivative of Gaussian (DOG) wavelet provides a poor frequency resolution but a good time localisation on the other hand we can expect that Morlet (a plane sine wave with amplitude windowed in time by a Gaussian function) wavelet choice gives a high frequency resolution. Sophisticated wavelets are more powerful in revealing hidden detailed structures for example Morlet wavelet has been used to examine the processes, models and structures behind the variability of solar activity [3].

I. COSMIC RAYS : EARLY RESEARCH AND OTHER WORKS

Victor Hess [16] has discovered Cosmic rays in 1912, when he found that an electroscope discharged more rapidly as he ascended in a balloon. In 1936 he was awarded the Nobel prize for his discovery. Cosmic rays are high energy charged particles, originating in outer space, that travel at nearly the speed of light and strike the Earth from all directions. This radiation was believed to be electromagnetic in nature for some time [23]. However, during the 1930's it was found that cosmic rays must be electrically charged because they are affected by the Earth's magnetic field.

Cosmic rays also include high energy electrons, positrons. From the 1930s to the 1950s, cosmic rays served as a source of particles for high energy physics investigations, and led to the discovery of subatomic particles that included the positron and muon.

Detectors of innovative charge measurement techniques have been developed. They can be divided into three general categories: recording detectors, such as photographic emulsions, detectors visual detectors, such as cloud chambers, and electronic detectors, such as Geiger-Muller counters.

In the late 1940s, [17,18] group of cosmic ray investigators at the University of Minnesota and the University of Rochester employed photographic emulsions carried to high altitudes, frequently above 27,000 meters, by balloons to determine the charge and energy of the cosmic rays. indeed collisions between incoming cosmic rays and air molecules can cause the cosmic rays to fragment into several lighter nuclei, thus altering their composition. At high altitudes, the probability of such a collision is low; therefore, the balloon detectors measure the primary composition of the particles in space. These early experiments demonstrated that of the nuclei in the cosmic rays, about 87 percent are hydrogen, or protons; 12 percent are helium; and the remaining 1 percent are nuclei heavier than helium. It is the composition of these heavier nuclei that contain the clues to the nucleosynthesis processes.

Before they have been slowed down and broken up by the atmosphere, cosmic rays were measured directly by instruments carried on spacecraft and high altitude balloons, using particle detectors similar to those used in nuclear and high energy physics experiments.

In 1956 [19,20], the physicist Frank McDonald, at Iowa State University, developed a combination of two electronic detectors- a scintillation counter and a Cherenkov counter- to determine the charge and velocity of the cosmic rays. Thus good measurements of the elemental abundance for elements up to iron are then provided by the combination of these detectors. The elements heavier that iron were so rare that their identification required a new technique.

In mid 1960 [21,22] Robert Fleischer, Buford Price, and Robert Walker, researchers at the General Electric Research and Development Center, found that the trails of ionizing particles were recorded in certain types of plastics, then these trails could be revealed by etching the plastic in an appropriate chemical agent. They proved that if the rate at which the trail was etching as well as the total etchable length were both measured; the charge and energy of the particle could be determined. In association to these plastic detectors Balloon flights provided information on the composition of the heavier elements in the cosmic rays.

Because cosmic rays are electrically charged The magnetic fields of the Galaxy, the solar system, and the Earth have scrambled the flight paths of these particles so much that we can no longer point back to their sources in the Galaxy. However, cosmic rays in other regions of the Galaxy can be traced by the electromagnetic radiation they produce. Supernova remnants such as the Crab Nebula are known to be a source of cosmic rays from the radio synchrotron radiation emitted by cosmic ray electrons spiraling in the magnetic fields of the remnant.

Galactic cosmic rays (GCRs) are the high-energy particles that flow into our solar system from far away in the Galaxy. GCRs are mostly pieces of atoms: protons, electrons, and atomic nuclei which have had all of the surrounding electrons stripped during their high-speed passage through the Galaxy. Cosmic rays provide one of our few direct samples of matter from outside the solar system.

The "composition" of cosmic rays describes what fraction of cosmic rays are protons, what fraction are helium nuclei, etc. All of the natural elements in the periodic table are present in cosmic rays, in roughly the same proportion as they occur in the solar system. about 89% of the nuclei are hydrogen (protons), 10% helium (alpha particles), and about 1% heavier elements. The common heavier elements (such as carbon, oxygen, magnesium, silicon, and iron) are present in about the same relative abundances as in the solar system, but there are important differences in elemental and isotopic composition that provide information on the origin and history of galactic cosmic rays. These require large detectors to collect enough particles to say something meaningful about the "fingerprint" of their source.

The HEAO Heavy Nuclei Experiment, launched in 1979 [16, 17], collected only about 100 cosmic rays between element 75 and element 87 (the group of elements that includes platinum, mercury, and lead), in almost a year and a half of flight, and it was much bigger than most scientific instruments flown by NASA today.

To make better measurements requires an even larger instrument, and the bigger the instrument, the greater the cost.

The term "cosmic rays" usually refers to galactic cosmic rays, which originate in sources outside the solar system, distributed throughout our Milky Way galaxy. However, this term has also come to include other classes of energetic particles in space, including nuclei and electrons accelerated in association with energetic events on the Sun (called solar energetic particles), and particles accelerated in interplanetary space.

The origin of cosmic radiation has been a more or less open question since its discovery.

From the more common supernovae to radio galaxies, quasi-stellar objects and now pulsars, all have been associated with the sources of cosmic rays. If pulsars are sources then the cosmic radiation must propagate to the earth. This problem has received extensive study [16]

Looking for the contribution of significant periods, different research teams have analysed time variability in cosmic ray fluxes records [1, 2]. Particularly continuous wavelet transform (WT) was recently used, and is useful for data series with non-stationary processes when dealing in term of time-frequency decomposition [3-6].

II. PARTICULAR CASES OF COSMIC RAYS TRANSPORT EQUATION: A PULSAR SOURCE AND WAVELETS ANALYSIS

1. Cosmic Rays Transport Equation Form

Simplifying the complete transport equation keeping an energy –dependent diffusion coefficient for the case of Kolmogorov type spectrum of turbulence [8] $D_0(E) = D_0 E^\alpha$ and considering also the case with continuous energy loss by synchrotron radiation and inverse Compton scattering.

The CR diffusion equation becomes:

$$\frac{\partial N}{\partial t} - \frac{\partial}{\partial E}\left(b_0 E^2 N\right) - D_0 E^\alpha \Delta N = q \tag{1}$$

where $N_i(E,r,t))$ is the particle number density of type i and $\dfrac{\partial}{\partial E}\left[b_0 E^2 N\right] = \dfrac{\partial}{\partial E}\left[\left\langle\dfrac{\partial E}{\partial t}\right\rangle^{ion} N\right]$

is the partial loss of energy due to the production of the final states (of multipion) at highest energies, it is roughly a constant [14], $-\dfrac{1}{E}\left(\dfrac{dE}{dt}\right)_\pi = C = (2.42 \pm 0.03)x10^{-8}\,yr^{-1}$

From the values determined for the fractional energy loss, one can directly derive the energy degradation of ultra high energy cosmic rays [14] in terms of their flight time. This is given by,

$$E(t) = E_0 \exp\left[-Ct\right] = E_0 \exp\left[-t/a\right]; \qquad for\ E \geq 10^{21} eV$$
$$a = 1/C = 41.10^6\ years$$

At energies corresonding to Larmor radii smaller than the scale of magnetic cells ($r_L(E) \leq l_{cell}$), we assume that the particles experience a Kolmogorov spectrum of magnetic field fluctuations, thus the corresponding diffusion coefficient [9] is $D(E) \propto E^{1/3}$.

For energies such that ($r_L(E) \geq l_{cell}$), the diffusion coefficient is linear in energy, $D(E) \propto E$. The Kolmogorov spectrum can be justified, if the magnetic field in the local magnetized supercluster LSC has structure similar to a cascade on the largest scales lcell to small scales. This can be the case if outflows from galaxies and peculiar velocities of plasma in the LSC have experienced enough turbulent motion as in reference [10, 15]. However for the limited range of energies where UHECRs are expected to be extragalactic, the resulting spectrum does not depend strongly on the specific choice of the energy power index in the diffusion coefficient. More precisely the diffusion coefficient is written as follows:

$$D(E) = D_{low} = D_0\left(cm^2/s\right)\left(\frac{E}{10^{19} eV}\right)^{1/3}\left(\frac{B}{5.10^{-8} G}\right)^{-1/3}\left(\frac{l_{cell}}{0.5 Mpc}\right)^{2/3}$$
$$for\ E \leq E_c = 2.3.10^{19}\left(\frac{B}{5.10^{-8} G}\right)\left(\frac{l_{cell}}{0.5 Mpc}\right)eV$$
$$and\ D(E) = D_{high}(E) = D_{low}(E_c)\left(\frac{E}{E_c}\right),\ pour\ E > E_c$$
$$D_{low}(E_c) = 1.452.10^{34}\ cm^2/s \quad (for\ E_c = 23 EeV\ and\ D_0 = 1{,}1.10^{34}\ cm^2/s)$$

We adopt in our work an intermediate model for the dependence in energy of the diffusion coefficient; we take an index of power for energy equal to 0.6 which correspond to a mean value for the energy index power between both modes for diffusion coefficient. We thus try within this framework to solve the cosmic rays transport equation. Different cases of sources for cosmic ray diffusion equation will be examined.

2. Time Dependant Descrete Source

We choose in this step a time dependant source term. In the case of discrete sources, the term of injection q is the sum on all the sources. For an individual source appearing at time t_0, we can write:

$$q(E, r, t) = B.f(t) = q_0 \delta(r) E^{-\alpha} \theta(t' - t_0)$$
$$\text{with } B = q_0 E^{-\alpha}$$

Generally the solution of Eq.1 is written as:

$$N = \int dE' \, dt' \, dr' \, q_1(E', t', r') G(r - r', t - t', E - E')$$

G is the Green's function and the propagator of particles from position and time injection r'=0 and t'=0 to position r and at time t, G must then satisfy the equation

$$\frac{\partial G}{\partial t} - \frac{\partial (b_0 E^2 G)}{\partial E} - D_0 E^a \Delta G = \delta(r - r') \delta(t - t') \delta(E - E')$$

The solution of equation (1) is then given by:

$$N = q_0 E^{-\alpha} \int_{-1/bE}^{0} dt' \frac{e^{-\frac{r^2}{4\Lambda}} \theta(t' - t_0)}{(4\pi\Lambda)^{3/2} (1 + b_0 Et')^{2-\alpha}} \tag{2}$$

where the length scale Λ is given by:

$$\Lambda = \int_{E}^{E_0} \frac{D_0 E'^a}{b_0 E'^2} dE' = \frac{D_0}{b_0(1-a)} E^{a-1} \left(1 - (1 + b_0 Et')^{1-a}\right)$$

We simulated numerically the evolution of particles number density N during the elapsed time. The resolution of cosmic rays diffusion equation and found shapes for CR particles density function N were performed for energy of 1 EeV. The curves (Figure 1) describing the evolution of CR density in the atmospheric shells located at lowest distances (in the vicinity of the source) present two branches, an increase corresponding to the creation of particles and a decrease describing the loss of particles. At long distances while moving away from the source it is especially the decrease of particles density corresponding only to the phenomena of particles losses which appears. We then studied the variation of the CR particles density function in time and for different values from energy for a total number of initial particles of 10^5 particles and located at r=10^6pc. The corresponding curves (Figure 2) show that the CR particles density reach a maximum corresponding to a maximum of particles creation for 5.10^9 years for an energy of 10^{-2} EeV . For energy of 1 EeV the corresponding time of this maximum decreases to $0.7.10^8 \sim 1.2.10^8$ years.

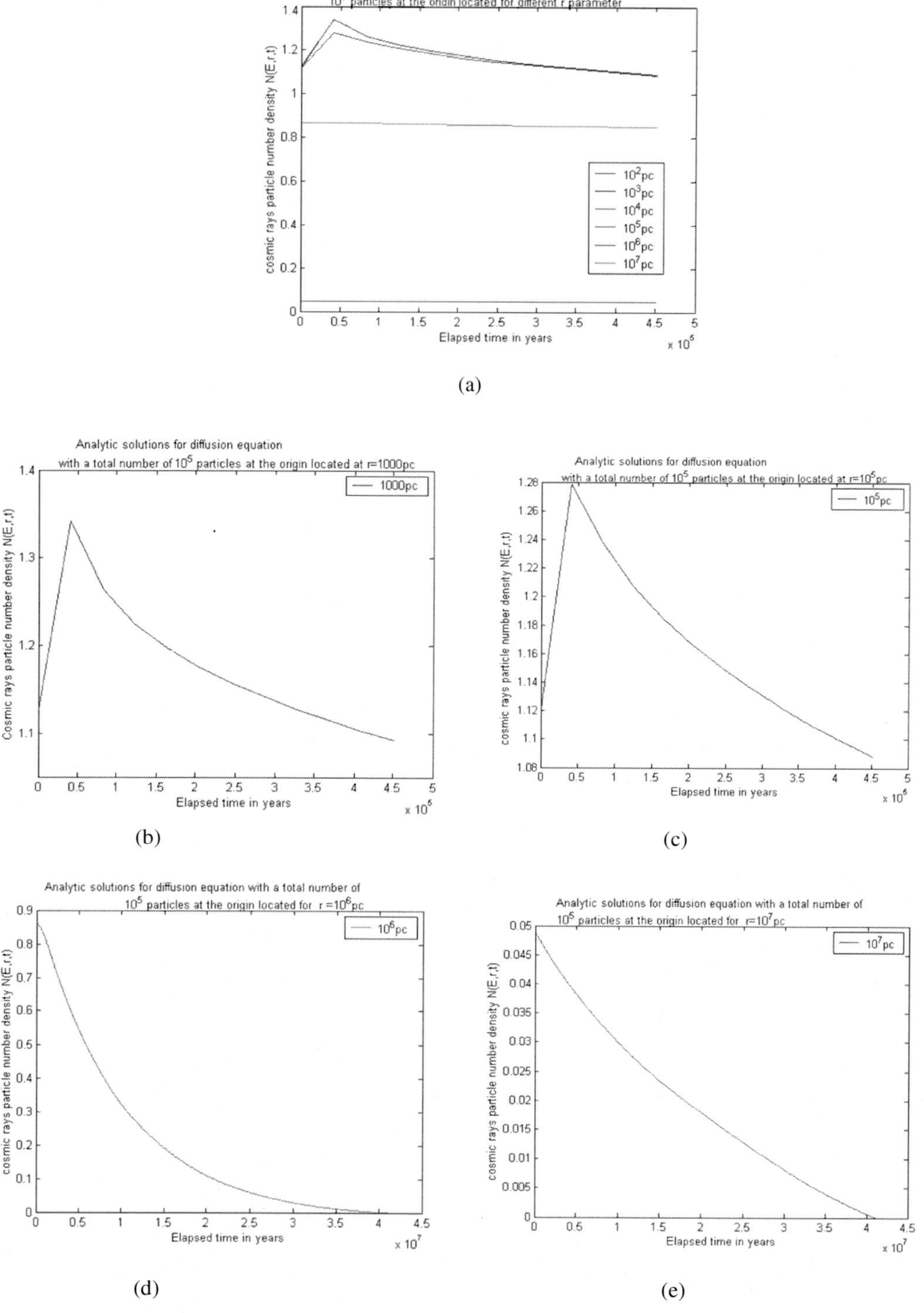

Figure 1. (a) Variation of the density of CR particles number in elapsed time for a total initial number of particles of 10^5 and for various values of the parameter r for the atmospheric shell. (b).r=10^3pc,(c) r=10^5pc, (d)r=10^6 pc(e) r=10^7 pc.

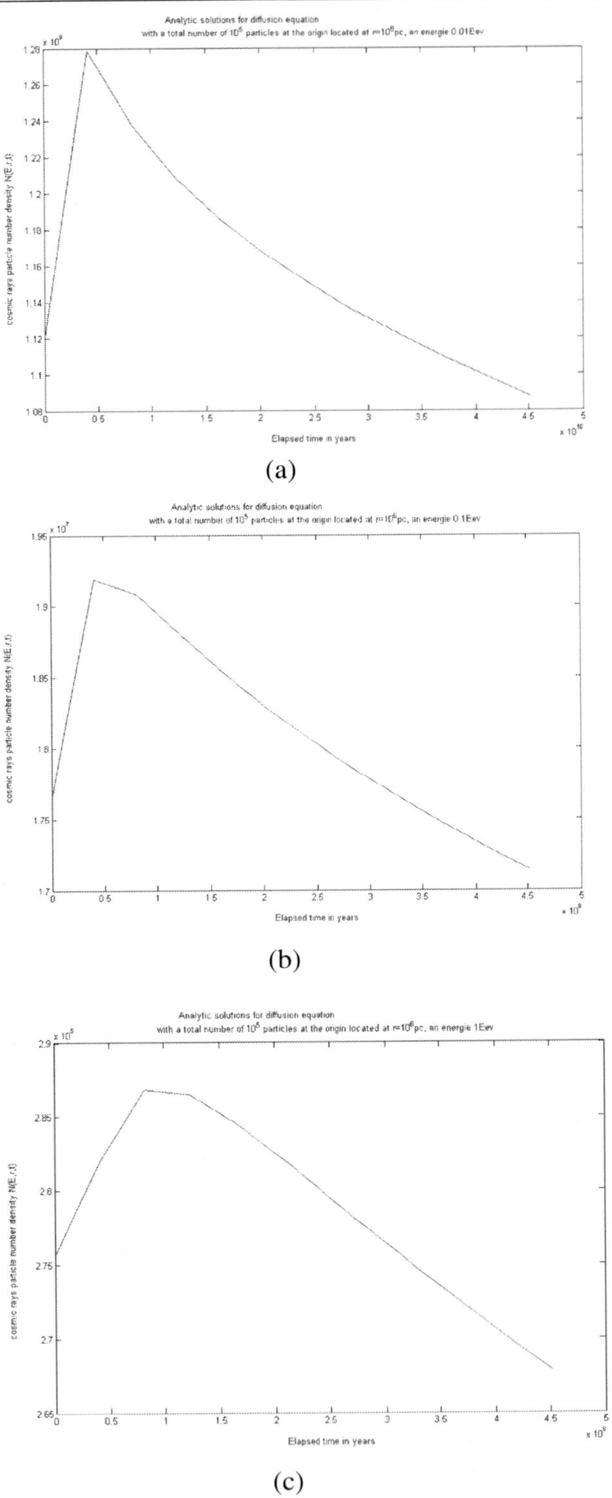

Figure 2. Variation of the CR particles density in time for a total initial number of particles of 10^5 located at r=10^6 pc and for various values of the parameter energy. (a).E=0.01EeV,(b) E=0.1EeV, (c) E=1EeV.

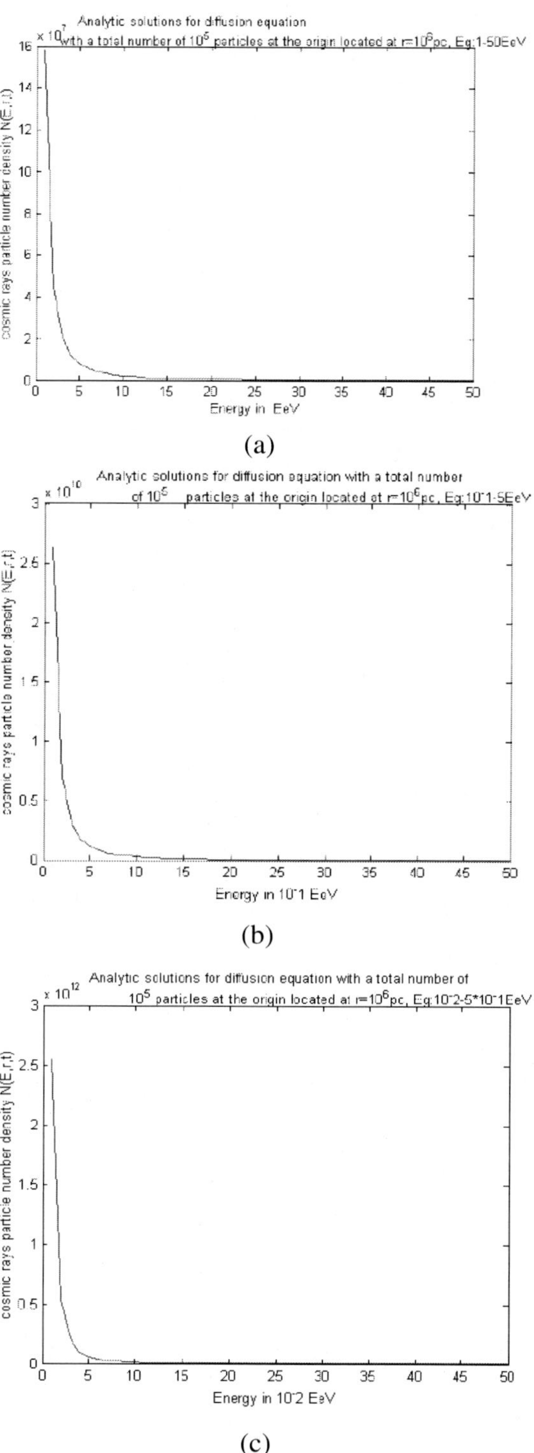

(a)

(b)

(c)

Figure 3. Variation of the CR particles density versus energy for a total initial number of particles of 10^5 located at r=10^6 pc and for various intervals of energy: (a).E: 1-50EeV, (b). E: 0.1-5EeV, (c). E: 0.01-0.5EeV.

The evolution of CR particles density versus energy (Figure 3) proves a decrease of the density while going towards highest energies. The highest CR particles densities are found for lowest energies for all energy intervals 10^{-2} to 50 EeV. For the first energy interval 1-50 EeV there are no more particles for an energy equal to cut-off energy E_c=23 EeV. For the second class of energy 0.1-5 EeV, the CR particles density reach zero particles for 1.7 EeV energy. For energy varying from 0.01 to 0.5 EeV the CR particles density decrease to zero for 0.1 EeV .

3. Morlet Wavelets Decomposition and Reconstruction of Cosmic Rays Particles Density

We have decomposed in Morlet wavelets the series of values for the CR density particles already derived in previous section and corresponding to 1EeV energy. The scale parameter aj is fixed to 1.

We have used discrete series in time for the function density of cosmic particles for a period study of 1000 years with one year step. The Morlet decomposition coefficient are given by:

$$w_k^R = a_j^{-1/2} \sum_{l=1}^{1000} y(t_l) g_k^R(t_l)$$

where $g(t) = \exp\left(i\omega_0 t - \frac{t^2}{2} \right)$ and $\omega_0 = 2\pi$, the period $T = 1\,year$

The curve describing the variation of decomposition coefficients in time shows a structure in the vicinity of the origin (Figure 4). The maximum decomposition coefficients is around a period of 4-5 years.

The Morlet reconstruction shows also a maximum for the CR particles density around the period 4-5 years.

4. Pulsar Modelled Source Cases

The discovery of RP J037-3039A with the period 23 ms around an other compact object with the orbital period 2.4h in a low eccentricity orbit (e=0.0878) was reported in reference [11]. The binary parameters suggested another neutron star as the companion. In reference [12], this companion was reported to be radio pulsar J0737f-3039B with the spin period 2.8s. Thus for the first time a binary system composed of two pulsars, a recycled one and a normal one was discovered. Since pulsar A rotates much faster and shines more intensely all the time except for a ~27s eclipse, the arrival times of pulses for pulsar A are measured much more accurately than for the pulsar B.

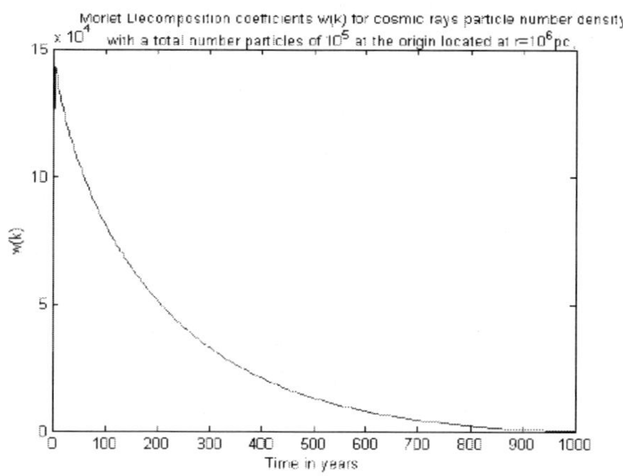

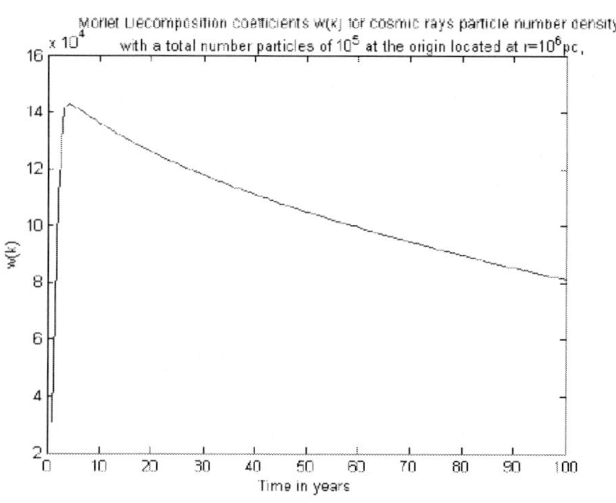

Figure 4. Morlet decomposition coefficients for the CR particles density.

A Single Morlet Wavelet to Describe the Pulsar Source

We have chosen the source in the shape of a pulsar PSR J0737-3039A of period T_0. We propose as a first step to describe the source pulsar oscillations by a single Morlet wavelet. This is given in terms of scale and translation parameters by:

$$q\left(E, r, \frac{t-k}{a_j}\right) = q_0 E^{-\alpha}\, \delta(r) f\left(t_{kj}\right) = B.\delta(r) f\left(t_{kj}\right), \text{ with } t_{kj} = \frac{t-k}{a_j},\ B = q_0 E^{-\alpha},\ q_0 = 10^5,\ E = 1 EeV, \alpha = 2$$

In case where we approximate $f(t)$ by a Morlet wavelet $\quad f\left(t_{kj}\right) = \cos\left(\omega_0.\left(\frac{t-k}{a_j}\right)\right).\exp-\left(\frac{1}{2}.\left(\frac{t-k}{a_j}\right)^2\right)$

$$q\left(E, r, \frac{t-k}{a_j}\right) = B.\delta(r)\cos\left(\omega_0.\left(\frac{t-k}{a_j}\right)\right).\exp-\left(\frac{1}{2}.\left(\frac{t-k}{a_j}\right)^2\right)$$

with $\omega_0 = 2\pi/T_0$, $T_0 = 22.7\,.ms$

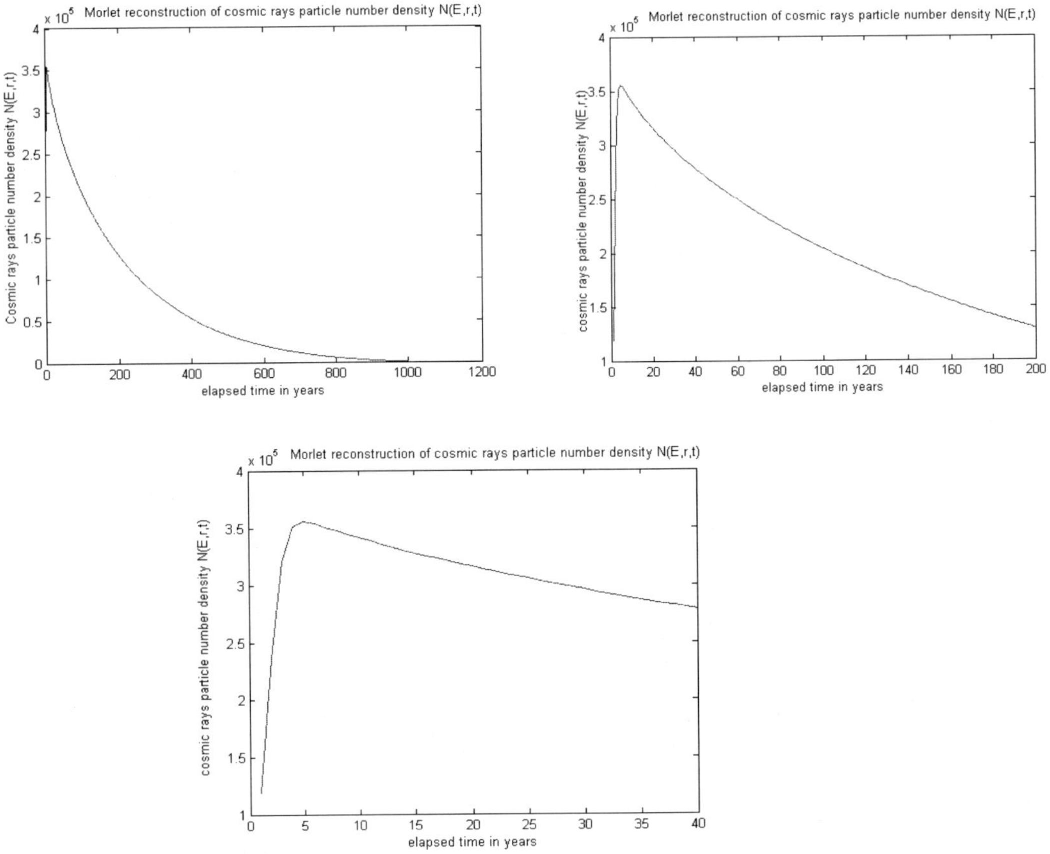

Figure 5. Morlet reconstruction of the CR particles density showing a maximum around 4-5 years.

For a translation parameter k=0 and a scale parameter aj =1, we obtain the following curves describing the oscillation of the source for short times (just after injection of particles) and long times far from the injection. This model reveals already the clean period of the pulsar ~20-23 ms (Figure 6a) which ceases emitting after~ 3s (~132 oscillations) (Figure 6b, 6c) We examined the oscillation of this pulsar source modelled in Morlet wavelets for different values from the scale parameter aj =10, 100, 1000 we obtain the following curves describing the oscillation of the source for short times (just after injection of particles) and for relatively long times.

The use of various values of scales parameters reveals in the oscillation of this pulsar two periods of fluctuations of ~2-3s, and ~5s (Figure 7a, b) for times far from the injection, the period of emission is then of 40 s. For aj =100, an oscillation of period 8s is added and the pulsar source ceases emitting after ~350s, for large values aj =1000 one period of 20s appears and the period of emission of the source is prolonged to ~3500-4000s. For short times in the vicinity of the source and for a parameter of scale aj =10, one characteristic period of 0.2s is found. For aj =100 the period of oscillation of the source is of ~2.2s (Figure 8 a,b,c)

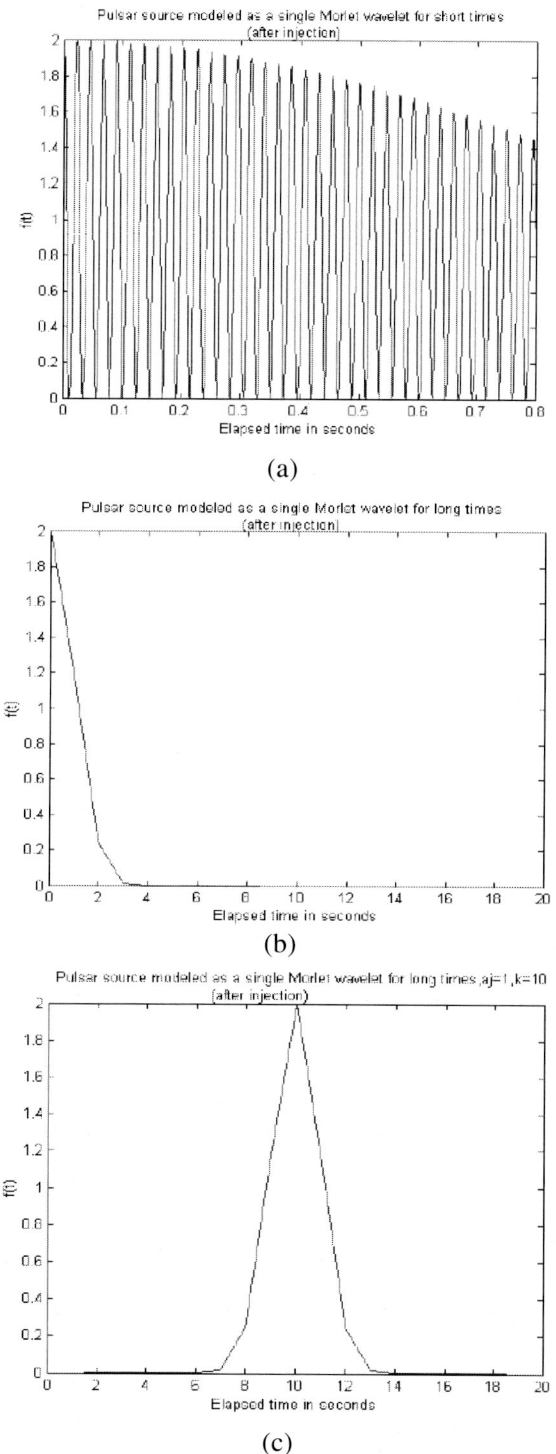

Figure 6. Oscillation of the pulsar source modelled by a single Morlet: wavelet (a) for short times (just after injection), (b), (c) far from injection.

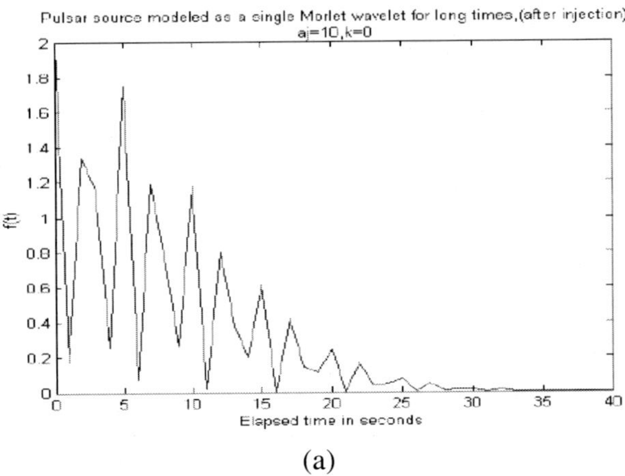

(a)

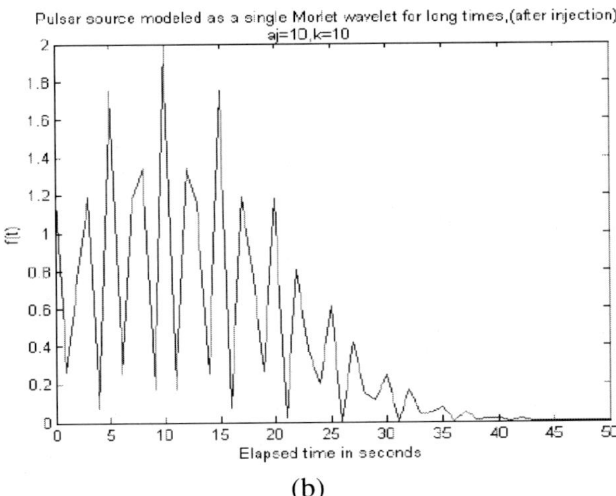

(b)

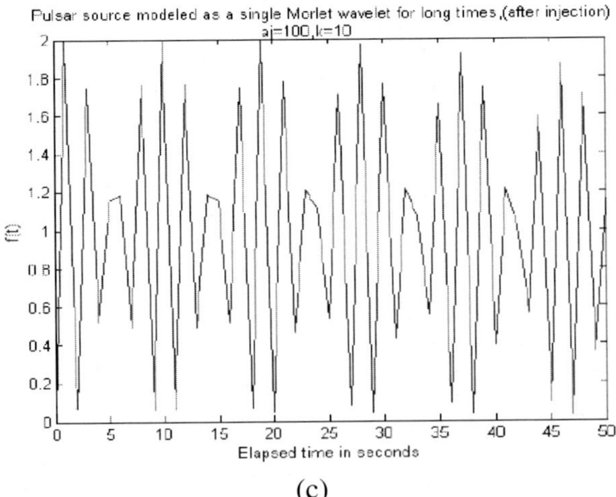

(c)

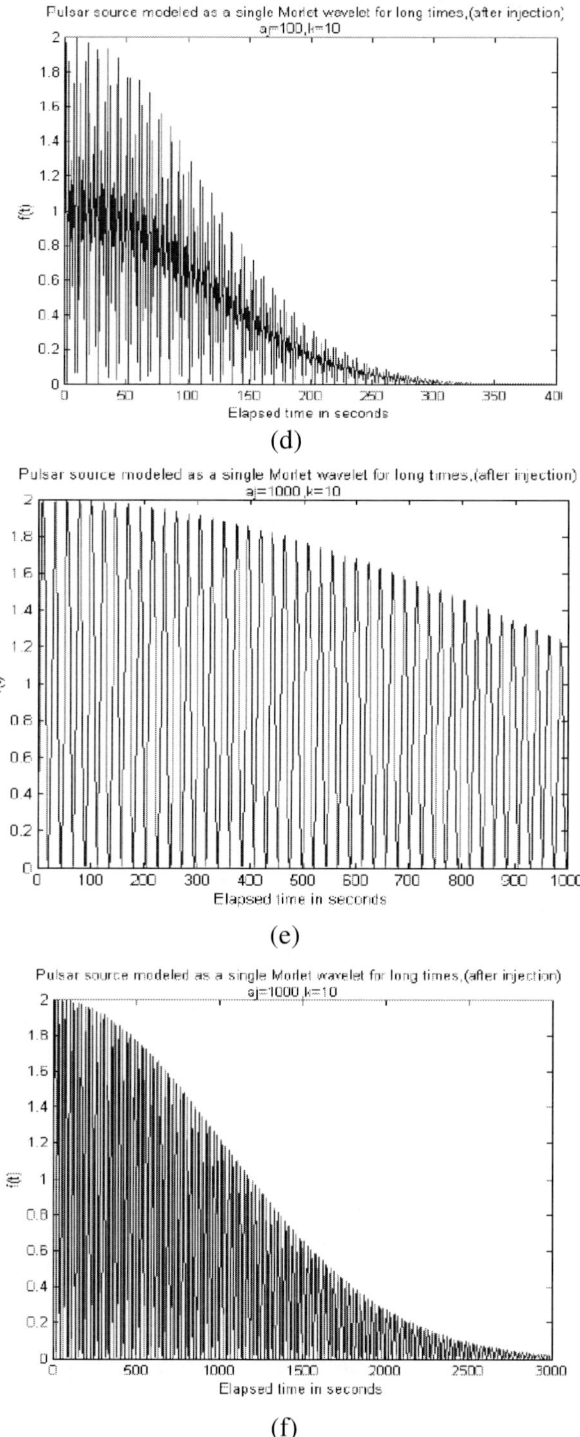

Figure 7. Oscillation of the single Morlet wavelet modelled pulsar for long times (far from the injection), for various parameters of scale: (a), (b): aj=10 (k=0,10); (c), (d): aj=100 (k=10); (e), (f): aj=1000 (k=10).

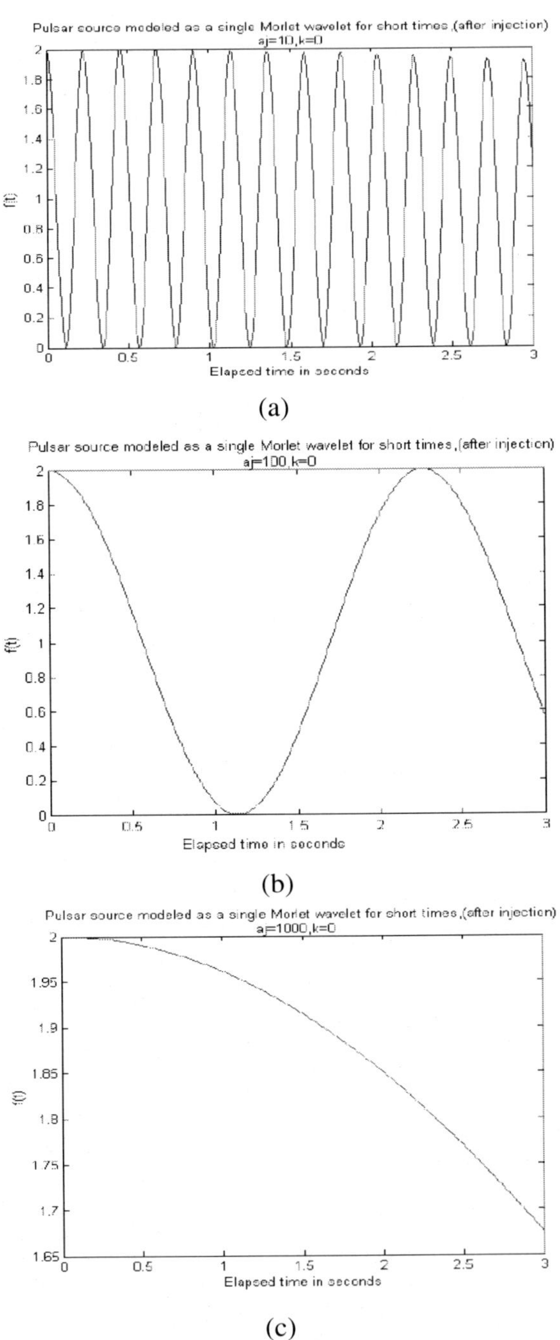

(a)

(b)

(c)

Figure 8. Oscillation of the pulsar source modelled by a single Morlet wavelet for short times (just after the injection) for different values from parameter of scale: (a): aj =10, (b): aj=100, (c): aj=1000.

Cosmic Ray Particles Density for a Single Wavelet Modelled Source

In this sequence we examine the diffusion equation of cosmic rays as an answer to a pulsar modelled source described by a single Morlet wavelet.

The analytic solution of equation (1) is then given by

$$N(E,r,t) = q_0 E^{-\alpha} \int\limits_{-1/bE}^{0} dt' \frac{e^{-\frac{r^2}{4\Lambda}} \cos(\omega_0 t').\exp-\left(\frac{1}{2}t'^2\right)}{(4\pi\Lambda)^{3/2}(1+b_0 Et')^{2-\alpha}}$$

with $\Lambda = \int\limits_{E}^{E_0} \frac{D_0 E'^a}{b_0 E'^2} dE' = \frac{D_0}{b_0} \int\limits_{E}^{E_0} E'^{a-2} dE' = \frac{D_0}{b_0(1-a)} E^{a-1}\left(1-(1+b_0 Et')^{1-a}\right)$

We thus examined the variation in time of the CR particles number density compared to the source oscillation model (Figure 9). We were interested in paces changes of evolution for the CR particles density and the characteristic periods versus the scale parameter aj. For aj =1, the oscillation period of the CR particles density as an answer to this model of pulsar source shows one characteristic period of ~20s. Thus the CR particles density is not cancelled after the end of source emission and continuous to oscillate according to the derived model.

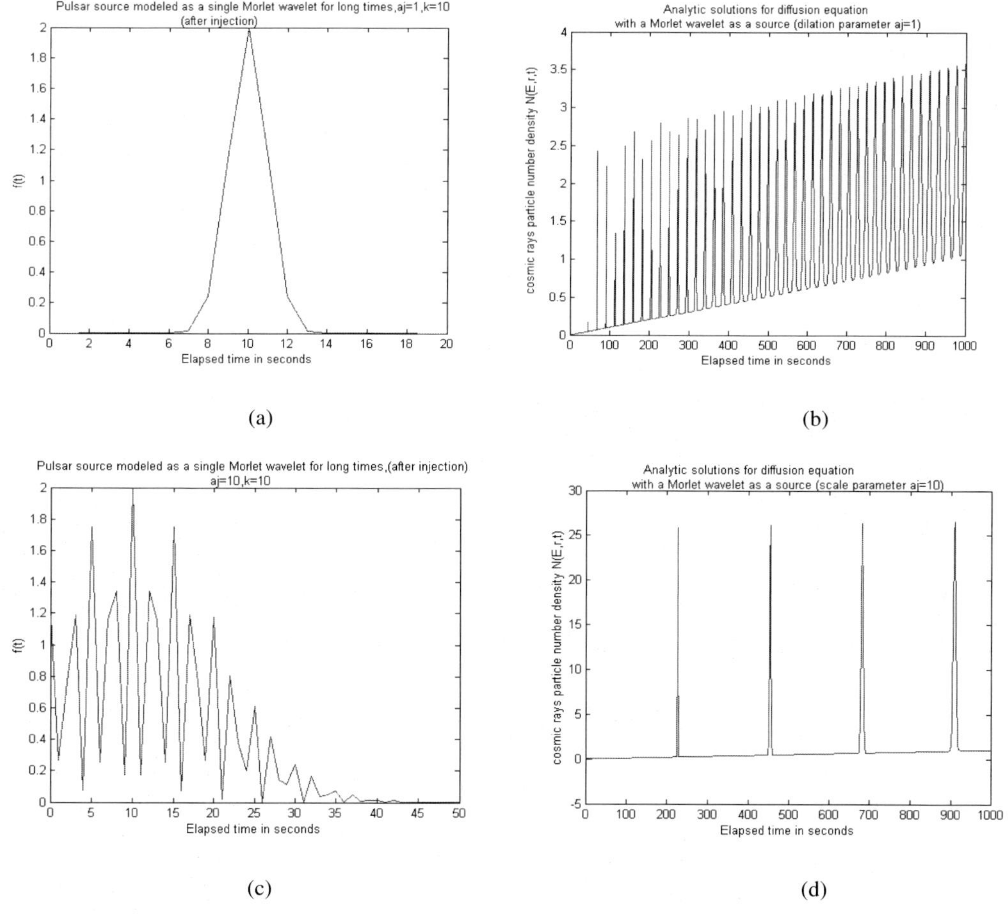

(a)

(b)

(c)

(d)

Figure 9. (Continued).

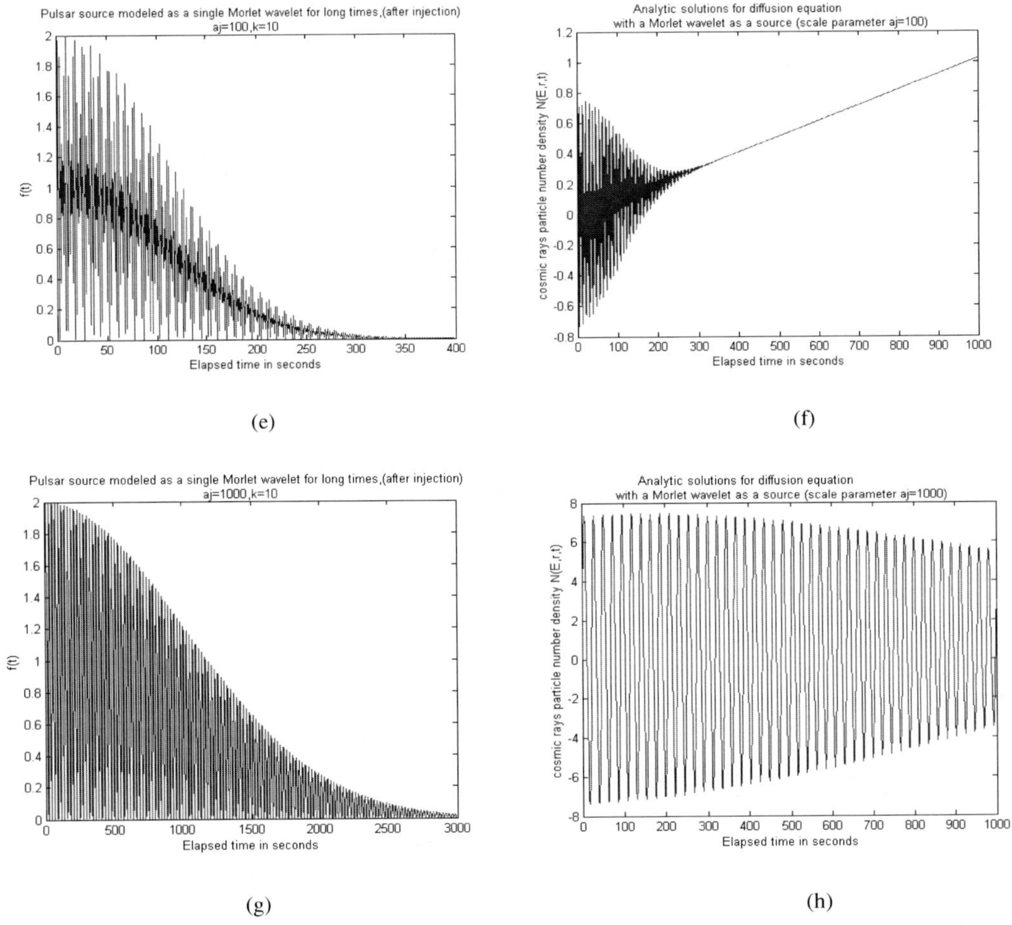

Figure 9. Variation of the CR particles density in time for different values from the scale parameter : (b) compared to the source (a): aj=1, (d) compared to the source (c) aj=10, (f) compared to the source (e) : aj=100, (h) compared to the source (g) aj=1000.

A period about 220s~240s appears for aj =10. For aj=100 the characteristic period of the source pulsar 8s is also present in the answer described by the density of CR particles. The amplitudes of oscillations of the latter are cancelled just at the end of emission period of ~350s from the source, thus the density of CR particles continue to increase quasi linearly with light undulations. The value of scale parameter aj =1000 shows for the cosmic particles density the same period 20s as the pulsar source which induced this answer.

Morlet Decomposition and Reconstructions of the Response N (E, R, T) to Pulsar Modelled Source

The decomposition period is taken equal to pulsar period;

$$\omega_0 = 2\pi/T_0 \ , \ T_0 = 22.7.10^{-3}\,s$$

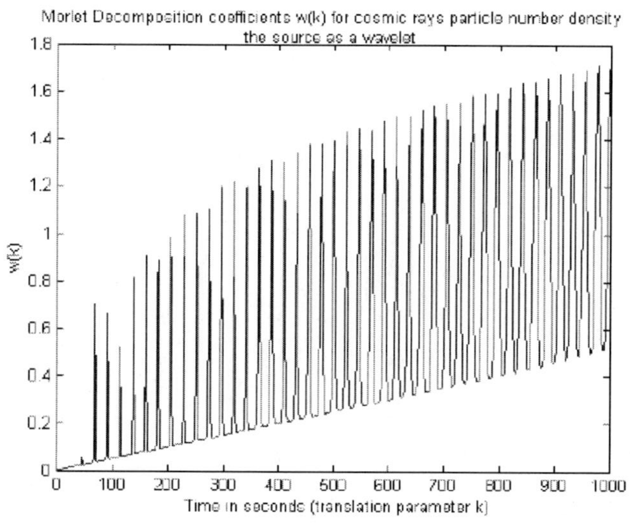

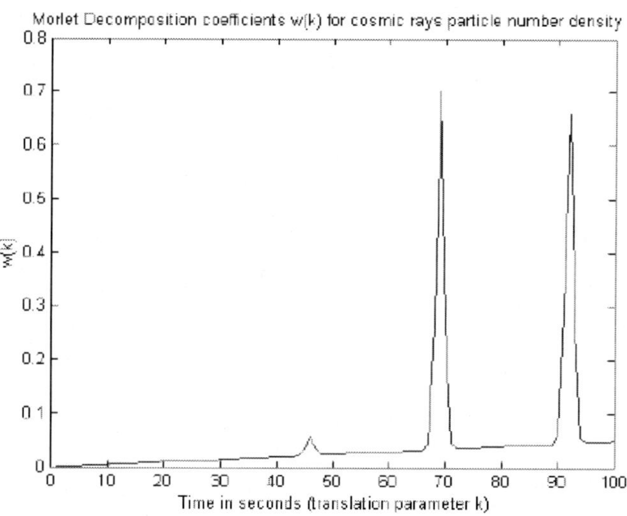

Figure 10. Variation in time of the Morlet decomposition coefficients for CR particles density where the source is a pulsar described by a Morlet wavelet.

Morlet reconstruction as well as decomposition coefficients (Figure 10,11) reveal the same original period of the CR particles density as a response to a pulsar modelled source. The density of particles of RC in the vicinity of the origin, decreased compared with the original density. it is divided by two for the hundred seconds period right after the injection.

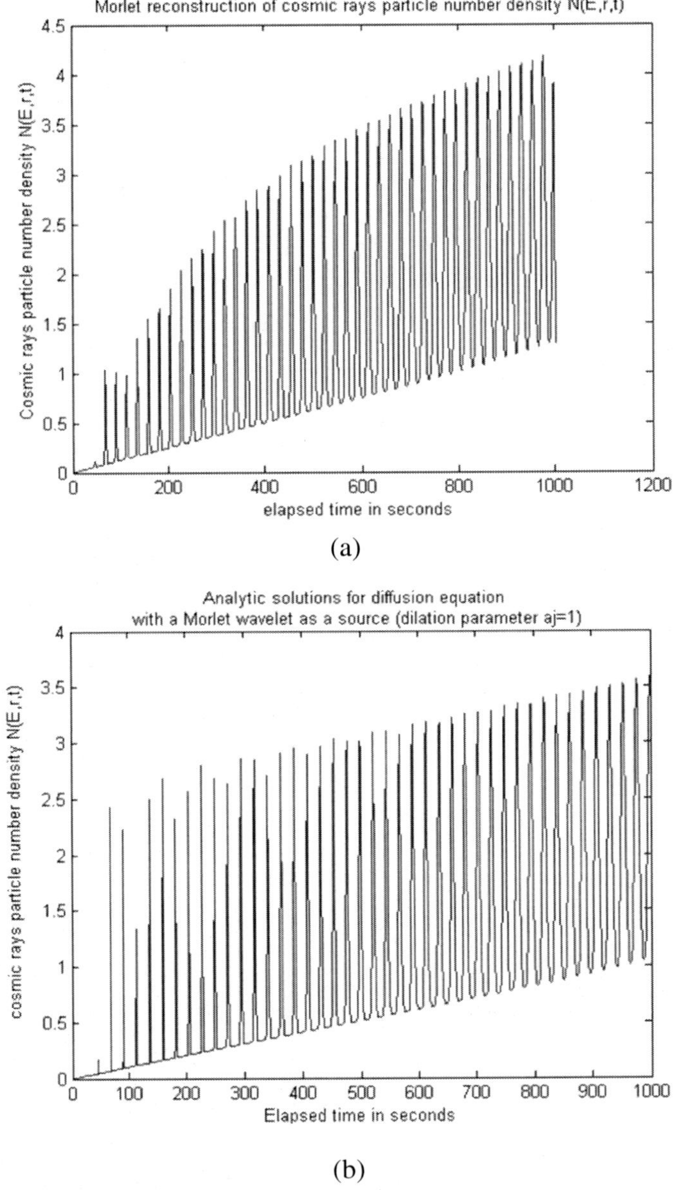

Figure 11.(a) Morlet reconstruction of the CR particles density where the source is a pulsar described by a Morlet wavelet compared with (b) the original curve.

5. A Sum of Morlet Wavelets Modelled Pulsar Source

Source Oscillation

The source always selected in the form of a pulsar PSR J0737-3039A of T_0 period. In this stage the oscillations of this pulsar are described in a more realistic way by a family thus a sum of ten Morlet wavelets shifted through various translation parameter values k, we examined various cases of scale parameter aj = 1, 10, 100 and 1000 (Fig12).

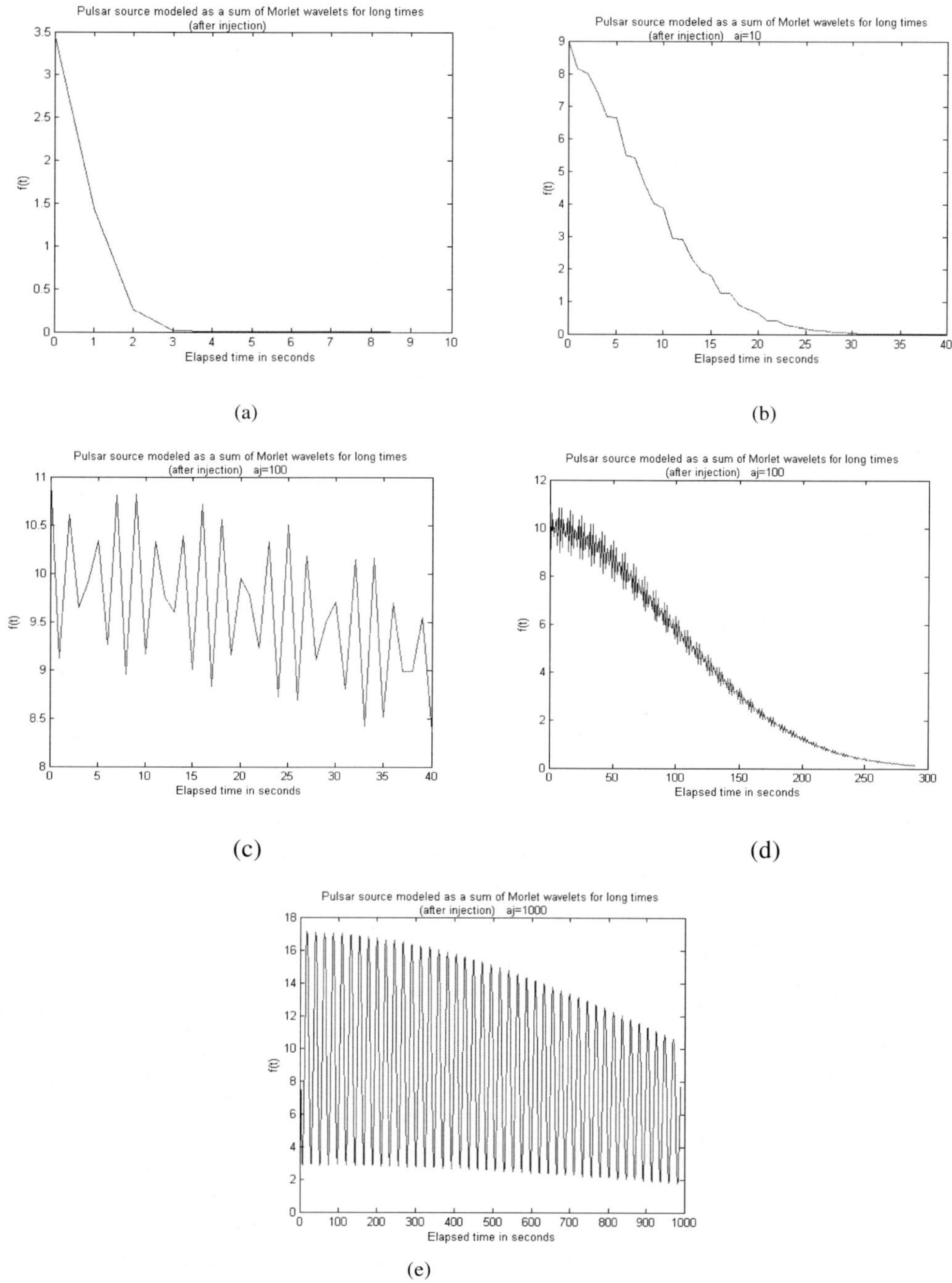

Figure 12. Oscillation of the pulsar source modelled by a sum of Morlet wavelets for long times (far from the injection) for various parameters of scale (a): aj=1, (b): aj=10, (c), (d): aj=100, (e): aj=1000.

$$q\left(E,r,\frac{t-k}{a_j}\right)=q_0 E^{-\alpha}\,\delta(r)\,f\left(t_{kj}\right)=B.\delta(r)\,f\left(t_{kj}\right)$$

$$\text{avec } t_{kj}=\frac{t-k}{a_j},\ B=q_0 E^{-\alpha},\ q_0=10^5,\ E=1\,EeV,\ \alpha=2$$

we approximate $f(t)$ by a sum of Morlet wavelets (all of weight 1):

$$f\left(t_{kj}\right)=\sum_k \cos\left(\omega_0.\left(\frac{t-k}{a_j}\right)\right).\exp-\left(\frac{1}{2}.\left(\frac{t-k}{a_j}\right)^2\right)$$

$$q\left(E,r,\frac{t-k}{a_j}\right)=B.\delta(r)\sum_k \cos\left(\omega_0.\left(\frac{t-k}{a_j}\right)\right).\exp-\left(\frac{1}{2}.\left(\frac{t-k}{a_j}\right)^2\right)$$

with $\omega_0=2\pi/T_0$, $T_0=22.7\,.ms$

The sum of Morlet wavelets taken as model of pulsar source naturally makes increase the maximum value of the function f (t) describing the source.

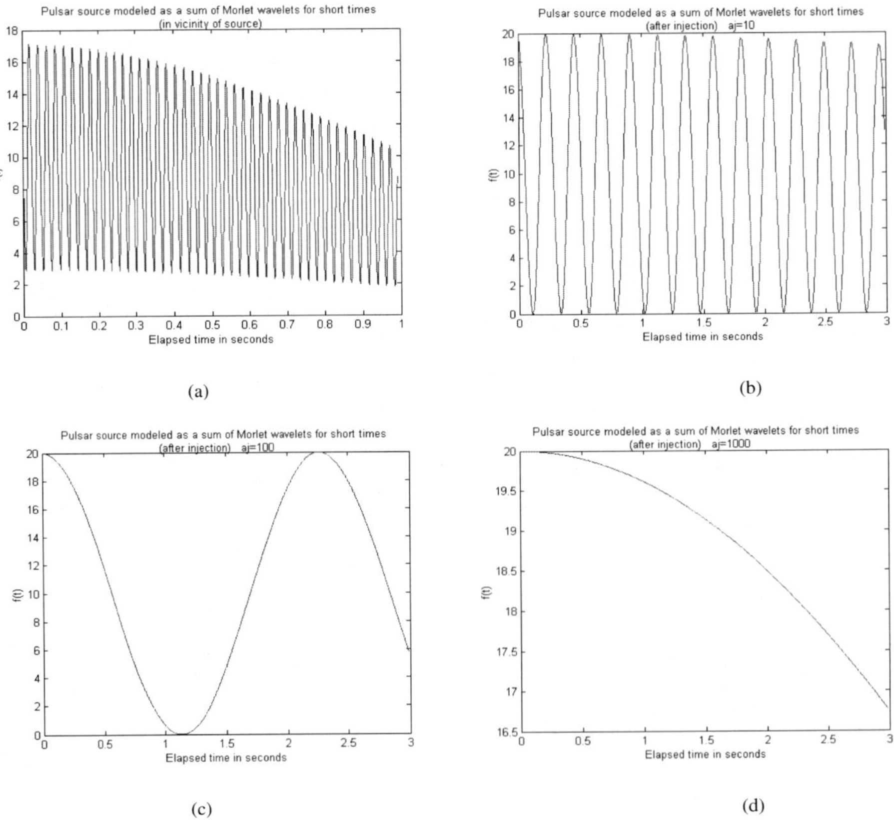

Figure 13. Oscillation of the pulsar source modelled by a sum of Morlet wavelets for short times (in the vicinity of the source) for various scale parameters (a): aj=1, (b): aj=10, (c): aj=100, (d):aj=1000.

For relatively long times (after the injection) and for aj =10 the source presents light undulations and cease to emit after ~40s. For aj =100, we find the same periods of oscillations for a single wavelet modelled source but the amplitudes of oscillations are relatively small compared to the values of function f (t) describing the new model of source.

We also find for aj =1000 the periods of oscillation of 20s found for the model with only one Morlet wavelet with an increase in the maximum value of the CR density.

For short times (in the vicinity of the source) (Figure 13), we find the characteristic period of pulsar of ~20-23ms for the scale parameter aj =1, with a relative increase in the intensity for the function f(t) describing the source. The periods 0.2 S and 2.2s are also present respectively for aj =10 and aj =100. As for the source with only one Morlet wavelet the function f (t) decrease for great values of the scale parameter aj =1000, thus showing the phenomenon of particles losses for the period of emission of the source. The maximum value for the function f(t) increased.

Cosmic Ray Particles Number Density as a Response to Morlet Wavelets Sum Modelled Pulsar Source

In this sequence the CR particles density is thus the answer to a pulsar source described by a sum of Morlet wavelets. The analytical solution of (Eq.1) is given then by

$$N\left(E,r,t_{kj}\right)= q_0 E^{-\alpha}\sum_{k}\int_{-1/bE}^{0} dt'\, \frac{e^{-\frac{r^2}{4\Lambda}}\cos\left(\omega_0 .t'_{kj}\right)\exp-\left(\frac{1}{2}.t'^2_{kj}\right)}{\left(4\pi\Lambda\right)^{3/2}\left(1+b_0 Et'_{kj}\right)^{2-\alpha}}$$

The variation in time of the CR particles density as a response to this model of source shows very reduced peaks at elapsed times 45s, 65s 90s 110s just after the injection of particles in the vicinity of the source. The appearing harmonics are of various amplitudes and show mainly two parts of the signal in opposition of phases and two classes of amplitudes for the CR density. CR densities of greater amplitudes especially show periods of ~25s at the end of signal, 150s, 200s, the 250s. CR particles densities of lower amplitudes oscillate mainly in opposition of phases to those of large amplitudes and during time of ~25s, 70s. Weak fluctuations also about 2~3s are detected around main peaks (Figure 14). CR particles density for aj =10 shows reduced peaks appearing each ~220-250s with a very slow variation of the CR density starting from the maximum which relatively increased. For aj =100, the CR density is almost constant, elapsed time to reach a maximum of CR density increased. A quasilinear growth wrapping a light undulation of the CR density appears for the scale parameter aj =1000 (Figure 15).

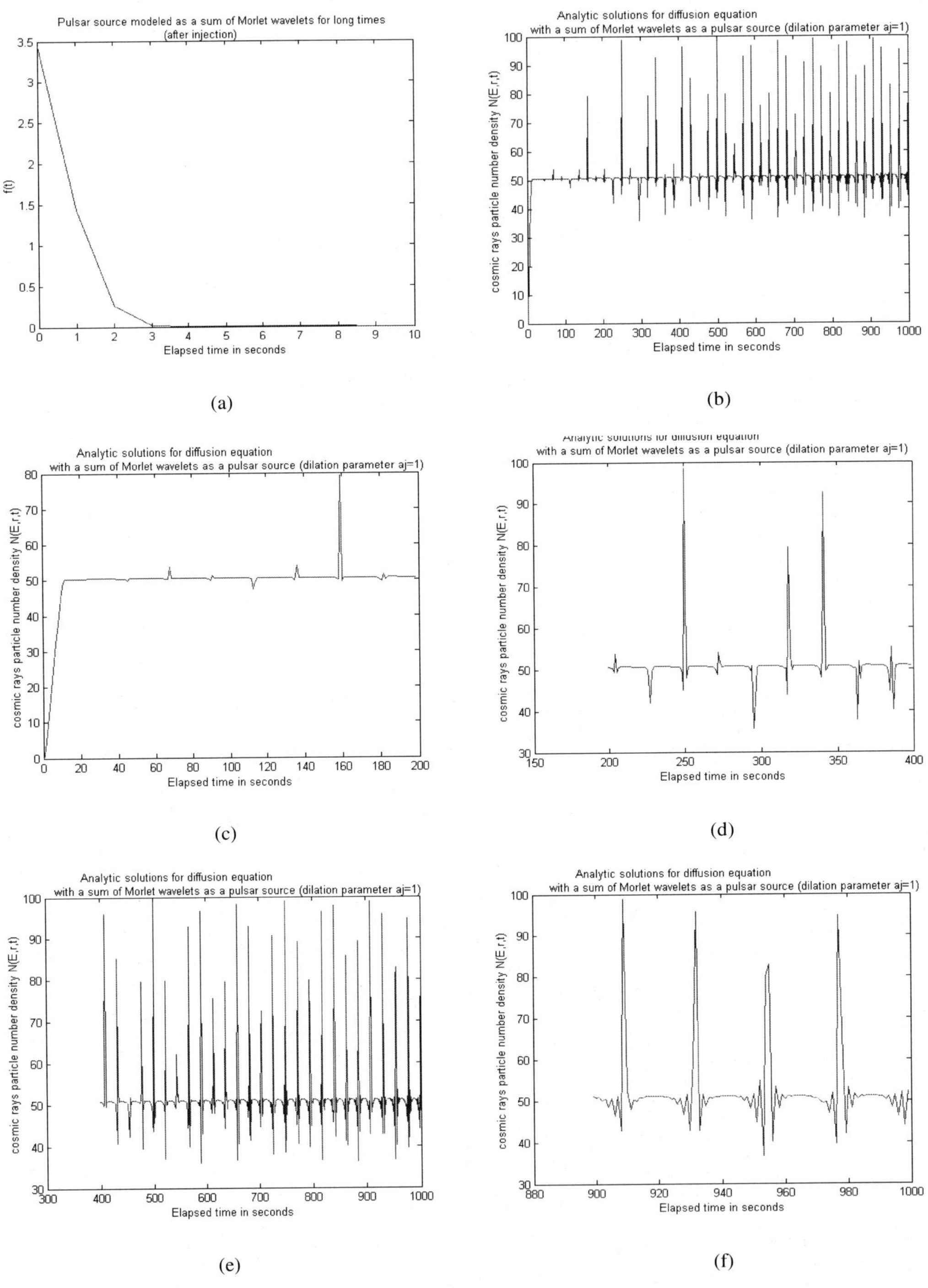

Figure 14. Variation in time of the CR density as a response to a pulsar source modelled by the oscillation of ten Morlet wavelets (a). Various zooms are shown according to the time interval of study: (b) signal showing the variation of N (E, r, t) for the total period of elapsed time, (c), (d): beginning of the signal,(e), (f): end of the signal of N (E, r, t).

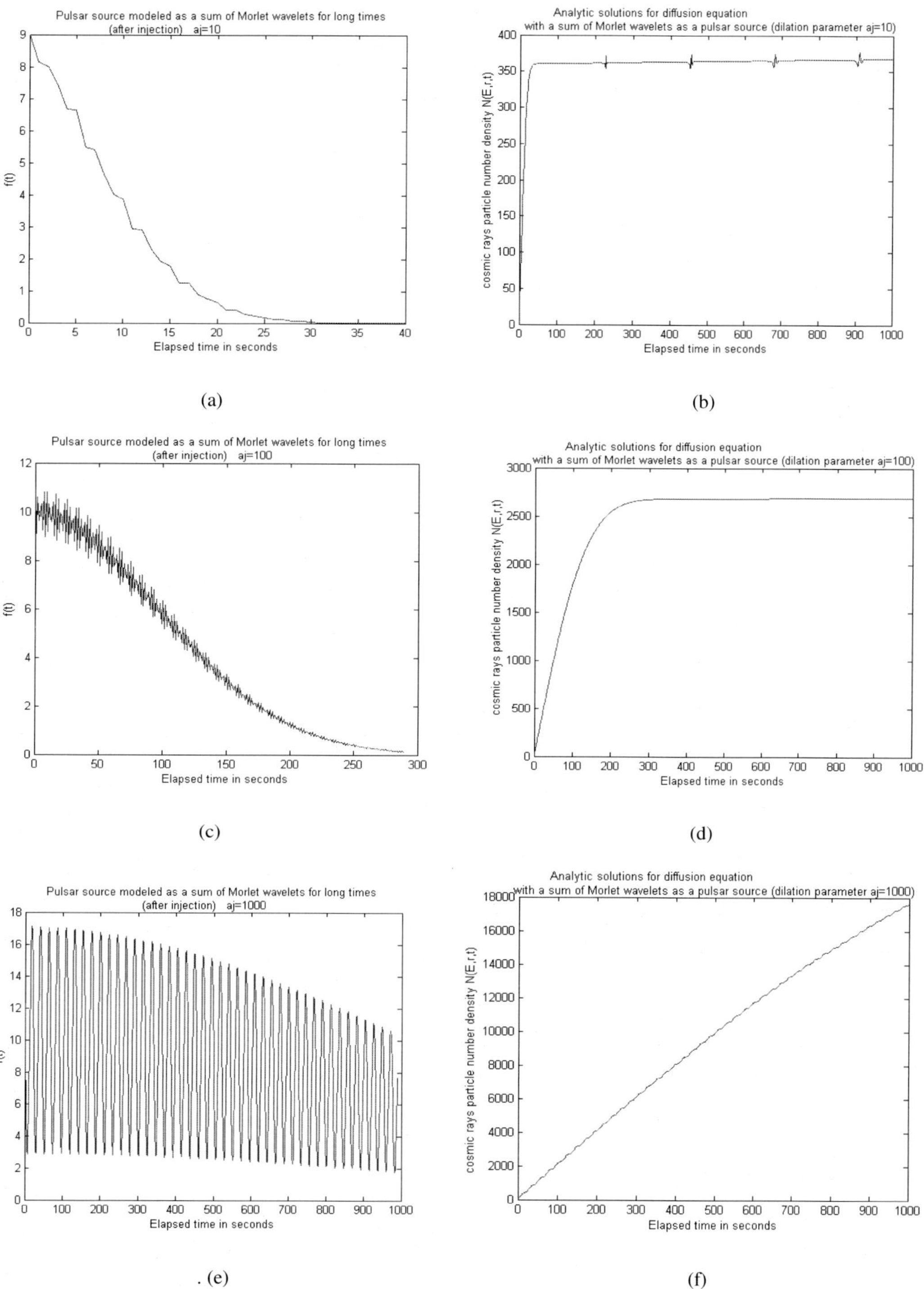

Figure 15. Variation in time of the CR particles density compared to the pulsar source modelled by the oscillation of ten Morlet wavelets. (a) and (b) are respectively the source and the answer N (E, r, t) for aj=10, (c) and (d) for aj=100 and (e), (f) for aj =1000.

Décomposition and Reconstruction of Réponse (N(E,R,T) : CR Particles Density) in Morlet Wavelets

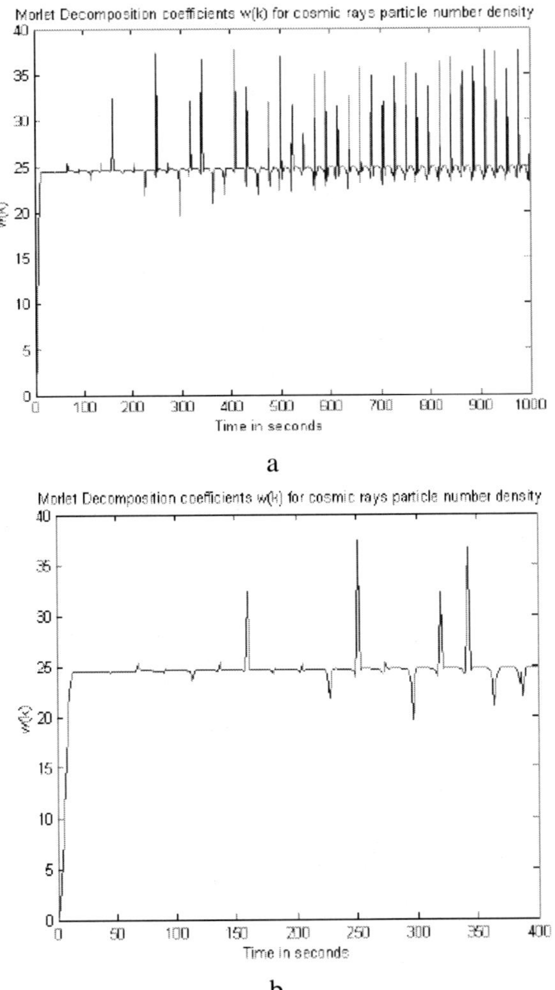

Figure 16. Variation in time of the decomposition coefficients in Morlet wavelets of the CR particles density with oscillation of pulsar source described by a sum of Morlet wayelts.

The T_0 period of decomposition is maintained equal to pulsar period;

$$\omega_0 = 2\pi/T_0 \ , \ T_0 = 22.7 \, ms$$

Information concerning the most important periods contained in curves of decomposition coefficients (Figure 16) especially show in the first part of decomposition signal a maximum around the period 250s, a secondary peak around 340s and other peaks around 160s and 320s. Thus all harmonics found in original curve of CR density as response to this type of pulsar modelled source are revealed with others in Morlet decomposition .the peaks in opposition of phase are located mainly at 300s.

The Morlet reconstruction curves (Figure 17) for the CR particles number density show periods of 90s, 160s and those of 25 S especially at the end of the signal found already in the original curve. The amplitudes of various harmonics describing the reconstructed signal decreased especially at 550~600s.

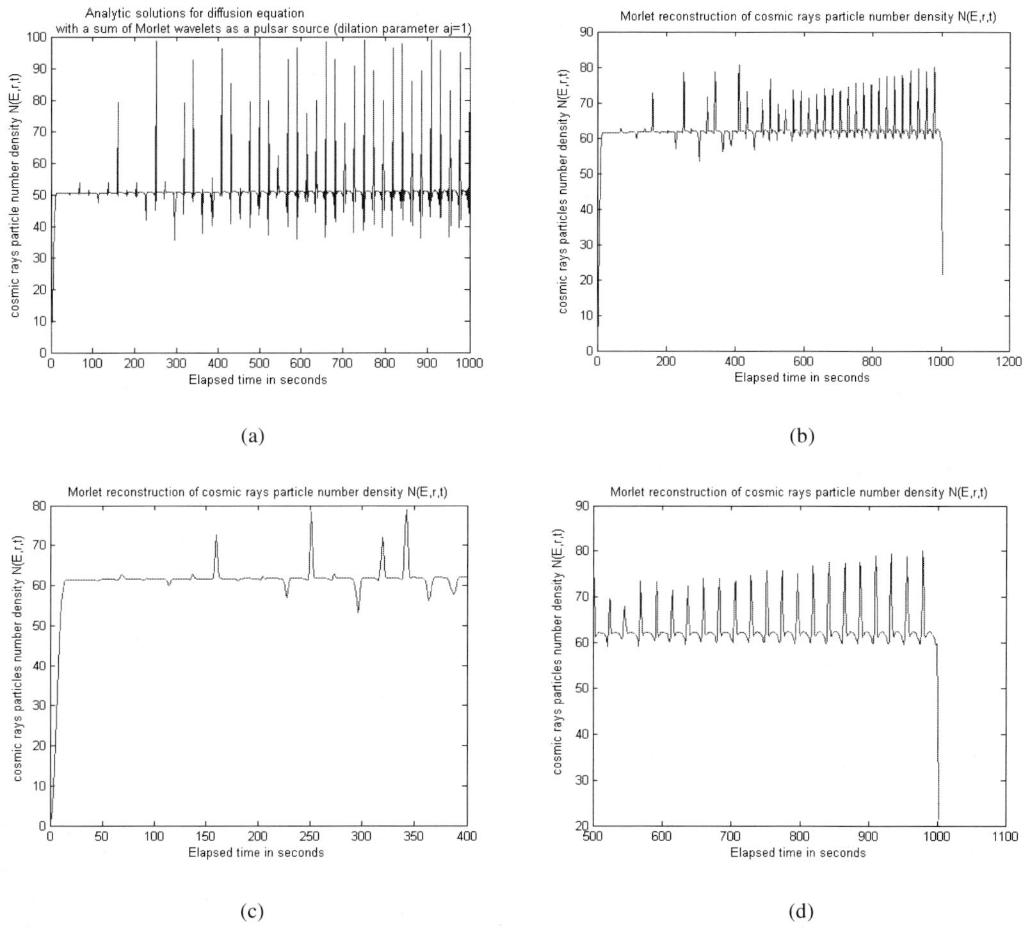

(a)

(b)

(c)

(d)

Figure 17. (b),(c),(d) Morlet reconstruction of the CR density where pulsar source oscillations are described by a sum of Morlet wavelets (zooms for various parts of the reconstructed signal), compared with (a) the original curve describing the CR density N (E, r, t) as response to this model of pulsar source oscillations.

CONCLUSION

One of the indirect observations we can make is the "composition" of GCRs. This can tell us a lot about the GCR sources and the cosmic rays' trip through the Galaxy. In this work particular attention was given to time dependence properties of source and CR densities oscillations. We have studied in a first step the evolution of the cosmic rays particles density function in time for very high energies of about $\sim 10^3$ EeV and taking into account of continuous energy losses by synchrotron radiation and inverse Compton scattering, and a diffusion coefficient depending on energy in the case of the Kolmogorov type spectrum of turbulence. We have adopted an intermediate model for an index of energy power of 0.6 for diffusion coefficient. The curves describing the evolution of the CR density particles for a total number of initial particles of 10^5 particles and at different distances from the source show the two phenomena of creation and losses of particles especially at short distances in the vicinity of the source. At $r=10^6$pc from the source, the curves analysing the CR particles density in time when varying the energy parameter from 0.01 EeV to 1 EeV show that the CR particles density reach a maximum corresponding to a maximum of particles creation for 5.10^9 years for an energy of 10^{-2} EeV , the corresponding time of this maximum decrease with energy. The decrease of CR particles density with energy at $r=10^6$pc keeps the same pace with exception that the energy corresponding to disappearance of particles is ranging from E= E_c=23 EeV for interval 1-50 EeV to 0.1 EeV for interval 0.01 to 0.5 EeV .

We then decomposed the function of CR particles density in Morlet wavelets and we detected the presence of a maximum for the decomposition coefficients around a period from 4 to 5 years around the time of injection for which also the CR particles density presents a maximum and describes the two phenomena creation and losses of particles.

In another part the source is described by a pulsar whose motion is approximated by the oscillation of Morlet wavelet. The CR particles density as a response to this model of source shows periods of 20s, 220s~240s, 8s, 20s respectively for the values of scale parameters aj=1,10,100 and 1000. Then we have decomposed in Morlet wavelets the CR particles density. The curve describing the decomposition coefficients reveals peaks around 70s, 90s, a very reduced peak around 42s, all the other peaks are increasing in amplitudes and are located around 110s, 130,150s etc... The Morlet reconstruction shows the same original period of the CR particles density as an answer to a pulsar also described by Morlet wavelet. The CR particles density around the origin has decreased by half right after the injection.

However, a more realistic manner to describe the oscillations of the pulsar modelled source is to approximate its oscillation by a superposition of Morlet wavelets shifted in time. Indeed the deadened aspect of cosmic particles emission given by a single Morlet wavelet constitutes a handicap for the real description of a pulsar emitting in a continuous way.

The appearing harmonics show primarily two parts of the signal in phase opposition and two classes of amplitudes for the CR density and for aj=1. The CR densities of greater amplitudes have periods of ~25s, 150s, 200s, 250s. The CR particles densities of lower amplitudes oscillate mainly in phase opposition to those of large amplitudes and during time of ~25s, 70s. Weak fluctuations about 2~3s are also detected.

For aj =10 the CR density shows a very slow variation and reduced peaks appearing each ~220-250s recalling the period of large peaks in the case of one wavelet modelling for aj=10. The CR density is almost constant for aj=100 and after having to reach a maximum at the end of emission period of source. A quasi-linear growth wrapping a light undulation of the CR density appears for the scale parameter aj =1000.

The most important periods given by the curves of Morlet decomposition coefficients for aj=1, especially show a maximum around the period 250s, a secondary peak around 320-340s and other peaks around 160s, the peaks in phase opposition are located mainly at 300s.

Morlet reconstructions show periods of 90s, 160s and those of 25 s especially at the end of the signal and already found in the original curve.

We project to extrapolate this study to long times far from injection of particles by the source then to reveal longer periods , and to see on the other hand the other effects related to source and atmospheric depth on the response.

REFERENCES

[1]   M.R. Attolini et al., *Planetary Space Sci.* 23, 1603 (1975).

[2]   K. Kudela et al., *J. Geophys. Res.* 96, 15871(1991).

[3]   V.I., Kozlov and V.V. Markov, Wavelet image of the fine structure of the 11-year cycle based on studying cosmic ray fluctuations during cycles 20-23., *Geomagnetism and Aeronomy.* 47, 43-51, 2007.

[4]   V.I., Kozlov and V.V. Kozlov : A new index of solar activity : An index of Cosmic Ray Scintillation., Geomagnetism and Aeronomy, 2008, vol.48, No.4, pp.463-471, *Pleiades Publishing,* Ltd., 2008.

[5]   V.I., Kozlov and V.V. Markov,wavelet image of a Heliospheric Storm in Cosmic rays., *Geomagnetism and Aeronomy,* 2007, vol 47, No1,pp.56-53.

[6]   Andreas Prokoph and R Timothy Patterson., Application of Wavelet and regression Analysis in assessing temporal and geographic climate variability: Eastern, Ontario, Canada as a case study., *Atmosphere Ocean* 42 (3) 2004,201-212.

[7]   J Morlet ,G. Arehs, I , Forgeau and D,Giard. 1982. Wave propagation and sampling theory. *Geophys.* 47:203.

[8]   C. Y Huang and M.Pohl., Monte Carlo study of Cosmic ray propagation in the galaxy and diffuse gamma ray production., 30th International cosmic ray conference.

[9]   P Blasi. et al., A magnetized local supercluster and the origin of the highest energy cosmic rays., arxiv:astro-ph/980624v1 19 juin 1998.

[10] R Kulsrud, D Ryu. R,Cen.,and J.P Ostriker.., *Astrophys J.* (1997).

[11] M Burgay et al . *Nature* 426 531 (2003).

[12] A G Lyne et al. *Science* 303 1153 (2004).

[13] K. Kudela , M.Storini , A.Antalova and J.Rybak . On the wavelet approach to cosmic ray variability., proceeding of ICRC 2001: 3773 c *Copernicus Gesellschaft* 2001.

[14] L. Anchordoqui, T. Paul, S. Reucroft, and J. Swain., Ultrahigh Energy Cosmic Rays:The state of the art before the Auger Observatory., arxiv.org/abs/hep-ph/0206072v3

[15] D.Ryu, H.Kang, and P. Biermann, e-print astro-ph/9803275.

[16] V.L Guinzburg and S.I. Syrovatskii, *"The Origin of Cosmic Rays"* Macmillan 1964.

[17] Longair, S Malcolm. Particles, Photons, and Their Detection, Vol. 1 of High Energy Astrophysics, 2nd ed. , Cambridge, *Cambridge University Press,* 1992.

[18] Gaisser. K Thomas., Cosmic Rays and Particle Physics, Cambridge, *Cambridge University Press,* 1990.

[19] J. A Simpson,., "Elemental and Isotopic Composition of the Galactic Cosmic Rays, *Annual Reviews of Nuclear and Particle Science*, Vol. 33, 323-381. 1983,

[20] J. R Jokipii,., and F. B. McDonald, "Quest for the Limits of the Heliosphere", Scientific American 272, 58-63, 1994.

[21] Friedlander, W Michael. *"Cosmic Rays"*. Harvard University Press, 1989.

[22] Rossi, Bruno. *"Cosmic Rays"* McGraw Hill 1964.

[23] N. Zarrouk, R. Bennaceur. Estimates Of Cosmic Radiation Exposure on Tunisian Passenger Aircraft Radiation Protection Dosimetry (2008), pp. 1–8.

[24] N.Zarrouk, R. Bennaceur. A Wavelet Based Analysis of Cosmic Rays Modulation, Science Direct , *Acta astronautica,* 65 (2009) 262-272.

[25] N.Zarrouk, R. Bennaceur. Neural Network and Wavelets In Prediction of Cosmic Ray Variability: The North Africa as Study Case, Science Direct, *Acta astronautica,* 66 (2010) 1008–1016.

[26] N.Zarrouk, R. Bennaceur. Extrapolating cosmic ray variations and impacts on life : Morlet wavelet analysis. *International Journal of Astrobiology* 8(3): 169-174 (2009). Cambridge University

[27] N.Zarrouk, R. Bennaceur. Link nature between low cloud amounts and cosmic rays through wavelet analysis. *Acta* astronautica, 66 (2010) 1311–1319.

Chapter 12

ANISOTROPY OF ULTRAHIGH-ENERGY COSMIC RAYS

A. A. Mikhailov

Yu.G. Shafer Institute of Cosmophysical Research and Aeronomy,
Yakutsk, Russia

ABSTRACT

The problem associated with the origin of ultrahigh energy cosmic rays is one of priority in high-energy astrophysics. There are two main hypotheses about the origin of cosmic rays - galactic and extragalactic. In a galactic model, possible origins of cosmic rays are supernovae stars, pulsars, or a galactic center; in extragalactic models – nuclei active galactics, the spinning supermassive black holes, quasars, etc.

We analyzed arrival directions of extensive air showers (EAS) or particles which originated these EAS by Yakutsk, AGASA, and HiRes arrays, and P. Auger observatory data. Analyses results show: distributions associated with the number of particles in galactic latitudes differs relative to the the center/anticenter of the Galaxy; arrival directions of particles according toYakutsk, AGASA, P.Auger data correlate with pulsars of the Local arm Orion; and the rotatation period of pulsars which correlate with arrival directions of particles have more short periods than other pulsars.

We show that useing a harmonic analysis for the search anisotropy for arrival directions of particles is ineffective. Therefore, we suggest a new method to search sources and anisotropy arrival directions of particles. This method differs from other methods: the region of search sources and anisotropy of cosmic rays can vary on a celestial sphere. By this method we found fluxes of particles from the galactic plane and the side of a galactic center which exceed the expected fluxes of particles than 4σ (\sigma) and $5\sigma\square$ in the case of isotropy more. It is shown that the arrival directions of particles according to PAO data correlate to both pulsars and active galaxies. The problem associated with the origin of cosmic rays is also discussed.

INTRODUCTION

The Yakutsk EAS array is designed for detecting cosmic rays with energy $>10^{15}$ eV. It provides measurements associated with main components of extensive air showers (EAS): electrons, muons and Cerenkov light emissions. There are experiments on radio-emission detection. The array itself consists of several instruments, combined in a single system: the main array; small Cherenkov array; Cherenkov tracing detector, based on a camera-obscure, large muon detector; and and a laser lidar to measure atmosphere parameters.

The Yakutsk array began taking data in 1974. The array consists of 49 surface scintillation plastic detectors on an area 2×2 m^2; 9 surface plastic detectors at 2 m^2; 5 underground scintillation plastic detectors, each 20 m^2 for muons with an energy threshold of $E_\mu > 1 \times \sec\theta$ GeV (where θ - the zenith angle); and 45 Cherenkov light detectors. Due to the Cherenkov light measurements, the energy dissipated in the atmosphere by the shower and the energy of EAS could be estimated without a calorimetry-method model. Now, the surface detectors form a uniform net of 10 km^2 with a distance of 500 m between the neighboring units. In the center of the array are 192 m^2 muon detectors with a threshold $E_\mu > 0.5 \sec\theta$ GeV. Signals for timing at each detector are distributed over a microwave link. A full description of the array has given by [1, 2].

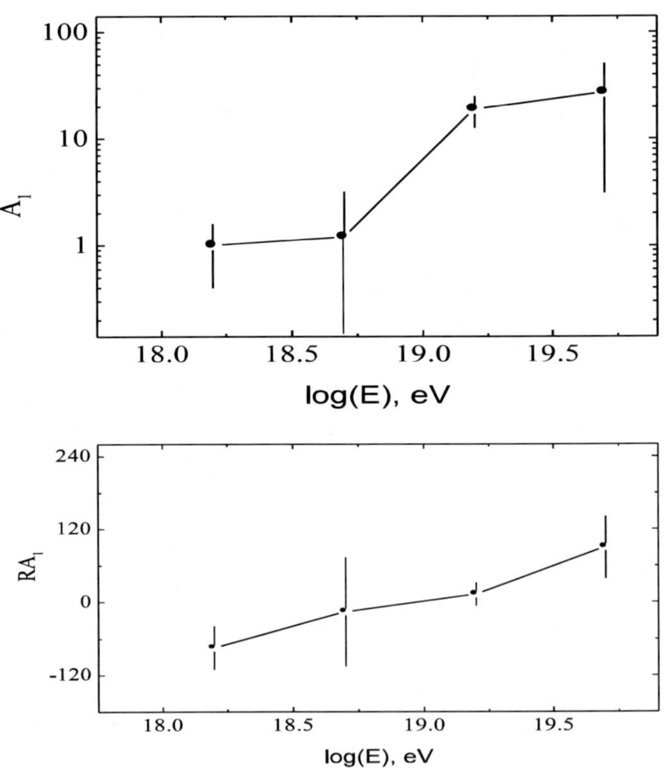

Figure 1. Amplitudes A$_1$ and phases RA$_1$ of the 1st harmonic Fourier.

1. THE HARMONIC ANALYSIS

We have analyzed arrival directions of particles from data from the Yakutsk array. Showers with energy $E>10^{18}$ eV, with zenith angles $\theta<60°$ and axes lying inside of perimeter of the array are considered. We have analyzed arrival directions of extensive air showers (EAS) from Yakutsk EAS array data for 1974-2010. The accuracy definition for solid angles of arrival directions of EAS at $E \sim 5.10^{18}$ is 5-7°, $E>4.10^{19}$ eV $\sim 3°$; the accuracy of energy estimation - $\sim 30\%$.

The latter divided an observed region of energy $>10^{18}$ eV into 4 intervals: 1) $10^{18} - 5.10^{18}$ eV, 2) $5.10^{18} - 8.10^{19}$ eV, 3) $8.10^{19} - 4.10^{19}$ eV, 4) $>4.10^{19}$ eV. Results from the harmonic analysis distribution of particles according to harmonic functions of the Fourier in right ascension are shown in Figure 1. In interval energy 10^{19} - 4.10^{19} eV, the amplitude of the 1st harmonic is at a value of $A_1=18.8 \pm 6.3\%$ and its phase RA=14°; the probability of chance of amplitude is equal $P \sim 1.2\times10^{-2}$ according to [3]. The number of EAS is equal to 502. The amplitude of the 2nd harmonic is - 5 %. Note that the phase of the 1st harmonic at $E\sim10^{18}$ eV from the Local Arm of Galaxy at RA$\sim$300° varies gradually with energy to RA $\sim$ 90° at $E\sim4.10^{19}$ eV.

Nonetheless, we suppose that the harmonic analysis only gives an average anisotropy in right ascension. Declinations of the celestial sphere are observed by different array angles and areas. Therefore, the isotropy of cosmic rays will be observed from different declinations at several different events. The observed number of particles at different declinations from the harmonic analysis is summarized. For this summarization, any local maximum or minimum at declinations smooth out and, as a result, we receive an average anisotropy on a right ascension. For us, we had greater interest in finding the anisotropy arrival directions of particles in different declinations and fluxes of particles in separate regions of the celestial sphere. Therefore, we suggest a new method for analyzing arrival directions of ultrahigh-energy particles.

2. PARTICLES FROM LOCAL ARM ORION AND PULSARS

As we consider arrival directions of particles with energy $E>4\times10^{19}$ eV we ask, "Do they correlate with pulsars of the Local arm Orion?"

Earlier we found a correlation [4, 5] between arrival directions of particles and pulsars [6] along the magnetic field of the Local arm Orion. The angular distance between the arrival direction of each particle and each pulsar was calculated. If this distance was less than the definite angle θ from the given pulsar then this particle was considered to be related to this pulsar. Each particle was considered only one time and to the nearest pulsar. We consider particles with energies $E = (0.8-4)\cdot10^{19}$ eV. A maximum correlation between arrival directions of particles and position of pulsars was found at $\vartheta<6°$. The observed number of particles was $n_{obs} = 157$ and the number of expected particles was $n_{exp} = 130.1$

According to the method in [7, 8] we estimated the chance probability to be - $P = 2.10^{-4}$. Particles moving along magnetic field lines are deviated minimally and the probability of correlation between shower arrival directions and pulsars increases.

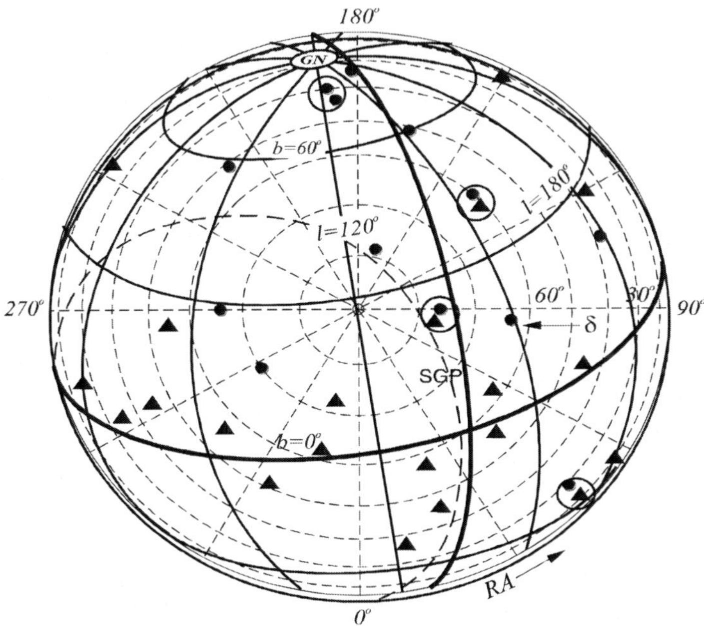

Figure 2. Yakutsk: distribution particles with E>4×10^{19} eV, ▲, • - particles which correlated and uncorrelated with pulsars accordingly. Shaped circle – boundary Local arm. δ - declination, RA – right ascension, b – galactic latitude, l – longitude.

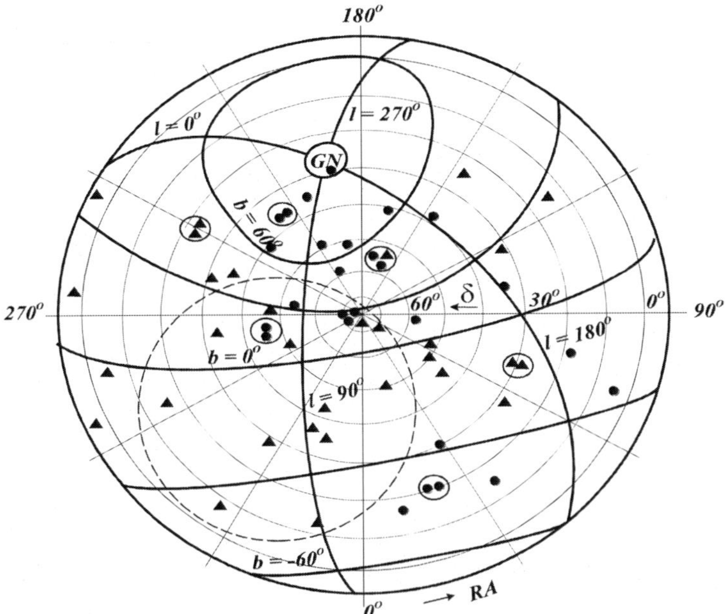

Figure 3. AGASA: distribution particles with E>4×10^{19} eV, ▲, • - particles which correlated and uncorrelated with pulsars accordingly. Shaped circle – boundary Local arm. Large circles – clusters.

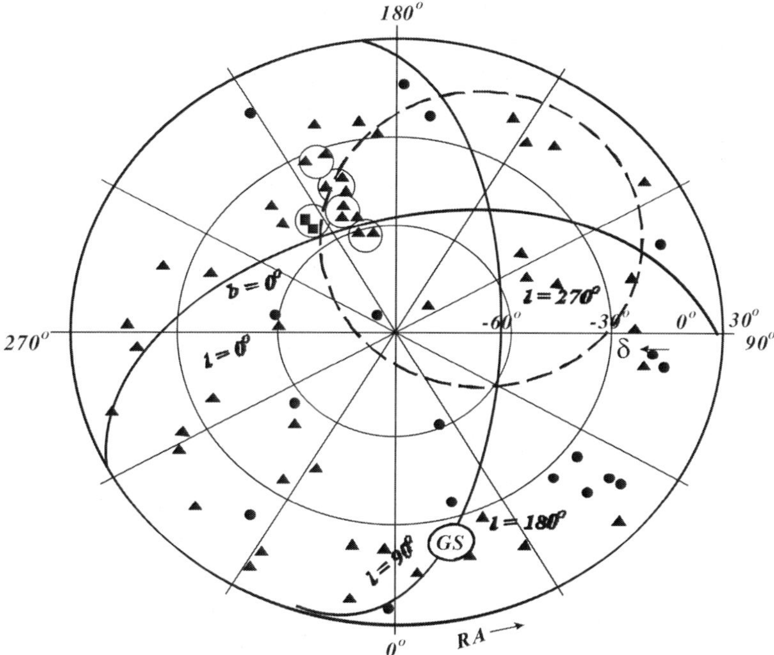

Figure 4. PAO: distribution particles with E>5.7×10^{19} eV, ▲, • - particles which correlate and uncorrelate with pulsars accordingly.

Figure 2 presents the distribution of 34 particles according to the Yakutsk array data with an energy E>4.10^{19} eV on the map of equal exposition of the celestial sphere for the period of 1974-2010 (the method to construct this map is based on the estimation of the expected number of showers in the celestial sphere [8]). On the map of equal exposition, an equal number of particles from the equal parts of the sphere is expected. According to the Yakutsk data we have found 19 particles, from which 34 correlate with pulsars (these particles are noted by triangles, Figure 2); according to array AGASA [9] there are 21 particles from 57 (these particles are noted by triangles, Figure 3); and according to array P. Auger [10] there are 49 particles from 69 (these EAS are noted by triangles, Figure 4).

Below we consider particles only inside the Local arm Orion or within a cone with angles <45° from galactic coordinates b=0° and l=90° for Yakutsk and AGASA arrays, and from galactic coordinates b=0° and l=270° for the P. Auger observatory.

In Figure 2 only distribution particles are shown inside the Local arm Orion as according to theYakutsk data. We find only 13 particles from 35 (such number of particles was registered the whole time of operation during 1974-2009). Inside the Local arm Orion 11 particles from 13 correlate with pulsars (if particles are <6° from pulsars we consider this to be a correlation between particles and pulsars [7]). The chance probability in 4 dimensions system coordinat according to [8] is P ~ 0.1.

Figure 3 shows distribution particles only inside the Local arm Orion as according to AGASA array data [9], in operation during 1990 – May 2000. Inside the Local arm Orion 12 particles from 18 correlate with pulsars. The chance probability to find 12 particles from 18 at R<6° according to [8] is P ~ 10^{-4}.

Figure 4 shows the distribution particles only inside the Local arm Orion according to P. Auger data registered during 2004-2009. Inside the Local arm Orion 16 particles from 18 correlate with pulsars. The chance probability according to [8] is $P\sim10^{-3}$.

According to [2], the largest array data arrival directions of particles with energy $E>4.10^{19}$ eV correlate with pulsars: at the Input (AGASA) and Output (P. Auger) Local arm Orion of the Galaxy. We did not find any correlation between arrival directions of particles and pulsars outside the Local arm by this array data.

3. FLUX PARTICLES FROM THE LOCAL ARM

Figure 5 shows the distribution of 34 particles with energy $E>4.10^{19}$ eV as according to the Yakutsk data in the map of equal exposition of the celestial sphere. We consider particles from the side of the Input of the Local Arm Orion of the Galaxy at a galactic latitude $3.3°<b<29.7°$ and longitude $60.1°<l<116.8°$ (this region is noted by dash quadrangles). There are 9 particles inside these coordinates. We divided the whole period of observation into 2 periods, equal to the exposition of observation: 1974 -1985 and 1986-2007. Four particles (circles) have been registered for the first period; 5 particles have been registered for the second period −. We take the first period as a sample, and for second the period 5 particles (triangles) were registered; the probability of chance to find of 34 particles in the above-mentioned coordinates is $P\sim0.03$ according to [8].

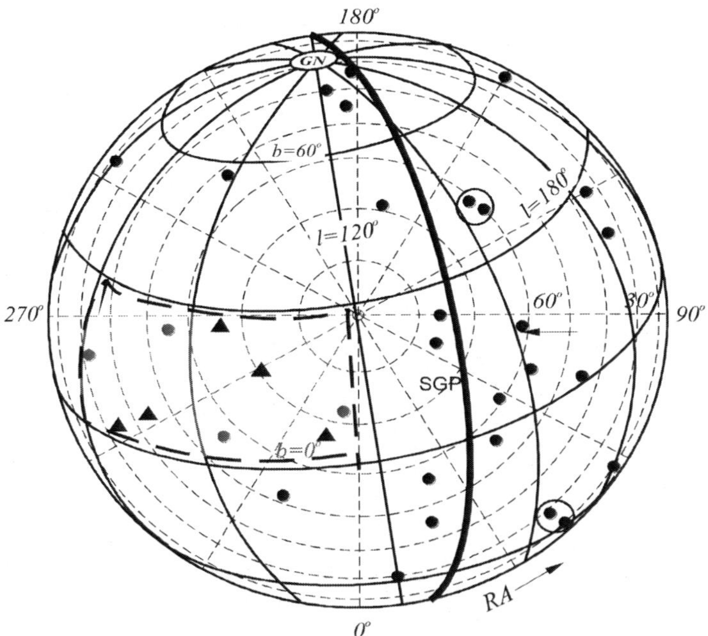

Figure 5. On the map of equal exposition, particles with energy $E>4.10^{19}$ eV are shown as according to Yakutsk EAS array data. SGP – Super Galactic Plane. Dashed quadrangles on the left – a considered region of a celestial sphere. Big circles – clusters.

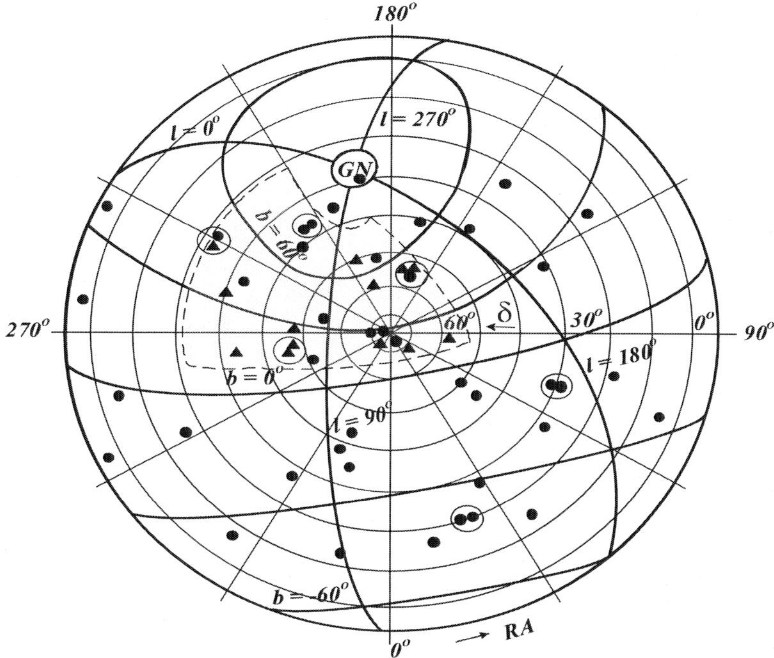

Figure 6. The same as in Figure 1 for the AGASA. Dashed quadrangles on the upper - a considered region of a celestial sphere. Inside quadrangles: ● -1984-1994, ▲ - 1995-2001.

We consider particles with energy E>4.10^{19} eV from the AGASA array data [9]. As it is seen from this map, almost a half of the events (25 particles of 58) are within coordinates 11.2°<b<69.3° and 38.9°<l<154° towards a side of the Input of the Local Arm (Figure 6). Also, we have divided the whole period of observation into 2 periods, equal to the exposition of observation: 1984 - 1994 and 1995 - 2001. For the first period, 13 particles (circles) were registered; for the second period 12 particles were registered. We take the first period as a sample, and the probability of chance to find 12 of 58 particles in the above-mentioned coordinates is P~10^{-3}.

Thus, the statistically significant particle flux in the case of the AGASA array is observed from the side of the Input of Local Arm as well as from data of the Yakutsk EAS array.

The last time the P. Auger observatory increased their data [10], we analyzed this data suggestion by using the new method (you can see it at the end of this paper, titled "Anisotropy of arrival direction of particles").

As follows from the distribution of particles on the celestial sphere according to the data of Yakutsk and AGASA (Figures 5, 6), there is a large-scale anisotropy in the arrival direction of particles connected to the Local Arm Orion of the Galaxy.

Earlier we found that if sources of particles are distributed uniformly in the Galaxy disc, then protons with the energy E~10^{18} eV mainly move along the magnetic field lines of the Galaxy arms [11]. Therefore, one can suggest that the observed particles with the energy E>4.10^{19} eV come from the side of the Input and Output of the Local Arm Orion of the Galaxy along the magnetic field lines and have a rigidity R~10^{18} eV or particles ultrahigh energy have heavily charged nuclei. A similar conclusion about the composition of cosmic rays was made by us on a basis associated with the analysis of the distribution of particles in the galactic latitude and the muon content of extensive air showers at energy $E=10^{19}$-10^{20} eV according to the Yakutsk EAS array data [5]. Also, independently from us, our colleagues using another method of muon data of showers analysis have made a conclusion [12] that the iron nuclei portion at energy $E>2.10^{19}$ eV can be from ~29% up to 68% of the total number of particles. The last data of the P. Auger array indicates that the particles at energy $E>4.10^{19}$ eV are not only protons, but, most likely, iron nuclei [13, 14].

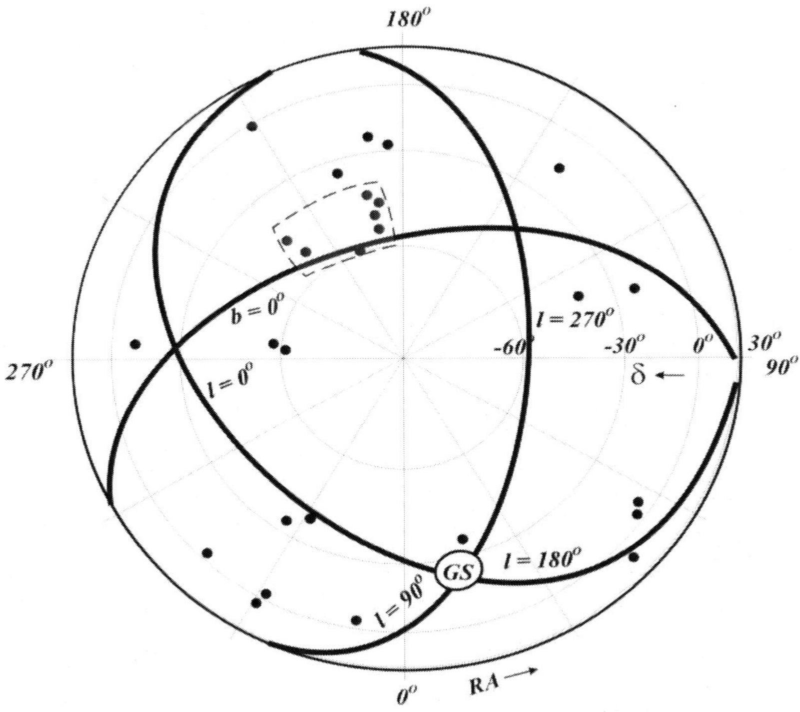

Figure 7. On the map of equal exposition, particles with energy $E>4.10^{19}$ eV are shown as according to P.Auger EAS array data for 2004-2007. Dashed quadrangles – a considered region of a celestial sphere. δ - declination, RA – right ascension, b, l – a galactic latitude and longitude. For the distribution of particles with the energy $E>4.10^{19}$ eV for 2004-2007 of the P. Auger array data [10] the map of equal exposition of the celestial sphere has been constructed. As is seen from this map, 7 particles of 27 are within of coordinates towards the side of the Output of the Local Arm near a galactic plane: $1.7°≤b≤19.2°$ and $-52.4°≤l≤-34.4°$. Also, we have divided the whole period of observation into 2 periods: 2004 – 2005 and 2006 - 2007. For the first period, 3 particles were registered; the probability of chance to find 3 of 27 particles in above0mentioned coordinates is P~10^{-3}. We take the first period as a sample, and for the second period, 4 particles were registered; the probability of chance to find 4 of 27 particles in the above-mentioned coordinates is P~3.10^{-5}. Unfortunately, for 2007 – 2009, we did not find more particles inside the above coordinates.

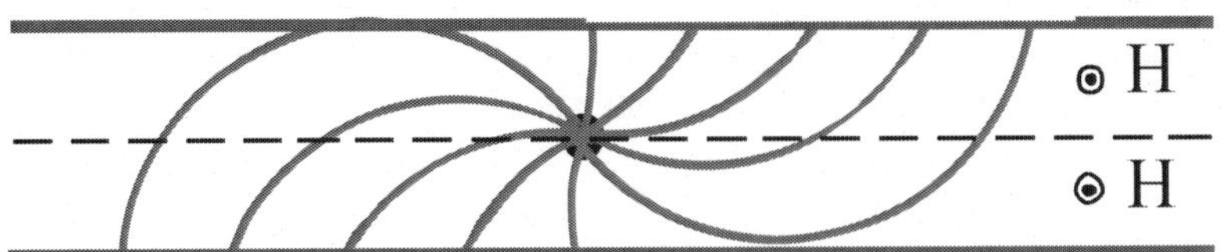

Figure 8. Trajectories of particles in a magnetic field H of a disc, ⊙ - a magnetic field.

4. SIGN OF THE GALACTIC ORIGIN

In the paper [15], a galactic model for the cosmic ray origins of ultrahigh energy has been considered. It is supposed that sources of particles are uniformly distributed over the Galaxy disc and the regular magnetic field of the Galaxy mainly has azimuth directions. If a magnetic field doesn't change its direction above and below a galactic plane (near the Sun a magnetic field is directed in galactic longitudes l ~ 90°), then as a result of the influence of a large-scale regular magnetic field, characteristic trajectories of particles are possible (Figure 8).

Thus, a flux of particles from the side of the Local Arm Orion of the Galaxy has been found.

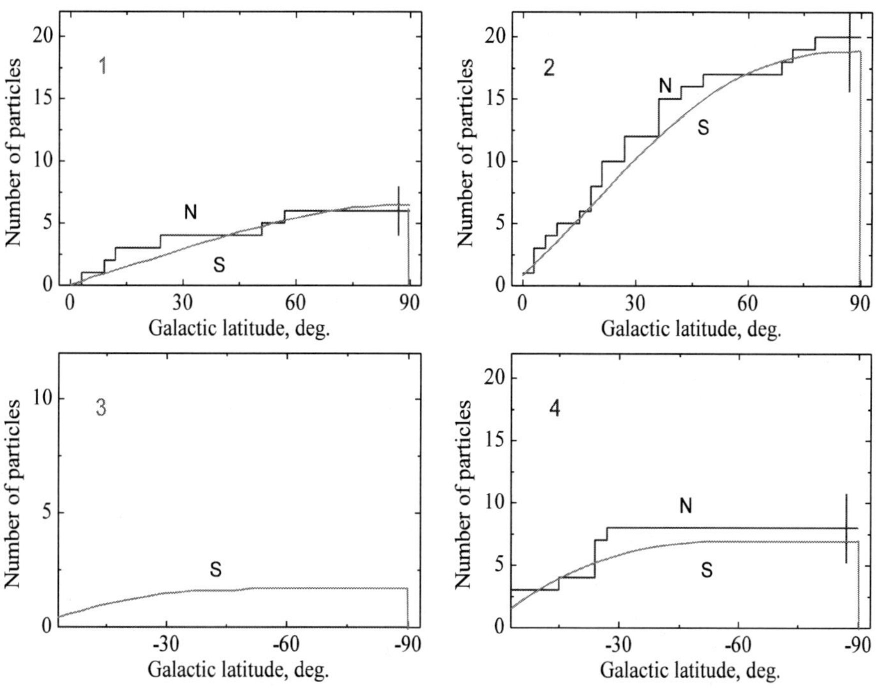

Figure 9. Yakutsk: distribution of particles n in galactic latitude at center/anticenter directions by Yakutsk array data. A solid curve S - the expected number of events in a case of isotropy. Directions: 1,3 - center, 2,4 - anticenter.

When there is a uniform distribution of cosmic ray sources in the Galaxy disc an expected number of particle fluxes from different directions of the celestial sphere will be proportional to the lengths of particle trajectories in this direction. As follows from Figure 8, increased numbers of particles are expected from the center of the Galaxy with positive latitudes and from the anticenter with negative latitudes. Therefore, from the side of the Galaxy center (longitudes - 90°<l<90°), the ratio of the number of particles above the galactic plane $n_c(b>0°)$ to the number of particles below the plane will be $n_c(b>0°)/n_c(b<0°)>1$; from the side of anticenter, the ratio number of particles $n_a(b>0°)$ will be $n_a(b>0°)/n_a(b<0°)<1$. This is a sign that cosmic rays are of a galactic origin.

We have considered the distribution of particles in galactic latitudes from data according to Yakutsk, AGASA, and P. Auger for the sides: the center and the anticenter of the Galaxy. The particle distribution according to the Yakutsk data of the northern arrays is considered in Figure 9. It presents the distribution of particles in galactic latitude above/below a galactic plane relative on the Galaxy center/anticenter directions according to data of the Yakutsk array. The interval of angles of 0 - 90° in galactic latitude b is divided though 3°; the number of particles in every subsequent interval of angles has been summed up (integral distribution). The expected number of particles, marked as S (the curve), is also shown in the case of isotropy.

The expected number of particles has been obtained in terms of the exposition of the elementary part of the celestial sphere for the array of EAS according to [8].

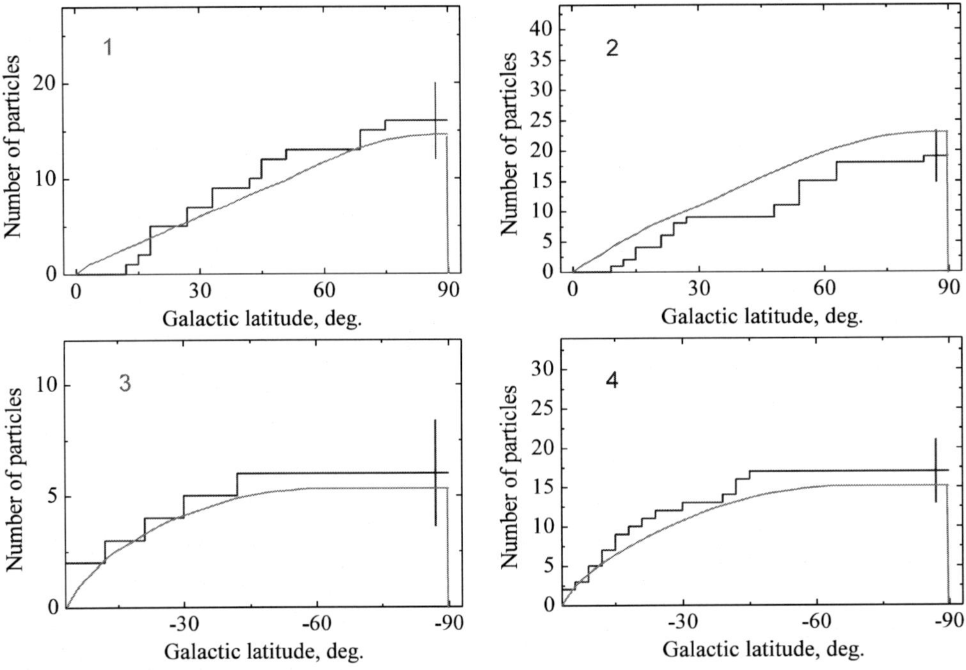

Figure 10. Distribution of particles in galactic latitude relative to the the center/anticenter directions according to data from the AGASA array.

Figure 10 presents the distribution of particles in galactic latitudes relative to the directions of the center/anticenter of the Galaxy according to AGASA data. Figure 11 illustrates the distribution of particles in galactic latitudes relative to the directions of the center/anticenter of the Galaxy according to data from the P.Auger observatory [16].

In a certain case, the Yakutsk array agrees with the suggestion of galactic model origin of particles (Figure 8), executing at (Figure 9, 3,4); the same is true of AGASA - (Figure 10, 1,2,3), and in a case for P. Auger (Figure 11, 1,2,3). The ratio between the number of particles above/below the galactic plane both from the side of the center R_c and from the side of galactic anticenter R_a has been obtained (Figure 12) in terms of the expected number of events, S, in the case of isotropy $R_c=n_c(b=90°)S_c(b=-90°)/(_c(b=-90°)S_c(b=90°))$ and $R_a=n_a(b=90°)S_a(b=-90°)/(n_a(b=-90°)S_a(b=90°))$.

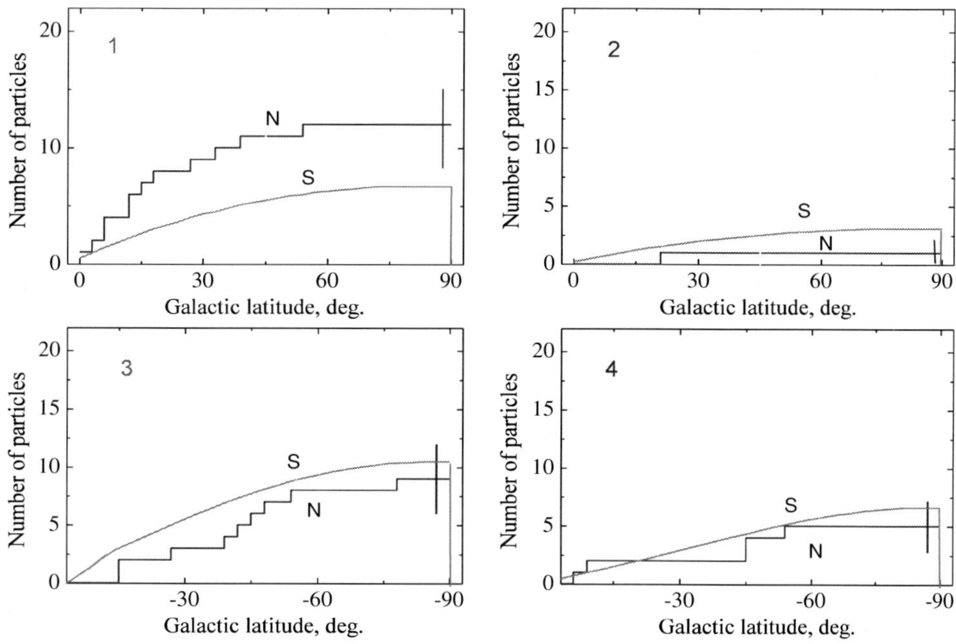

Figure 11. Distribution of particles in galactic latitude relative to the the center/anticenter directions according to data from the P.Auger observatory. The designation is the same as in Figure 1.

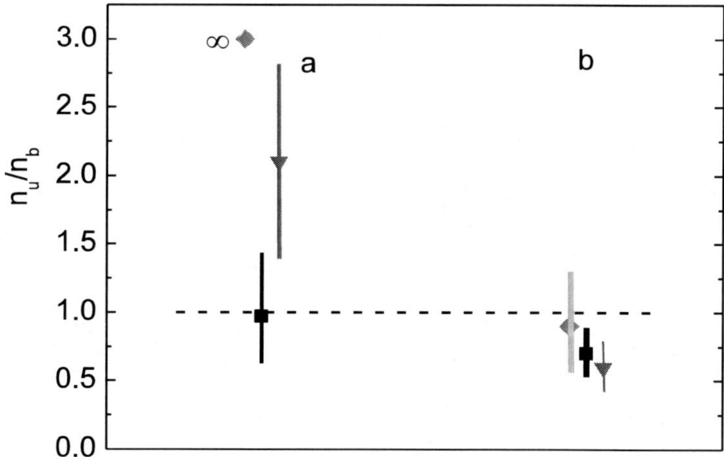

Figure 12. It is shown ratios R_c, R_a according to data from 3 EAS arrays ◆-Yakutsk, ■ - AGASA ▼- PAO.

In a case, the value R_c at latitude s equals $R_c = \infty$ (Figure 12) according to the Yakutsk array, because particles have not been registered at b<0° from the center of the Galaxy. The value $R_a = 0.9 \pm 0.4$ also shows ratios for R_c, R_a latitudes |b|=90° according to data from

AGASA and P. Auger EAS arrays. The independent data of the southern array confirms the particle distribution in the galactic latitude of northern arrays.

Note, the chance probability, P, to observe particles according to the common data of 3 arrays by the following sides: ≥ 34 particles (center, b>0°), ≥ 30 particles (anticenter, b<0°), ≤ 40 particles (anticenter, b>0°), ≤ 15 particles (center, b<0°) is P ~ 10^{-2}. This probability was found by simulating 119 events over the whole celestial sphere in terms of exposuring it to arrays according to [8].

We show ratio numbers of particles from the side of the galactic center/anticenter $R_c \geq 1$ and $R_a \leq 1$ according to common data of the arrays. According to [15], cosmic rays of ultrahigh energy most likely have galactic origin.

5. EXTENSIVE AIR SHOWERS WITH A DEFICIT MUON COMPONENT

We have analyzed the muon content in extensive air showers (EASs) detected at the Yakutsk array with energies E > 5.10^{18} eV, zenith angles <60°, and axes lying within the array perimeter. In this paper, data from the years 1987-2008 have been considered.

Earlier, we considered EAS without muons [17, 18] etc. In [18] we showed that ultrahigh-energy showers can be conventionally classified by 4 classes in terms of the muon content:

1) showers with usual muon (μ) content - ~ 97%,
2) showers pour in muons - ~1%,
3) showers without muons (E>1 GeV) - ~1%,
4) showers with high muon content - ~<1%.

Showers with high muon content are observed only for the highest energies. This fact can be important for determining the composition and origin of cosmic rays with extremely high energies.

The detection time of EASs observed at the Yakutsk array was separated into 6-hour time intervals. The time intervals when the muon detectors did not operate were excluded from the analysis. First, we selected showers without muon components (i.e., showers for which the readings of the muon detectors were absent (equal to zero) within the limits of the detection threshold, which was equal to 1 GeV).

For a zero reading of muon detectors, the probability that this muon detector would not trigger for the expected particle number N is estimated as P= $\Pi(P_1+P_2)$, where P_1 is the probability that no muon reaches the i-th detector and P_2 is the probability that one particle reaches the detector but it does not trigger. At P>10^{-3}, this shower was excluded from consideration.

Each case of registering showers without muon components was carefully checked. We demanded that detectors of muons which had zero indications had shown registered muons from other EAS before and after 100 minutes from the given event. For example, EAS with usual muons come before 8 minutes (t_1) and after = 50 minutes (t_2), then the shower number without muons (t_0) (see Figure 13).

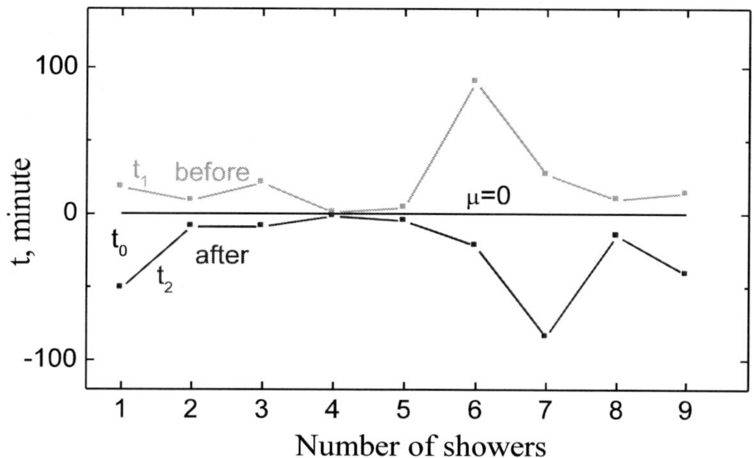

Figure 13. Relativearrival times in a minute of usual and with deficit muons EAS : t_1, t_2 – arrival time of EAS with muons, t_0 (line) – arrival time EAS without muons.

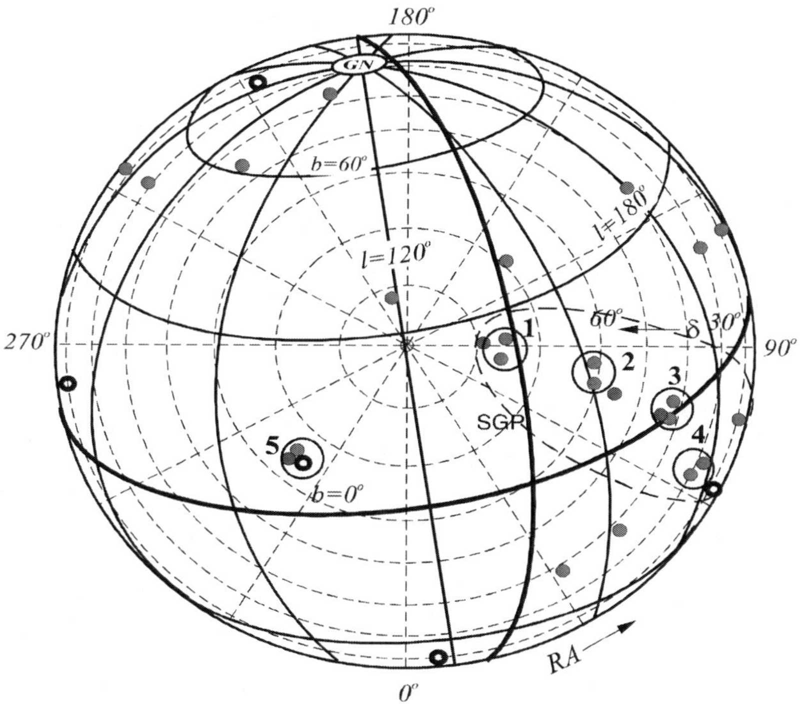

Figure 14. Distribution of showers without and pour muons , Large circles denote clusters 1-7, EAS without μ, o – EAS with pour μ, Galactic and Supergalactic (SGP) planes – black curves.

After such a procedure we found 21 showers without muons and 5 EAS with poor muons - the muon density at a distance >100 m from the axis was less than expected, anticipated to have more at 3σ. Above energy E>4.10^{19} eV we did not find EAS without muons. The portion of EAS with without muons to the number of showers with the normal muon composition is ~1%.

Figure 14 presents the distribution of EAS with deficit muons. We have found 5 doublets which consist of fEAS without and with poor muons.

Also, distributions of EAS are at a maximum from the site a galactic a plane (Figure 15): number of EAS is n(|b|<30°)/n(|b|>30°)=1.7±0.6 instead of 1.2 in cases associated with the isotropy of cosmic rays [1].

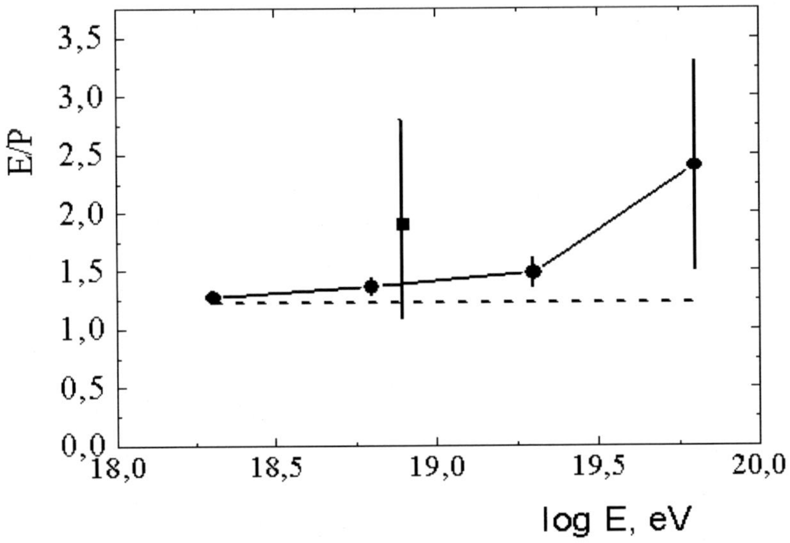

Figure 15. Ratio number of particles equator/pole ,(E/P), which originated EAS with deficit μ - ■, and usual μ - •, a shaped line – ratio particles in a case of isotropy.

It is shown in [19] that the muon content of showers reflects the mass composition of the particles which have formed the EAS. Probably, showers with the usual muon content are formed by the charged particles, showers without and pour muons by by neutral particles, and showers with high muon content by heavy particles. From 26 EAS with deficit muons (Figure 14) only 17 EAS correlate with pulsars [6] or arrival directions of EAS <6° from pulsars.

The arrival directions of 4 EASs of the 1st cluster are within an angular range of 9° around the PSR 0809+74 pulsar [6], which is located at a distance of 0.3 kpc from the Earth. The probability that arrival directions of 4 EAS from 26 are within <9° from this pulsar is equal P~10^{-3}. The EAS arrival directions of triplets are within an angular range of 7.6° from pulsar PSR 0450+55. The EAS arrival directions of doublets 3,4,5 are within a 9° angle from pulsars PSR 2217+47, 2241+69, 0458+46 pulsars, respectively, which are located at a distance of less than 2.4 kpc. We did not find pulsars near doublet 6 at <9°. Doublets 6,7 consist of EAS without and pour muons. Arrival directions for doublet 7 are located <2.6° from pulsar 2045+56.

Earlier, we found out the correlation of arrival directions for EAS with the usual muon content at energy E~10^{19} eV with positions of pulsars located along magnetic force lines of the Local Arm of the Galaxy [5]. Here, arrival directions of 18 EAS without muons from 26 are located within 7° from pulsar positions [6] (this angle accurately determines the arrival direction EAS with energy E>5.10^{18} eV). The chance probability to find 18 events from 26 within angular distances $<7°$ from pulsars is $P\sim10^{-3}$. Thus, we find some a correlation between EAS arrival directions and deficit muons and pulsars.

We consider the ratio number of EASs which originated with a deficit and usual μ relative to the equator/pole - $n(|b|<30°)/n(|b|>30°)$ above energy $E>10^{17}$ eV (Figure 15). The ratio of particles with usual EASs in cases of isotropy is equal to 1.2 according to [1, 20]. Here, we observe a gradual increase associated with the flux of particles which originated the EAS with usual μ with the energy, but statistical data is not enough. The ratio number of particles which originate the EAS with a deficit μ is higher (but error is high too) than the ratio number of particles (EAS with usual μ). This difference could possibly be caused by charged particles (usual EAS) that are tangled by a magnetic field, while neutral particles (EAS with particles and their sources are pulsars) [18].

Furthermore, we consider the EAS distribution on the right ascension (Figure 16). We have divided a region of energy into 3 intervals: 1) $5\times10^{18}-10^{19}$ eV; 2)$10^{19}- 4\times10^{19}$ eV; 3)$E>4\times10^{19}$ eV. We observe a maxima of the EAS distribution at coordinates $60°<RA<90°$ at the first two intervals of energy where doublets of EAS with usual μ are situated (Figure 14).

For the entire period of the operation, the Yakutsk EAS array has registered 3 particles with energy $E>10^{20}$ eV (Table 2). Note that from them 2 particles come from region $60°<RA<80°$ where there are doublets of EAS with deficit muons (Figure 14, 16).

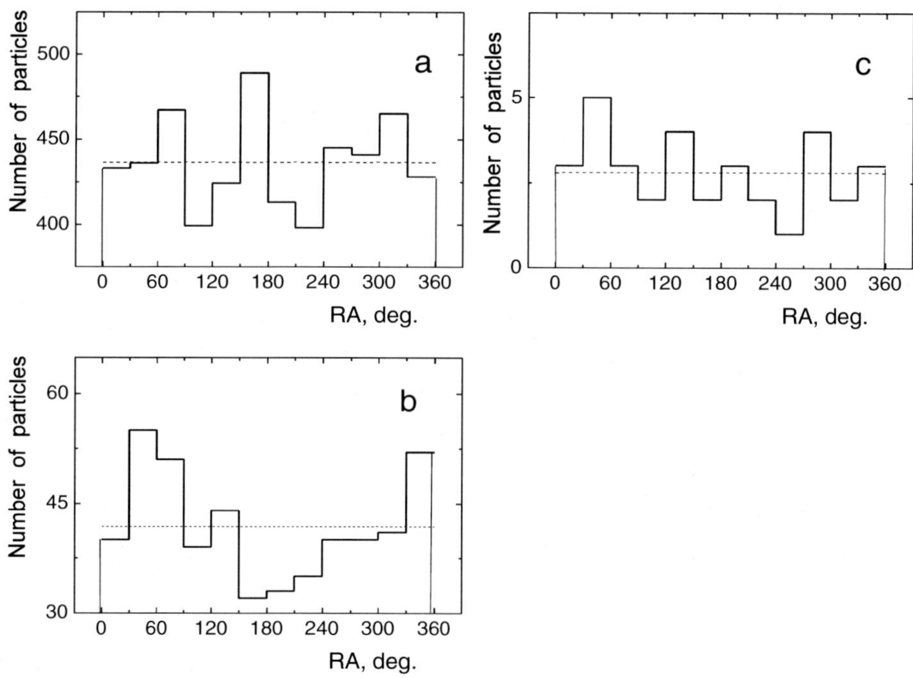

Figure 16. Distribution particles in energy intervals: a) $E=5\times10^{18} - 10^{19}$ eV; b)$10^{19} - 4\times10^{19}$ eV; c) $E>4\times10^{19}$ eV. Dot linear - the expected number of particles in a case of isotropy.

Note that arrival directions for EAS without muons correlate with pulsars irrespectively of their position concerning force lines of a magnetic field of the Local Arm. The detection of correlating arrival directions of EASs with deficit muons with pulsars irrespectively of their position testifies that these EAS are originated by a neutral deficit μ.

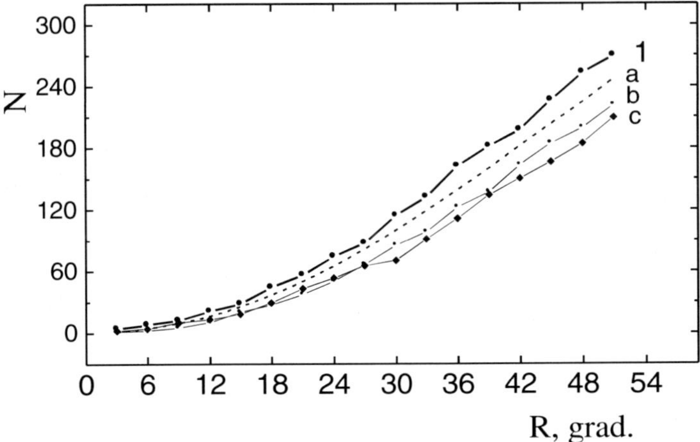

Figure 17. A distribution number of particles around pulsars (Table 1) at radius R: 1 - PSR 0458+46, a - the expected number of particles in cases of isotropy around pulsar PSR 0458+46; b, c – observed number of particles around points B,C accordingly (Table 2).

In Figure 17 the observed and expected number of particles around pulsar PSR 0458+46 (Table 2) correlates with most energy particles according to the Yakutsk array data. Also, distribution particles around 2 points B,C are shown (with same δ but other RA or other galactic latitude b, latitudes points B,C higher than latitude PSR 0458+46, Table 2). In cases of isotropy, the expected number of particles around pulsar PSR 0458+4 will be the same as around points B, C. However, the number of particles around PSR 0458+46 is higher then the number of particles around B,C points (Figure 17). This fact points to the fact that the number of particles increase with decreases of galactic latitude.

Table 1. EAS with E>10^{20} eV, which correlated with pulsars, D – distance, T – period of pulsar and points B,C

Date	E, 10^{20} eV	Pulsar, PSR	D, kpc	P, sec.	B, C
12.21.1977	1.1	0940+16	1.7	1.08	B - δ=46°, RA=254.5°; b=38.2°
05.07.1989	1.5	0458+46 (b=2.5°)	1.6	0.63	C - δ=46°, RA=180.5°, b=69°
12.18.2004	1.6	0410+69	1.6	0.39	

Note that earlier we found a correlation between particles with energy E>~10^{19} eV and pulsars [20-22]. In this papers we show that the maximum correlation between particles and pulsars was at an angular distance <6°. Therefore, we have selected pulsars below the latter as correlates with EAS which are situated at an angular radius <6° from arrival directions of the EAS.

6. ORIGIN OF COSMIC RAYS OF ULTRAHIGH ENERGY

Introduction

The origin problem of ultrahigh-energy cosmic rays is one of priority in high energy astrophysics. There are two main hypotheses about the origin of cosmic rays with $E > 4.10^{19}$ eV - galactic [15, 21] and extragalactic [10, 23-26]. In a galactic model, the origin of cosmic rays are from more probably sources: supernova stars, pulsars [15, 27-29], etc.

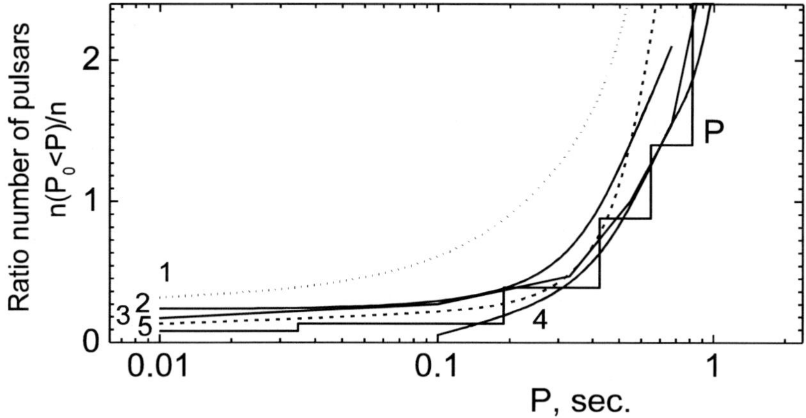

Figure 18. Ratio number of pulsars with period P_0 - $n(P_0 < P)/n$:
1 - pulsars, which correlated with EAS of PAO;
2 - pulsars, which correlated with EAS of AGASA;
3 - pulsars, which correlated with EAS of HiRes;
4 - pulsars, which correlated with usual EAS, Yak. 1;
5 - pulsars, which correlated with EAS with deficit muon, Yak. 2;
P - pulsars according catalogue.

6.1. Pulsars with Shorter Rotation Periods

Here, data associated with extensive air showers (EAS), in terms of their exposition of the celestial sphere to arrays, are analyzed from Yakutsk and AGASA arrays, and the P. Auger observatory. What are the sources of particles with $E > 4.10^{19}$ eV and do they correlate with pulsars? Note that earlier we found particles that correlated with $E \sim 10^{19}$ eV and pulsars at angular distances $R < 6°$ [4, 22]. Here, we select particles of global array data with $E > 4.10^{19}$ eV and search pulsars at angular distances $R < 6°$. We consider P. Auger observatory data with energy $E > 5.5 \times 10^{19}$ eV [10] and find that 43 particles correlate from 69 correlates with pulsars or 68% of particles correlate with pulsars; according to AGASA array data at $E > 4 \times 10^{19}$ eV - 31 EAS correlate from 57 or 53% particles; according to HiRes array data [30] at $E > 4 \times 10^{19}$ eV - 80% particles; according to Yakutsk array data with usual muons at $E > 8 \times 10^{19}$ eV - correlate 61.7%, and with deficit muons − 65.4 %, at $E > 4 \times 10^{19}$ eV - 80% [31]. Thus, according to data from constantly operating global arrays (Yakutsk, P. Auger, AGASA), the portion of particles which correlated with pulsars is ~ 60 %.

  A. A. Mikhailov

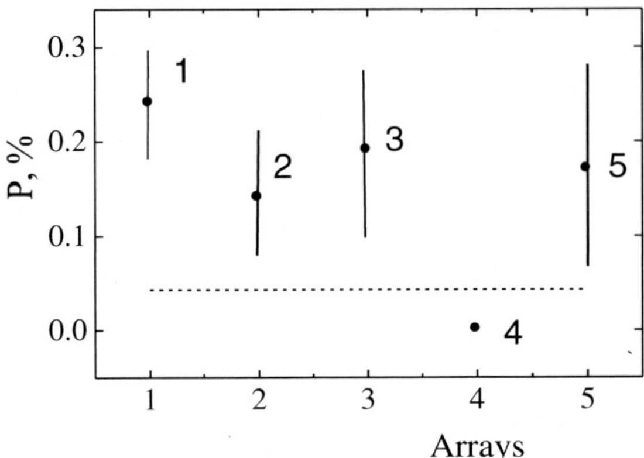

Figure 19. Potion pulsars with periods P<0.01 sec. which correlate with particles to the total number of pulsars: 1 –PAO, 2 – HiRes, 3 – AGASA, 4 – Yak.1, 5 - Yak.2. Dash line - potion pulsars with periods P<0.01 sec. to total number of pulsars.

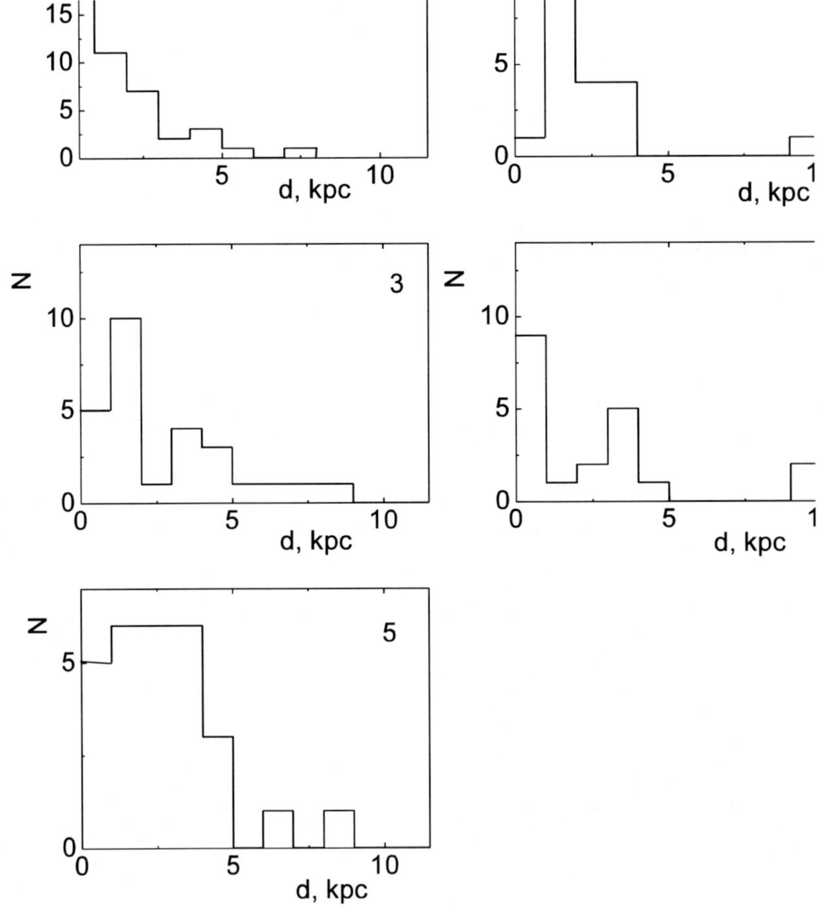

Figure 20. Distances of pulsars which correlate with particles: 1 –PAO, 2 – HiRes, 3 – AGASA, 4 – Yak.1, 5 - Yak.2.

We consider the rotation periods of pulsars [6] which correlated with EAS. We chose those pulsars which have periods with $P_0 < 0.01$. The ratio number of pulsars with periods $P < P_0$ which correlate with particles to all the number of pulsars is shown in Figures 18 and 19 and Table 2 (in cases of the Yakutsk array we have considered EAS with usual and deficit muons, marked as Yak.1, and with deficit muons, marked as Yak.2).

A portion of short-period pulsars is higher (P = 20 – 30%) than the expected number of pulsars according to the catalog at 4%, but statistical data is not enough. In Figure 20 pulsar distances correlating with particles are shown. A majority of pulsars situated at distances <2 kpc or those which contribute the most to correlations associated with the arrival directions of particles give the nearest pulsars.

Table 2. Particles and pulsar with short period rotation

Array	N_1	N_2	P_1,chance probability	P, chance Probability
PAO	47	12	0.01	
HiRes	28	4	0.47	
AGASA	30	6	0.16	4.8×10^{-4}
Yakutsk, usual μ	34	0	1	
Yakutsk, deficit μ	17	2	0.64	

where: N_1 – number of particles which correlated with pulsars, N_2 – number of particles which correlated with msec pulsars, P_1 – chance probability that a definite number of particles correlated with msec pulsars, P – common chance probability of 4 arrays.

6.2. Particles Flux of Ultrahigh Energy from a Galactic Plane

According to the Pierre Auger Observatory data, a correlation was shown between the highest energy arrival directions of ultrahigh energy particles and nearby extragalactic matter distribution [10]. The correlation was established from a comparison of the arrival directions of EAS and positions of AGNs. By performing a scan over the maximum angular separation, the largest significance was obtained for angles of 3.1° and, distances of 71 Mpc and energies above 57 eV. The HiRes array and P. Auger observatory establish the energy spectrum cut-off energy above $E > 4.10^{19}$ eV [30, 32]. This indicates that the highest particles are most likely extragalactic, supporting the conclusion that the observed suppression in the cosmic ray spectrum at the highest energies is indeed due to the GZK (Greizen-Zatzepi-Kuzmin [33, 34]) effect rather than to the exhaustion of the acceleration power of the sources.

Data from the AGASA array (where 11 showers with energies above 10^{20} eV were registered) indicates the absence of a cutoff in the energy spectrum [35]. The data of the Yakutsk array is not enough to form any conclusion about the energy spectrum $E > 4.10^{19}$ eV [36]. The spectra below 10^{20} eV obtained have a similar shape in all these experiments, but significantly different intensities. The contradiction is most likely to be caused by the systematic difference in the energy estimates for individual showers in different experiments.

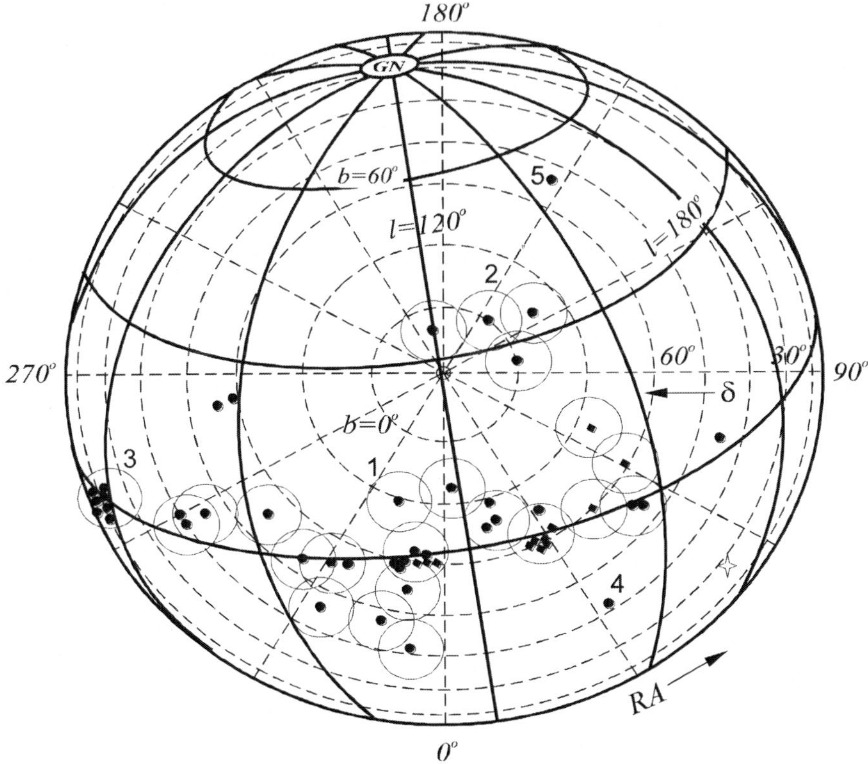

Figure 21. Distribution of pulsars in the celestial sphere, Yakutsk, E>8.10^{18} eV: ● - pulsars around which ≥10 particles were observed at R <6°, also it is shown in a circle around pulsars with radius R=6°.

6.3. Method Search Sources and Anisotropy

Usually, to search for an anisotropy of cosmic rays in the region of ultrahigh energy a method of harmonic functions is applied. First, the number of declination events are summarized, then amplitudes and phases of function Furies are found. The given method can be not effective in analyzing arrival directions of ultrahigh-energy particles because in result of summarizing the number of particles on declinations local maximum and minimum of distribution particles in some declinations can smooth them out. Therefore, we offer a new method of analysis. Our method changes the region of search sources and the anisotropy of cosmic rays on a celestial sphere. For each proposed source of particles we shall define a number of events inside an angular distance R from it. We selected sources when the number of particles at an angular distance <R from sources was above a certain threshold (≥n). If the distance between two nearest sources is less than <2R we will accept the crossed regions as one region and determine the observed number of events inside of this region. Then, we estimate the chance probability for the observed number of events in the given region from the common number of registered particles in 4-dimensions system coordinat [8]. If the observed number of events around the proposed sources is not chance then we consider these sources as their probable sources.

6.4. Particles Ultrahigh Energy and Pulsars

Note that in papers [5, 20]and other sources of information we found a correlation between arrival directions of ultrahigh particle energy according to Yakutsk EAS array data and pulsars located along the lines of a large-scale regular magnetic field of the Orion arm. The correlation between them was observed at an angular distance R<6°. Furthermore, we searched for a correlation between the arrival directions of particles and pulsars at this angular distance. We analyzed arrival directions of ultrahigh energy particles according to Yakutsk EAS array data at energy $E>8\times10^{18}$ eV and P. Auger EAS array data at $E>5.7\times10^{19}$ eV [10].

First, we considered 898 particles with energy $E>8\times10^{18}$ eV according to Yakutsk EAS array data. According to our method, we selected pulsars from the catalog for the arrival direction of particles that were around an angular distance R<6° [6]. We selected pulsars if there were more n≥10 events at angular distances of R<6° from their number of particles. If the distance between pulsars was <2R or the angular distance around this was crossed we took it to be one region.

We have found 3 groups of pulsars after connecting crossed regions. There were 37 found in the 1st group, 4 in the 2nd group, and 6 pulsars in 3rd group. There are 178, 27 and 14 particles in these groups of pulsars accordingly. The chance probability to find out such number of events in these groups of pulsars from 898 particles is equal to $P_1 \sim 3.10^{-6}$, $P_2 \sim 0.8$ and $P_3 \sim 2.2\times10^{-3}$ according to [8].

The location of the 1st and 3rd groups of pulsars, from which the statistically significant number of particles is observed, are in a galactic plane. The maximal number of events $20\leq n\leq23$ was observed around the 1st groups of pulsars J0141+6009, 0146+6145, 0147+5922, 0157+6212, 0215+6218. The given pulsars have been considered earlier as probable sources of particles [5, 37].

In Figure 21 we observed among separate pulsar : only at two pulsars (4,5) were more particles found than expectedin the case of isotropy.. Around R<6° we found pulsar J0218+4232 – 10 particles with a chance probability P ~ 0.07; PSR J1012+5307 – 13 particles, P ~ 0.05.

We consider arrival directions of 34 particles with energy $E>4\times10^{19}$ eV according to Yakutsk EAS array data (Figure 22). Those pulsars are shown for which the number of particles around such pulsars at R<6° is n=1, 2 on the map of equal exposition of the celestial sphere where an equal number of particles is expected from an equal area (Figure 22). We have received 3 groups of pulsars after connecting crossed regions with an angular distance <2R between them.

Nineteen pulsars were found in the 1st group, 14 particles in the 2nd group, and 7 in the 3rd group. There are 3, 4 and 2 particles in these groups of pulsars accordingly. The chance probability to find out such a number of events in these groups of pulsars from 34 particles is equal to $P_1 \sim 4.10^{-2}$, $P_2 \sim 0.2$ and $P_3 \sim 0.2$ according to [8]. Note, a flux of particles from pulsars from the side center of the Galaxy (Figure 22, number 1) was observed (Figure 21, number 3). The chance probability from another group of pulsars and separate pulsars is P>0.1.

Note that excess particles were found at $E\sim10^{18}$ eV near the center of the Galaxy according to AGASA array data [38] and Sydney [39] where we observed fluxes from the center of the Galaxy.

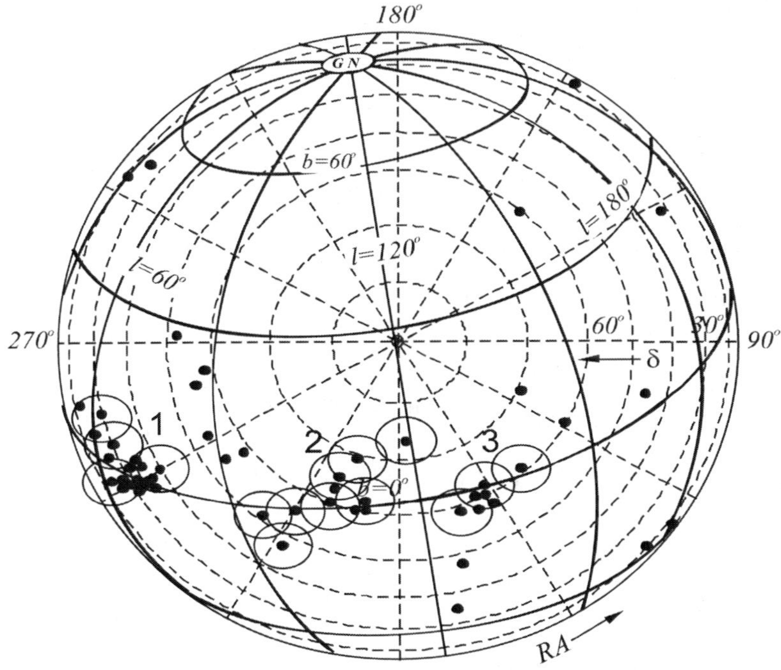

Figure 22. Distribution of pulsars in the celestial sphere, Yakutsk, E>4.10^{19} eV: ● - pulsars around which ≥10 particles were observed at R <6°.

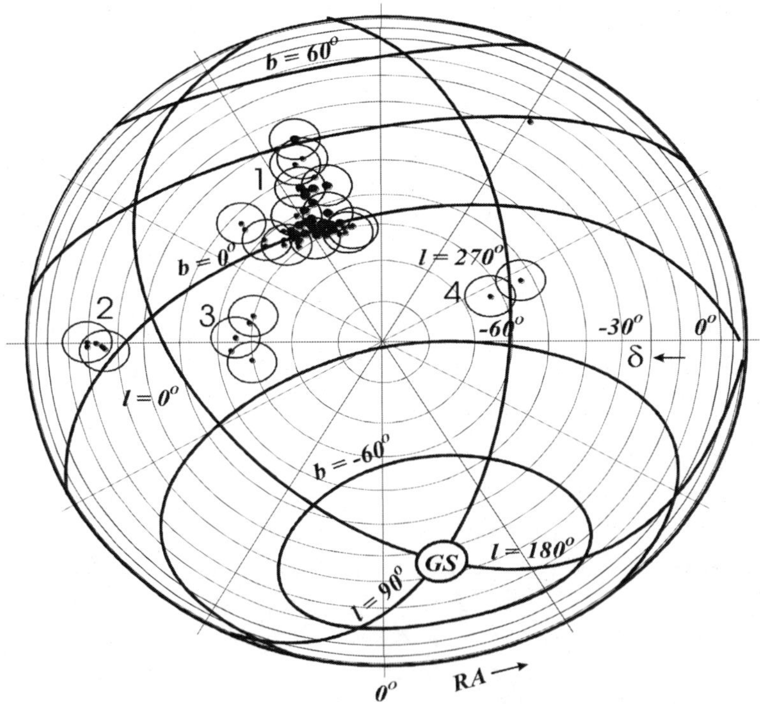

Figure 23. PAO, E>5.7×10^{19} eV: distribution pulsars in the celestial sphere: at radius R<6°, ·● - 2, ● - 3 particles. The 3rd group of pulsars is shown along with circles with radius R<6°.

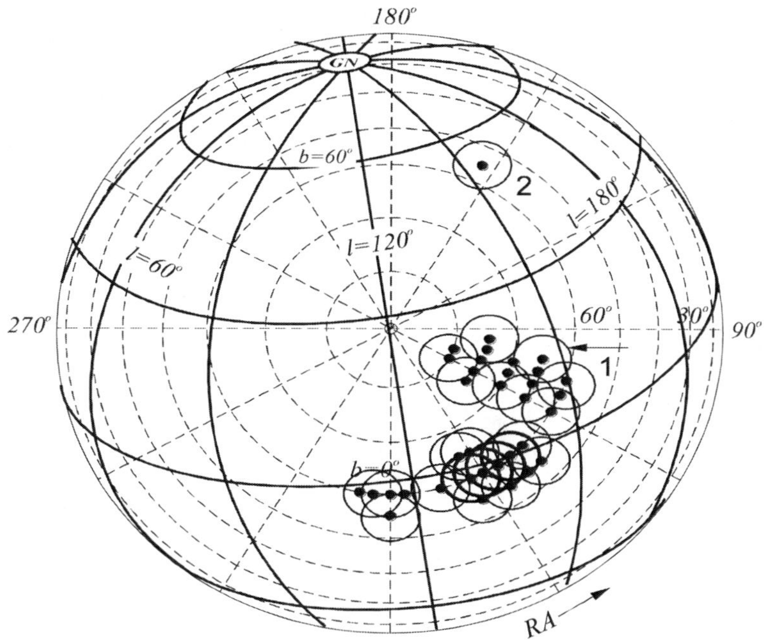

Figure 24. Yakutsk, $E>8.10^{18}$ eV, anisotropy: distribution points in the celestial sphere ● within which at radius $R<6°$ ≥ 15 particles were observed.

In Figure 23 those pulsars are shown for which there are 24 particles with energy $E>5.7\times10^{19}$ eV at an angular distance R $<6°$, according to PAO observatory data [10]. We have selected those pulsars around which ≥ 3 particles were observed (Figure 25, large black circles). Twenty-three pulsars are inside at $R<6°$ from each other, except for pulsar J1332-3032. The chance probability to find 9 events around the given groups of pulsars from 69 particles is equal $P\sim10^{-4}$. These pulsars also are near a galactic plane.

Next, we selected those pulsars around which ≥ 2 particles were observed at $R<6°$ (Figure 23, small black circles). Thus, we found 48 pulsars within $<6°$ from each other (mark number 1, Figure 23, large black points mean 3 particles, small black points – 2 particles) between a right ascension $180°<RA<240°$. The chance probability to find 14 particles in the given group of pulsars from 69 particles is equal $P \sim10^{-6}$ according to [8]. These pulsars are near a galactic plane output of the Orion arm. Also, we found 2 groups of pulsars marked 2,3 on Figure 23 around pulsars at $R<6°$ that have 2 particles. The chance probability to find 2 particles around the 2nd group of pulsars for the side of center of the Galaxy is $P\sim7\times10^{-2}$, to find 2 particles around the 3rd group of pulsars is P ~0.3.

6.5. Anisotropy of Particle Arrival Directions

We have estimated the anisotropy of arrival directions of particles. . For this it is the drawn line on a coordinate grid with a declination δ and a right ascension RA through 5°×5° on all celestial sphere, beginning with the declination δ=0° and the right ascension RA=0°

then have counted up the number of particles around these points and compare number of particles around these points.

We have chosen areas with R<6° around given points where the observed number of particles n was n≥15 events according to the Yakutsk array (Figure 24). We have found 130 particles after connecting crossed regions into one region . The chance of probability to find out such number of events from ageneral number 898 of particles is $P_1 \sim 2.10^{-5}$. The region where arrival directions of particles with a number of events n≥15 has the coordinates 45°<δ<80° и 60°<RA<95°, and this region is larger than the region occupied by pulsars of the 1st group. The maximums for this distribution of particles is around coordinates δ=60°, RA=25° - 40° or b ~ 0°, l ~ 125° - 140° (more black circles, Figure 24), number of particles at R<6° - n≥18. This maximum number of particles coincide with positions of pulsars J0141+6009, 0146+6145, 0147+5922, 0157+6212, 0215+6218 [6].

The peak of the given maximum of distribution is at δ=60° and RA=25°. Twenty-three particles around this point were observed at R<6° (an expected number of events in case of isotropy is n~9.5). The chance probability to observe such a number of events from 898 particles is $P \sim 10^{-4}$. The latter is caused by the small number of pulsars in the given region – there are only 4 pulsars according to [6]; 3 pulsars from them are shown in Figure 23; around the 4th pulsar the number of particles is not enough.

Note that pulsar J0205+6449 has a short period of rotation, P_0=0.06 seconds at distance of 3.2 kpc on an angular distance of 5.6°,near this maximum, which is the most probable source of particles according to [40, 41].

We considered regions with radius R<6° around the given points in cases associated with P. Auger observatory data where the number of observed particles was n≥2 events. After the connection of crossed regions we found 2 regions (Figure 25). We found 14 particles in the 1st group of pulsars. The chance probability to find 14 events from 69 particles is equal $P_1 < 10^{-6}$. The maximum of this distribution is around coordinates: δ ~ - 0°, RA ~ 200°; δ ~ - 55°, RA~210°, δ ~ -55°, RA ~ 205° (b ~0 ÷ 15°, l ~310°). The peak of the given maximum of distribution is around a point at δ=-55°, RA=210°. Around this point at R <6°, 5 particles were observed (an expected number of events in case of isotropy is n ~ 0.38). The chance of probability to observe 5 events from 69 particles is equal $P \sim 4.10^{-5}$. Note that from the given maximum of distribution particles on distance 3.1° and 5.6° are pulsars J0205+6449, J1439-5501 with short periods of rotation, P_0 ~ 0.034 and 0.028 sec at distances 1.7 and 0.7 kpc from the Earth.

6.6. HiRes Array Data

Furthermore, we consider arrival directions of particles according to HiRes data $E > 4.10^{19}$ eV [30]. The HiRes experiment consisted of two sites (HiRes I and II) 12.6 km apart, located at Dugway Proving Ground in the state of Utah (USA). Each site consisted of telescope units (22 at HiRes I and 42 at HiRes II) pointing at different parts of the sky. The detectors observed the full 360 degrees in the azimuth, but cover from 3 to 16.5 (HiRes I) and from 3 to 30 degrees (HiRes II) in elevation angles. HiRes detectors operate only 10% of the time due to the requirement of a dark, clear, moonless night. We selected pulsars around which ≥2 events have been observed at R<6°. We found 3 pulsars J0922+0638, J0946+0951,

J0953+0755 (Figure 26, 1), around which 3 events were observed. The chance probability to find 3 events around these pulsars at R<6° from 35 particles is P ~ 2.10^{-3}. Unfortunately, we did not find any correlating arrival directions of particles and positions of active galaxies where the chance probability was >3σ.

For anisotropy we considered regions with a radius R<6° around the given points in cases according to HiREs data where the number of observed number of particles was n≥2 events. After connecting crossed regions we found one region (Figure 26, 2). We found 2 particles in this region. The chance probability to find 2 events from 35 particles is equal P~0.06. The maximum of particles distribution is in the galactic plane.

6.7. AGASA Array Data

The Akeno Giant Air Array (AGASA) operated in a stable condition since 1990 until the end of May 2000. Here, we consider arrival directions for 57 particles with energy E>4.10^{19} eV [9]. The zenith angle of particles is limited to less then 45°.

First, we searched for a correlation between arrival directions of particles and pulsar positions. We found a group of 5 pulsars in region 1 at δ~40° and RA ~ 55° (Figure 27, 1, shade circle), but the chance probability to find 3 events from 57 around the inside this region is P ~ 5.10^{-2}.

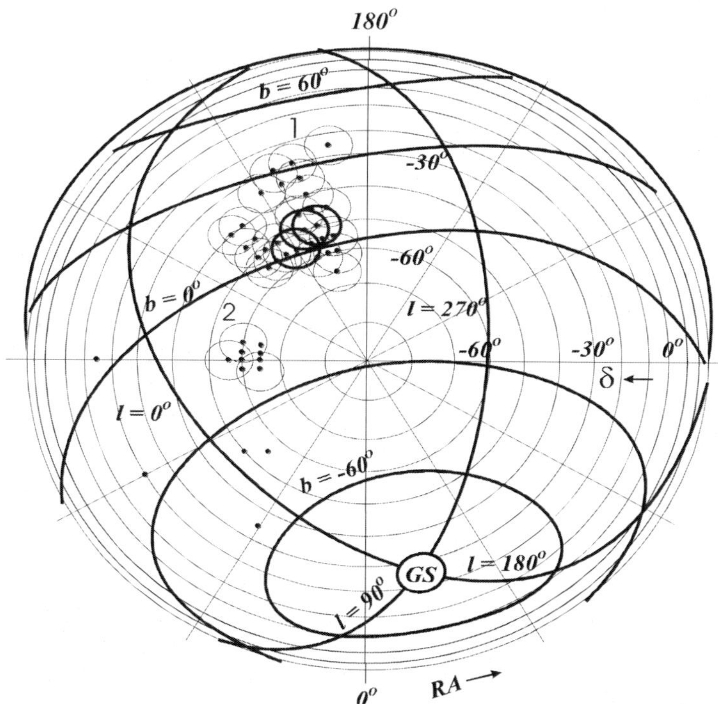

Figure 25. PAO, E>5.7×10^{17} eV, anisotropy: distribution points in the celestial sphere - at radius R<6° - n>2 particles.

A. A. Mikhailov

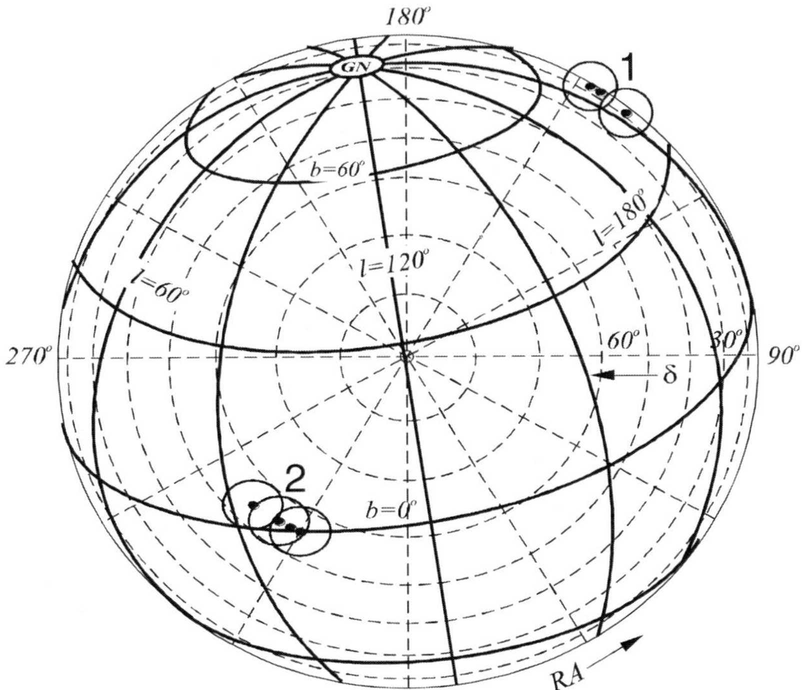

Figure 26. HiRES, E>4.10^{19} eV, 1 – distribution pulsars, 2 – distribution points.

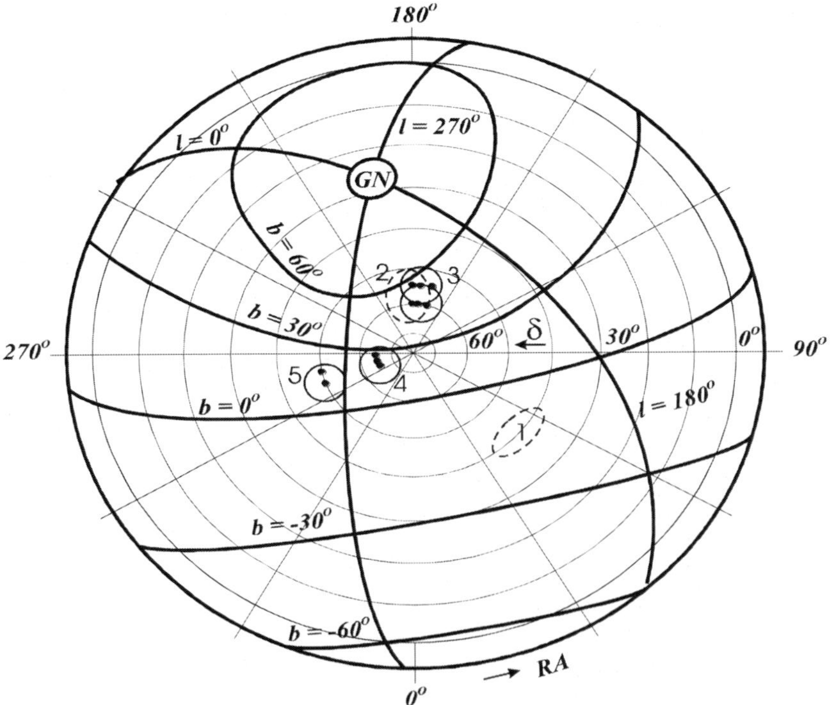

Figure 27. AGASA, E>4.10^{19} eV, 1 - pulsars, 2 – active galaxies, anisotropy: 3,4,5 – distribution points.

We found some correlation arrival directions of particles with the position of 8 active galaxies, situated at region $\delta \sim 58°$ and RA $\sim 175°$ (Figure 27, 2, shade circle). The number particles around every galaxy at R<6° is n≥3 events. The chance probability to find 3 events inside this region is P~4.10^{-2}.

For the search anisotropy we considered points around which at R<6° number of particles was for n≥3 events. Three regions were found, and inside every 3 regions we found 3 particles (Figure 27, 3-5). The chance probability to find 3 events in region number 1 is $P_1 \sim$ 0.14, in region 2 – $P_2 \sim 3.10^{-3}$, and in region $P_3 \sim 2.10^{-2}$. The maximum of particles distribution is near the galactic plane.

7. PARTICLES WITH ENERGIES E>8.10^{19} EV AND AN ACTIVE GALAXIES

Data collected by the Pierre Auger Observatory through August 31, 2007 showed evidence for anisotropy in the arrival directions of cosmic rays above the Greisen- Zatsepin- Kuzmin [34, 35] energy threshold, 6.10^{19} eV. The anisotropy was measured by the fraction of arrival directions that is less than 3.1° from the position on an active galaxy's nucleus within 75 Mpc (using the Veron-Getty and Veron 12th ed. catalog [42]).

We searched for a correlation between the arrival directions of UHECRs with positions of active galaxies in the Veron-Getty and Veron (VCV catalog) with z<0.018.

In Figure 28 we consider arrival directions of 898 particles according to the Yakutsk EAS array data at energy E>8.10^{18} eV. Active galaxies are shown around which there are pulsars at R<6° and the number of particles is n≥10. We have received one group of active galaxies after connecting crossed regions with an angular distance <2R between them. Sixteen active galaxies have been found in this region. There are 53 particles in this group. The chance probability to find 53 events from 898 particles is equal to P_1 ~0.98. The chance probabilities around other galaxies is P>0.1.

In Figure 2931 we consider arrival directions of 34 particles according to Yakutsk EAS array data at energy E>4.10^{19} eV. Active galaxies are shown around which there are pulsars at R<6° and the number of particles is n≥2. We have received one group of 5 active galaxies after connecting crossed regions with an angular distance <2R between them. There are 3 particles in this group. The chance probability to find 3 events from 34 particles is equal $_{to}$ P_1 ~0.016. The chance probabilities around another galaxy is P>0.1.

In Figure 30 we consider arrival directions of particles according to PAO observatory data at energy E>5.7×10^{19} eV. Active galaxies are shown around which there are pulsars at R<6° and the number of particles is n≥3. We have received one group of active galaxies after connecting crossed regions with an angular distance <2R between them. There were 7 active galaxies found. There are 9 particles inside this group. The chance probability to find 9 events from 69 particles is equal to P_1 ~3.10^{-5}. However, the character of this distribution contradicts the distribution of maximum of particles at E>4.10^{19} eV (Figure 25). Note that these 9 particles also correlate with pulsars (Figure 23).

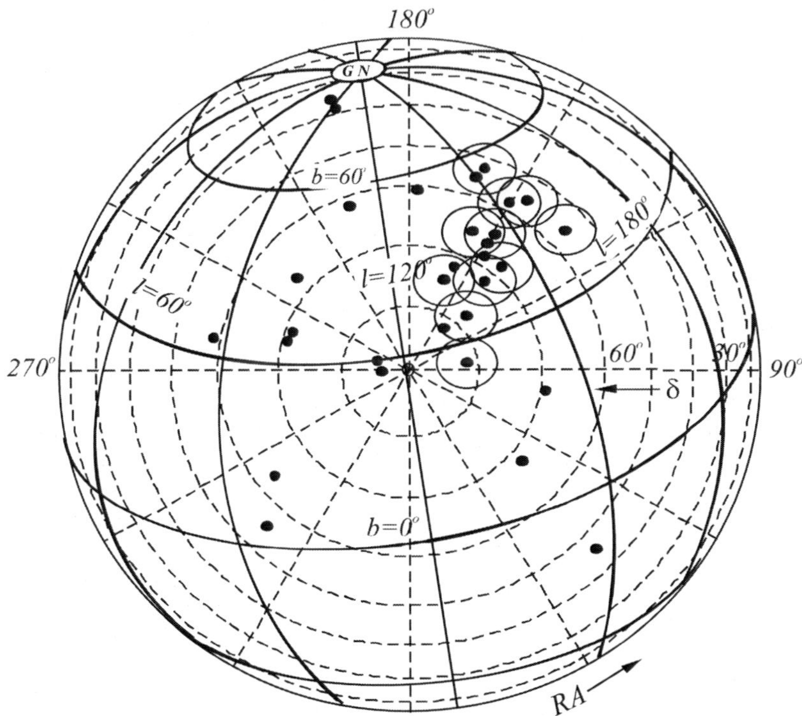

Figure 28. Yakutsk, E>8.10^{18} eV: distribution of active galaxies: around galaxies at radius R<6° - n> 10 particles.

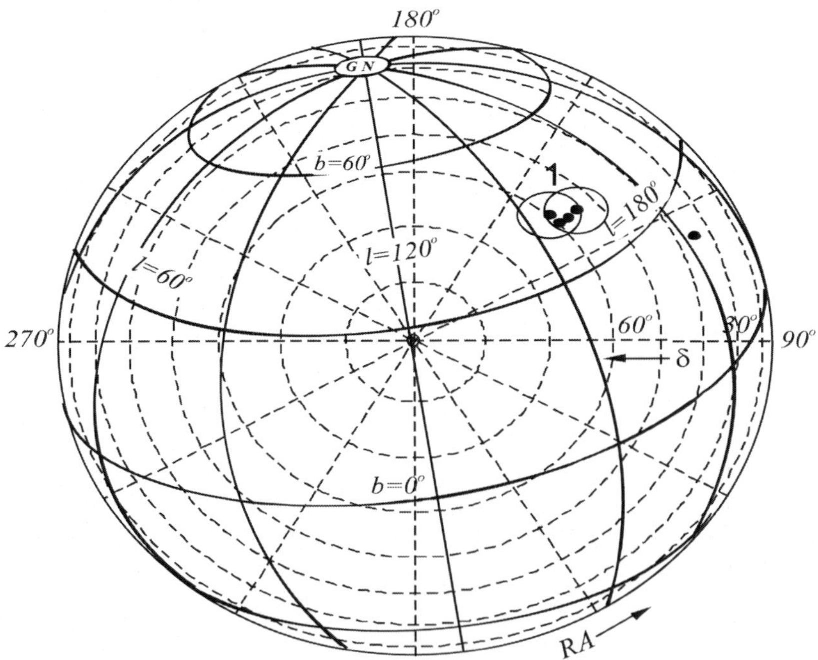

Figure 29. Yakutsk, E>4.10^{19} eV: distribution active galaxies: around galaxies at radius R<6° - n> 2 particles.

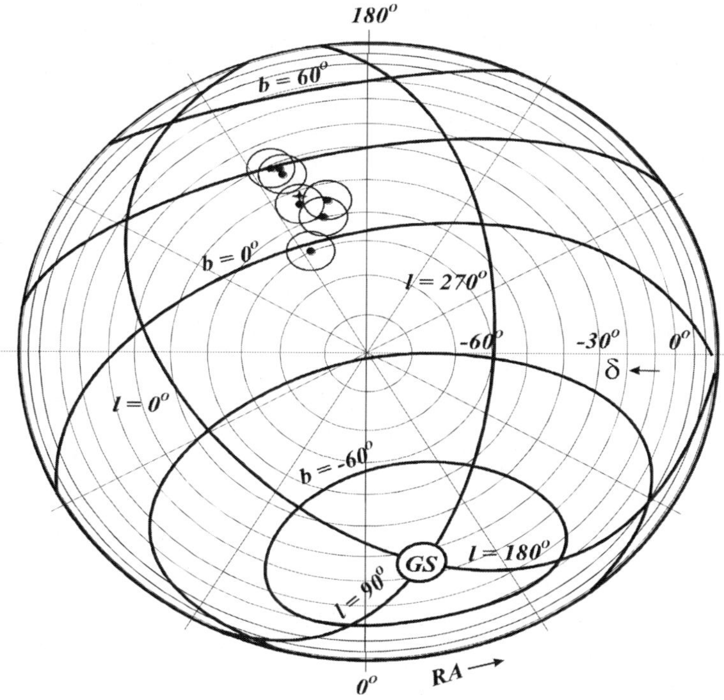

Figure 30. PAO, E>4.10^{19} eV, distribution active galaxies, around which at R<6° there are 3 particles.

Note, around radiogalaxy Cen A at R<6° there are 3 particles with a chance probability to find 3 events from 69 particles at P~5.7×10^{-3}. We have not found any statistical values for excess particles around radio galaxy Cen A.

CONCLUSION

We found a flux of particles from pulsars of the galactic plane and from pulsars near the galactic center. There was a correlation between arrival directions of particles by AGASA and P.Auger data and the position of pulsars of the Local arm Orion at E>4.10^{19} eV. Pulsars which correlate with particles have rotation periods shorter than other pulsars. There was an anisotropy of particle arrival directions from the side of a galactic plane according to Yakutsk, PAO, HiRES, AGASA array data at E>4.10^{19} eV. All observed cosmic rays are galactic and their sources are pulsars.

REFERENCES

[1]   Dyakonov M.N., Efimov N.N., A.V. Gluskov, Mikhailov A.A. et al., Kosmicheskoe izluchenie predelno visokich energii, *Monograph,* Novosivisk, Nauka, 1991, 252p.

[2]   Afanasiev B.N., Glushkov A.V. et al. *Proceedings of the Tokyo Workshop on Techniques for the Study of the Extremely High Energy Cosmic Rays,* edited by M. Nagano, Tokyo, 1993. 32.

[3]   Linsley J. *Phis. Rev. Lett.,* 1975, V.34, 1530-1533.

[4]   Mikhailov A.A. *Proc. 16-th European cosm. rays symp.* Alkala, 1998. 245.

[5]   Becker P.A., Bisnovatyi-Kogan G.S., Casadei D., …, Mikhailov A.A. et al. Frontiers in cosmic ray research, *Monograph,* New York, Nova Sc. Publ., 2007, ch. 6, 161.

[6]   Manchester R.N., Hobbs G.B., Teoh A., & Hobbs M., AJ., 2005, 129, 1993.

[7]   Mikhailov A.A. Bulletin Nauchno-Technical informaion. *Problems of Cosmophysics and Aeronomy.* Yakutsk, 1982, December, 10.

[8]   N.N. Efimov, A.A. Mikhailov. *Astrop. Phys.* 1994, 2, 329.

[9]   Hayashida N., Honda K., Inoue N. et al. *Astrophys. J.,* 1999, 522, 225.

[10]   The P. Auger Collaboration. *Astropart. Phys.,* 2010, 34, 314.

[11]   Berezinsky V.S., Mikhailov A.A. *Proc.20-th ICRC.* Moscow, 1987, 2, 54.

[12]   A.A.Glushkov, I.T. Makarov, M.I. Pravdin et al, *Pisma v Zhetf.,* 2008, 87, 220.

[13]   A.A. Watson. *Proc. 30-th ICRC,* Merida, 2007, e-disk; arXiv/astro-ph: 0801.2321.

[14]   The P. *Auger Collaboration.* arXiv/astro-ph:1107.4804.

[15]   S.I. Syrovatsky. *"O vozmoshnosti galacticheskogo proiskozdenia kosmicheskich luchei sverchvisokich energii.",* Preprint of P. N. Lebedev Physical institute of AS USSR, M., 1969, p.7.

[16]   The P. Auger Collaboration. *Astrop. Phys.* 2008, 29, 188.

[17]   A.A. Mikhailov, E.S. Nikiforova. *JETP Lett.,* 2000, 72, 229.

[18]   A.A. Mikhailov, N.N. Efremov, E.S. Nikiforova. *JETP Lett.,* 2006, 83, 265.

[19]   G.B. Khristiansen, G.V. Kulikov, Yu.A. Fomin. *Kosmicheskie luchi sverchvisokich energi,* Moscow. Atomizdat, 1975, 241.

[20]   Malov I.E., Mikhailov A.A., Avramenko A. et al. Pulsars: Theory, Categories and Application. *Monograph,* NY, Nova Science Pulishers, Nova Science Publishers, Inc., NY, 2010, 240.

[21]   A.A. Mikhailov. *Nucl. Phys. B., Proc. Suppl.,* 2008, 175-176, 537.

[22]   A.A. Mikhailov, *Izv. AN, s. fiz.,* 1999, 63, 556.

[23]   V.S. Berezinsky, A. Gazizov, S. Grigoreva. *Phys. Review,* 2006, 74, 4305.

[24]   V.S. Berezinsky, A. Gazizov, S. Grigoreva. *Phys. Lett. B.,* 2005, 147, 612.

[25]   E. G. Berezhko, *The Astroph. J.* 2008, 684, L69.

[26]   V.S. Ptuskin, V.N. Zirakashvili, *Astron. Astrophys.,* 2005, 429, 755.

[27]   Hong-Bo Hu, Hu H., Yuan Q., Wang B. et al., *Astroph. J.,* 2009, 700, L170.

[28]   Giler M., Lipski M., *Proc. 27-th ICRC,* Hamburg, 2001, 6, 2092.

[29]   P. Blasi, R.I. Epstein, A.V. Olinto. *Astroph. J.,* 2000, 533, L123.

[30]   Abbasi R.U, Abu-Zayad T., Allen M., et al., arXiv/astro-ph:0804.0382v2.

[31]   A. A. Mikhailov, N.N. Efremov, N.S. Gerasimova et al. *Proc. 29-th ICRC,* Pune, 2005, 7, 227.

[32]   The P. *Auger Collaboration.* ArXiv: 11074809.

[33]   K. Greisen. *Phys. Rev. Lett.* 1966, 16, 748.

[34]   G.T. Zatsepin and V.A. Kuzmin. *Pis'ma v Zh Eksp. Teor. Fiz.,* 1966, 4, 114.

[35]   N. Hayashida, K. Honda, N. Inoue et al., ArXiv/astro-ph: 0008102v1.

[36]   V.P. Egorova, A.V. Glushkov, S.P. Knurenko et al. ArXiv/astro-ph: 0411685v1.

[37] Erlykin A.D., Mikhailov A.A. and Wolfendale A.W. *J. Phys. G: Nucl. Part. Phys.* 2002, 28, 2225.

[38] Hayashida N., Honda K., Inoue N. et al. *Astronomical Journal*, 2000, 120, 2190.

[39] Winn M.M., Ulrichs J., Peak L.S. et al. *J. Phys. G: Nucl. Phys.* 1986, 12, 653.

[40] Giller M., Lipski M. Proc. 27-th ICRC, Hamburg, 2001, 6, 2092.

[41] Olinto A.V., Epstein R.I., *Blasi P. Proc. 26-th ICRC*, 1999, 4, 361.

[42] M.P. Veron-Getty, P.Veron. *Astron.Astrophys.* 2006, 455, 773.

In: Homage to the Discovery of Cosmic Rays …
Editor: Jorge A. Perez-Peraza

ISBN: 978-1-63483-084-3
© 2015 Nova Science Publishers, Inc.

Chapter 13

COSMIC RAYS AND OTHER SPACE WEATHER FACTORS THAT INFLUENCE SATELLITE OPERATION AND TECHNOLOGY, PEOPLE'S HEALTH, CLIMATE CHANGE, AND AGRICULTURE PRODUCTION

Lev Dorman[1,2,], Lev Pustil'nik[1], Gregory Yom Din[1,3] and David Shai Applbaum[1]*

[1]Israel Cosmic Ray and Space Weather Centre with Emilio Segré Israel-Italian Observatory, affiliated to Tel Aviv University, Golan Research Institute and Israel Space Agency, Israel
[2]IZMIRAN of Russian Academy of Sciences, Moscow, Russia
[3]Open University of Israel, Israel

ABSTRACT

This paper is an example of how fundamental research in Cosmic Ray (CR) Astrophysics and Geophysics can be applied to several very important modern practical problems. Monitoring of space weather using online cosmic ray data allows for the prediction of space phenomena which are not only dangerous for satellites' electronics and astronauts' health in space, but also for crew and passengers' health on commercial jets in the atmosphere. In some rare cases this even extends to technology and people on the ground; CR and other space weather factors may even influence climate change and agriculture production.

It is well known that during great SEP (Solar Energetic Particle) events, fluxes can be so big that they may destroy the memory of computers and other electronics in space. Furthermore, satellites and other spacecraft may die. Each year, insurance companies pay more than 500,000,000 dollars for these failures; if there were ever an event as large as the one that happened on February 23, 1956, almost all satellites would be destroyed in 1-2 hours, costing 10-20 billion dollars. This is in addition to the obvious downsides of the destruction of satellite telecommunications. In these periods, it is necessary to switch off some part of the electronics for a short time to protect computer memories. These periods

---

[*] Contact e-mail: lid@physics.technion.ac.il (Lev Dorman).

are also dangerous for astronauts on space-ships, and passengers and crew in commercial jets (especially during S5-S7 radiation storms). The problem is, how exactly does one forecast these dangerous phenomena? We show that exact forecasts can be made by using high-energy particles (those having an energies of about 2-10 GeV/nucleon and higher), whose transportation from the Sun is characterized by a much bigger diffusion coefficient than that of small and middle energy particles. Because of this, high energy particles arrive from the Sun much earlier (8-20 minutes after their acceleration and escape into the solar wind) than the majority of the smaller energy particles that cause dangerous situations for electronics and people's health (these arrive 30-60 minutes later).

We describe here the principles and experience of the automatically-working programs "SEP-Search-1 min", "SEP-Search-2 min","SEP-Search-5 min", developed and checked in the Emilio Segre' Observatory of the Israel Cosmic Ray and Space Weather Center (Mt. Hermon, 2050 m above sea level). The second step is automatic determination of the flare energetic particle spectrum, and then automatic determination of the diffusion coefficient in interplanetary space, time of ejection and energy spectrum of SEP at the source, and then – the forecast of expected SEP flux and radiation hazard for space-probes in space, satellites in the magnetosphere, jets and various objects in the atmosphere and on the ground. We will describe also the theory and experience of high energy cosmic ray data use for forecasting major geomagnetic storms accompanied by Forbush effects (which influence very much satellites, communications, navigation systems, people's health, and high-level technology systems in space, in the atmosphere, and on the ground).

1. Introduction

CR are one of the most important factors in space weather. This is namely because CR of galactic and solar origin determine radiation storms—and therefore radiation hazard—for people and technology, computer and memory upsets and failures, solar cell damage, radio wave propagation disturbances, and failures in communication and navigation systems. Besides this, CR can be used as an effective instrument for space weather monitoring and the forecast of dangerous phenomena, especially, great magnetic storms. In Figure 1 are shown CR and other space weather factors' influence on satellites, communication, navigation systems, people's health, and others.

2. Satellite Anomalies in Connection with CR and Other Space Weather Factors

2.1. The Matter of the Problem

Satellite anomalies (or malfunctions), including total distortion of electronics and loss of some satellites, cost Insurance Companies billions of dollars each year. During especially active periods, the probability of big satellite anomalies and their loss increases greatly. Now, when a great number of civil and military satellites are continuously working for our day to day life, the problem of satellite anomalies becomes very important. Many years ago, about half of satellite anomalies were caused by technical reasons (for example, in the Russian Kosmos satellites), but with the increase of production quality with time, this fraction has become smaller and smaller. The other part, which now is dominant, is caused by different space weather effects.

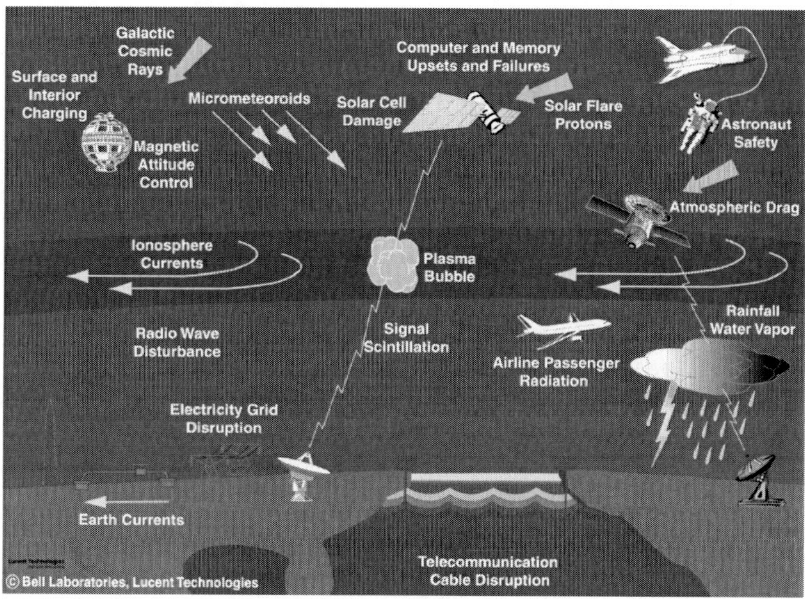

Figure 1. Space weather effects (from Bell Laboratories web-site in Internet).

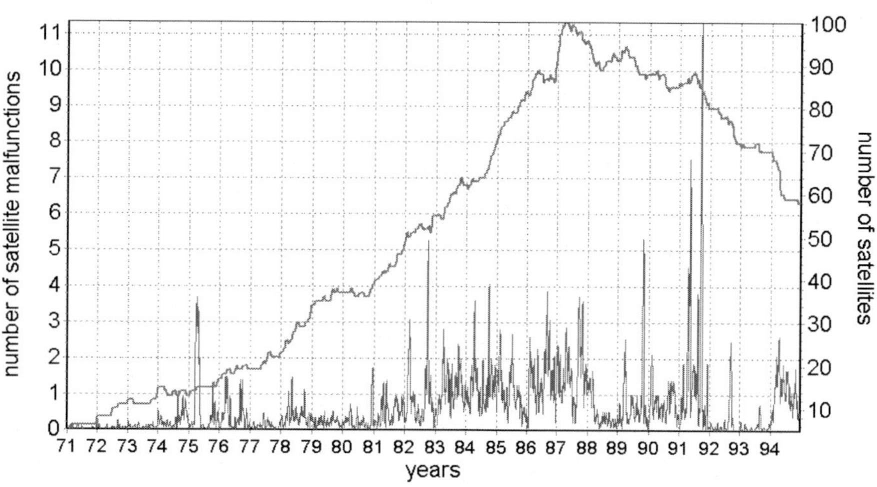

Figure 2. Time variations of monthly data for ~ 300 satellites and ~ 6000 anomalies.

2.2. Formation of a Database of Satellite Anomalies

The main contribution was from NGDC satellite anomaly database, created by J. Allen and D. Wilkinson [5] + "Kosmos" data (circular orbit at 800 km altitude and 74° inclination) + 1994 year anomalies − W. Thomas report [6] + The satellites characteristics - from:

http://spacescience.nasa.gov/missions/index.htm; http://www.skyrocket.de/space/index2.htm; http://hea-www.harvard.edu/QEDT/jcm/space/jsr/jsr.html; http://www.astronautix.com/index. htm.

2.3. Time Variations of Satellite Number and Its Anomaly Total Numbers

Results are shown in Figure 2.

2.4. Three Groups of Satellites

In Figure. 3 is shown the separation of satellites into three groups depending on altitude and inclination of their orbits (as it will be shown below, space weather effects sufficiently depend on satellite orbit). The biggest group of satellites – GEO (135 satellites) – *high altitude – low inclination* (in Figure. 3 this group is shown as *green*). In the group *low altitude – high inclination* (in Figure. 3 is shown as *blue*) are 68 satellites (mostly Russian Kosmos satellites). In the group *high altitude – high inclination* (in Figure 3 is shown as *red*) are only 14 satellites (mostly MEO satellites), but they reported about 1000 anomalies. There is also a fourth group *low altitude – low inclination* (in Figure. 3 is not shown) with only 5 satellites and very a low number of anomalies (not considered in this paper).

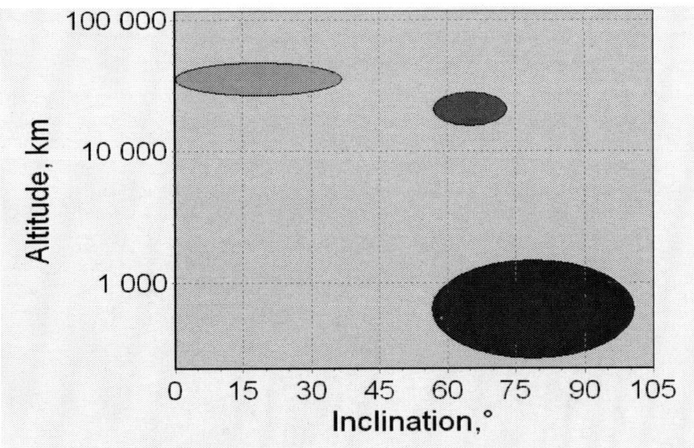

Figure 3. The separation of satellites on three groups.

2.5. Periods with Big Numbers of Satellite Anomalies

Two examples are shown in Figures 4 and 5.

In Figure 4 is shown the well known period – October 1989, characterized by very bad space weather. We see here three big solar proton events and very strong magnetic storm. We have three clusters of satellite anomalies and they coincide with maximal solar proton fluxes.

In Figure 5 the majority of anomalies coincide with a magnetic storm and enhancements of electron fluxes from radiation belts. The anomalies are absent entirely in the high altitude -

high inclination group, which played the main role in the preceding example. Only a few anomalies were reported in the GEO group and a huge majority – in the blue group (low altitude - high inclination).

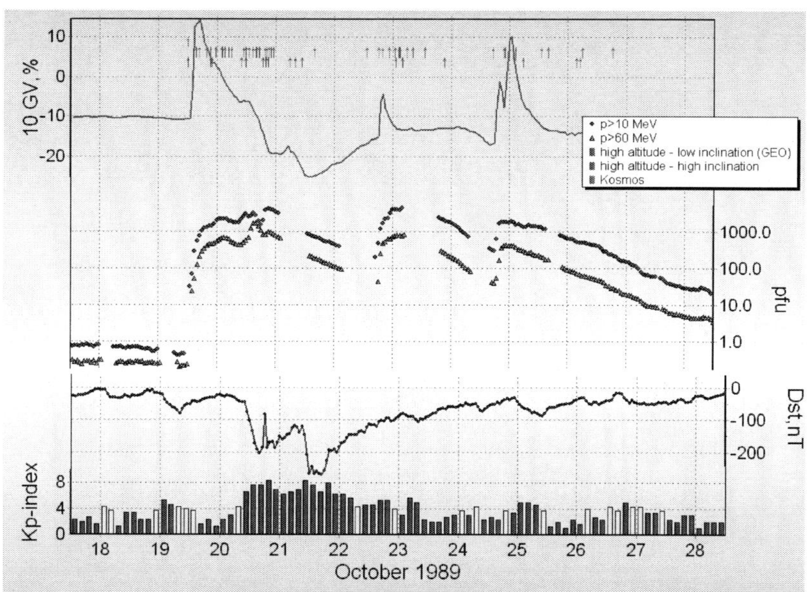

Figure 4. Period in October 1989. Upper panel: vertical arrows - moments of satellite anomalies, CR variations at 10 GV, solar protons (> 10 MeV and >60 MeV) fluxes. Lower panel: Kp- and Dst-indices.

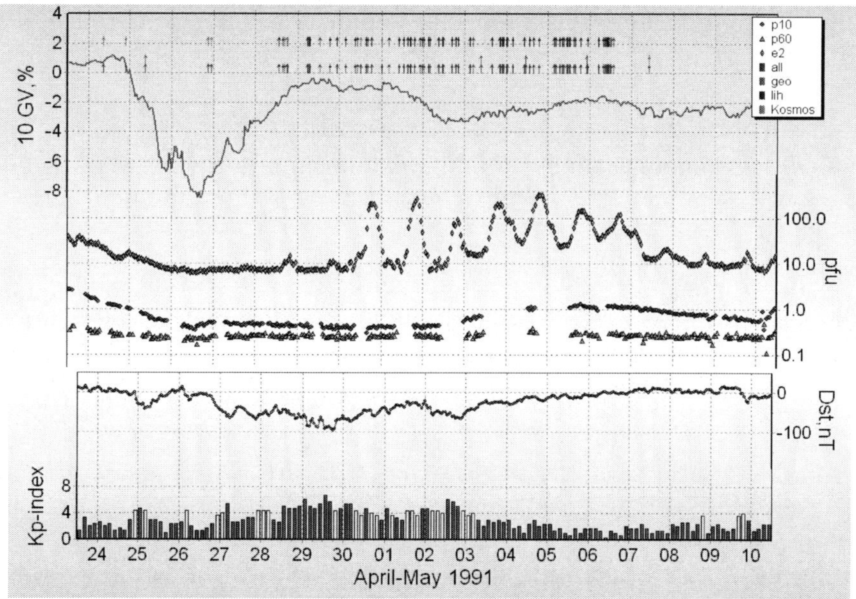

Figure 5. The same as in Figure 4, but for the period in April-May 1991 and in the upper panel are added fluxes of electrons with energy > 2 MeV.

2.6. Seasonal Variations of Anomalies of All Satellites in Comparison with Ap Index

Results are shown in Figure 6.

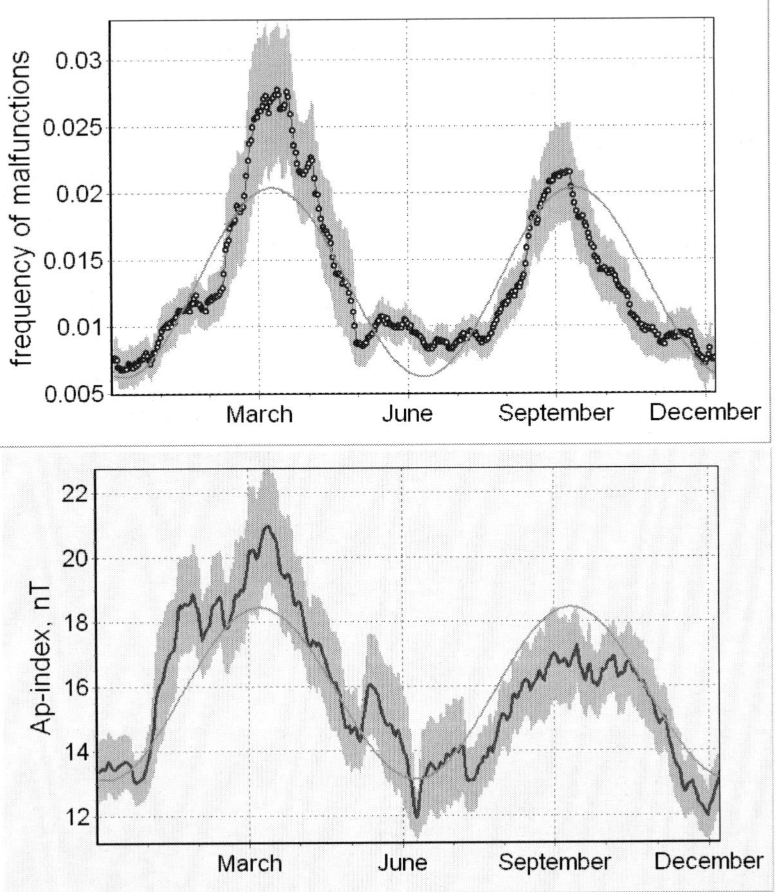

Figure 6. Satellite anomalies frequency (per day and per one satellite) and Ap-index averaged over the period 1975-1994. The curve with points is the 27-day running mean of values; the grey band corresponds to the 95% confidence interval. The sinusoidal curve is a semiannual wave with maxima in equinoxes best fitting the frequency data.

2.7. Seasonal Variations of Anomalies for Different Satellite Orbits

Results are shown in Figures 7 – 9.

From Figures 7 – 9 it can be seen that the biggest seasonal variation is in the GEO green group (high altitude – low inclination, Figure 7). The red group (high altitude – high inclination, Figure 8) did not demonstrate visible spring – autumn preference. The blue group (low altitude – high inclination, Figure 9) demonstrated some spring – autumn preference, but not as clearly as the green group (compare with Figure 7).

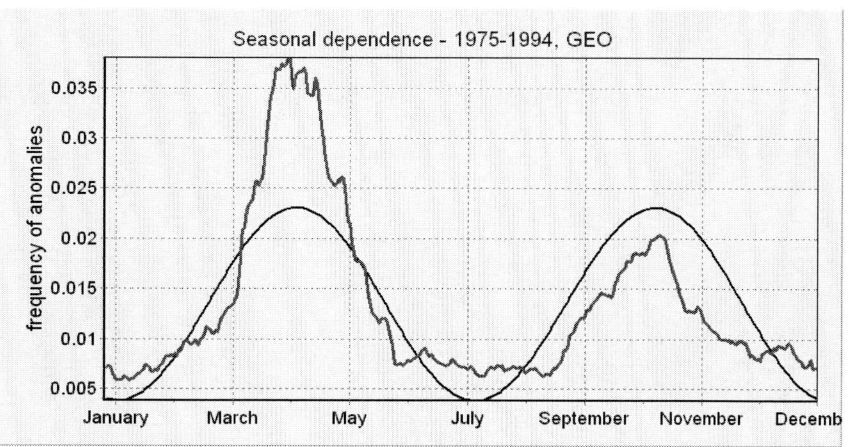

Figure 7. 27-day averaged anomaly frequencies (per day and per one satellite) and corresponding half year wave for high altitude – low inclination (green) group.

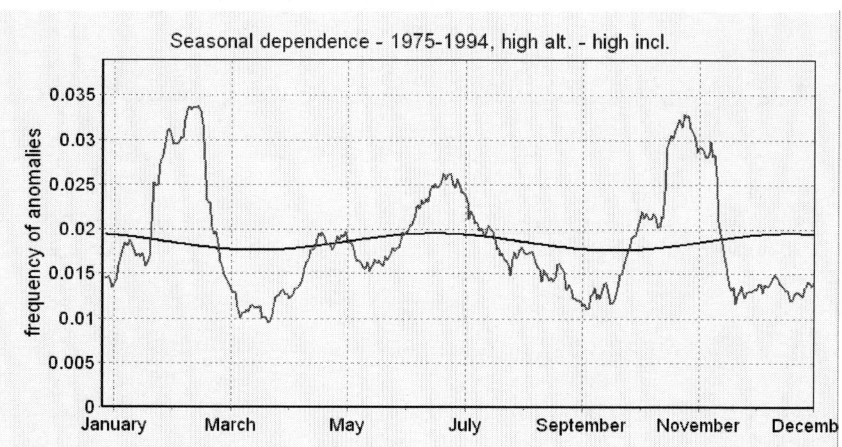

Figure 8. The same as in Figure 7, but for high altitude – high inclination satellite (red) group.

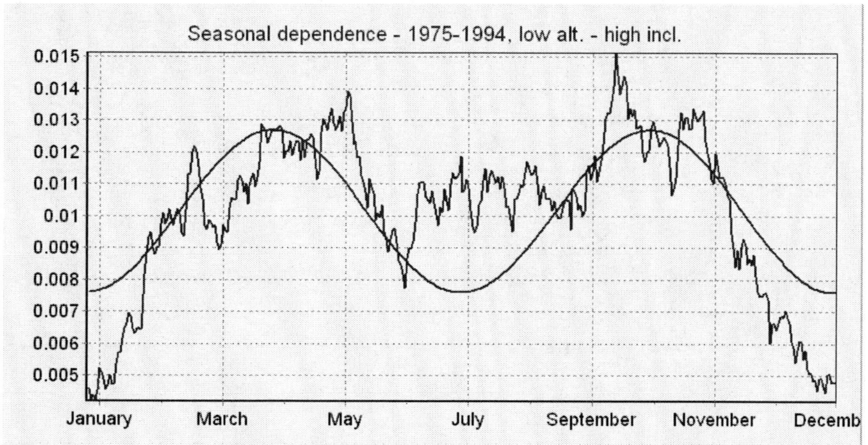

Figure 9. The same as in Figure 7, but for low altitude – high inlination satellite (blue) group.

2.8. Clustering of Satellite Anomalies

Figure 10 shows that satellite anomalies have a tendency to clustering in small groups of few days.

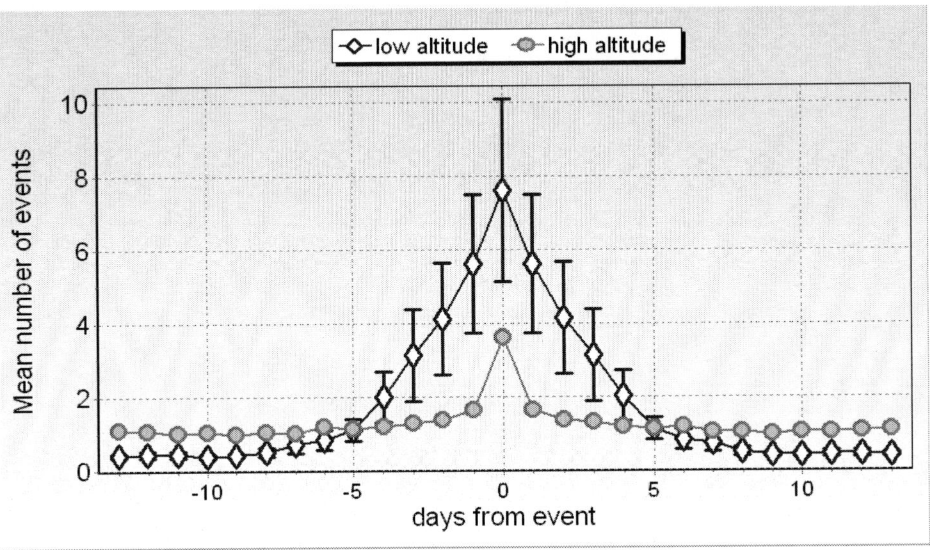

Figure 10. The tendency towards clustering of satellite anomalies in small groups of few days for low altitude (blue) and high altitude (green) satellites.

2.9. Influence of Geomagnetic Storms on Satellite Anomalies

In Figures 11 and 12 are shown the connection of geomagnetic storms SSC with satellite anomalies.

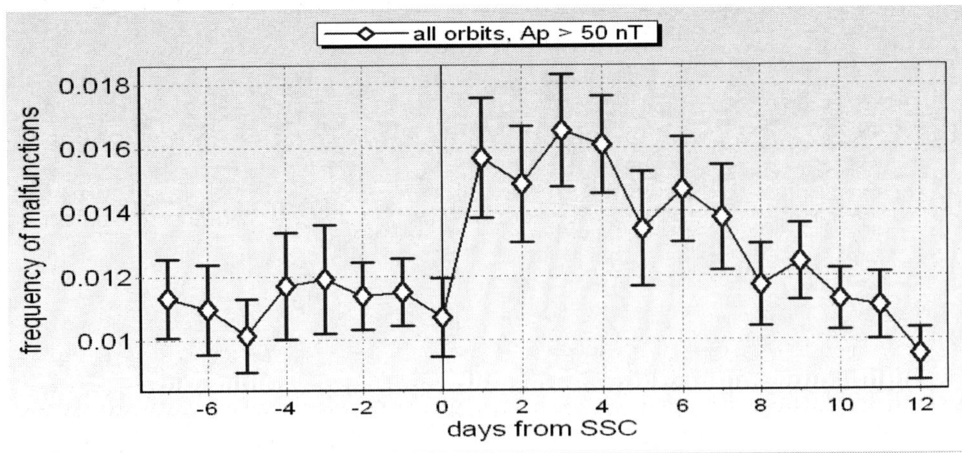

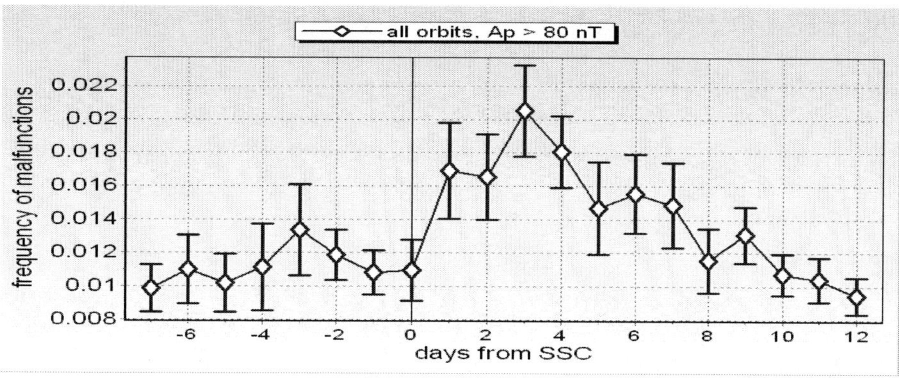

Figure 11. SSC and satellite anomalies for all orbits and for Ap > 50 nT and > 80 nT (upper and low panels).

Figure 11 shows that there is an effective connection between magnetic storms and satellite anomalies (probability of anomaly increases in about 1.5 times), and from Figure 12 it can be seen that on average, anomalies start later and lasted longer than magnetic storms.

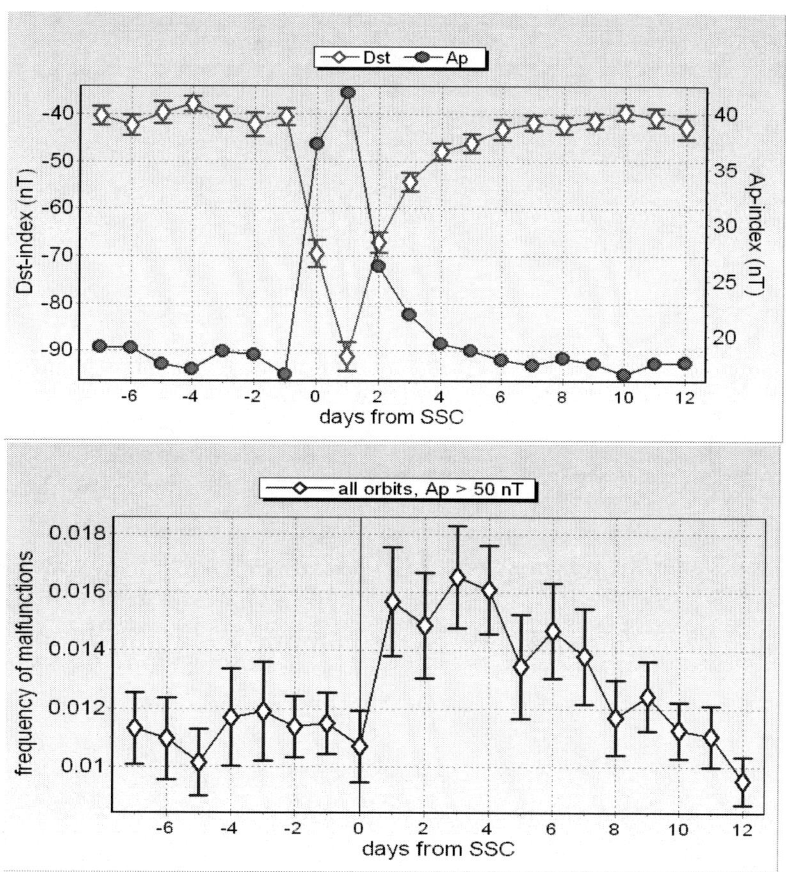

Figure 12. Variations of Ap and Dst indexes (upper panel) in comparison with satellite anomalies (low panel).

2.10. Solar Cosmic Ray Events and Satellite Anomalies

Results are shown in Figures 13-15.

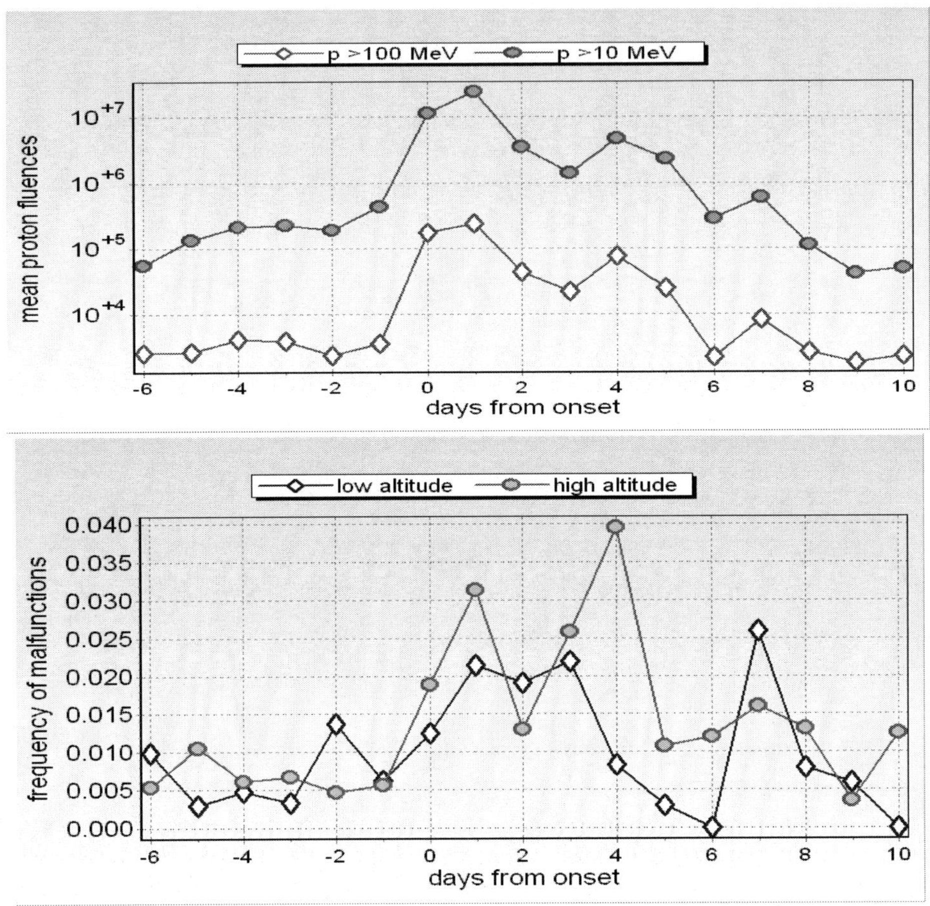

Figure 13. Solar proton events and satellite anomalies.

In Figure 13 the 0-day is proton event onset day. Hear we used only enhancements in which the average hour flux was bigger than 300 pfu. We see increasing of satellite anomalies in the 0-day and in 1-day. The increase is bigger at high altitudes than at low altitudes.

From Figure 14 it can be seen that the biggest effect is in the red group (high altitude, high inclination). The smaller effect is in the green group (high altitude, low inclination), and nothing for the blue group (low altitudes). The differences between satellite groups are visible.

From Figure 15 it can be seen that usually we have 10% probability of anomaly, but in special solar proton days this probability rises up to 100%. If we consider only data when there were only more than 6 satellites in the high altitude—high inclination group, the correlation coefficient between fluence of solar protons with energy > 10 MeV and frequency of anomalies increased up to 0.83.

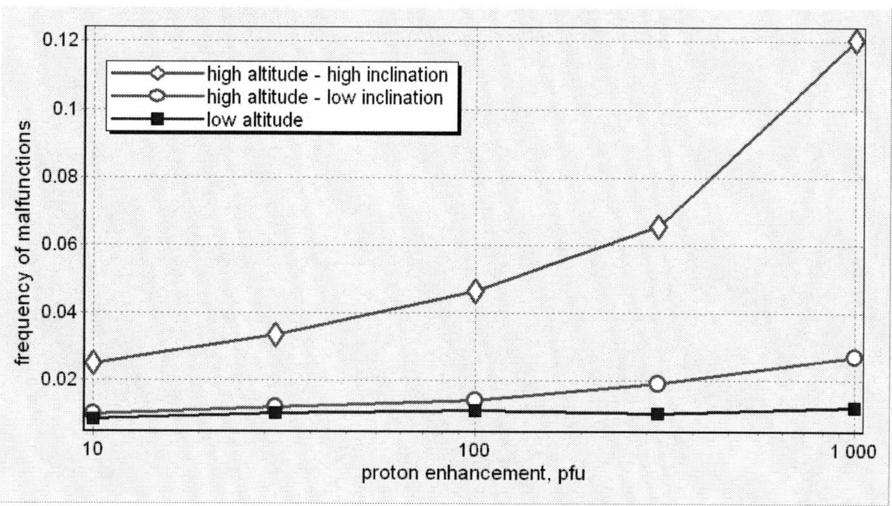

Figure 14. Dependence of the frequency of satellite anomaly (per day and per one satellite) on the flux of protons with energy ≥ 10 MeV for different groups of satellites.

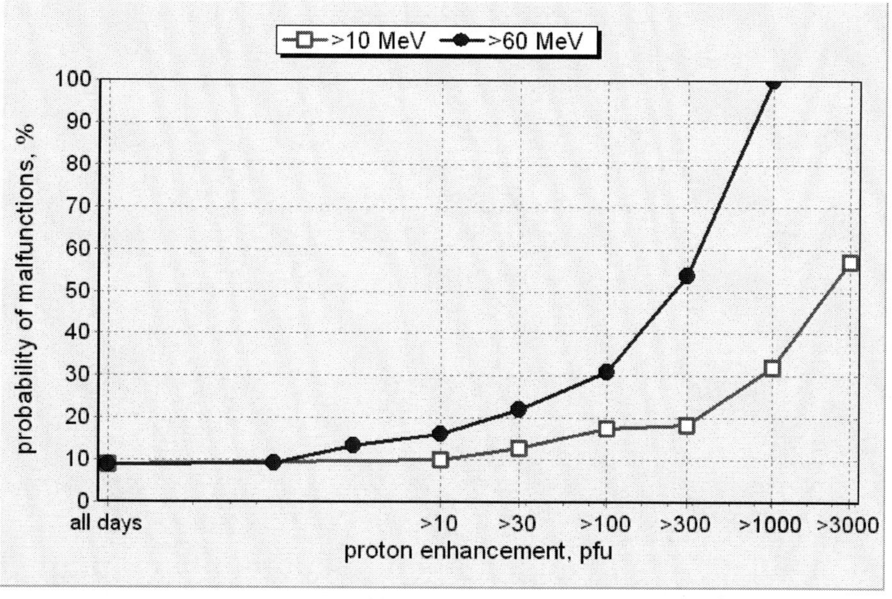

Figure 15. Dependence of the probability of satellite anomaly (per one day) from the flux of protons with energies > 10 MeV and > 60 MeV for high altitude - high inclination group of satellites.

2.11. Hazards of Energetic Protons from Solar Flares and Relativistic Electrons from Radiation Belts for Satellites on Different Orbits

Results of statistical analysis are shown in Figure 16.

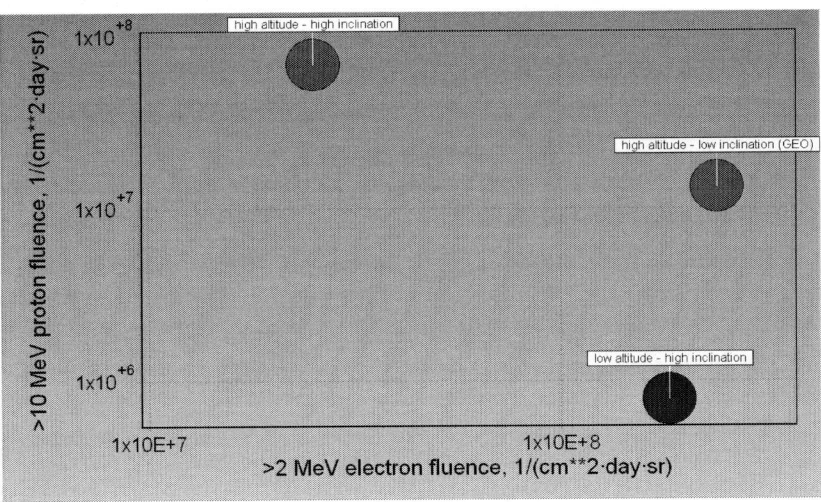

Figure 16. Mean proton and electron fluencies on the anomaly day for the different satellite groups.

From Figure 16 it can be seen that anomalies in the red group (high altitude – high inclination) are caused mostly by energetic protons from solar flares. In the blue group (low altitude – high inclination) – they are caused mostly by energetic electrons from radiation belts, and in the green group (high altitude – low inclination) it is mixed – they are caused by both energetic protons and electrons.

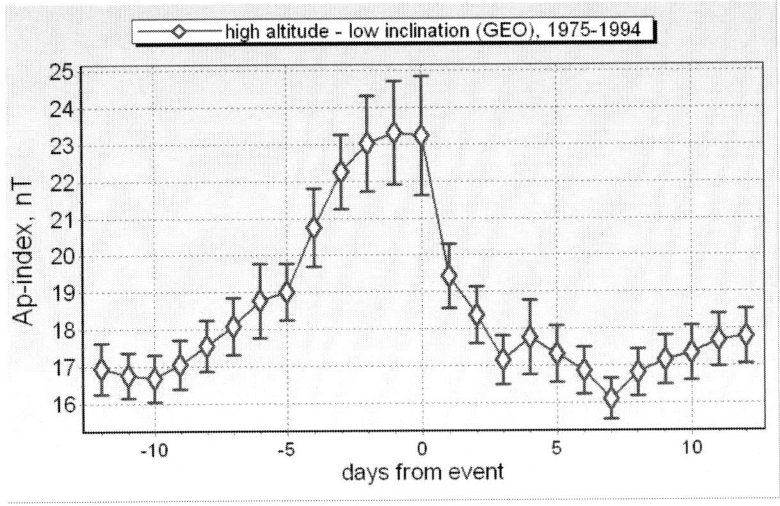

Figure 17. Mean behaviour of Ap-index in anomaly periods for high altitude – low inclination group (GEO satellites).

2.12. Possible Precursors for Satellite Anomalies

We found interesting behavior of some parameters of space weather related to satellite anomalies.

2.12.1. Ap Index

Results of statistical analysis for Ap index are shown in Figure 17.

From Figure 17 we see the enhanced geomagnetic activity not only on the day of satellite anomaly (zero-day) but also several days before.

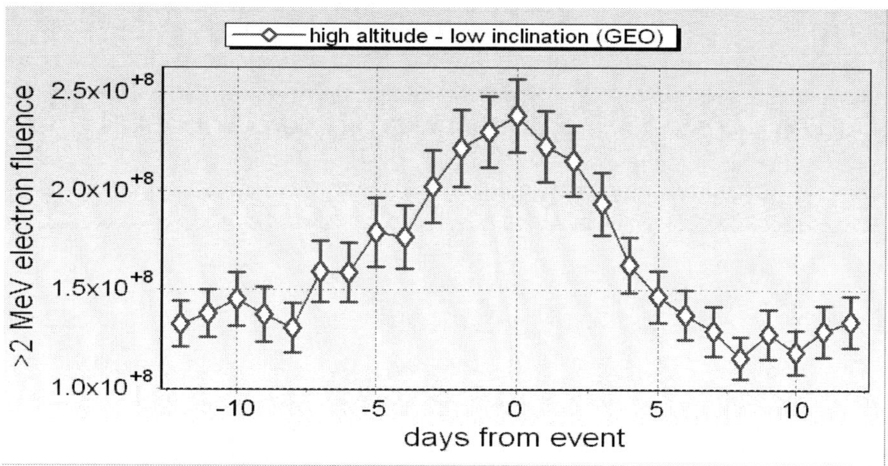

Figure 18. The mean behavior of >2 MeV electron fluence in anomaly periods for high altitude – low inclination group (GEO satellites).

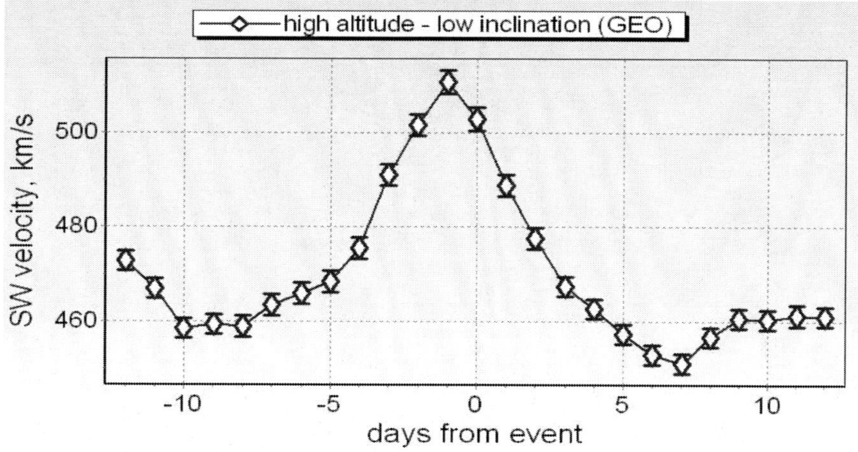

Figure 19. The mean behavior of solar wind velocity in anomaly periods for high altitude – low inclination group (GEO satellites).

2.12.2. Fluencies of Electrons from Radiation Belts with Energy > 2 MeV

From Figure 18 can be seen that >2 MeV electron fluence also may be used as a several day precursor for high altitude – low inclination satellite anomalies.

2.12.3. Velocity of Solar Wind

Results are shown in Figure 19.

From Figure 19 it can be seen that solar wind velocity also may be used as a several days precursor for high altitude – low inclination satellite anomalies. In principle, all three factors considered in this Section may be used as combined precursors for anomalies in the high altitude – low inclination satellite group.

2.13. Models for Satellite Anomaly Probabilities and Forecasting

Results schematically are shown in Figure 20.

From Figure 20 it can be seen that the main role in the formation of anomalies is played by energetic electrons (generated mostly in the Earth's magnetosphere during magnetic storms) in the green and blue groups of satellites (especially – in the green group), but in the red group solar energetic protons are much more important than other indices. Therefore, for forecasting satellite anomalies it is important to forecast SEP events (see below Sections 4-8) and magnetic storms (see Sections 9-10).

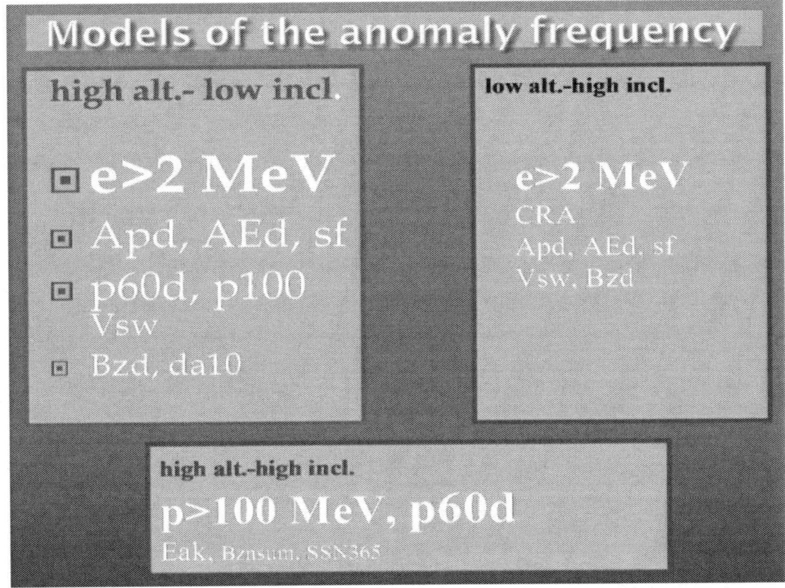

Figure 20. Results for three groups of satellites. Bigger symbols correspond bigger to regression coefficients.

3. DATA FROM THE PAST AND CLASSIFICATION OF DANGEROUS SPACE WEATHER PHENOMENA (NOAA CLASSIFICATION AND ITS MODERNIZATION)

As is well known, the NOAA Space Weather Scale establishes 5 gradations of SEP events, called Solar Radiation Storms: from S5 (the highest level of radiation, corresponding to flux of solar protons with energy >10 MeV about 10^5 proton.cm^{-2}.s^{-1}) up to S1 (the lowest level, flux of about 10 proton.cm^{-2}.s^{-1} for protons with energy >10 MeV). In our opinion, using ground level CR neutron monitors and muon telescopes, it is possible to monitor and forecast (using much higher energy particles than smaller energy particles which cause the main radiation hazard) SEP events of levels S5, S4 and S3. With the increase of an SEP event, the level of radiation will increase the accuracy of forecasting. Let us note that in our opinion, for satellite damage and influence on people's health and technology, on communications by HF radio-waves the total fluence of SEP during the event is more important than the protons flux that is used now in NOAA Space Weather Scale. The second note is that the level S5 (corresponds to the flux 10^5 proton.cm^{-2}.s^{-1}, or fluency F ~ 10^9 proton.cm^{-2} for protons with $E_k \geq 10$ MeV) is not maximal (as it is supposed by NOAA Solar Radiation Storms Scale), but can be much higher and with much smaller probability than S5 (Dorman et al., 1993; Dorman and Venkatesan, 1993; Dorman and Pustil'nik, 1995, 1999). As it was shown by McCracken et al. (2001), the dependence of event probability on fluence can be prolonged at least up to $F = 2 \times 10^{10}$ proton.cm^{-2} for protons with $E_k \geq 30$ MeV, which was observed in SEP of September 1859 according to data of nitrate contents in polar ice and on other SEP events (see Figures 21 - 23).

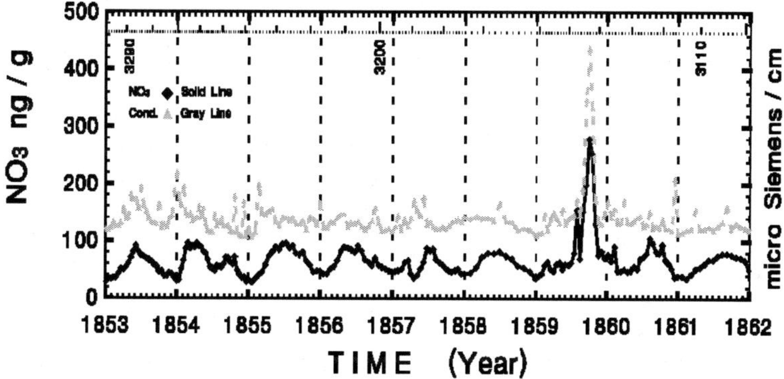

Figure 21. Example on a great SEP event in September 1859 with fluency F = 2×10^{10} proton.cm^{-2} for protons with $E_k \geq 30$ MeV. According to McCracken et al., 2001.

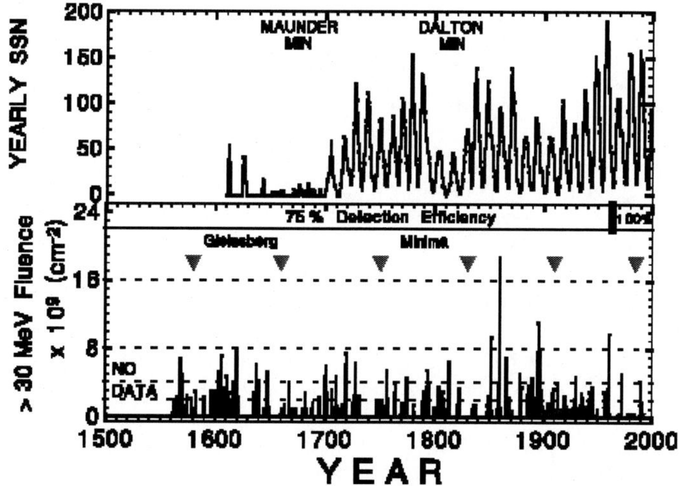

Figure 22. Great SEP events in the last 450 years according to nitrate data (McCracken et al., 2001).

Table 1. Extended SEP events radiation hazard scale (based on NOAA Space Weather Scale for Solar Radiation Storms)

		SEP events radiation hazard	Fluence level ≥ 30 MeV protons	Number of events per one year
S7	Especially extreme	**Biological:** Lethal doze for astronauts, for passengers and crew on commercial jets; great influence on people's health and gene mutations on the ground **Satellite operations:** very big damages to satellites' electronics and computer memory, damage to solar panels, loosing of many satellites **Other systems:** complete blackout of HF (high frequency) communications through polar and middle-latitude regions, big position errors make navigation operations extremely difficult.	$10^{(11\text{-}12)}$	**One in several thousand years**
S6	Very extreme	**Biological:** About lethal doze for astronauts, serious influence on passengers and crew health on commercial jets; possible influence on people's health and gene mutations on the ground **Satellite operations:** big damages to satellites electronics and computer memory, damage to solar panels, loss of several satellites **Other systems:** complete blackout of HF communications through polar regions, some position errors make navigation operations very difficult.	$10^{(10\text{-}11)}$	**One in a few hundred years**
S5	Extreme	**Biological:** unavoidable high radiation hazard to astronauts on EVA (extra-vehicular activity); high radiation exposure to passengers and crew in commercial jets at high latitudes (approximately 100 chest X-rays) is possible. **Satellite operations:** satellites may be rendered useless, memory impacts can cause loss of control, may cause serious noise in image data, star-trackers may be unable to locate sources; permanent damage to solar panels possible. **Other systems:** complete blackout of HF (high frequency) communications possible through the polar regions, and position errors make navigation operations extremely difficult.	10^9	**One in 20-50 years**
S4	Severe	**Biological:** unavoidable radiation hazard to astronauts on EVA; elevated radiation exposure to passengers and crew in commercial jets at high latitudes (approximately 10 chest X-rays) is possible. **Satellite operations:** may experience memory device problems and noise on imaging systems; star-tracker problems may cause orientation problems, and solar panel efficiency can be degraded. **Other systems:** blackout of HF radio communications through the polar regions and increased navigation errors over several days are likely.	10^8	**One in 3-4 years**
S3	Strong	Biological: radiation hazard avoidance recommended for astronauts on EVA; passengers and crew in commercial jets at high latitudes may receive low-level radiation exposure (approximately 1 chest X-ray). Satellite operations: single-event upsets, noise in imaging systems, and a slight reduction of efficiency in solar panel are likely. Other systems: degraded HF radio propagation through the polar regions and navigation position errors likely.	10^7	**One per year**

This type of great dangerous events is very rare (occurring about once every few hundred years). According to Figure 23, it is not excluded that in principle very great SEP events with fluency 10 and even 100 times bigger (correspondingly one in a few thousand and one in several tens of thousands of years) can occur. So, we suppose to correct the very important classification, developed by NOAA, in two directions: to use fluency F of SEP during all event (in units proton.cm^{-2}) instead of flux I, and to extend levels of radiation hazard. As a result, the modernized classification of SEP events is shown in Table 1.

Let us note that the expected frequency of SEP events in the last column of Table 1 is averaged over the solar cycle. Really, this frequency is much higher in periods of high solar activity than in periods of low solar activity (Dorman et al., 1993; Dorman and Pustil'nik, 1995, 1999).

4. ONLINE SEARCH FOR THE START OF GREAT SEP EVENTS, AUTOMATIC FORMATION OF ALERTS, ESTIMATION OF PROBABILITY OF FALSE ALERTS AND PROBABILITY OF MISSING ALERTS (REALIZED IN THE EMILIO SEGRÉ OBSERVATORY AT MT. HERMON)

4.1. Why Do We Need to Use Solar High Energetic Particles for SEP Forecasting and to Determine the Time of Their Arrival?

It is well known that in periods of great solar energetic particle (SEP) ground events, fluxes of energetic particles can be so big that memory of computers and other electronics in space may be damaged, and satellite and spacecraft operations can be seriously degraded. In these periods it is necessary to switch off some part of the electronics for a few hours to protect computer memories. The problem is how to forecast exactly these dangerous phenomena. We show that exact forecasts can be made by using high-energy particles (few GeV/nucleon and higher) whose transportation from the Sun is characterized by much bigger diffusion coefficients than lower energy particles. High-energy particles arrive from the Sun much earlier (8-20 minutes after acceleration and escaping into the solar wind) than the lower energy particles that damage electronics (about 30-60 minutes later). We describe here the principles and operation of the automated programs "SEP-Search-1 min", "SEP-Search-2 min," and "SEP-Search-5 min", developed and checked in the Emilio Segre' Observatory (ESO) of the Israel Cosmic Ray Center (2025 m above sea level, R_c =10.8 GV). The determination of increasing flux is made by comparison with the intensity, averaged from 120 to 61 minutes, prior to the current one-minute data. For each minute of data the program "SEP-Search-1 min" is run. If the result is negative (no simultaneous increase in both channels of total intensity $\geq 2.5\sigma_1$, where σ_1 is the standard deviation for one minute of observation in one channel [for ESO σ_1 =1.4 %]), start the program "SEP-Search-2 min", using two minute averages with $\sigma_2 = \sigma_1/\sqrt{2}$, and so on. If any positive result is obtained, the "SEP-Search" programs check the next minute of data. If the result is again positive, the online the programs "SEP-Collect" and "SEP-Research" that determine the expected flux and spectrum and generate automatic alerts automatically run. These programs are described in Dorman and Zukerman (2001).

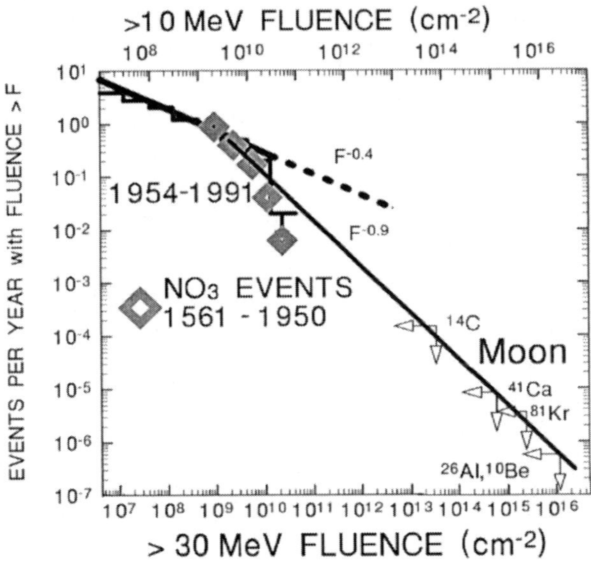

Figure 23. The dependence of SEP events probability (number of events per year) on the value of fluence according to direct satellite and NM data, nitrate in polar ice data and cosmogenic nuclide data on the moon (McCracken et al., 2001).

4.2. Short Description of ICR&SWC and Emilio Segre' Observatory

The Israel Cosmic Ray & Space Weather Center (ICR&SWC) and Israel-Italian Emilio Segre' Observatory (ESO) were established in 1998, with affiliation to Tel Aviv University, to the Technion (Israel Institute of Technology, Haifa) and to the Israel Space Agency (under the aegis of the Ministry of Science). The Mobile Cosmic Ray Neutron Monitor of the Emilio Segre' Observatory was prepared in collaboration with scientists of the Italian group in Rome, and transferred in June 1998 to the site selected for the Emilio Segre' Observatory (33°18'N, 35°47.2'E, 2050 m above sea level, vertical cut-off rigidity R_c =10.8 GV). The results of measurements (data taken at one-minute intervals of cosmic ray neutron total intensities from two separate 3NM-64, as well as similar one-minute data about the intensities relating to neutron multiplicities $m \geq 1$, ≥ 2, ≥ 3, ≥ 4, ≥ 5, ≥ 6, ≥ 7 and ≥ 8) have been computer-stored. Similar one-minute data relating to the atmospheric electric field, wind speed, air temperature outside, and humidity and temperature inside the Cosmic Ray Observatory have also been recorded and archived. Each month one-hour data of the Emilio Segre' Observatory (short title ESO) are sent to the World Data Center in Boulder (USA, Colorado), to the WDC C-2 for cosmic rays (Japan) and to many Cosmic Ray Observatories in the world as well as uploaded to our website. We established the automatic system of electric power supply using a diesel generator for providing continuous power for the Emilio Segre' Observatory. We finished the foundation of direct radio-connection in real time scale of the Emilio Segre' Observatory with our Laboratory in Qazrin, and with the Internet. In Figure 24 we show a block-scheme of the main components of the Emilio Segré Observatory (ESO) and their connection with the Central Laboratory of Israel Cosmic Ray & Space Weather Center in Qazrin and with the Internet.

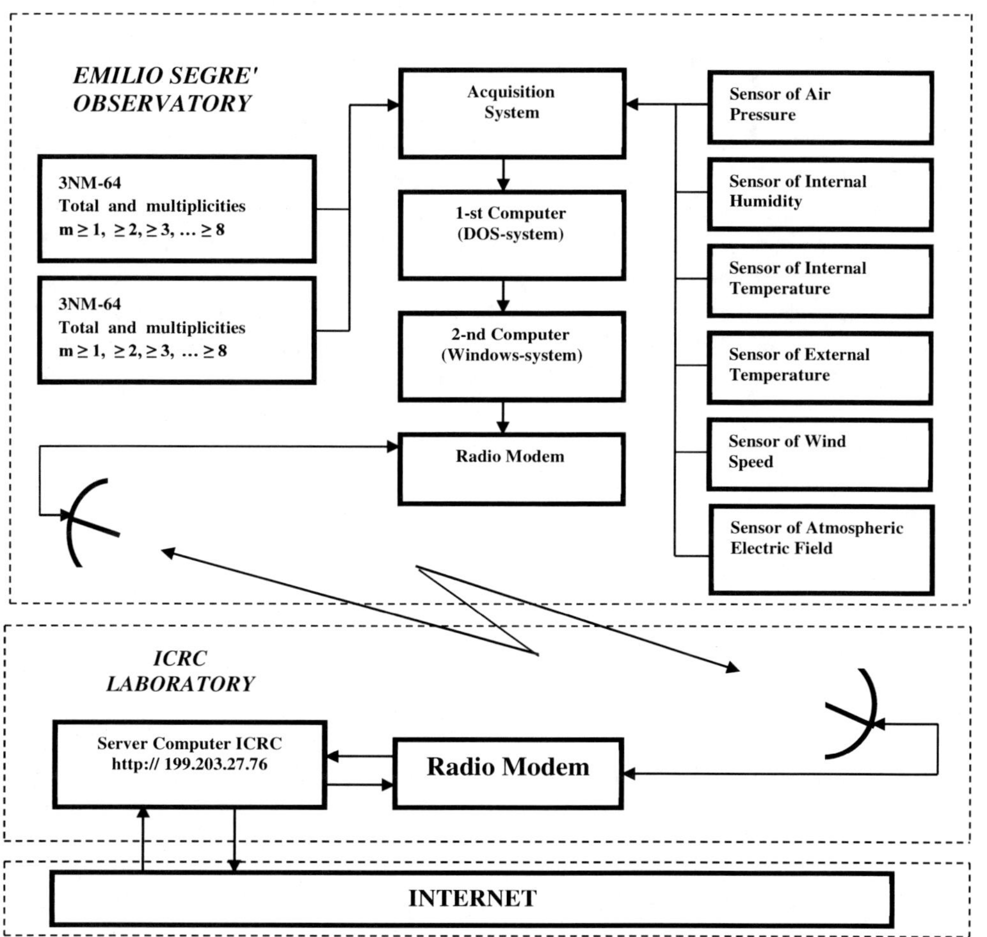

Figure 24. Schematic of the main components of the Emilio Segre' Observatory (ESO) and their connection with the Central Laboratory of Israel Cosmic Ray Center (ICRC) in Qazrin and with the INTERNET.

4.3. The Method of Automatically Searching for the Start of Ground SEP Events

Let us consider the problem of automatically searching for the start of ground SEP events. Of course, the patrol of the Sun and forecast of great solar flares are very important, but not enough: only a very small part of great solar flares produce dangerous SEP events. In principal this exact forecast can be made by using high-energy particles (few *GeV/nucleon* and higher) whose transportation from the Sun is characterized by much bigger diffusion coefficient than for small and middle energy particles. Therefore high-energy particles arrive from the Sun much earlier (8-20 minutes after acceleration and escaping into the solar wind) than the lower energy particles that cause a dangerous situation for electronics (at least about

30-60 minutes later). The flux of high-energy particles is very small and cannot be dangerous for people and electronics.

The problem is that this very small flux cannot be measured with enough accuracy on satellites to use for forecast (it needs very large effective surfaces of detectors and thus large weight). High-energy particles of galactic or solar origin are measured continuously by ground-based neutron monitors, ionization chambers and muon telescopes with very large effective surface areas (many square meters) that provide very small statistical errors. It was shown on the basis of data in periods of great historical SEP events (as the greatest of February 23, 1956 and many tens of others), that one-minute online data of high energy particles could be used for forecasting of incoming dangerous fluxes of particles with much smaller energy. The method of coupling (response) functions (Dorman, 1957; Dorman et al., 2000; Clem and Dorman, 2000; Dorman, 2004) allows us to calculate the expected flux above the atmosphere, and out of the Earth's magnetosphere from ground based data. Let us describe the principles and online operation of programs "SEP-Search-1 min", "SEP-Search-2 min","SEP-Search-5 min", developed and checked in the Emilio Segre' Observatory of ICRC.

The determination of increasing flux is made by comparison with intensity averaged from 120 to 61 minutes before the present Z-th one-minute data. For each Z minute data, start the program "SEP-Search-1 min" (see Figure 25). The program for each Z-th minute determines the values

$$D_{A1Z} = \left[\ln(I_{AZ}) - \sum_{k=Z-120}^{k=Z-60} \ln(I_{Ak})/60 \right] \Big/ \sigma_1 \tag{1}$$

$$D_{B1Z} = \left[\ln(I_{BZ}) - \sum_{k=Z-120}^{k=Z-60} \ln(I_{Bk})/60 \right] \Big/ \sigma_1 \,, \tag{2}$$

where I_{Ak} and I_{Bk} are one-minute total intensities in the sections of neutron super- monitor A and B. If simultaneously

$$D_{A1Z} \geq 2.5, \; D_{B1Z} \geq 2.5, \tag{3}$$

the program "SEP-Search-1 min" repeats the calculation for the next Z+1-th minute and if Equation (3) is satisfied again, the onset of a great SEP is determined and programs "SEP-Collect" and "SEP-Re-search," described in the next section, are started.

If Equation (3) is not satisfied, the program "SEP-Search-2 min" starts by using two-min data characterized by $\sigma_2 = \sigma_1/\sqrt{2}$. In this case, the program "SEP-Search-2 min" will calculate values \

$$D_{A2Z} = \left[\left(\ln(I_{AZ}) + \ln(I_{A,Z-1}) \right) \Big/ 2 - \sum_{k=Z-120}^{k=Z-60} \ln(I_{Ak}) \Big/ 60 \right] \Big/ \sigma_2 \,, \tag{4}$$

$$D_{B2Z} = \left[\left(\ln(I_{BZ}) + \ln(I_{B,Z-1})\right)\bigg/ 2 - \sum_{k=Z-120}^{k=Z-60} \ln(I_{Bk})\bigg/ 60\right]\bigg/ \sigma_2 \tag{5}$$

If the result is negative (no simultaneous increase in both channels of total intensity ≥ 2.5 σ_2, i.e. the condition $D_{A2Z} \geq 2.5$, $D_{B2Z} \geq 2.5$ fails), then "SEP-Search-3 min" uses the average of three minutes Z-2, Z-1 and Z with $\sigma_3 = \sigma_1/\sqrt{3}$. If this program also gives a negative result, then the program "SEP-Search-5 min" uses the average of five minutes Z-4, Z-3, Z-2, Z-1 and Z with $\sigma_5 = \sigma_1/\sqrt{5}$. If this program also gives a negative result, i.e. all programs "SEP-Search-K min"(where K=1,2,3,5) give a negative result for the Z-th minute, it means that in the next 30-60 minutes there will be no radiation hazard (this information can be also very useful). After obtaining this negative result, the procedure repeats for the next, Z+1-th minute, and so on. If any positive result is obtained for some Z=Z', the "SEP-Search" programs check the next Z'+1-th minute data. If the result obtained is again positive, then the programs, described in Section 4, are started to determine the expected flux and spectrum and to send automated alerts.

4.4. Monitoring and Preliminary Alert for Starting of Great SEP Events on the Web-site of ICR&SWC and ESO

Figure 26 shows an example of the image in the website of ICR&SWC and ESO for the moment 20:27:03 UT at 13 May 2001 (it was obtained 3 seconds after registration by NM in ESO, at 20:27:00 UT). On this image are shown online one-minute data of total neutron intensity for the last 6 hours (360 circles in the upper part) and one-hour data during the last 6 days (144 circles in the down part). One minute data are upgraded each minute in ASCII form. In the bottom of the image is given information on variations in A and B channels on the basis of one-, two-, and three-minute data (A1, A2, A3, B1, B2 and B3 in units of standard deviations).

In the last line in Figure 26 is shown the Alert for starting of great SEP (based on data of both channels and checked by the situation in the previous minute). It can be seen that in the considered moment of time there is the following situation, SEP ALERT: No. This means that at least for one hour there will be no dangerous radiation hazard.

4.5. The Probability of False Alarms

Because the probability function $\Phi(2.5) = 0.9876$, the probability of an accidental increase with amplitude of more than 2.5σ in one channel will be $(1 - \Phi(2.5))/2 = 0.0062\,\mathrm{min}^{-1}$, meaning one in 161.3 minutes (in one day we expect 8.93 accidental increases in one channel). The probability of accidental increases simultaneously in both channels will be $((1 - \Phi(2.5))/2)^2 = 3.845 \times 10^{-5}\,\mathrm{min}^{-1}$ meaning one in 26007 minutes ≈ 18 days. The probability that the increases of 2.5σ will be accidental in both channels in two successive minutes is equal to $((1 - \Phi(2.5))/2)^4 = 1.478 \times 10^{-9}\,\mathrm{min}^{-1}$, meaning one in 6.76×10^8 minutes ≈ 1286 years. If this false alarm (one in about 1300 years) is sent, it is not dangerous, because the first alarm is preliminary and can be cancelled if in the third successive minute there is no increase in both channels bigger than 2.5σ (it is not excluded that in the third minute there will be also an accidental increase, but the probability of this false alarm is negligible: $((1 - \Phi(2.5))/2)^6 = 5.685 \times 10^{-14}\,\mathrm{min}^{-1}$, meaning one in 3.34×10^7 years). Let us note that the false alarm can be sent in the case of a solar neutron event (which really is not dangerous for electronics in spacecrafts or for astronauts health), but these events are usually very short (lasting only a few minutes) and this alarm will be automatically canceled in the successive minute after the end of the solar neutron event.

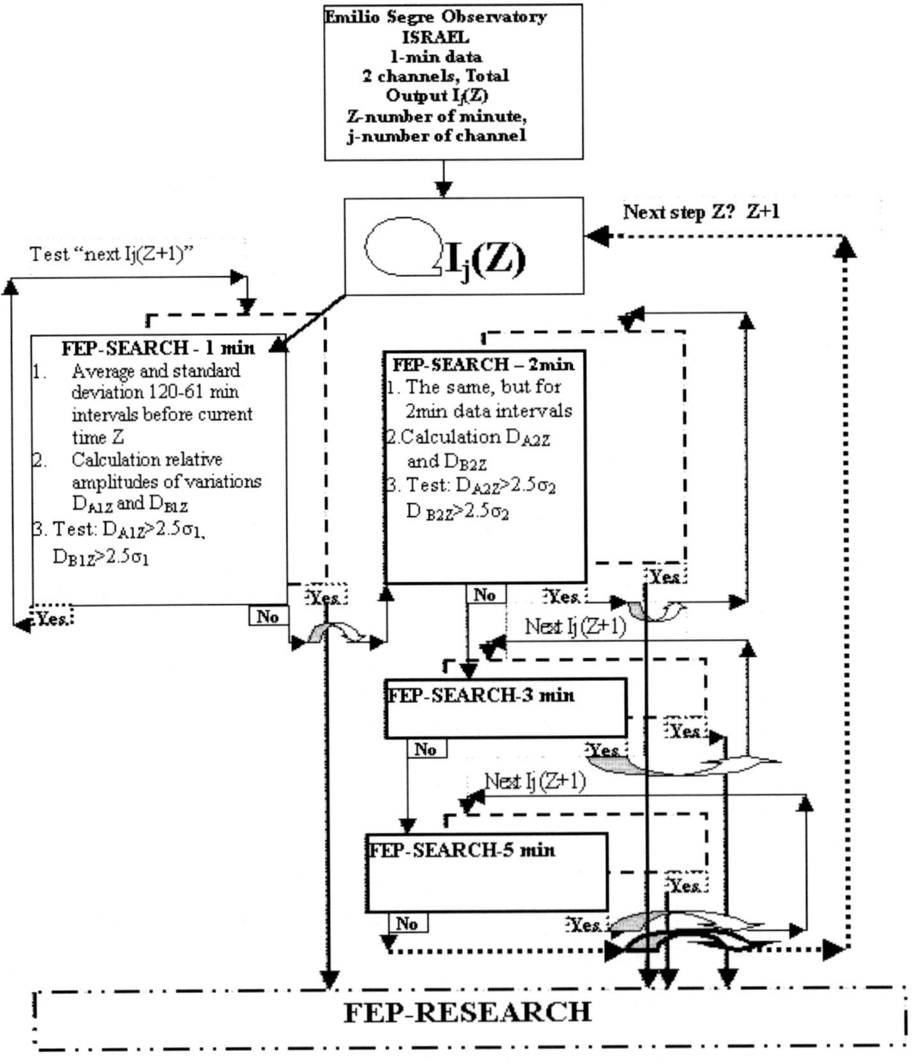

Figure 25. Schematic of programs "SEP-Search".

4.6. The Probability of Missed Triggers

The probability of missed triggers depends very strongly on the amplitude of the increase. Let us suppose for example that we have a real increase of 7σ (that for ESO corresponds to an increase of about 9.8 %). The trigger will be missed in any of both channels and in any of both successive minutes if as a result of statistical fluctuations the increase of intensity is less than 2.5σ. For this the statistical fluctuation must be negative, with amplitude of more than 4.5σ. The probability of this negative fluctuation in one channel in one minute is equal to $(1 - \Phi(4.5))/2 = 3.39 \times 10^{-6}\ \mathrm{min}^{-1}$, and the probability of missed trigger for two successive minutes of observation simultaneously in two channels is 4 times larger: 1.36×10^{-5}. This means that a missed trigger is expected only once per about 70000 events. In Table 2 are listed probabilities P_{mt} of missed triggers for ESO (where standard deviation for one channel for one minute σ=1.4%) as a function of the amplitude of increase A.

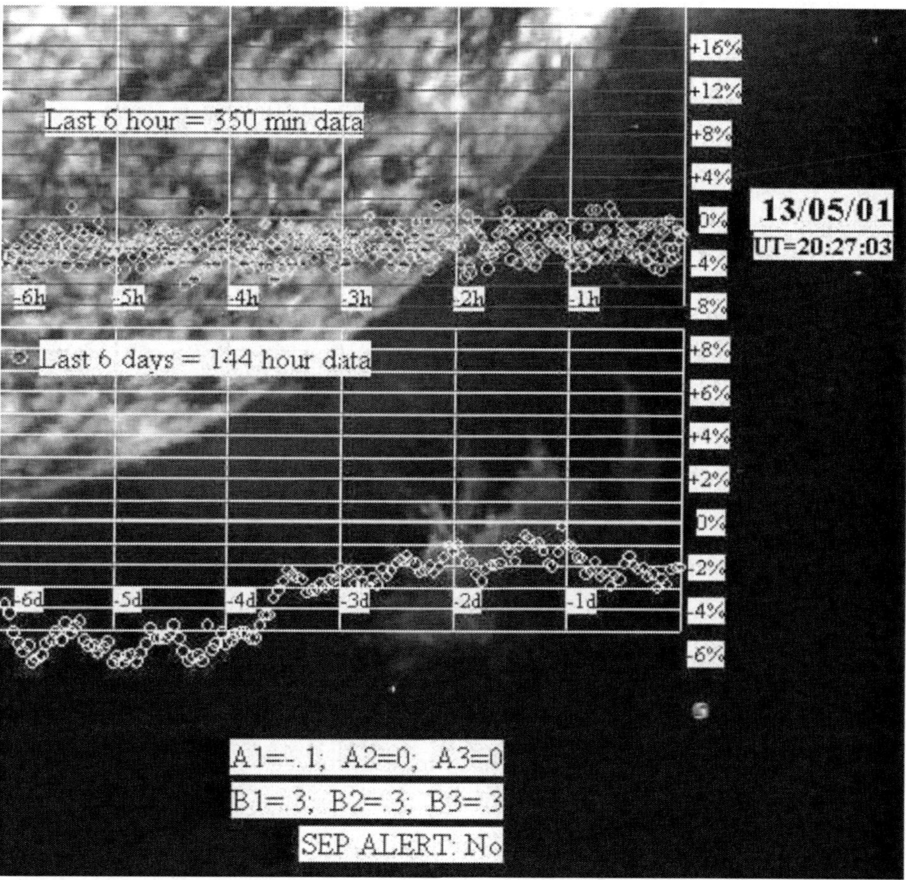

Figure 26. Monitoring and preliminary alert for starting of great SEP events on the website of ICR&SWC.

Table 2. Probabilities P_{mt} of missed triggers as a function of amplitude of increase A (in % and in σ)

A, %	5.6	6.3	7.0	7.7	8.4	9.1	9.8	10.5
A, σ	4.0	4.5	5.0	5.5	6.0	6.5	7.0	7.5
P_{mt}	2.7×10^{-1}	9.1×10^{-2}	2.5×10^{-2}	5.4×10^{-3}	9.3×10^{-4}	1.3×10^{-4}	1.4×10^{-5}	1.1×10^{-6}

4.7. Discussion on the Supposed Method

The results obtained show that the considered method of automatically searching for the onset of great, dangerous SEP on the basis of one-minute NM data rarely gives false alarms (the probability of a false preliminary alarm is one in about 1300 years, and for a false final alarm- one in 3.34×10^{7} years). Non-dangerous solar neutron events also can be separated automatically. We estimated also the probability of missed triggers; it was shown that for events with amplitude of increase more than 10% the probability of a missed trigger for successive two minutes NM data is smaller than 1.36×10^{-5} (this probability decreases sufficiently with increased amplitude A of the SEP increase, as shown in Table 2). Historical ground SEP events show very fast increase of amplitude in the start of event (Dorman, 1957; Carmichael, 1962; Dorman and Miroshnichenko, 1968; Duggal, 1979; Dorman and Venkatesan, 1993; Stoker, 1995). For example, in the great SEP event of February 23, 1956 amplitudes of increase in the Chicago NM were at 3.51 UT - 1%, at 3.52 UT – 35%, at 3.53 UT – 180%, at 3.54 UT – 280 %. In this case the missed trigger can be only for the first minute at 3.51 UT. The described method can be used in many Cosmic Ray Observatories where one-minute data are detected. Since the frequency of ground SEP events increases with decreasing cutoff rigidity, it will be important to introduce described method in high latitude Observatories. For low latitude Cosmic Ray Observatories the SEP increase starts earlier and the increase is much faster; this is very important for forecasting of dangerous situation caused by great SEP events.

5. ONLINE DETERMINATION OF SEP SPECTRUM BY THE METHOD OF COUPLING FUNCTIONS

5.1. Principles of SEP Radiation Hazard Forecasting

The problem is that the time-profiles of solar cosmic ray increases are very different for different great SEP events. It depends on the situation in interplanetary space. If the mean free path of high-energy particles is large enough, the initial increase will be sharp, very short, only a few minutes, and in this case one or two-minute data will be useful. Conversely when the mean free path of high energy particles is much smaller, the increase will be gradual, possibly prolonged for 30-60 minutes, and in this case 2-, 3- or 5-minute data will be useful. Moreover, for some very anisotropic events (as February 23, 1956) the character of increase on different stations can be very different (sharp or gradual depending on the station location and anisotropy). Dorman et al. (2001) described the operation of programs "SEP Search-K min" (where $K=1$, 2, 3, and 5). If any of the "SEP Search-K min" programs gives a positive result for any Cosmic Ray Observatory the online program "SEP Collect" is started and collects all available data on the SEP event from Cosmic Ray Observatories and satellites. The many "SEP Research" programs then analyze these data. The real-time research method consists of:

- Determination of the energy spectrum above the atmosphere from the start of the SEP-event (programs "SEP Research-Spectrum");
- Determination of the anisotropy and its energy dependence (program "SEP Research-Anisotropy");
- Determination of the propagation parameters, time of SEP injection into the solar wind and total source flux of SEP as a function of energy (programs "SEP Research-Propagation", "SEP Research-Time Ejection", "SEP Research-Source");
- Forecasting the expected fluxes and the spectrum in space, in the magnetosphere and in the atmosphere (based on the results obtained from steps 1-3 above) using programs "SEP Research-Forecast in Space", "SEP Research-Forecast in Magnetosphere", and "SEP Research-Forecast in Atmosphere";
- Issuing of preliminary alerts if the forecast fluxes are at dangerous levels (space radiation storms *S5, S4* or *S3* according to the classification of NOAA. These preliminary alerts are from the programs "SEP Research-Alert 1 for Space", "SEP Research-Alert 1 for Magnetosphere", and "SEP Research-Alert 1 for Atmosphere".

Then, based on further online data collection, more accurate Alert 2, Alert 3 and so on are sent. Here we will consider three modes of the research method:

1. a single station with continuous measurements and at least two or three cosmic ray components with different coupling functions for magnetically quiet and disturbed periods;
2. two stations with continuous measurements at each station and at least two cosmic ray components with different coupling functions; and
3. an International Cosmic Ray Service (ICRS), as described in Dorman et al. (1993), that could be organized in the near future based on the already existing world-wide network of cosmic ray observatories (especially important for anisotropic SEP events).

These programs can be used with real-time data from a single observatory (very roughly), with real-time data from two observatories (roughly), with real-time data from several observatories (more exactly), and with an International Cosmic Ray Service (much more exactly). Here we consider how to determine the spectrum of SEP and with a simple model of SEP propagation in the interplanetary space, and how to determine the time of injection, diffusion coefficient, and flux in the source. Using this simple model we can calculate expected fluxes in space at 1/2, 1, 3/2, 2 and more hours after injection. The accuracy of the programs can be checked and developed through comparison with data from the historical large ground SEP events described in detail in numerous publications (see for example: Elliot (1952), Dorman (1957), Carmichael (1962), Dorman and Miroshnichenko (1968), Duggal (1979), Dorman and Venkatesan (1993), Stoker (1995)).

5.2. Main Properties of Online CR Data Used for Determining SEP Spectrum in Space; Available Data for Determining the SEP Energy Spectrum

Let us suppose that we have online one-minute data from a single observatory with at least 3 cosmic ray components with different coupling functions (if the period is magnetically disturbed) or at least 2 components (in a magnetically quiet period). For example, our Emilio Segre' Observatory (ESO) at 2020 m altitude, with a cut-off rigidity of 10.8 GV, currently has the following different components (Dorman et al., 2001): total and multiplicities $m=1$, 2, 3, 4, 5, 6, 7, and ≥ 8. We started to construct a new multidirectional muon telescope that includes 1441 single telescopes in vertical and many inclined directions with different zenith and azimuth angles. From this multidirectional muon telescope we will have more than a hundred components with different coupling functions. Now many cosmic ray stations have data of several components with different coupling functions, where the method described below for determining online energy spectrum can be used. Also available are real-time satellite data, which are very useful for determining the SEP energy spectrum (especially at energies lower than detected by surface cosmic ray detectors).

5.3. Analytical Approximation of Coupling Functions

Based on the latitude survey data of Aleksanyan et al. (1985), Moraal et al., 1989; and Dorman et al. (2000) the polar normalized coupling functions for total counting rate and different multiplicities m can he approximated by the function (Dorman, 1969):

$$W_{om}(R) = a_m k_m R^{-(k_m+1)} \exp\left(-a_m R^{-k_m}\right), \tag{6}$$

where m = tot, 1, 2, 3, 4, 5, 6, 7, ≥ 8. Polar coupling functions for muon telescopes with different zenith angles θ can be approximated by the same type of functions (6) determined only by two parameters $a_m(\theta)$ and $k_m(\theta)$. Let us note that functions (6) are normalized:

$$\int_0^\infty W_{om}(R)dR = 1$$ at any values of a_m and k_m. The normalized coupling functions for point

with cut-off rigidity R_c, will be

$$W_m(R_c, R) = a_m k_m R^{-(k_m+1)}\left(1 - a_m R_c^{-k_m}\right)^{-1} \exp\left(-a_m R^{-k_m}\right), \text{ if } R \geq R_c,$$

$$W_m(R_c, R) = 0, \text{ if } R < R_c \tag{7}$$

5.4. The First Approximation of the SEP Energy Spectrum

In the first approximation the spectrum of primary variation of SEP event can be described by the function

$$\Delta D(R)/D_o(R) = bR^{-\gamma} ,$$
(8)

where $\Delta D(R) = D(R,t) - D_o(R)$, $D_o(R)$ is the differential spectrum of galactic cosmic rays before the SEP event and $D(R,t)$ is the spectrum at a later time t. In Equation 8 parameters b and γ depend on t. Approximation (8) can be used for describing a limited interval of energies in the sensitivity range detected by the various components. Historical SEP data show that, in the broad energy interval, the SEP spectrum has a maximum, and the parameter γ in Equation 8 depends on particle rigidity R (usually γ increases with increasing R) that can be described by the second approximation.

5.5. The Second Approximations of the SEP Energy Spectrum

The second approximation of the SEP spectrum can be determined if it will be possible to use online 4 or 5 components with different coupling functions. In this case the form of the spectrum will be

$$\Delta D(R)/D_o(R) = bR^{-\gamma_o - \gamma_1 \ln(R/R_o)} ,$$
(9)

with 4 unknown parameters $b, \gamma_o, \gamma_1, R_o$. The position of the maximum in the SEP spectrum will be at

$$R_{\max} = R_o \exp(-\gamma_o/\gamma_1) ,$$
(10)

which varies significantly from one event to another and changes very much with time: in the beginning of the SEP event it is great (many GV), but, with time during the development of the event, $R_{\max}$ decreases very much.

5.6. Online Determining of the SEP Spectrum from Data of Single Observatory

5.6.1. The Case of Magnetically Quiet Periods ($\Delta R_c = 0$); Energy Spectrum in the First Approximation

In this case the observed variation $\delta I_m(R_c) \equiv \Delta I_m(R_c)/I_{mo}(R_c)$ in some component m can be described in the first approximation by function $F_m(R_c, \gamma)$:

$$\delta I_m(R_c) = b F_m(R_c, \gamma) \tag{11}$$

where m = tot, 1, 2, 3, 4, 5, 6, 7, ≥ 8 for neutron monitor data (but can denote also the data obtained by muon telescopes at different zenith angles and data from satellites), and

$$F_m(R_c, \gamma) = a_m k_m \left(1 - \exp\left(-a_m R_c^{-k_m}\right)\right)^{-1} \int\limits_{R_c}^{\infty} R^{-(k_m+1+\gamma)} \exp\left(-a_m R^{-k_m}\right) dR \tag{12}$$

is a known function. Let us compare data for two components m and n. According to Equation 11 we obtain

$$\delta I_m(R_c)/\delta I_n(R_c) = \Psi_{mn}(R_c, \gamma), \text{ where } \Psi_{mn}(R_c, \gamma) = F_m(R_c, \gamma)/F_n(R_c, \gamma) \tag{13}$$

as calculated using Equation 12. Comparison of experimental results with function $\Psi_{mn}(R_c, \gamma)$ according to Equation 13 gives the value of γ, and then from Equation 11 the value of the parameter b. The observed SEP increase for different components allows the determination of parameters b and γ for the SEP event beyond the Earth's magnetosphere.

5.6.2. The Case of Magnetically Quiet Periods ($\Delta R_c = 0$); Energy Spectrum in the Second Approximation

Let us suppose that available online are at least 4 components $i = k, l, m, n$ with different coupling functions. Then for the second approximation taking into account Equation 9, we obtain

$$\delta I_i(R_c) = b \Phi_i(R_c, \gamma_o, \gamma_1, R_o), \tag{14}$$

where $i = k, l, m, n$ and

$$\Phi_i(R_c, \gamma_o, \gamma_1, R_o) = a_i k_i \left(1 - \exp\left(-a_i R_c^{-k_i}\right)\right)^{-1} \int\limits_{R_c}^{\infty} R^{-(k_i+1+\gamma_o+\gamma_1 \ln(R/R_o))} \exp\left(-a_i R^{-k_i}\right) dR \tag{15}$$

By comparison data of different components we obtain

$$\frac{\delta I_k(R_c)}{\delta I_l(R_c)} = Y_{kl}(R_c, \gamma_o, \gamma_1, R_o),$$

$$\frac{\delta I_l(R_c)}{\delta I_m(R_c)} = Y_{lm}(R_c, \gamma_o, \gamma_1, R_o), \quad \frac{\delta I_m(R_c)}{\delta I_n(R_c)} = Y_{mn}(R_c, \gamma_o, \gamma_1, R_o), \tag{16}$$

where $(i, j = k, l, m, n)$

$$Y_{ij}(R_c, \gamma_o, \gamma_1, R_o) = \Phi_i(R_c, \gamma_o, \gamma_1, R_o)/\Phi_j(R_c, \gamma_o, \gamma_1, R_o) \tag{17}$$

is calculated using Equation 15. The solving of the system of Equation 16 gives the values of γ_0, γ_1, R_0, and then parameter b can be determined by Equation 14:

$b = \delta I_i(R_c)/\Phi_i(R_c, \gamma_0, \gamma_1, R_0)$ for any $i = k, l, m, n$.

5.6.3. The Case of Magnetically Disturbed Periods ($\Delta R_c \neq 0$); Energy Spectrum in the First Approximation

For magnetically disturbed periods the observed cosmic ray variation, instead of Equation 11, will be described by

$$\delta I_k(R_c) = -\Delta R_c W_k(R_c, R_c) + b F_k(R_c, \gamma), \tag{18}$$

where ΔR_c is the change of cut-off rigidity due to change of the Earth's magnetic field, and $W_k(R_c, R_c)$ is determined by Equation 7 at $R = R_c$. Now for the first approximation of the SEP energy spectrum we have unknown variables γ, b, ΔR_c, and for their determination we need data from at least 3 different components $k=l, m, n$ in Equation 18. In accordance with the spectrographic method (Dorman, 1975) let us introduce the function

$$\Psi_{lmn}(R_c, \gamma) = \frac{W_l F_m(R_c, \gamma) - W_m F_l(R_c, \gamma)}{W_m F_n(R_c, \gamma) - W_n F_m(R_c, \gamma)}, \tag{19}$$

where

$$W_l \equiv W_l(R_c, R_c), \quad W_m \equiv W_m(R_c, R_c), \quad W_n \equiv W_n(R_c, R_c). \tag{20}$$

Then from

$$\Psi_{lmn}(R_c, \gamma) = \frac{W_l \delta I_m(R_c) - W_m \delta I_l(R_c)}{W_m \delta I_n(R_c) - W_n \delta I_m(R_c)}, \tag{21}$$

the value of γ can be determined. Using this value of γ, for each time t, we determine

$$\Delta R_c = \frac{F_l(R_c, \gamma)\delta I_m(R_c) - F_m(R_c, \gamma)\delta I_l(R_c)}{F_m(R_c, \gamma)\delta I_n(R_c) - F_n(R_c, \gamma)\delta I_m(R_c)}, \quad b = \frac{W_l \delta I_m(R_c) - W_m \delta I_l(R_c)}{W_l F_m(R_c, \gamma) - W_m F_l(R_c, \gamma)}, \tag{22}$$

In magnetically disturbed periods, the observed SEP increase for different components again allows the determination of parameters γ and b, for the SEP event beyond the Earth's magnetosphere, and ΔR_c, giving information on the magnetospheric ring currents.

5.6.4. The Case of Magnetically Disturbed Periods ($\Delta R_c \neq 0$); Energy Spectrum in the Second Approximation

In this case, for magnetically disturbed periods we need at least 5 different components, and the observed cosmic ray variation, instead of Equation 18, will be described by

$$\delta I_i(R_c) = -\Delta R_c W_i(R_c, R_c) + b\Phi_i(R_c, \gamma_o, \gamma_1, R_o), \tag{23}$$

where $i=j,\ k,\ l,\ m,\ n$. Here $W_i(R_c, R_c)$ are determined by Equation 7 at $R = R_c$ and $\Phi_i(R_c, \gamma_o, \gamma_1, R_o)$ are determined by Equation 15. By excluding from the system of Equation 23 linear unknown variables ΔR_c and b we obtain three equations for determining three unknown parameters γ_o, γ_1, R_o of SEP energy spectrum of the type

$$\frac{W_l \delta I_m(R_c) - W_m \delta I_l(R_c)}{W_m \delta I_n(R_c) - W_n \delta I_m(R_c)} = Y_{lmn}(R_c, \gamma_o, \gamma_1, R_o), \tag{24}$$

where

$$Y_{lmn}(R_c, \gamma_o, \gamma_1, R_o) = \frac{W_l \Phi_m(R_c, \gamma_o, \gamma_1, R_o) - W_m \Phi_l(R_c, \gamma_o, \gamma_1, R_o)}{W_m \Phi_n(R_c, \gamma_o, \gamma_1, R_o) - W_n \Phi_m(R_c, \gamma_o, \gamma_1, R_o)}. \tag{25}$$

In Equation 24 the left side is known from experimental data for each moment of time t, and the right side are known functions from γ_o, γ_1, R_o, what is calculated by taking into account Equation 15 and Equation 7 (at $R = R_c$). From the system of three equations of the type of Equation 24 we determine online γ_o, γ_1, R_o, and then parameters b and ΔR_c:

$$b = \frac{W_l \delta I_m(R_c) - W_m \delta I_l(R_c)}{W_l \Phi_m(R_c, \gamma_o, \gamma_1, R_o) - W_m \Phi_l(R_c, \gamma_o, \gamma_1, R_o)}, \tag{26}$$

$$\Delta R_c = \frac{\Phi_l(R_c, \gamma_o, \gamma_1, R_o)\delta I_m(R_c) - \Phi_m(R_c, \gamma_o, \gamma_1, R_o)\delta I_l(R_c)}{\Phi_m(R_c, \gamma_o, \gamma_1, R_o)\delta I_n(R_c) - \Phi_n(R_c, \gamma_o, \gamma_1, R_o)\delta I_m(R_c)}. \tag{27}$$

5.7. Using Real-time Cosmic Ray Data from Two Observatories

For determining the online SEP energy spectrum in the first and in the second approximations (Equations 8 and 9, correspondingly) it is necessary to use data from pairs of observatories in the same impact zone (to exclude the influence of the anisotropy distribution of the SEP ground increase), but with different cut-off rigidities R_{c1} and R_{c2}.

5.7.1. Magnetically Quiet Periods ($\Delta R_c = 0$); The SEP Energy Spectrum in the Interval $R_{c1} \div R_{c2}$

If it is the same component m for both observatories (e.g., total neutron component on about the same level of average air pressure h_o), the energy spectrum in the interval $R_{c1} \div R_{c2}$ can be determined directly for any moment of time t (here $W_{om}(R)$ is determined by Equation 6):

$$\frac{\Delta D}{D_o}(R_{c1} \div R_{c2}) = \left(\delta I_m(R_{c1})\left(1 - a_m R_{c1}^{-km}\right) - \delta I_m(R_{c2})\left(1 - a_m R_{c2}^{-km}\right) \right) \Big/ \int_{R_{c1}}^{R_{c2}} W_{om}(R) dR , \qquad (28)$$

5.7.2. Magnetically Quiet Periods ($\Delta R_c = 0$); the SEP Energy Spectrum in the First Approximation

In this case for determining b and γ in Equation 8 we need on line data at least of one component on each Observatory. If it is the same component m on both Observatories, parameter γ will be found from equation

$$\delta I_m(R_{c1}) \big/ \delta I_m(R_{c2}) = \Psi_{mm}(R_{c1}, R_{c2}, \gamma),$$

where

$$\Psi_{mm}(R_{c1}, R_{c2}, \gamma) = F_m(R_{c1}, \gamma) \big/ F_m(R_{c2}, \gamma), \qquad (29)$$

and then can be determined

$$b = \delta I_m(R_{c1}) \big/ F_m(R_{c1}, \gamma) = \delta I_m(R_{c2}) \big/ F_m(R_{c2}, \gamma). \qquad (30)$$

If there are different components m and n, the solution will be determined by equations:

$$\delta I_m(R_{c1}) \big/ \delta I_n(R_{c2}) = \Psi_{mn}(R_{c1}, R_{c2}, \gamma),$$

where

$$\Psi_{mn}(R_{c1}, R_{c2}, \gamma) = F_m(R_{c1}, \gamma) \big/ F_n(R_{c2}, \gamma), \qquad (31)$$

$$b = \delta I_m(R_{c1}) \big/ F_m(R_{c1}, \gamma) = \delta I_n(R_{c2}) \big/ F_n(R_{c2}, \gamma). \qquad (32)$$

5.7.3. Magnetically Quiet Periods ($\Delta R_c = 0$); the SEP Energy Spectrum in the Second Approximation

In this case we need at least 4 components: there can be 1 and 3 or 2 and 2 different components in both observatories with cut-off rigidities R_{c1} and R_{c2}. If there are 1 and 3 components, parameters γ_o, γ_1, R_o will be determined by the solution of the system of equations

$$\frac{\delta I_k(R_{c1})}{\delta I_l(R_{c2})} = \frac{\Phi_k(R_{c1},\gamma_o,\gamma_1,R_o)}{\Phi_l(R_{c2},\gamma_o,\gamma_1,R_o)}, \quad \frac{\delta I_l(R_{c2})}{\delta I_m(R_{c2})} = Y_{lm}(R_{c2},\gamma_o,\gamma_1,R_o),$$

$$\frac{\delta I_m(R_{c2})}{\delta I_n(R_{c2})} = Y_{mn}(R_{c2},\gamma_o,\gamma_1,R_o), \tag{33}$$

and then we determine

$$b = \frac{\delta I_k(R_{c1})}{\Phi_k(R_{c1},\gamma_o,\gamma_1,R_o)} = \frac{\delta I_l(R_{c2})}{\Phi_l(R_{c2},\gamma_o,\gamma_1,R_o)} = \frac{\delta I_m(R_{c2})}{\Phi_m(R_{c2},\gamma_o,\gamma_1,R_o)} = \frac{\delta I_n(R_{c2})}{\Phi_n(R_{c2},\gamma_o,\gamma_1,R_o)}. \tag{34}$$

If there are two and two components in both observatories with cut-off rigidities R_{c1} and R_{c2}, the system of equations for determining parameters γ_o, γ_1, R_o will be

$$\frac{\delta I_k(R_{c1})}{\delta I_l(R_{c1})} = Y_{kl}(R_{c1},\gamma_o,\gamma_1,R_o), \quad \frac{\delta I_l(R_{c1})}{\delta I_m(R_{c2})} = \frac{\Phi_l(R_{c1},\gamma_o,\gamma_1,R_o)}{\Phi_m(R_{c2},\gamma_o,\gamma_1,R_o)},$$

$$\frac{\delta I_m(R_{c2})}{\delta I_n(R_{c2})} = Y_{mn}(R_{c2},\gamma_o,\gamma_1,R_o) \tag{35}$$

and then we determine

$$b = \frac{\delta I_k(R_{c1})}{\Phi_k(R_{c1},\gamma_o,\gamma_1,R_o)} = \frac{\delta I_l(R_{c1})}{\Phi_l(R_{c1},\gamma_o,\gamma_1,R_o)} = \frac{\delta I_m(R_{c2})}{\Phi_m(R_{c2},\gamma_o,\gamma_1,R_o)} = \frac{\delta I_n(R_{c2})}{\Phi_n(R_{c2},\gamma_o,\gamma_1,R_o)}. \tag{36}$$

5.7.4. Magnetically Disturbed Periods ($\Delta R_c \neq 0$); The SEP Energy Spectrum in the First Approximation

In this case we have 4 unknown variables: $\Delta R_{c1}, \Delta R_{c2}, b, \gamma$. If there are 1 and 3 components in both observatories with cut-off rigidities R_{c1} and R_{c2}, the system of equations for determining $\Delta R_{c1}, \Delta R_{c2}, b, \gamma$ is

$$\delta I_k(R_{c1}) = -\Delta R_{c1} W_k(R_{c1},R_{c1}) + b F_k(R_{c1},\gamma), \tag{37}$$

$$\delta I_l(R_{c2}) = -\Delta R_{c2} W_l(R_{c2},R_{c2}) + b F_l(R_{c2},\gamma), \tag{38}$$

$$\delta I_m(R_{c2}) = -\Delta R_{c2} W_m(R_{c2},R_{c2}) + b F_m(R_{c2},\gamma), \tag{39}$$

$$\delta I_n(R_{c2}) = -\Delta R_{c2} W_n(R_{c2}, R_{c2}) + b F_n(R_{c2}, \gamma) . \tag{40}$$

In this case we determine, from Equations 38-40, γ, b, and ΔR_{c2} as it was described above by Equations 19-22. From Equation 37 we then determine

$$\Delta R_{c1} = \left(b F_k(R_{c1}, \gamma) - \delta I_k(R_{c1}) \right) / W_k(R_{c1}, R_{c1}). \tag{41}$$

If there are two and two components in both observatories, we will have the same system as Equations 37-40; however, Equation 38 is replaced by

$$\delta I_l(R_{c1}) = -\Delta R_{c1} W_l(R_{c1}, R_{c1}) + b F_l(R_{c1}, \gamma) \tag{42}$$

From the system of Equations 37, 39, 40, and 42, we exclude linear unknown variables b, ΔR_{c1}, ΔR_{c2} and finally obtain a equation for determining γ:

$$\frac{W_k \delta I_l(R_{c1}) - W_l \delta I_k(R_{c1})}{W_m \delta I_n(R_{c2}) - W_n \delta I_m(R_{c2})} = \Psi_{klmn}(R_{c1}, R_{c2}, \gamma) , \tag{43}$$

where

$$\Psi_{klmn}(R_{c1}, R_{c2}, \gamma) = \frac{W_k(R_{c1}, R_{c1}) F_l(R_{c1}, \gamma) - W_l(R_{c1}, R_{c1}) F_k(R_{c1}, \gamma)}{W_m(R_{c2}, R_{c2}) F_n(R_{c2}, \gamma) - W_n(R_{c2}, R_{c2}) F_m(R_{c2}, \gamma)} \tag{44}$$

can be calculated for any pair of stations using known functions $F_k(R_{c1}, \gamma)$, $F_l(R_{c1}, \gamma)$, $F_m(R_{c2}, \gamma)$, $F_n(R_{c2}, \gamma)$ (calculated from Equation 12), and known values $W_k(R_{c1}, R_{c1})$, $W_l(R_{c1}, R_{c1})$, $W_m(R_{c2}, R_{c2})$, and $W_n(R_{c2}, R_{c2})$ (calculated from Equation 7). After determining γ we can determine the other 3 unknown variables:

$$\Delta R_{c1} = \frac{F_k(R_{c1}, \gamma) \delta I_l(R_{c1}) - F_l(R_{c1}, \gamma) \delta I_k(R_{c1})}{W_k F_l(R_{c1}, \gamma) - W_l F_k(R_{c1}, \gamma)} , \tag{45}$$

$$\Delta R_{c2} = \frac{F_m(R_{c2}, \gamma) \delta I_n(R_{c2}) - F_n(R_{c2}, \gamma) \delta I_m(R_{c2})}{W_m F_n(R_{c2}, \gamma) - W_n F_m(R_{c2}, \gamma)} , \tag{46}$$

$$b = \frac{W_k \delta I_l(R_{c1}) - W_l \delta I_k(R_{c1})}{W_k F_l(R_{c1}, \gamma) - W_l F_k(R_{c1}, \gamma)} = \frac{W_m \delta I_n(R_{c2}) - W_n \delta I_m(R_{c2})}{W_m F_n(R_{c2}, \gamma) - W_n F_m(R_{c2}, \gamma)} \tag{47}$$

5.7.5. Magnetically Disturbed Periods ($\Delta R_c \neq 0$); Energy Spectrum in the Second Approximation

In this case we have 6 unknown variables $\Delta R_{c1}, \Delta R_{c2}, b, \gamma_o, \gamma_1, R_o$, thus from both observatories we need a total of at least 6 components in any combination. This problem also can be solved by the method described above, only instead of functions F and Ψ the functions Φ and Y will be used. Here we have not enough space to describe these results; it will be done in another paper.

5.8. Using Real-time Cosmic Ray Data from Many Observatories (ICRS)

Using the International Cosmic Ray Service (ICRS) proposed by Dorman et al. (1993) and the above technique, much more accurate information on the distribution of the increased cosmic ray flux near the Earth can be found. We hope that for large SEP events it will be possible to use the global-spectrographic method (reviewed by Dorman, M1974) to determine, in real-time, the temporal changes in the anisotropy and its dependence on particle rigidity. This will allow a better determination of SEP propagation parameters in interplanetary space, and of the total flux and energy spectrum of particles accelerated in the solar flare, in turn improving detailed forecasts of dangerous large SEP events.

5.9. Determining of Coupling Functions by Latitude Survey Data

The coefficients a_m and k_m for the coupling functions in Equations 6 and 7 were determined from a latitude survey by Aleksanyan et al. (1985) and they are in good agreement with theoretical calculations of Dorman and Yanke (1981), and Dorman et al. (1981). Improved coefficients were determined on the basis of the recent Italian expedition to Antarctica (Dorman et al., 2000). The dependence of a_m and k_m on the average station pressure h (in bar) and solar activity level is characterized by the logarithm of CR intensity (we used here monthly averaged intensities from the Climax, USA neutron monitor as $\ln(I_{Cl})$; however, the monthly averages of the Rome NM or monthly averages of the ESO NM (with some recalculation coefficients) can be approximated by the functions:

$$a_{tot} = \left(-2.9150h^2 - 2.2368h - 8.6538\right)\ln(I_{Cl}) + \left(24.5842h^2 + 19.4600h + 81.2299\right), \quad (48)$$

$$k_{tot} = \left(0.1798h^2 - 0.8487h + 0.7496\right)\ln(I_{Cl}) + \left(-1.4402h^2 + 6.4034h - 3.6975\right), \quad (49)$$

$$a_m = \left[\left(-2.915h^2 - 2.237h - 8.638\right)\ln(I_{Cl}) + \left(24.584h^2 + 19.46h + 81.23\right)\right] \\ \times \left(0.987m^2 + 0.2246m + 6.913\right)/9.781 \quad (50)$$

$$k_m = \left[\left(0.1798h^2 - 0.8487h + 0.7496\right)\ln(I_{Cl}) + \left(-1.4402h^2 + 6.4034h - 3.6975\right)\right] \\ \times (0.0808m + 1.8186)/1.9399. \quad (51)$$

Instead of Climax, other stations can be used in Equations 48-51 with the appropriate coefficients.

5.10. Calculation of Integrals $F_m(R_c,\gamma)$ and $\Phi_m(R_c,\gamma_0,\gamma_1,R_o)$

The integrals $F_m(R_c,\gamma)$ of Equation 12 are calculated for values of γ from -1 to +7, for different average station air pressure, h, cut-off rigidities, R_c, and different levels of solar activity characterized by the value $\ln(I_{Cl})$ according to

$$F_m(R_c,\gamma) = a_m k_m \left(1 - \exp\left(-a_m R_c^{-k_m}\right)\right)^{-1} R_c^{-k_m - \gamma} \sum_{n=0}^{\infty} (-1)^n a_m^n R_c^{-nk_m} \left(n!((n+1)k_m + \gamma)\right)^{-1}$$

(52)

Results of calculations of Equation 52 at different values of γ from -1 to +7 show that integrals $F_m(R_c,\gamma)$ can be approximated, with correlation coefficients between 0.993 and 0.996, by the function

$$F_m(R_c,\gamma,h,\ln(I_{Cl})) = A_m(R_c,h,\ln(I_{Cl}))\exp(-\gamma B_m(R_c,h,\ln(I_{Cl}))).$$

(53)

Results of calculations of functions $\Phi_m(R_c,\gamma_0,\gamma_1,R_o)$ (for the second approximation of the SEP energy spectrum) will be presented in another paper.

5.11. Calculation of Functions $\Psi_{mn}(R_c,\gamma)$, $\Psi_{lmn}(R_c,\gamma)$ and $\Psi_{klmn}(R_{c1},R_{c2},\gamma)$

Functions $\Psi_{mn}(R_c,\gamma)$ determined by Equation 13 now by using Equation 53 becomes

$$\Psi_{mn}(R_c,\gamma) = F_m(R_c,\gamma)/F_n(R_c,\gamma) = (A_m/A_n)\exp(-\gamma(B_m - B_n)),$$

(54)

Equation 19 becomes

$$\Psi_{lmn}(R_c,\gamma) = \frac{W_l(R_c,R_c)A_m \exp(-B_m\gamma) - W_m(R_c,R_c)A_l \exp(-B_l\gamma)}{W_m(R_c,R_c)A_n \exp(-B_n\gamma) - W_n(R_c,R_c)A_m \exp(-B_m\gamma)},$$

(55)

and Equation 44 becomes

$$\Psi_{klmn}(R_{c1},R_{c2},\gamma) = \frac{W_k(R_{c1},R_{c1})A_l \exp(-B_l\gamma) - W_l(R_{c1},R_{c1})A_k \exp(-B_k\gamma)}{W_m(R_{c2},R_{c2})A_n \exp(-B_n\gamma) - W_n(R_{c2},R_{c2})A_m \exp(-B_m\gamma)}$$

(56)

5.12. Examples of Determination of the SEP Energy Spectrum

For example let us compare Equation 54 and Equation 13: the spectral index is

$$\gamma = \left(\ln\left(A_m / A_n\right) - \ln\left(\delta I_m\left(R_c\right) / \delta I_n\left(R_c\right)\right)\right) / \left(B_m - B_n\right)$$

(57)

Then from Equation 11 we determine

$$b = \delta I_m\left(R_c\right)\left[A_m \exp\left(-B_m\left(\left(\ln\left(A_m / A_n\right) - \ln\left(\delta I_m\left(R_c\right) / \delta I_n\left(R_c\right)\right)\right) / \left(B_m - B_n\right)\right)\right)\right]^{-1}.$$

(58)

So for magnetically quite times the inverse problem can be solved online for each one-minute data interval during the rising phase. For magnetically disturbed times the inverse problem can also be solved automatically in real-time, using the special function described in Equation 55 (for observations by at least three components at one observatory), or by using the function described in Equation 56 (for observations by two observatories). Note that for the first approximation we have assumed a two-parameter form of the energy spectrum. In reality the SEP energy spectrum could be more complicated; γ may also change with energy.

If the spectrum can only be described by three or four-parameters (considered in the second approximation given previously) then observations from a range of neutron monitors (including those with low cut-off rigidities) and from satellites will be needed. We conclude that in these cases solutions can also be obtained for the energy spectrum, its change with time and the change of cut-off rigidities. The details of these solutions will be reported in another paper.

5.13. Special Program for Online Determination of Energy Spectrum for Each Minute after SEP Start (for the Case of Single Observatory)

Let us consider data from a neutron monitor with independent registration of total flux and 7 multiplicities as input flux of data as vector $I_m(i)$, where i –is number of time registration index, m- number of channel ($m = 0$ –total flux, $m = 1 \div 7$ –different multiplicities). For estimation of the statistical properties of input flux we calculate mean and standard deviation for each channel $\langle I_m(i)\rangle$ and $\sigma_m(i)$ on the 1-hour interval, preceded to i-time moment on 1-hour. During the next step we calculate vector of relative deviations:

$$\delta I_m(i) = \left(I_m(i) - \langle I_m(i)\rangle\right) / \langle I_m(i)\rangle$$

(59)

and corresponding standard deviation $\sigma\left(\delta I_m(i)\right)$. On the basis of deviation's vector $\delta I_m(i)$ we calculate matrix

$$R_{mn}(i) = \delta I_m(i) / \delta I_n(i)$$

(60)

of ratios deviations for different energetic channels m, n at the time moment i and its standard deviation $\sigma(R_{mn}(i))$. On the basis of Equation 57 and Equation 58, by using Equation 60 we calculate matrix of the spectral slope $\gamma_{mn}(i)$, matrix of amplitude $b_{mn}(i)$ and estimations of correspondent standard deviations $\sigma(\gamma_{mn}(i))$ and $\sigma(b_{mn}(i))$.

For conversation of matrix output of estimated slope $\gamma_{mn}(i)$ to the scalar's form $\gamma(i)$ we use estimation of the average values taking into account the correspondent weights W of the individual values $\gamma_{mn}(i)$ determined by the correspondent standard deviations

$$W(\gamma_{mn}(i)) = (\sigma(\gamma_{mn}(i)))^{-2}\left(\sum_{m,n}(\sigma(\gamma_{mn}(i)))^{-2}\right)^{-1}, \tag{61}$$

that

$$\gamma(i) = \sum_{m,n}\gamma_{mn}(i) \times W(\gamma_{mn}(i)). \tag{62}$$

The same we made for parameter $b(i)$:

$$b(i) = \sum_{m,n} b_{mn}(i) \times W(b_{mn}(i)), \tag{63}$$

where

$$W(b_{mn}(i)) = (\sigma(b_{mn}(i)))^{-2}\left(\sum_{m,n}(\sigma(b_{mn}(i)))^{-2}\right)^{-1} \tag{64}$$

By founding values of $\gamma(i)$ and $b(i)$ can be determined SEP energy spectrum by using Equation 8 for each moment of time:

$$\Delta D(R,i) = b(i)R^{-\gamma(i)} \times D_o(R) \tag{65}$$

We checked this procedure on the basis of neutron monitor data on Mt. Gran-Sasso in Italy for the event September 29, 1989. We used one minute data of total intensity and multiplicities from 1 to 7. The number of minutes is start from 10.00 UT at September 29, 1989. Upper limit of indexes m, n we choose about 3÷4, since for higher multiplicities accuracy level decreases fast and we may decrease confidence of the estimation.

6. ONLINE DETERMINATION OF DIFFUSION COEFFICIENT IN INTERPLANETARY SPACE, TIME OF EJECTION AND ENERGY SPECTRUM OF SEP AT THE SOURCE BY NM DATA (THE CASE WHEN DIFFUSION COEFFICIENT DEPENDS ONLY ON PARTICLE RIGIDITY)

6.1. Online Determination of SEP Spectrum at the Source When Diffusion Coefficient in the Interplanetary Space and Time of Ejection are Known

According to observation data of many events for about 60 years the time change of SEP flux and energy spectrum can be described in the first approximation by the solution of isotropic diffusion from the pointing instantaneous source described by function

$$Q(R,r',t') = N_o(R)\delta(r')\delta(t'). \tag{66}$$

Let us suppose that the time of ejection and diffusion coefficient are known. In this case the expected SEP rigidity spectrum on the distance r from the Sun in the time t after ejection will be

$$N(R,r,t) = N_o(R) \times \left[2\pi^{1/2} \left(K(R)t \right)^{3/2} \right]^{-1} \times \exp\left(-\frac{r^2}{4K(R)t} \right), \tag{67}$$

where $N_o(R)$ is the rigidity spectrum of total number of SEP in the source, t is the time relative to the time of ejection and $K(R)$ is the known diffusion coefficient in the interplanetary space in the period of SEP event. At $r = r_1 = 1$ AU and at some moment t_1 the spectrum determined in Section 4, will be described by the function

$$N(R,r_1,t_1) = b(t_1)R^{-\gamma(t_1)}D_o(R), \tag{68}$$

where $b(t_1)$ and $\gamma(t_1)$ are parameters determined the observed rigidity spectrum in the moment t_1, and $D_o(R)$ is the spectrum of galactic cosmic rays before event (see Section 5). From the other side, the SEP spectrum will be determined at $r = r_1$, $t = t_1$ according to Equation 66. Then we obtain equation

$$b(t_1)R^{-\gamma(t_1)}D_o(R) = N_o(R)\left[2\pi^{1/2}(K(R)t_1)^{3/2} \right]^{-1} \times \exp(-r_1^2/(4K(R)t_1)) \tag{69}$$

If the diffusion coefficient $K(R)$ and time of ejection (i.e., time t_1 relative to the time of ejection) are known, then from Equation 69 we obtain

$$N_O(R) = 2\pi^{1/2} b(t_1) R^{-\gamma(t_1)} D_o(R) \times \left(K(R)t_1\right)^{3/2} \exp\left(r_1^2 / (4K(R)t_1)\right). \tag{70}$$

6.2. Online Determination of Diffusion Coefficient in the Interplanetary Space When SEP Spectrum in Source and Time of Ejection Are Known

Let us consider the case when SEP spectrum in source and time of ejection may be known (e.g. from direct solar gamma-ray and solar neutron measurements what gave information on the time of acceleration and spectrum of accelerated particles, see detail review in Dorman, M2010). In this case, from Equation 69 we obtain the following iteration solution

$$K(R) = \frac{r_1^2}{4t_1}\left(\ln(N_O(R)) - \ln\left(b(t_1)R^{-\gamma(t_1)}D_O(R)\right) - \ln\left(2\pi^{1/2}\right) - \frac{3}{2}\ln\left(K(R)t_1\right)\right)^{-1}. \tag{71}$$

As the first approximation in the right hand of Equation 71 can be used $K(R)$ for galactic cosmic rays obtained from investigation of hysteresis phenomenon (Dorman, 2001, Dorman et al., 2001a, b). The iteration process in Equation 71 is very fast and only few approximations are necessary.

6.3. Online Determination of the Time of Ejection When Diffusion Coefficient in the Interplanetary Space and SEP Spectrum in Source Are Known

In this case we obtain from Equation 69

$$t_1 = \frac{r_1^2}{4K(R)}\left(\ln(N_O(R)) - \ln\left(b(t_1)R^{-\gamma(t_1)}D_O(R)\right) - \ln\left(2\pi^{1/2}\right) - \frac{3}{2}\ln\left(K(R)t_1\right)\right)^{-1}. \tag{72}$$

As the first approximation in the right hand of Equation 72 can be used t_1 obtained from measurements of the start of increase in very high energy region where the time of propagation from the Sun is about the same as for light. The iteration process in Equation 72 is very fast and only few approximations is necessary.

6.4. Online Determination Simultaneously of SEP Diffusion Coefficient and SEP Spectrum in Source, if the Time of Ejection Is Known

Let us suppose that $K(R)$ and $N_O(R)$ are unknown functions, but the time of ejection is known. In this case we need to use data at least for two moments of time t_1 and t_2 relative to the time of ejection (what is supposed as known value). In this case we will have system from two equations:

$$b(t_1)R^{-\gamma(t_1)}D_O(R) = N_O(R) \times \left[2\pi^{1/2}(K(R)t_1)^{3/2}\right]^{-1} \times \exp\left(-r_1^2/(4K(R)t_1)\right), \tag{73}$$

$$b(t_2)R^{-\gamma(t_2)}D_O(R) = N_O(R) \times \left[2\pi^{1/2}(K(R)t_2)^{3/2}\right]^{-1} \times \exp\left(-r_1^2/(4K(R)t_2)\right), \tag{74}$$

By dividing Equation 73 on Equation 74, we obtain

$$\frac{b(t_1)}{b(t_2)}(t_1/t_2)^{3/2} R^{-[\gamma(t_1)-\gamma(t_2)]} = \exp\left(-\frac{r_1^2}{4K(R)}\left(\frac{1}{t_1}-\frac{1}{t_2}\right)\right), \tag{75}$$

from which follows

$$K(R) = \left(-\frac{r_1^2}{4}\left(\frac{1}{t_1}-\frac{1}{t_2}\right)\right) \times \left(\ln\left(\frac{b(t_1)}{b(t_2)}(t_1/t_2)^{3/2} R^{-[\gamma(t_1)-\gamma(t_2)]}\right)\right)^{-1}. \tag{76}$$

The found result for $K(R)$ will be controlled and made more exactly on the basis of Equation 76 by using the next data in moments t_2 and t_3, as well as for moments t_1 and t_3, then by data in moments t_3, and t_4, and so on. By introducing result of determining of diffusion coefficient $K(R)$ on the basis of Equation 76 in Equation 69 we determine immediately the expected flux and spectrum SEP in the source:

$$\begin{aligned} N_O(R) &= 2\pi^{1/2}b(t_1)R^{-\gamma(t_1)}D_O(R) \times (K(R)t_1)^{3/2}\exp\left(r_1^2/(4K(R)t_1)\right) \\ &= 2\pi^{1/2}b(t_2)R^{-\gamma(t_2)}D_O(R) \times (K(R)t_2)^{3/2}\exp\left(r_1^2/(4K(R)t_2)\right) \end{aligned} \tag{77}$$

where $K(R)$ is determined by Equation 76.

6.5. Online Determination Simultaneously of the Time of Ejection and SEP Spectrum at the Source if the Diffusion Coefficient Is Known

Let us suppose that the diffusion coefficient $K(R)$ is known, but the time of ejection T_e is unknown. In this case we need measurements of SEP energy spectrum at least in two moments of time. Let us suppose that in two moments T_1 and T_2 were made measurements of energetic spectrum as it was described in Section 4. Here times T_e, T_1 and T_2 are in UT scale. Let us suppose that

$$t_1 = T_1 - T_e = x, \quad t_2 = T_2 - T_e = T_2 - T_1 + x, \tag{78}$$

where t_1 and t_2 are times relative to the moment of SEP ejection into solar wind, and T_1 and T_2 are known UT and $T_e = T_1 - x$ is unknown value what can be determined from the following system of equations:

$$b(T_1)R^{-\gamma(T_1)}D_O(R) = N_O(R) \times \left[2\pi^{1/2}(K(R)x)^{3/2} \right]^{-1} \times \exp\left(-r_1^2 / (4K(R)x)\right), \tag{79}$$

$$b(T_2)R^{-\gamma(T_2)}D_O(R) = N_O(R) \times \left[2\pi^{1/2}(K(R)(T_2 - T_1 + x))^{3/2} \right]^{-1} \times \exp\left(-r_1^2 / (4K(R)(T_2 - T_1 + x))\right) \tag{80}$$

By dividing of Equation 79 on Equation 80, we obtain

$$\frac{b(T_1)}{b(T_2)}(x/(T_2 - T_1 + x))^{3/2} R^{-[\gamma(T_1)-\gamma(T_2)]} = \exp\left(-\frac{r_1^2}{4K(R)}\left(\frac{1}{x} - \frac{1}{T_2 - T_1 + x} \right) \right), \tag{81}$$

From Equation 81 unknown value x can be found by iteration:

$$t_1 = x = \Omega^{2/3}(T_2 - T_1)(1 - \Omega^{2/3})^{-1}, \tag{82}$$

where

$$\Omega(x) = \frac{b(T_2)}{b(T_1)} R^{[\gamma(T_1)-\gamma(T_2)]} \times \exp\left(-\frac{r_1^2}{4K(R)}\left(\frac{1}{x} - \frac{1}{T_2 - T_1 + x} \right) \right). \tag{83}$$

As the first approximation we can use $x_1 = T_1 - T_e \approx 500$ sec what is a minimum time of relativistic particles propagation from the Sun to the Earth's orbit. Then by Equation 83 we determine $\Omega(x_1)$ and by Equation 82 determine the second approximation x_2. To put x_2 in Equation 83 we compute $\Omega(x_2)$, and then by Equation 82 determine the third approximation x_3, and so on. After determining $x = t_1$ and UT time of ejection $T_e = T_1 - x$ we can determine the SEP spectrum in the source:

$$N_O(R) = 2\pi^{1/2}\big(K(R)(T_1 - T_e)\big)^{3/2} b(T_1)R^{-\gamma(T_1)} D_O(R)\exp\big(r_1^2 /(4K(R)(T_1 - T_e))\big) =$$

$$= 2\pi^{1/2}\big(K(R)(T_2 - T_e)\big)^{3/2} b(T_2)R^{-\gamma(T_2)} D_O(R)\exp\big(r_1^2 /(4K(R)(T_2 - T_e))\big) \tag{84}$$

6.6. Online Determination Simultaneously Time of Ejection, Diffusion Coefficient and SEP Spectrum at the Source

Let us suppose that time of ejection T_e, diffusion coefficient $K(R)$ and SEP spectrum in the source $N_O(R)$ are unknown. In this case for determining online simultaneously time of ejection T_e, diffusion coefficient $K(R)$ and SEP spectrum in the source $N_O(R)$ we need information on SEP spectrum at least in three moments of time T_1, T_2 and T_3 (all times T are in UT scale). In this case instead of Equation 6.13 we will have for times after SEP ejection into the solar wind:

$$t_1 = T_1 - T_e = x, \; t_2 = T_2 - T_1 + x, \; t_3 = T_3 - T_1 + x \tag{85}$$

where T_2-T_1 and T_3-T_1 are known values, and x is unknown value what we need to determine by solution of the system of equations

$$b(T_1)R^{-\gamma(T_1)} D_O(R) = N_O(R) \times \Big[2\pi^{1/2}\big(K(R)x\big)^{3/2}\Big]^{-1} \times \exp\big(-r_1^2 /(4K(R)x)\big), \tag{86}$$

$$b(T_2)R^{-\gamma(T_2)} D_O(R) = N_O(R) \times \Big[2\pi^{1/2}\big(K(R)(T_2 - T_1 + x)\big)^{3/2}\Big]^{-1} \times \exp\big(-r_1^2 /(4K(R)(T_2 - T_1 + x))\big) \tag{87}$$

$$b(T_3)R^{-\gamma(T_3)} D_O(R) = N_O(R) \times \Big[2\pi^{1/2}\big(K(R)(T_3 - T_1 + x)\big)^{3/2}\Big]^{-1} \times \exp\big(-r_1^2 /(4K(R)(T_3 - T_1 + x))\big) \tag{88}$$

From this system of Equations 86-88 by dividing one equation on other (to exclude $N_O(R)$) we obtain

$$\frac{T_2 - T_1}{x(T_2 - T_1 + x)} = -\frac{4K(R)}{r_1^2} \times \ln\left\{ \frac{b(T_1)}{b(T_2)} (x/(T_2 - T_1 + x))^{3/2} R^{-[\gamma(T_1) - \gamma(T_2)]} \right\}, \tag{89}$$

$$\frac{T_3 - T_1}{x(T_3 - T_1 + x)} = -\frac{4K(R)}{r_1^2} \times \ln\left\{\frac{b(T_1)}{b(T_3)}(x/(T_3 - T_1 + x))^{3/2} R^{-[\gamma(T_1) - \gamma(T_3)]}\right\}. \tag{90}$$

By dividing Equation 89 on Equation 90 (to exclude $K(R)$) we obtain

$$x = [(T_2 - T_1)\Psi(x) - (T_3 - T_1)]/(1 - \Psi(x)), \tag{91}$$

where $\Psi(x)$ is function very weekly depended from x:

$$\Psi(x) = \frac{T_3 - T_1}{T_2 - T_1} \times \frac{\ln\left\{\frac{b(T_1)}{b(T_2)}\left(\frac{x}{T_2 - T_1 + x}\right)^{3/2} R^{\gamma(T_2) - \gamma(T_1)}\right\}}{\ln\left\{\frac{b(T_1)}{b(T_3)}\left(\frac{x}{T_3 - T_1 + x}\right)^{3/2} R^{\gamma(T_3) - \gamma(T_1)}\right\}}. \tag{92}$$

Equation 91 can be solved by the iteration method: as the first approximation we can use, as in Section 6.5, $x_1 = T_1 - T_e \approx 500\ \mathrm{sec}$ which is a minimum time of relativistic particles propagation from the Sun to the Earth's orbit. Then by Equation 92 we determine $\Psi(x_1)$ and by Equation 91 determine the second approximation x_2. To put x_2 in Equation 92 we compute $\Psi(x_2)$, and then by Equation 91 determine the third approximation x_3, and so on:

$$x_k = [(T_2 - T_1)\Psi(x_{k-1}) - (T_3 - T_1)]/(1 - \Psi(x_{k-1})), \tag{93}$$

After solving Equation 91 and determining the time of ejection $T_e = T_1 - x$, we can compute diffusion coefficient from Equation 89 or Equation 90:

$$K(R) = -\frac{\dfrac{r_1^2(T_2 - T_1)}{4(T_1 - T_e)(T_2 - T_e)}}{\ln\left\{\frac{b(T_1)}{b(T_2)}\left(\frac{T_1 - T_e}{T_2 - T_e}\right)^{3/2} R^{\gamma(T_2) - \gamma(T_1)}\right\}} = -\frac{\dfrac{r_1^2(T_3 - T_1)}{4(T_1 - T_e)(T_3 - T_e)}}{\ln\left\{\frac{b(T_1)}{b(T_3)}\left(\frac{T_1 - T_e}{T_3 - T_e}\right)^{3/2} R^{\gamma(T_3) - \gamma(T_1)}\right\}} \tag{94}$$

After determining time of ejection and diffusion coefficient we can determine the total flux and energy spectrum in the source from any of Equations 86-88:

$$\begin{aligned}
N_o(R) &= 2\pi^{1/2} b(T_1) R^{-\gamma(T_1)} D_o(R)(K(R)(T_1 - T_e))^{3/2} \exp(r_1^2/(4K(R)(T_1 - T_e))) = \\
&= 2\pi^{1/2} b(T_2) R^{-\gamma(T_2)} D_o(R)(K(R)(T_2 - T_e))^{3/2} \exp(r_1^2/(4K(R)(T_2 - T_e))) = \\
&= 2\pi^{1/2} b(T_3) R^{-\gamma(T_3)} D_o(R)(K(R)(T_3 - T_e))^{3/2} \exp(r_1^2/(4K(R)(T_3 - T_e))),
\end{aligned} \tag{95}$$

where $T_e = T_1 - x$ was determined by Equation 91, and $K(R)$ by Equation 94.

6.7. Controlling Online of the Used Model of SEP Generation and Propagation

The online controlling of use of the model can be made by data obtained in the next minutes T_4, T_5, T_6, and others. For example, we can determine the time of ejection T_e, diffusion coefficient $K(R)$ and SEP spectrum in the source $N_o(R)$ on the basis of data obtained in moments T_2, T_3, T_4, then in moments T_3, T_4, T_5, then in moments T_4, T_5, T_6, or for any other combinations of time moments, for example, at T_1, T_3, T_5, and so on. Obtained values of T_e, $K(R)$ and $N_o(R)$ in the frame of errors must be the same what obtained on the basis of data in time moments T_1, T_2, T_3. If this condition will be satisfied for any combinations of T_i, T_j, T_k, it will mean that the used model of SEP generation and propagation in the interplanetary space is correct and it can be used also for prediction of expected SEP fluxes in space, in the magnetosphere, and in the Earth's atmosphere. If this condition will be not satisfied, it will mean that the real model of SEP generation and propagation in the interplanetary space is more complicated. In this case, the using of data from only one observatory is not enough, it is necessary to use data from many cosmic ray observatories to determine anisotropy of SEP fluxes on the Earth and parameters of more complicated model. On the basis of consideration of many SEP events we expect that after some short time $T_{\min} - T_e$ SEP distribution became isotropic and considered above simple model of SEP generation and propagation in the interplanetary space became correct. This moment $T_{\min} - T_e$ can be determined automatically by the described above method of estimation of T_e, $K(R)$ and $N_o(R)$ for different combinations T_i, T_j, T_k.

6.8. Online Forecasting of Expected SEP Flux and Radiation Hazard for Space-probes in Space, Satellites in the Magnetosphere, Jets and Various Objects in the Atmosphere, and on the Ground in Dependence of Cut-off Rigidity

If the controlling described in the Section 6.7 will give a positive result, we can predict the expected SEP rigidity spectrum in the space at any moment T and at any distance r from the Sun:

$$D_S(R,r,T) = N_o(R) \times \left[2\pi^{1/2} \left(K(R)(T - T_e) \right)^{3/2} \right]^{-1} \times \exp\left(-r^2 / (4K(R)(T - T_e)) \right) \quad (96)$$

The expected flux inside any space-probe at distance r from the Sun with the threshold energy $E_{k\min}$ will be

$$I_S(r,T,E_{k\min}) = \int_{E_{k\min}}^{\infty} N_o(R(E_k)) \times \left[2\pi^{1/2}\left(K(R(E_k))(T-T_e)\right)^{3/2}\right]^{-1} \times \exp\left(-r^2/(4K(R(E_k))(T-T_e))\right)dE_k$$

$$(97)$$

and the fluency that a space-probe will receive during the entire time of the event (determined the radiation doze) will be

$$Y_S(r,E_{k\min}) = \int_{T_e}^{\infty} dT \int_{E_{k\min}}^{\infty} N_o(R(E_k)) \times \left[2\pi^{1/2}\left(K(R(E_k))(T-T_e)\right)^{3/2}\right]^{-1} \times \exp\left(-r^2/(4K(R(E_k))(T-T_e))\right)dE_k \quad (98)$$

6.9. Prediction of Expected SEP Energy Spectrum, SEP Flux and SEP Fluency (Radiation Doze) for Satellites at Different Cut-off Rigidities inside the Earth's Magnetosphere

Inside the Earth's magnetosphere the expected SEP energy spectrum will be determined by Equation 96 at $r = r_1 = 1 AU$. For satellites at different cut-off rigidities $R_c(T)$ inside the Earth's magnetosphere the expected SEP flux will be

$$I_S(T,R_c(T)) = \int_{R_c(T)}^{\infty} N_o(R) \times \left[2\pi^{1/2}\left(K(R)(T-T_e)\right)^{3/2}\right]^{-1} \times \exp\left(-r_1^2/(4K(R)(T-T_e))\right)dR$$

$$(99)$$

where $R_c(T)$ is determined by the orbit of a satellite. The expected fluency that a satellite will receive with orbit $R_c(T)$ during the entire time of the event (proportional to the radiation doze) will be

$$Y_S(R_c(T)) = \int_{T_e}^{\infty} dT \int_{R_c(T)}^{\infty} N_o(R) \times \left[2\pi^{1/2}\left(K(R)(T-T_e)\right)^{3/2}\right]^{-1} \times \exp\left(-r_1^2/(4K(R)(T-T_e))\right)dR$$

$$(100)$$

6.10. Prediction of Expected Intensity of Secondary Components Generated by SEP in the Earth's Atmosphere and Expected Radiation Doze for Planes and Ground Objects on Different Altitudes at Different Cut-off Rigidities

The expected intensity of secondary component of type i (electron-photon, nucleon, muon and others) generated in the Earth's atmosphere by SEP at a moment of time T will be in the some point characterized by pressure level h_o and cut-off rigidity R_c as following

$$I_{si}(T,R_c,h_o) = \int_{R_c}^{\infty} N_o(R) \times \left[2\pi^{1/2}\left(K(R)(T-T_e)\right)^{3/2}\right]^{-1} \times \exp\left(-r_1^2/(4K(R)(T-T_e))\right)W_i(R,h_o)dR \cdot (101)$$

where $W_i(R,h_o)$ is the coupling function. The expected total fluency (proportional to the radiation doze) for planes and ground objects on different altitudes at different cut-off rigidities will be determined by integration of Equation 7.6 over all time of the event and summarizing over all secondary components:

$$Y_S(R_c, h_o) = \sum_i \int_{T_e}^{\infty} dT \int_{R_c}^{\infty} N_o(R) \times \left[2\pi^{1/2} \left(K(R)(T - T_e) \right)^{3/2} \right]^{-1} \times \exp\left(-r_i^2 / (4K(R)(T - T_e)) \right) W_i(R, h_o) dR, \quad (102)$$

6.11. Alerts in Cases if the Radiation Dozes Are Expected to Be Dangerous

If the forecasted radiation doses described in Sections 6.8-8.10 are expected to be dangerous, there will be sent in the first few minutes of the event preliminary "SEP-Alert_1/Space", "SEP-Alert_1/ Magnetosphere", "SEP-Alert_1/ Atmosphere", "SEP-Alert_1 /Ground", and then by obtaining online more exact results on the basis of for coming new data there will be automatically sent more exact Alert_2, Alert_3 and so on. These Alerts will give information on the expected time and level of danger; experts must operatively decide what to do: for example, for space-probes in space and satellites in the magnetosphere were to switch-off the electric power for 1-2 hours to save the memory of computers and high level electronics, for jets to decrease their altitudes from 10-20 km to 4-5 km to protect crew and passengers from great radiation hazard, and so on.

6.12. The Checking of the Model When Diffusion Coefficient Does not Depend on the Distance from the Sun Using Real Data during the SEP Event of September 1989

We will use the data obtained during the great SEP event in September 1989 by the NM on the top of Gran-Sasso in Italy (Dorman et al., 2005a, b). This NM detects one-minute data not only of total neutron intensity, but also many of neutron multiplicities ($\geq 1, \geq 2, \geq 3$, up to ≥ 8), which gives the possibility of using the method of coupling functions to determine the energy spectrum in high energy range (≥ 6 GV) for each minute. On the basis of these data we determine at first the values of the diffusion coefficient $K(R)$. These calculations have been done according to the procedure described above, by supposing that $K(R)$ does not depend on the distance to the Sun. Results are shown in Figure 27.

From Figure 27 can be seen that at the beginning of the event the obtained results are not stable, due to large relative statistical errors. After several minutes the amplitude of CR intensity increase becomes many times bigger than statistical error for one minute data σ (about 1%), and we can see a systematical increase of the diffusion coefficient $K(R)$ with time. This result contradicts the conditions at which was solved the inverse problem in Section 6. Really the systematical increase of the diffusion coefficient with time reflects the increasing of $K(R)$ with the diffusion propagation of solar CR from the Sun, i.e. reflects the increasing of $K(R)$ with the distance from the Sun. This means that for the considered SEP event we need to apply the inverse problem, where it will be assumed increasing of the diffusion coefficient with the distance from the Sun.

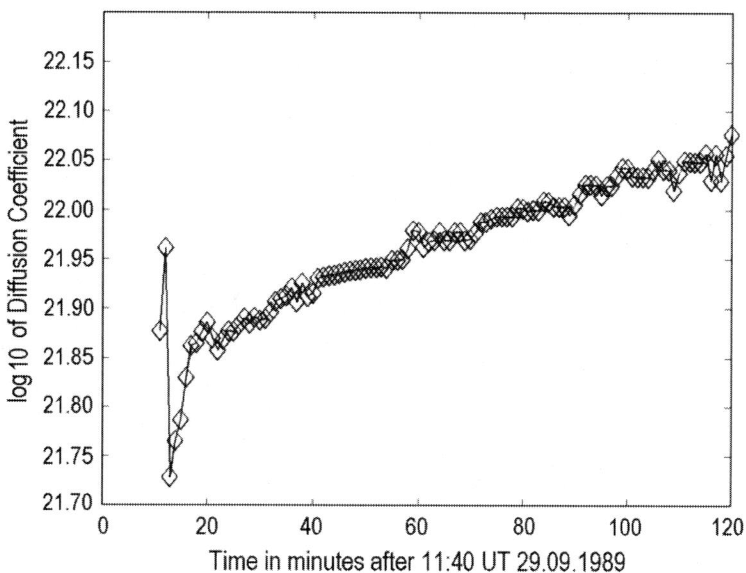

Figure 27. The time behavior of $K(R)$ for R ~ 10 GV for the SEP event 29 September, 1989. From Dorman et al. (2005a, b).

7. DETERMINATION OF THE DIFFUSION COEFFICIENT IN THE INTERPLANETARY SPACE, TIME OF EJECTION AND ENERGY SPECTRUM OF SEP AT THE SOURCE BY NM AND SATELLITE DATA (THE CASE WHEN DIFFUSION COEFFICIENT DEPENDS ON PARTICLE RIGIDITY AND DISTANCE FROM THE SUN)

7.1. The Inverse Problem for the Case When Diffusion Coefficient Depends from Particle Rigidity and from the Distance to the Sun

Let us suppose, according to Parker (M1963), that the diffusion coefficient

$$\kappa(R,r) = \kappa_1(R) \times (r/r_1)^\beta .$$

(103)

In this case the solution of the diffusion equation will be

$$N(R,r,t) = \frac{N_o(R) \times r_1^{3\beta/(2-\beta)}(\kappa_1(R)t)^{-3/(2-\beta)}}{(2-\beta)^{(4+\beta)/(2-\beta)}\Gamma(3/(2-\beta))} \times \exp\left(-\frac{r_1^\beta r^{2-\beta}}{(2-\beta)^2 \kappa_1(R)t}\right),$$

(104)

where t is the time after SEP ejection into the solar wind. So now we have four unknown parameters: time of SEP ejection into the solar wind T_e, β, $\kappa_1(R)$, and $N_o(R)$. Let us assume that according to the ground and satellite measurements at the distance $r = r_1 = 1$ AU from the Sun we know $N_1(R)$, $N_2(R)$, $N_3(R)$, $N_4(R)$ at UT times T_1, T_2, T_3, T_4. In this case

$$t_1 = T_1 - T_e = x, \quad t_2 = T_2 - T_1 + x, \quad t_3 = T_3 - T_1 + x, \quad t_4 = T_4 - T_1 + x, \tag{105}$$

For each $N_i(R, r = r_1, T_i)$ we obtain from Equation 104 and Equation 105:

$$N_i(R, r = r_1, T_i) = \frac{N_o(R) \times r_1^{3\beta/(2-\beta)} (\kappa_1(R)(T_i - T_1 + x))^{-3/(2-\beta)}}{(2-\beta)^{(4+\beta)/(2-\beta)} \Gamma(3/(2-\beta))} \times \exp\left(-\frac{r_1^2 (2-\beta)^{-2}}{\kappa_1(R)(T_i - T_1 + x)}\right), \tag{106}$$

where i = 1, 2, 3, and 4. To determine x let us step by step exclude unknown parameters $N_o(R)$, $\kappa_1(R)$, and then β. In the first we exclude $N_o(R)$ by forming from four Equation 106 for different i three equations for ratios

$$\frac{N_1(R, r = r_1, T_1)}{N_i(R, r = r_1, T_i)} = \left(\frac{x}{T_i - T_1 + x}\right)^{-3/(2-\beta)} \times \exp\left(-\frac{r_1^2}{(2-\beta)^2 \kappa_1(R)}\left(\frac{1}{x} - \frac{1}{T_i - T_1 + x}\right)\right), \tag{107}$$

where i = 2, 3, and 4. To exclude $\kappa_1(R)$ let us take the logarithm from both parts of Equation 107 and then divide one equation on another; as a result we obtain the following two equations:

$$\frac{\ln(N_1/N_2) + (3/(2-\beta))\ln(x/(T_2 - T_1 + x))}{\ln(N_1/N_3) + (3/(2-\beta))\ln(x/(T_3 - T_1 + x))} = \frac{(1/x) - (1/(T_2 - T_1 + x))}{(1/x) - (1/(T_3 - T_1 + x))}, \tag{108}$$

$$\frac{\ln(N_1/N_2) + (3/(2-\beta))\ln(x/(T_2 - T_1 + x))}{\ln(N_1/N_4) + (3/(2-\beta))\ln(x/(T_4 - T_1 + x))} = \frac{(1/x) - (1/(T_2 - T_1 + x))}{(1/x) - (1/(T_4 - T_1 + x))}. \tag{109}$$

After excluding from Equation 108 and Equation 109 unknown parameter β, we obtain an equation for determining x:

$$x^2(a_1 a_2 - a_3 a_4) + xd(a_1 b_2 + b_1 a_2 - a_3 b_4 - b_3 a_4) + d^2(b_1 b_2 - b_3 b_4) = 0, \tag{110}$$

where

$$d = (T_2 - T_1)(T_3 - T_1)(T_4 - T_1), \tag{111}$$

$$a_1 = (T_2 - T_1)(T_4 - T_1)\ln(N_1/N_3) - (T_3 - T_1)(T_4 - T_1)\ln(N_1/N_2), \tag{112}$$

$$a_2 = (T_3 - T_1)(T_4 - T_1)\ln(x/(T_2 - T_1 + x)) - (T_2 - T_1)(T_3 - T_1)\ln(x/(T_4 - T_1 + x)), \tag{113}$$

$$a_3 = (T_2 - T_1)(T_3 - T_1)\ln(N_1/N_4) - (T_3 - T_1)(T_4 - T_1)\ln(N_1/N_2), \tag{114}$$

$$a_4 = (T_3 - T_1)(T_4 - T_1)\ln(x/(T_2 - T_1 + x)) - (T_2 - T_1)(T_4 - T_1)\ln(x/(T_3 - T_1 + x)), \tag{115}$$

$$b_1 = \ln(N_1/N_3) - \ln(N_1/N_2), \quad b_2 = \ln(x/(T_2 - T_1 + x)) - \ln(x/(T_4 - T_1 + x)), \tag{116}$$

$$b_3 = \ln(N_1/N_4) - \ln(N_1/N_2), \quad b_4 = \ln(x/(T_2 - T_1 + x)) - \ln(x/(T_3 - T_1 + x)). \tag{117}$$

As it can be seen from Equation 110 and Equations 111-117, coefficients a_2, a_4, b_2, b_4 very weakly (as logarithm) depend on x. Therefore, Equation 110 we solve by iteration method, as above we solved Equation 99: as a first approximation, we use $x_1 = T_1 - T_e \approx 500$ sec (which is the minimum time of propagation of relativistic particles from the Sun to the Earth's orbit). Then by Equation 113 and Equations 115-117 we determine $a_2(x_1)$, $a_4(x_1)$, $b_2(x_1)$, $b_4(x_1)$ and by Equation 110 we determine the second approximation x_2, and so on. After determining x, i.e. according Equation 105 determining t_1, t_2, t_3, t_4, the final solutions for β, $\kappa_1(R)$, and $N_o(R)$ can be found. Unknown parameter β in Equation 103 we determine from Equation 108 and Equation 109:

$$\beta = 2 - 3\left[\left(\ln(t_2/t_1)\right) - \frac{t_3(t_2 - t_1)}{t_2(t_3 - t_1)}\ln(t_3/t_1)\right] \times \left[\left(\ln(N_1/N_2)\right) - \frac{t_3(t_2 - t_1)}{t_2(t_3 - t_1)}\ln(N_1/N_3)\right]^{-1}. \tag{118}$$

Then we determine unknown parameter $\kappa_1(R)$ in Equation 103 from Equation 107:

$$\kappa_1(R) = \frac{r_1^2\left(t_1^{-1} - t_2^{-1}\right)}{3(2-\beta)\ln(t_2/t_1) - (2-\beta)^2\ln(N_1/N_2)} = \frac{r_1^2\left(t_1^{-1} - t_3^{-1}\right)}{3(2-\beta)\ln(t_3/t_1) - (2-\beta)^2\ln(N_1/N_3)}. \tag{119}$$

After determining parameters β and $\kappa_1(R)$ we can determine the last parameter $N_o(R)$ from Equation 106:

$$N_o(R) = N_i(2-\beta)^{(4+\beta)/(2-\beta)}\Gamma(3/(2-\beta))r_1^{-3\beta/(2-\beta)}\left(\kappa_1(R)t_i\right)^{3/(2-\beta)}\exp\left(\frac{r_1^2}{(2-\beta)^2\kappa_1(R)t_i}\right), \tag{120}$$

where index $i = 1$, 2 or 3.

Above we show that for some simple model of SEP propagation it is possible to solve an inverse problem on the basis of ground and satellite measurements at the beginning of the event. Obtained results we used in the method of great radiation hazard forecasting based on online CR one-minute ground and satellite data (Dorman et al., 2005b).

7.2. The Checking of the Model When Diffusion Coefficient Depends on the Distance to the Sun

On the basis of the inverse problem solution described in Section 7.1, by using the first few minutes of NM data of the SEP event we can determine the effective parameters β by Equation 118, $\kappa_1(R)$ by Equation 119, and $N_o(R)$ by Equation 120, corresponding to high rigidity, about 10 GV. In Figure 28 the values of the parameter $\kappa_1(R)$ are shown for the event in September 1989.

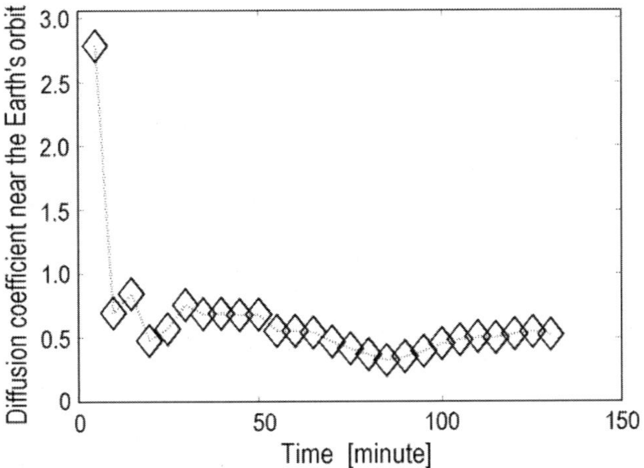

Figure 28. Diffusion coefficient $\kappa_1(R)$ near the Earth's orbit (in units 10^{23} cm^2 sec^{-1}) in dependence of time (in minutes after 11.40 UT of September 29, 1989).

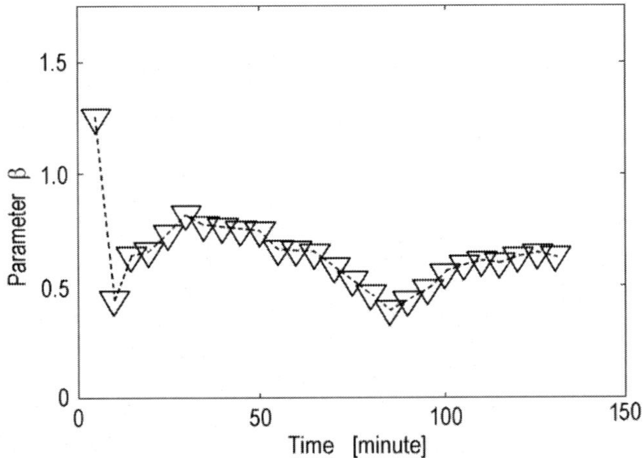

Figure 29. Values of parameter β in dependence of time (in minutes after 11.40 UT of September 29, 1989).

From Figure 28 it can be seen that at the very beginning of event (the first point) the result is unstable: in this period the amplitude of increase is relatively small, so the relative accuracy is too low, and we obtain very big diffusion coefficient. Let us note, that at the very beginning of the event the diffusion model can be very hardly applied (more correct would be the application of kinetic model of SEP propagation). After the first point we have about stable result with accuracy ± 20 % (let us compare with Figure 27, where the diffusion coefficient was found as effectively increasing with time). In Figure 29 are shown values of the parameter β.

It can be seen from Figure 29 that again the first point is anomalously big, but after the first point the result becomes almost stable with an average value $\beta \approx 0.6$ (with accuracy about $\pm$ 20%). Therefore, we can hope that the model of the inverse problem solution, described in Section 7.1 (the set of Equations 103–120) reflects adequately SEP propagation in the interplanetary space.

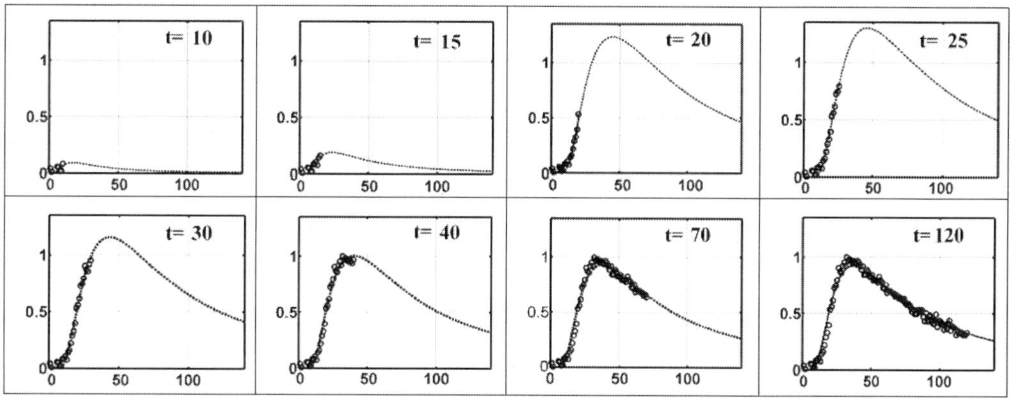

Figure 30. Calculation for each new minute of SEP intensity observations parameters β, $\kappa_1(R)$, $N_o(R)$ and forecasting of total neutron intensity (time t is in minutes after 11.40 UT of September 29, 1989; curves – forecasting, circles – observed total neutron intensity). From Dorman et al. (2005a, b).

7.3. The Checking of the Model by Comparison of Predicted SEP Intensity Time Variation with NM Observations

More accurate and exact checking of the solution of the inverse problem can be made by comparison of predicted SEP intensity time variation with NM observations. For this aim after determining of the effective parameters β, $\kappa_1(R)$, and $N_o(R)$ we may determine by Equation 104 the forecasting curve of expected SEP flux behavior for total neutron intensity. With each new minute of observations we can determine parameters β, $\kappa_1(R)$, and $N_o(R)$ more and more exactly. It means that with each new minute of observations we can determine more and more exactly the forecasting curve of expected SEP flux behavior. We compare this forecasting curve with time variation of observed total neutron intensity (see Figure 30 which contains 8 panels for time moments $t = 10$ min up to $t = 120$ min after 11.40 UT of 29 September, 1989). From Figure 30 it can be seen that it is not enough to use only the first few minutes of NM data ($t = 10$ min): the obtained curve forecasts too low intensity. For $t = 15$ min the forecast shows some bigger intensity, but also not enough. Only for $t = 20$ min (15 minutes of increase after beginning) and later (up to $t = 40$ min and more) we obtain about stable forecast with good agreement with observed CR intensity.

7.4. The Checking of the Model by Comparison of Predicted SEP Intensity Time Variation with NM and Satellite Observations

The results described above, based only on NM data, reflect the situation in SEP behavior in the high energy (more than 6 GeV) region. For extrapolation of these results to the low energy interval (dangerous for space-probes and satellites), we use satellite online data available through the Internet. The problem is how to extrapolate the SEP energy spectrum from high NM energies to very low energies detected by GOES satellite. The main idea of this extrapolation is the following: 1) the time of ejection for high and small energy ranges (detected by NM and by satellite) is the same, so it can be determined by using only NM data; 2) the source function relative to time is a δ–function, and relative to energy is a power function with an energy-dependent index $\gamma = \gamma_o + \ln(E_k/E_{ko})$ with maximum at $E_{k\max} = E_{ko}\exp(-\gamma_o)$:

$$N_o(R,T) = N_o\delta(T - T_e)R^{-(\gamma_o + \ln(E_k/E_{ko}))}. \tag{121}$$

Figure 31 shows results based on the NM and satellite data of forecasting of expected SEP fluxes also in small energy intervals and comparison with observation satellite data.

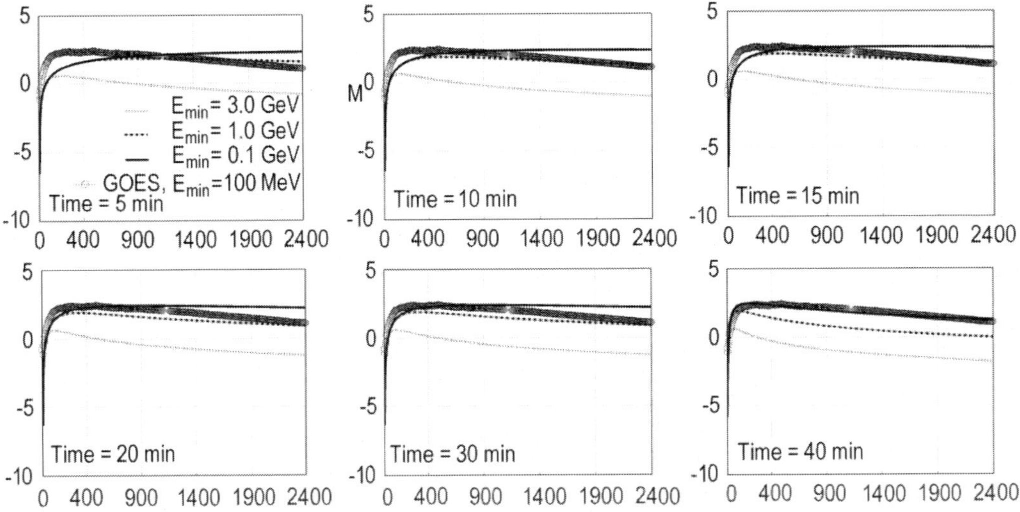

Figure 31. Predicted SEP integral fluxes for $E_k \geq E_{min} =$ 0.1, 1.0, and 3.0 GeV. The forecasted integral flux for $E_k \geq E_{min} =$ 0.1 GeV is compared with the observed fluxes for $E_k \geq 100$ MeV on GOES satellite. The ordinate is log10 of the SEP integral flux (in cm-2sec-1sr-1), and the abscissa is time in minutes from 11.40 UT of September 29, 1989. From Dorman et al. (2005a, b).

Results of the comparison presented in Figure 30 and Figure 31 show that by using online data from ground NM in the high energy range and from the satellite in the low energy range during the first 30-40 minutes after the start of the SEP event, it is possible by using only CR data to solve the inverse problem by formulas in Sections 7.2 and 7.3: to determine the properties of SEP source on the Sun (time of ejection into the solar wind, source SEP energy spectrum, and total flux of accelerated particles) and parameters of SEP propagation in the interplanetary space (diffusion coefficient and its dependence from particle energy and from the distance from the Sun).

7.5. The Inverse Problems for Great SEP Events and Space Weather

Let us note that the solving of inverse problems for great SEP events has important practical sense: to predict the expected SEP differential energy spectrum on the Earth's orbit and integral fluxes for different threshold energies up to many hours (and even up to few days) ahead. The total (event-integrated) fluency of the SEP event, and the expected radiation hazards can also be estimated on the basis of the first 30-40 minutes after the start of the SEP event and corresponding Alerts to experts operating different objects in space, in magnetosphere, and in the atmosphere at different altitudes and at different cut-off rigidities can be sent automatically. These experts should decide what to do operationally (for example, for space-probes in space and satellites in the magnetosphere to switch-off the electric power for few hours to save the memory of computers and high level electronics; for jets to decrease their altitudes from 10 km to 4-5 km to protect crew and passengers from great radiation hazard, and so on). From this point of view especially important are the solving of inverse problems for great SEP by using online data of many NM and several satellites in the frame of models in which CR propagation described by the theory of anisotropic diffusion or by kinetic theory. The solving of these inverse problems will make it possible on the basis of world-wide CR Observatories and satellite data (in real scale time, applicable from Internet) to make forecasting of radiation hazard for much shorter time after SEP event beginning.

8. FORECASTING OF RADIATION HAZARD AND THE INVERSE PROBLEM FOR SEP PROPAGATION AND GENERATION IN THE FRAME OF ANISOTROPIC DIFFUSION AND IN KINETIC APPROACH

8.1. Kinetic and Anisotropic Diffusion Cases

In this approach we will be based mainly on theoretical works Fedorov et al. (2002), Dorman et al. (2003), and Dorman (2008a), according to which the evolution of the particle distribution function $f(y, \tau, \mu)$ follows from the kinetic equation written in the drift approximation:

$$\frac{\partial f}{\partial \tau} + \frac{\mu \partial f}{\partial y} + f - \frac{1}{2} \int_{-1}^{1} f d\mu = \frac{v_s}{v} \delta(y) \delta(\tau) \varphi(\mu), \tag{122}$$

where y is a coordinate along regular IMF and τ is time in dimensionless units ($y = z v_s / v$, $\tau = v_s t$); $v_s = v/\Lambda$ is the collision frequency of SEP with magnetic inhomogeneities; $\mu = \cos\theta$, and θ is the particle pitch-angle. The right-hand side of Equation 122 describes an instantaneous injection of SEP with an initial angular distribution

$$\varphi(\mu) = \frac{a_\mu \Delta_\mu}{2\left(\Delta_\mu^2 + (\mu - \mu_o)^2\right)}. \tag{123}$$

For the finite time the injection in the right-hand side of Equation 122 instead of $\delta(\tau)$ will be

$$\chi(\tau) = v_o^2 \tau \exp(-v_o \tau), \tag{124}$$

where v_o^{-1} characterizes the mean duration of the injection. In this case the solution of Equation 122, obtained by the method of direct and inverse Fourier–Laplace transform, will be

$$G(y,\tau) = \int_0^\tau d\xi \int_{-1}^1 d\mu \chi(\tau - \xi) f(y,\xi,\mu)\psi(\mu). \tag{125}$$

This solution consists of three terms

$$G(y,\tau) = G_{us}(y,\tau) + G_s^o(y,\tau) + G_s^d(y,\tau). \tag{126}$$

The first component describes a contribution of the un-scattered particles which exponentially decreases with time τ:

$$G_{us}(y,\tau) = \frac{v_s v_o^2 \exp(-v_o \tau)}{v} \times \int_0^\tau \frac{d\xi}{\xi}(\tau - \xi)\varphi\left(\frac{y}{\xi}\right)\psi\left(\frac{y}{\xi}\right)\exp(\xi(v_o - 1)). \tag{127}$$

The contribution of the scattered particles can be divided into two parts. One, the non-diffusive term $G_s^o(y,\tau)$, also exponentially decreases with time, and another term, $G_s^d(y,\tau)$, has a leading meaning in the diffusive limit of $\tau \gg 1$. Namely, the non-diffusive term reads

$$G_s^o(y,\tau) = \frac{v_s v_o^2 \exp(-v_o \tau)}{8\pi v}\left\{\int_0^{y/\tau} d\eta \Psi(y,\tau,\eta)[S(\tau) - S(y)] + \int_{y/\tau}^1 d\eta \Psi(y,\tau,\eta)[S(y/\eta) - S(y)]\right\}, \tag{128}$$

where

$$\Psi(y,\tau,\eta) = \frac{\exp(y\Lambda(\eta)/2)}{\left((\mu_o - \eta)^2 + \Delta_\mu^2\right)\left((\lambda_o - \eta)^2 + \Delta_\lambda^2\right)}. \tag{129}$$

The last (diffusive non-vanishing) term in Equation 126 has a sense only for $|y| < \tau$ and reads as

$$G_s^d\left(y,\tau\right)=\frac{v_s v_o^2}{4\pi v}\int\limits_{-\pi/2}^{\pi/2}\frac{dk}{k^2}\Phi\left(y,k\right)\times\left\{e^{\tau\kappa}-e^{(y-\tau)v_o+y\kappa}\left[1+\left(\tau-y\right)\left(v_o+\kappa\right)\right]\right\}. \qquad (130)$$

where $\kappa \equiv k\cot k - 1$, and

$$\Phi(y,k)=\frac{\left(B_\mu B_\lambda-\Gamma_\mu\Gamma_\lambda\right)\cos\left(ky\right)+\left(B_\mu\Gamma_\lambda+\Gamma_\mu B_\lambda\right)\sin\left(ky\right)}{D_\mu D_\lambda\cos^2 k}. \qquad (131)$$

8.2. Expected Temporal Profiles for NM and Comparison with Observations

For example, some selected NM data for the 24 May 1990 are demonstrated in Figure 32.

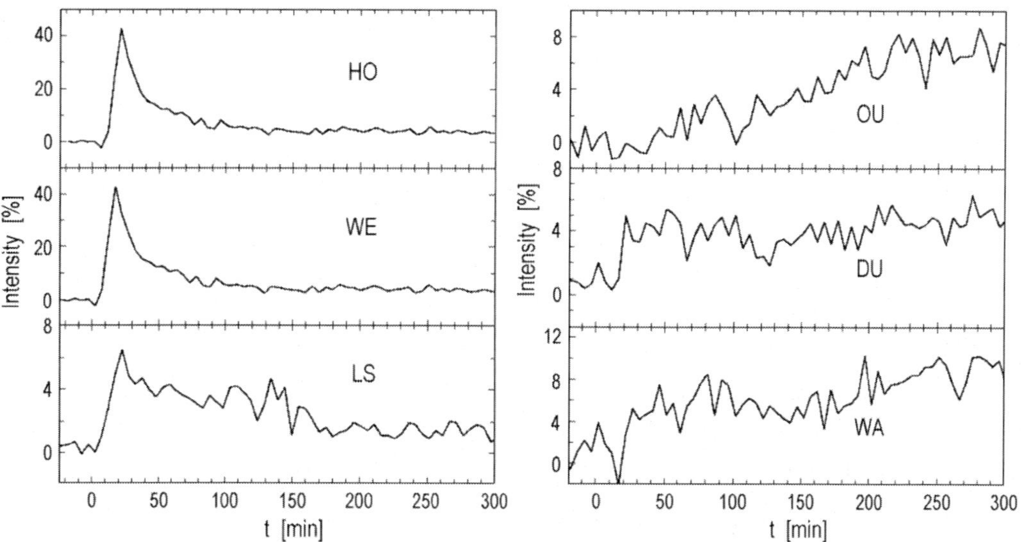

Figure 32. Two groups of NM records of the event on 24 May 1990. The left - have the narrow peak of the anisotropic stream of the first fast particles (HO – Hobart, WE– Mt. Wellington, LS – Lomnický Štít); the right - show a diffusive tail with a wide maximum at a later time (OU – Oulu, DU – Durham, WA– Mt.Washington). From Fedorov et al. (2002).

In Figure 32 the time (in min) is measured from the onset of particle injection taken as 20.50 UT of May 24, 1990. The theoretically predicted temporal profiles for the selected NM in the model described above are demonstrated in Figure 33 in the left and right panels, respectively, using the calculated asymptotic direction for each NM station.

This calculation shows that HO and WE have very similar characteristics, λ_o is 0.9 and 0.86, respectively, with $\Delta\lambda = -0.26$. Station LS has $\lambda_o = 0.34$, $\Delta\lambda = 0.4$. In the second group of NM, OU, DU, and WA have λ_o = −0.94, −0.9, −0.85 and $\Delta\lambda$ = 0.06, 0.1, 0.3, respectively. Oulu and Apatity give absolutely the same theoretical curves resulting from their similar characteristics and very similar temporal profiles of the event. The last two NMs (DU and WA) experienced small increases at 1–2 hours after onset, as the theory predicts, see Figure 32 (right panel), owing to smaller λ_o and larger $\Delta\lambda$ and larger mean $\overline{R}$.

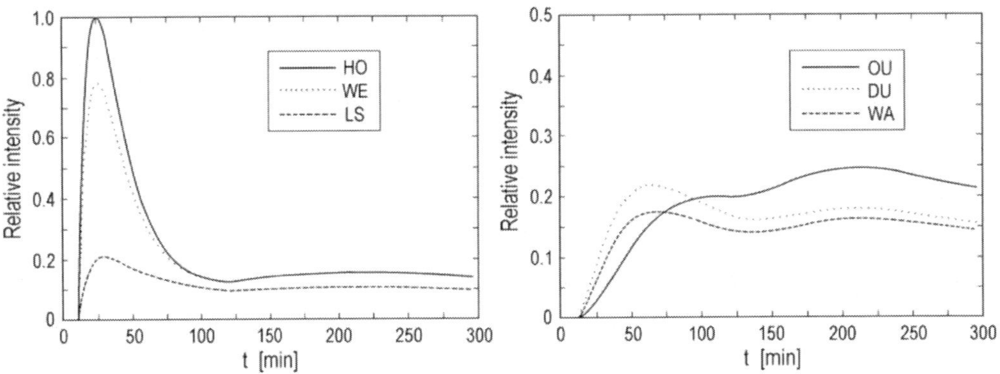

Figure 33. Theoretical prediction of temporal profiles for the selected NM using calculated parameters λ_o, $\Delta\lambda$ and mean $\overline{R}$ for each NM. On the ordinate axis are shown expected intensity relative to HO in maximum. According to Fedorov et al. (2002).

8.3. Seven Steps for Forecasting of Radiation Hazard during SEP Events

For the realization of the *first step* of forecasting we need one minute real-time data from about all NM of the world network. On each NM the program for the search of the start SEP events must work automatically, as it was described in Sections 1–3. This search will help to determine which NM from about 50 of the total number operated in the world network shows the narrowest peak of the anisotropic stream of the first arrived solar CR (NM of the 1-st type) and which shows a diffusive tail with a wide maximum at a later time (NM of the 2-nd type). In the *second step* we determine the rigidity spectrum of arrived SEP $I_s(R)$ above each NM outside of the atmosphere, and then separately for NM of the 1-st type and 2-nd type, by using the method of coupling functions as it was described above in Section 5 (in more detail see Chapter 3 in Dorman, M2004).

In the *third step* we should try to use the model of isotropic diffusion for a rough estimation of expected radiation hazard (see Sections 6 and 7). In the *fourth step* we should determine for different NM the mean $\overline{R}$, λ_o and $\Delta\lambda$ characterizing for this event. By using these parameters and experimental data on NM time profiles in the beginning time we cane determine parameters of SEP non-scattering and diffusive propagation, described in Sections 8.1 and 8.2 (the *fifth step*).

On the basis of determined parameters of SEP non-scattering and diffusive propagation we then determine expected SEP fluxes and pitch-angle distribution during total event in interplanetary space in dependence of time after ejection (the *sixth step*). In the *seventh step* by using again the method of coupling functions we should determine expected radiation doze which will be obtain during this event inside space probes in interplanetary space, satellites in the magnetosphere, aircrafts at different altitudes and cutoff rigidities, for people and technologies on the ground.

9. EFFECTS OF GREAT MAGNETIC STORMS (CAUSED BIG CR FORBUSH EFFECTS) ON THE FREQUENCY OF INFARCT MYOCARDIAL, BRAIN STROKES, CARE ACCIDENTS, AND TECHNOLOGY; EXTENDED NOAA CLASSIFICATION

There are numerous indications that natural, solar variability-driven time variations of the Earth's magnetic field can be hazardous in relation to health and safety. There are two lines of their possible influence: effects on physical systems and on human beings as biological systems. High frequency radio communications are disrupted, electric power distribution grids are blacked out when geomagnetically induced currents cause safety devices to trip, and atmospheric warming causes increased drag on satellites. An example of a major disruption on high technology operations by magnetic variations of large extent occurred in March 1989, when an intense geomagnetic storm upset communication systems, orbiting satellites, and electric power systems around the world. Several large power transformers also failed in Canada and United States, and there were hundreds of misoperations of relays and protective systems (Kappenman and Albertson, 1990; Hruska and Shea, 1993). Some evidence has been also reported on the association between geomagnetic disturbances and increases in work and traffic accidents (Ptitsyna et al. 1998 and refs. therein). These studies were based on the hypothesis that a significant part of traffic accidents could be caused by the incorrect or retarded reaction of drivers to the traffic circumstances, the capability to react correctly being influenced by the environmental magnetic and electric fields. The analysis of accidents caused by human factors in the biggest atomic station of former USSR, "Kurskaya", during 1985-1989, showed that ~70% of these accidents happened in the days of geomagnetic storms. In Reiter (1954, 1955) it was found that work and traffic accidents in Germany were associated with disturbances in atmospheric electricity and in geomagnetic field (defined by sudden perturbations in radio wave propagation). On the basis of 25 reaction tests, it was found also that the human reaction time, during these disturbed periods, was considerably retarded. Retarded reaction in connection with naturally occurred magnetic field disturbances was observed also by Koenig and Ankermueller (1982). Moreover, a number of investigations showed a significant correlation between the incidence of clinically important pathologies and strong geomagnetic field variations. The most significant results have been those on cardiovascular and nervous system diseases, showing some association with geomagnetic activity; a number of laboratory results on the correlation between the human blood system and solar and geomagnetic activity supported these findings (Ptitsyna et al. 1998 and refs. therein). Recently, the monitoring of cardiovascular function among cosmonauts on board the "MIR" space station revealed a reduction of heart rate variability during geomagnetic storms (Baevsky et al. 1996); the reduction in heart rate variability has been associated with a 550% increase in the risk of coronary artery diseases (Baevsky et al. 1997 and refs. therein). On the basis of great statistical data on several millions medical events in Moscow and in St. Petersburg it was found that a sufficient influence of geomagnetic storms accompanied with

CR Forbush decreases on the frequency of myocardial infarctions, brain strokes and car accident road traumas (Villoresi et al., 1994, 1995). Earlier we found that among all characteristics of geomagnetic activity, Forbush decreases are better related to hazardous effects of solar variability-driven disturbances of the geomagnetic field (Ptitsyna et al. 1998). Figure 34 shows the correlation between cardiovascular diseases, car accidents and different characteristics of geomagnetic activity (planetary index AA, major geomagnetic storms MGS, sudden commencement of geomagnetic storm SSC, occurrence of downward vertical component of the interplanetary magnetic field Bz and also decreasing phase of Forbush decreases (FD)).

The most remarkable and statistically significant effects have been observed during days of geomagnetic perturbations defined by the days of the declining phase of Forbush decreases in CR intensity. During these days the average numbers of traffic accidents, infarctions, and brain strokes increase by (17.4±3.1)%, (10.5±1.2)% and (7.0±1.7)% respectively.

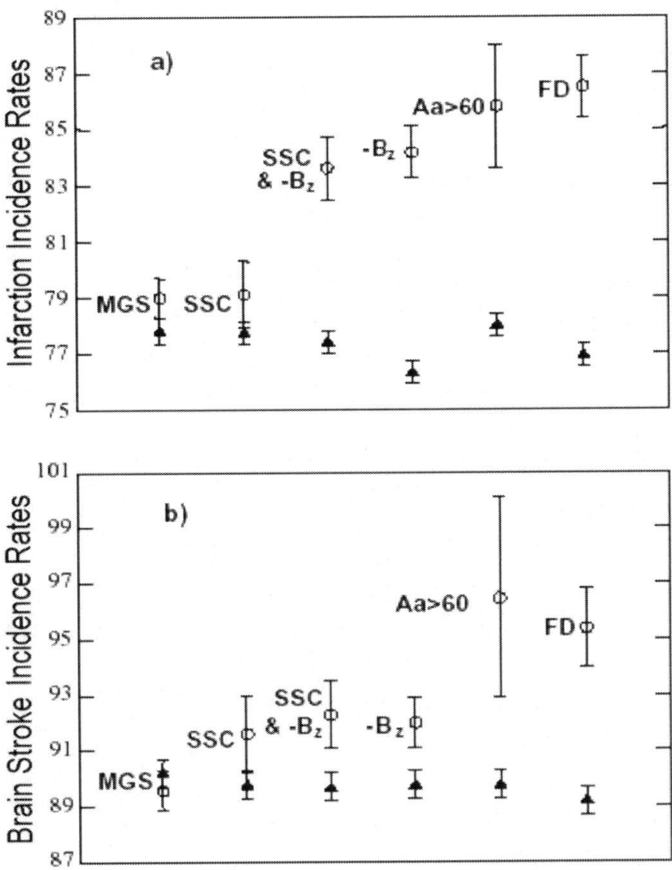

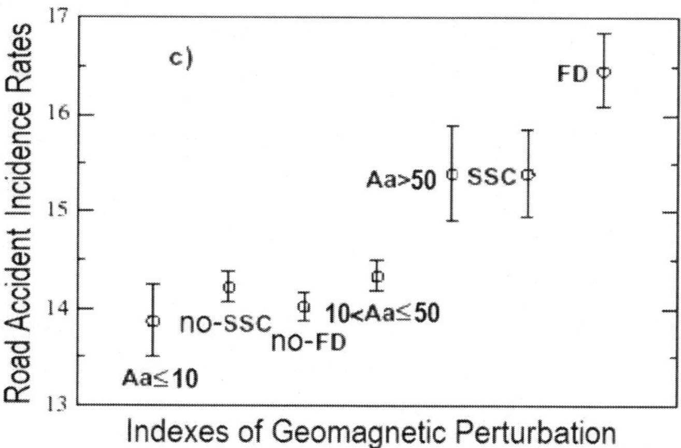

Figure 34. Myocardial infarction (a), brain stroke (b) and road accident (c) incidence rates per day during geomagnetic quiet and perturbed days according to different indexes of activity.

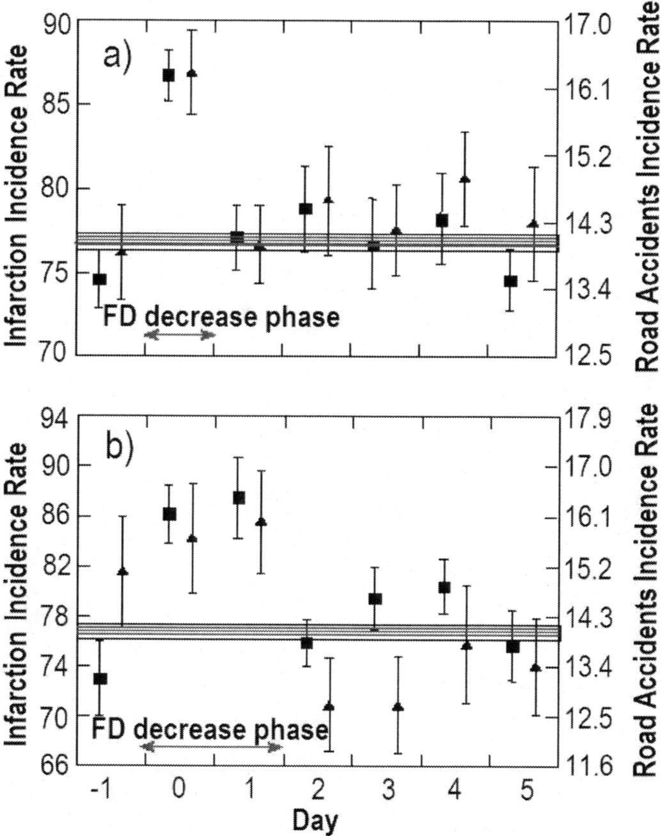

Figure 35. Infarction (full squares) and road accidents (full triangles) incidence during the time development of FD. (a): CR decrease phase T<1 day; (b): CR decrease phase 1 day<T<2 days.

In Figure 35 we show the effect on pathology rates during the time development of FD. All FD have been divided into two groups, according to the time duration T of the FD decreasing phase. Then, the average incidence of infarctions and traffic accidents was computed beginning from one day before the FD-onset till 5 days after. For the first group (T<1 day) the average daily incidence of infarctions and traffic incidence increases only in the first day of FD; no effect is observed during the recovery phase (that usually lasts for several days). Also for the second group (1 day<T<2 days) the increase in incidence rates is observed only during the 2-day period of the decreasing phase of FD.

Table 3. The extended NOAA scale of geomagnetic storms' influence on people's health, power systems, on spacecraft operations, and on other systems (the greatest three types, mostly accompanied with Forbush decreases). In addition to the original NOAA scale, biological effects are included according to results discussed above

Geomagnetic Storms			Kp values	Number per solar cycle
G5	Extreme	**Biological effects:** increase of more than 10-15% of the daily rate of myocardial infarctions, brain strokes and car road accident traumas for people population on the ground **Power systems:** widespread voltage control problems and protective system problems can occur, some grid systems may experience complete collapse or blackouts. Transformers may experience damage. **Spacecraft operations:** may experience extensive surface charging, problems with orientation, uplink/downlink and tracking satellites. **Other systems:** pipeline currents can reach hundreds of amps, HF (high frequency) radio propagation may be impossible in many areas for one to two days, satellite navigation may be degraded for days, low-frequency radio navigation can be out for hours.	Kp = 9	4 per cycle (4-8 days per cycle)
G4	Severe	**Biological effects:** increase of several percent (up to 10-15%) of the daily rate of myocardial infarctions, brain strokes and car road accident traumas for people population on the ground **Power systems:** possible widespread voltage control problems and some protective systems will mistakenly trip out key assets from the grid. **Spacecraft operations:** may experience surface charging and tracking problems, corrections may be needed for orientation problems. **Other systems:** induced pipeline currents affect preventive measures, HF radio propagation sporadic, satellite navigation degraded for hours, low-frequency radio navigation disrupted.	Kp = 8, including a 9-	100 per cycle (100-200 days per cycle)

Geomagnetic Storms			Kp values	Number per solar cycle
G3	Strong	**Biological effects:** increase of few percent of the daily rate of myocardial infarctions, brain strokes and car road accidents traumas for the population on the ground **Power systems:** voltage corrections may be required, false alarms triggered on some protection devices. **Spacecraft operations:** surface charging may occur on satellite components, drag may increase on low-Earth-orbit satellites, and corrections may be needed for orientation problems. **Other systems:** intermittent satellite navigation and low-frequency radio navigation problems may occur, HF radio may be intermittent.	Kp = 7	200 per cycle (200-400 days per cycle)

In Table 3 is shown NOAA scale of geomagnetic storms' influence on power systems, on spacecraft operations, and on other systems (the greatest three types, all mostly accompanied with Forbush decreases). We expect that for these three types of geomagnetic storms, online one-hour CR data of neutron monitors and muon telescopes for automatic online forecasting (at least before 15-20 hours of SSC) can be useful. In Table 3 we added some preliminary information on possible biological effects according to our results discussed above (see Figure 34 and Figure 35).

10. COSMIC RAY ONLINE ONE-HOUR DATA USING FOR FORECASTING DANGEROUS GEOMAGNETIC STORMS ACCOMPANIED WITH FORBUSH DECREASES; FORMATION OF THE INTERNATIONAL COSMIC RAY SERVICE (ICRS)

Thus, FD events can be used as reliable indicators of health- and safety-related harmful geomagnetic storms. For practical realization of forecasting hazardous geomagnetic storms by means of FD indicators, it will be necessary to get data from as many CR stations as possible in real-time. (now the majority data are available only after about one month). Therefore, it is necessary to found a special Real-Time Cosmic Ray World Data Center to transform the cosmic ray station network in a real-time International Cosmic Ray Service (ICRS) (Dorman et al. 1993). We present here basic ideas of the organization of such real-time data collection and processing, for providing a reliable forecast-service of FD and related dangerous disturbances of geomagnetic field. The main features observed in CR intensity before the beginning of FD that can be used for FD forecasting are shown schematically in Figure 36.

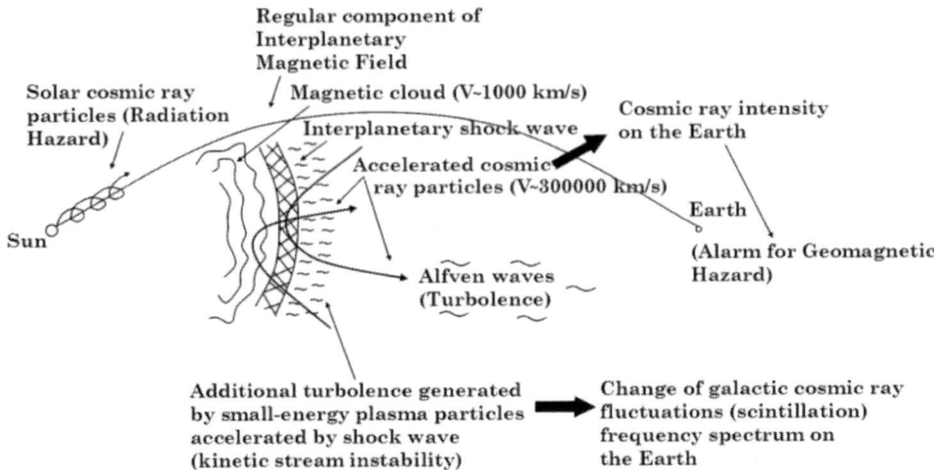

Figure 36. Scheme of mechanisms of possible precursory effects in CR.

They are following:

1. *CR pre-increase* (Blokh et al., 1959; Dorman, 1959; review in Dorman, M1963a, b). The discovery of this effect in 1959 (Blokh et al., 1959) stimulated the development of the mechanism of galactic CR interactions with interplanetary shock waves (Dorman, 1959; Dorman and Freidman, 1959) and further analyses (Dorman 1995, Belov et al. 1995) showed that this effect is related to particle interaction and acceleration by interplanetary shock waves;

2. *CR pre-decrease* (McCracken and Parsons, 1958; Fenton et al 1959; see review in Dorman, M1963a, b). This effect was analyzed recently both theoretically (Dorman et al., 1995) and experimentally on the basis of the network of CR stations (Belov et al. 1995). The pre-decrease effect can be due to a magnetic connection of the Earth with regions (moving from the Sun) with reduced CR density; this lower density can be observed at the Earth along the actual direction of IMF lines (Nagashima et al. 1990, Bavassano et al. 1994);

3. *CR fluctuations*. Many authors found some peculiarities in behavior of CR fluctuations before FD: changes in the frequency spectrum; appearance of peaks in the spectrum at some frequencies; variations in some special parameter introduced for characterizing the variability of fluctuations. Though the obtained results are often contradictory (Dorman et al. 1995), sometimes CR fluctuations appear as reliable phenomena for FD prediction, as expected from additional Alfven turbulence produced by kinetic stream instability of low-energy particles accelerated by shock waves (Berezhko et al. 1997);

4. *Change in 3-D anisotropy*. The CR longitudinal dependence changes abruptly in directions close to usual directions of interplanetary magnetic field and depends on the character and source of the disturbance. These effects, appearing much before Forbush decreases (up to 1 day) may be considered as predictors of FD. Estimation of CR anisotropy vector may be done by the global survey method described in Belov et al. (1997).

In Figure 37 we show an example of such an estimation done for the 9.09.1992 event represented the longitude-time CR intensity distribution. The grey circles mark the CR intensity decrease and white circles mark the CR intensity increase (in both cases bigger diameter of circle means bigger amplitude of intensity variation). The vertical line marks the time of Sudden Storm Commencement (SSC).

From Figure 37 one can see that the pre-increase, as well as the pre-decrease, occurs some hours (at least, 15-20 hours) before the SSC. As it was shown recently by Munakata et al. (2000), the CR pre-increase and pre-decrease effects can be observed very clearly also by the multidirectional muon telescope world network. They investigated 14 "major" geomagnetic storms characterized by $K_p \geq 8-$ and 25 large storms characterized by $K_p \geq 7-$ observed in 1992-1998. It was shown that 89% of "major" geomagnetic storms have clear precursor effects what can be used for forecasting (the probability of exact forecasting increased with increasing of the value of storm).

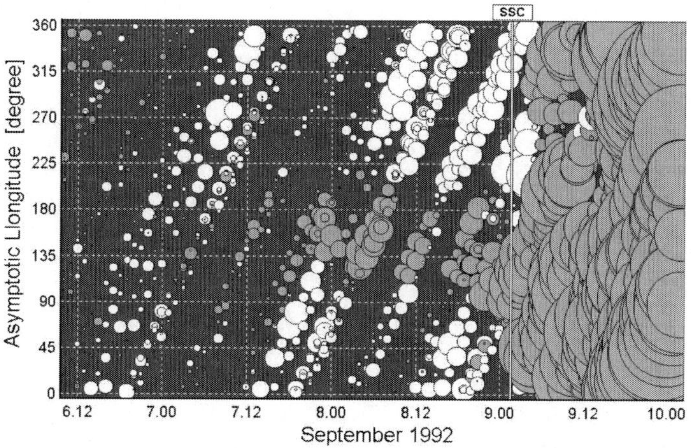

Figure 37. Galactic cosmic ray pre-increase (white circles) and pre-decrease (grey circles) effects before the Sudden Storm Commencement (SSC) of great magnetic storm in September 1992, accompanied with Forbush decrease.

We suppose that this type of analysis on the basis of online one-hour neutron monitor and muon telescope data from the world network CR Observatories can be made in the near future automatically with forecasting of great geomagnetic storms. This important problem can be solved, for example by the formation of the International Cosmic Ray Service (ICRS), what will be based on real-time collection and exchange through Internet of the data from about all cosmic-ray stations of the network (the possible scheme of ICRS working is shown in Figure 38 and Figure 39).

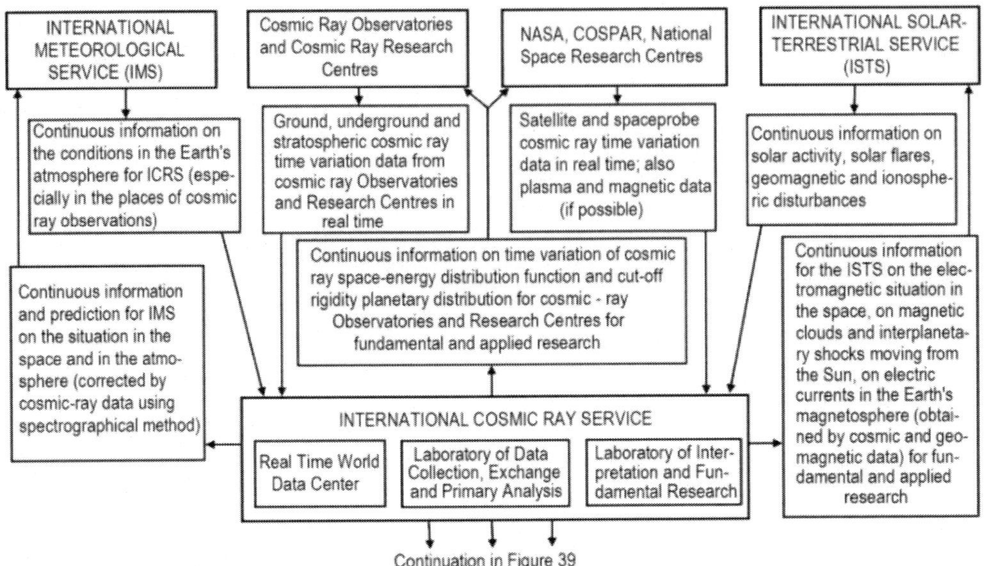

Figure 38. Supposed scheme of exchange data between collaborated CR Observatories in the frame of ICRS.

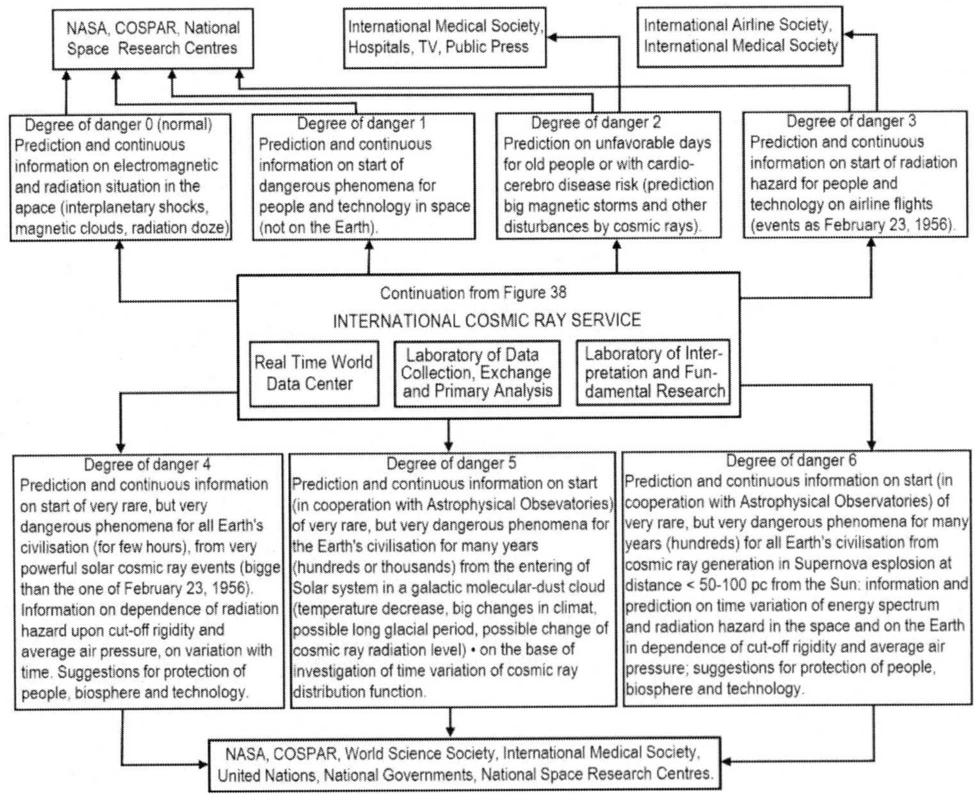

Figure 39. Output information on space weather and dangerous situations of different levels from ICRS to collaborated organizations.

Then, computerized data analysis and interpretation will be done on the basis of modern theories listed above and, as we hope, will be developed father in the near future. It will be necessary to use also related spacecraft data in real time: cosmic ray variations in small and very small energy regions, interplanetary magnetic field and solar wind data. For this purpose neutron monitor stations of the network should have a data collection time of 1 minute and 1 hour. The organization of ICRS and continue automatically forecast will provide necessary information to space agencies, health authorities, road police and other organizations to apply the appropriate preventive procedures for saving people and technologies from negative CR and other Space Weather factors negative effects.

11. CR AND OTHER SPACE WEATHER FACTORS INFLUENCED ON THE EARTH'S CLIMATE

11.1. Periodicities and Long-term Variations in Climate Change

About two hundred years ago, the famous astronomer William Herschel (1801) suggested that the price of wheat in England was directly related to the number of sunspots. He noticed that less rain fell when the number of sunspots was high (Joseph in the Bible, recognized a similar periodicity in food production in Egypt, about four thousand years ago). The solar activity level is known from direct observations over the past 450 years, and from data of cosmogenic nuclides (through CR intensity variations) for more than 10 thousand years (Eddy, 1976; see review in Dorman, M2004). Over this period there is a striking qualitative correlation between cold and warm climate periods and high and low levels of galactic CR intensity (low and high solar activity). As an example, Figure 40 shows the change in the concentration of radiocarbon during the last millennium (a higher concentration of ^{14}C corresponds to a higher intensity of galactic CR and to lower solar activity).

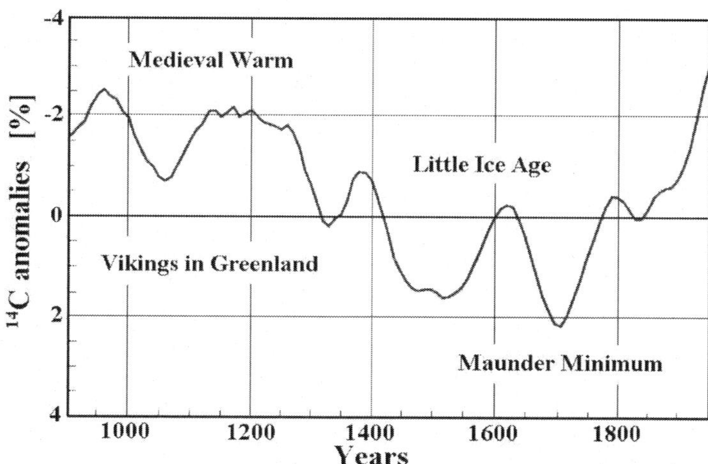

Figure 40. The change of CR intensity reflected in radiocarbon concentration during the last millennium. The Maunder minimum refers to the period 1645–1715, when sunspots were rare. From Swensmark (2000).

It can be seen from Figure 40 that during 1000–1300 AD the CR intensity was low and solar activity high, which coincided with the warm medieval period (during this period Vikings settled in Greenland). After 1300 AD solar activity decreased and CR intensity increased, and a long cold period followed (the so called Little Ice Age, which included the Maunder minimum 1645–1715 AD and lasted until the middle of 19th century).

11.2. The Possible Role of Solar Activity and Solar Irradiance in Climate Change

Friis-Christiansen & Lassen (1991) and Lassen & Friis-Christiansen (1995) found, from four hundred years of data, that the filtered solar activity cycle length is closely connected to variations of the average surface temperature in the northern hemisphere. Labitzke and Van Loon (1993) showed, from solar cycle data, that the air temperature increases with increasing levels of solar activity. Swensmark (2000) also discussed the problem of the possible influence of solar activity on the Earth's climate through changes in solar irradiance. But the direct satellite measurements of the solar irradiance during the last two solar cycles showed that the variations during a solar cycle was only about 0.1%, corresponding to about 0.3 $W·m^{-2}$. This value is too small to explain the present observed climate changes (Lean et al., 1995). Much bigger changes during solar cycle occur in UV radiation (about 10%, which is important in the formation of the ozone layer). High (1996) and Shindell et al. (1999) suggested that the heating of the stratosphere by UV radiation can be dynamically transported into the troposphere. This effect might be responsible for small contributions towards 11 and 22 years cycle modulation of climate but not to the 100 years of climate change that we are presently experiencing.

11.3. Cosmic Rays As an Important Link between Solar Activity and Climate Change

Many authors have considered the influence of galactic and solar CR on the Earth's climate. Cosmic radiation is the main source of air ionization below 40–35 km; only near the ground level, lower than 1 km, are radioactive gases from the soil also important in air ionization (see review in Dorman, M2004). The first to suggest a possible influence of air ionization by CR on the climate was Ney (1959). Swensmark (2000) noted that the variation in air ionization caused by CR could potentially influence the optical transparency of the atmosphere, by either a change in aerosol formation or influence the transition between the different phases of water. Many other authors considered these possibilities (Ney, 1959; Dickinson, 1975; Pudovkin and Raspopov, 1992; Pudovkin and Veretenenko, 1995, 1996; Tinsley, 1996; Swensmark and Friis-Christiansen, 1997; Swensmark, 1998; Marsh and Swensmark, 2000a, b; Dorman, 2009, 2012). The possible statistical connections between the solar activity cycle and the corresponding long term CR intensity variations with characteristics of climate change were considered in Dorman et al. (1987, 1988a, b). Dorman et al. (1997) reconstructed CR intensity variations over the last four hundred years on the basis of solar activity data and compared the results with radiocarbon and climate change data. Cosmic radiation plays a key role in the formation of thunderstorms and lightnings (see extended review in Chapter 11 in Dorman, M2004). Many authors (Markson, 1978; Price, 2000; Tinsley, 2000; Schlegel et al., 2001; Dorman et al., 2003; Dorman and Dorman, 2005) have considered atmospheric electric field phenomena as a possible link between solar activity and the Earth's climate. Also important in the relationship between CR and climate, is the influence of long term changes in the geomagnetic field on CR intensity through the changes of cutoff rigidity (see extended review in Chapter 7 in Dorman, M2009). One can consider the general hierarchical relationship to be: (solar activity cycles + long-term changes in the geomagnetic field) $\rightarrow$ (CR long term modulation in the Heliosphere + long term variation of cutoff rigidity) $\rightarrow$ (long term variation of clouds covering + atmospheric electric field effects) $\rightarrow$ climate change.

11.4. The Connection between Galactic CR Solar Cycles and the Earth's Cloud Coverage

Recent research has shown that the Earth's cloud coverage (observed by satellites) is strongly influenced by CR intensity (Swensmark, 2000; Tinsley, 1996; Swensmark, 1998; Marsh and Swensmark, 2000a, b). Clouds influence the irradiative properties of the atmosphere by both cooling through reflection of incoming short wave solar radiation, and heating through trapping of outgoing long wave radiation (the greenhouse effect). The overall result depends largely on the height of the clouds. According to Hartmann (1993), high optically thin clouds tend to heat while low optically thick clouds tend to cool (see Table 4).

Table 4. Global annual mean forcing due to various types of clouds, from the Earth Radiation Budget Experiment (ERBE), according to Hartmann (1993)

Parameter	High clouds		Middle clouds		Low clouds	Total
	Thin	Thick	Thin	Thick	All	
Global fraction /(%)	10.1	8.6	10.7	7.3	26.6	63.3
Forcing (relative to clear sky):						
Albedo (SW radiation)/(W·m^{-2})	-4.1	-15.6	-3.7	-9.9	-20.2	-53.5
Outgoing LW radiation /(W·m^{-2})	6.5	8.6	4.8	2.4	3.5	25.8
Net forcing /(W·m^{-2})	2.4	-7.0	1.1	-7.5	-16.7	-27.7

The positive forcing increases the net radiation budget of the Earth and leads to a warming; negative forcing decreases the net radiation and causes a cooling. (Note that the global fraction implies that 36.7% of the Earth is cloud free.)

From Table 4 it can be seen that low clouds result in a cooling effect of about 17 W·m^{-2}, which means that they play an important role in the Earth's radiation budget (Ohring and Clapp, 1980; Ramanathan et al., 1989; Ardanuy, 1991). The important issue is that even small changes in the lower cloud coverage can result in important changes in the radiation budget and hence has a considerably influence on the Earth's climate (let us remember that the solar irradiance changes during solar cycles is only about 0.3 W·m^{-2}). Figure 41 shows a comparison of the Earth's total cloud coverage (from satellite observations) with CR intensities (from the Climax neutron monitor) and solar activity data over 20 years.

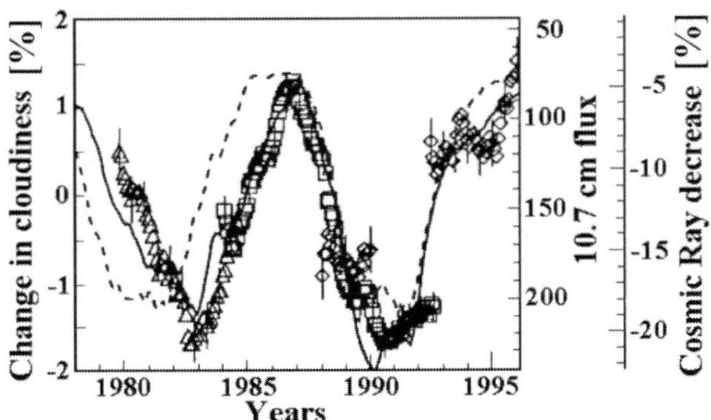

Figure 41. Changes in the Earth's cloud coverage: triangles - from satellite Nimbus 7, CMATRIX project (Stowe et al., 1988); squares - from the International Satellite Cloud Climatology Project, ISCCP (Rossow and Shiffer, 1991); diamonds – from the Defense Meteorological Satellite Program, DMSP (Weng and Grody, 1994; Ferraro et al., 1996). Solid curve – CR intensity variation according to Climax NM, normalized to May 1965. Broken curve – solar radio flux at 10.7 cm. All data are smoothed using twelve months running mean. From Swensmark (2000).

From Figure 41 it can be seen that the correlation of global cloud coverage with CR intensity is much better than with solar activity. Marsh and Swensmark (2000a) came to the conclusion that CR intensity relates well with low global cloud coverage, but not with high and middle clouds (see Figure 42).

It is important to note that low clouds lead, as a rule, to the cooling of the atmosphere. It means that with increasing CR intensity and cloud coverage (see Figure 41), we can expect the surface temperature to decrease. It is in good agreement with the situation shown in Figure 40 for the last thousand years, and with direct measurements of the surface temperature over the last four solar cycles (see Section 11.5, below).

11.5. The Influence of CR on the Earth's Temperature

Figure 43 shows a comparison of eleven year moving average Northern Hemisphere marine and land air temperature anomalies for 1935–1995 with CR intensity (constructed for Cheltenham/Fredericksburg for 1937–1975 and Yakutsk for 1953–1994, according to Ahluwalia, 1997) and Climax NM data, as well as with other parameters (unfiltered solar cycle length, sunspot numbers, and reconstructed solar irradiance). From Figure 43 one can see that the best correlation of global air temperature is with CR intensity, in accordance with the results described in sections 2.1–2.4 above. According to Swensmark (2000), the comparison of Figure 43 with Figure 41 shows that the increase of air temperature by 0.3 °C corresponds to a decrease of CR intensity of 3.5% and a decrease of global cloudiness of 3%; this is equivalent to an increase of solar irradiance on the Earth's surface of about 1.5 W.m^{-2} (Rossow and Cairns, 1995) and is about 5 times bigger than the solar cycle change of solar irradiance, which as we have seen, is only 0.3 W.m^{-2}.

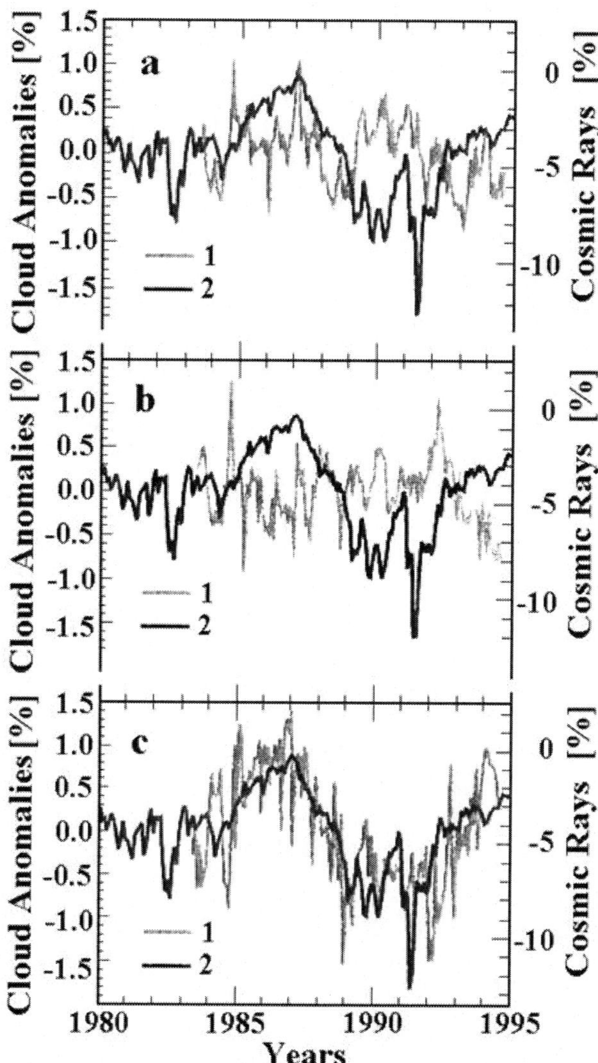

Figure 42. CR intensity obtained at the Neutron Monitor Huancayo/Haleakala (normalized to October 1965, curve 2) in comparison with global average monthly cloud coverage anomalies (curves 1) at heights, H, for: a – high clouds, H > 6.5 km, b – middle clouds, 6.5 km >H > 3.2 km, and c – low clouds, H < 3.2 km. From Marsh and Swensmark (2000a).

11.6. CR Influence on Climate during Maunder Minimum

Figure 44 shows the situation in the Maunder minimum (a time when sunspots were very rare) for: solar irradiance (Lean et al., 1992, 1995); concentration of the cosmogenic isotope ^{10}Be according to Beer et al., 1991 (as a measure of CR intensity, see extended review in Chapter 10 in Dorman, M2004); and reconstructed air surface temperature for the northern hemisphere (Jones et al., 1998).

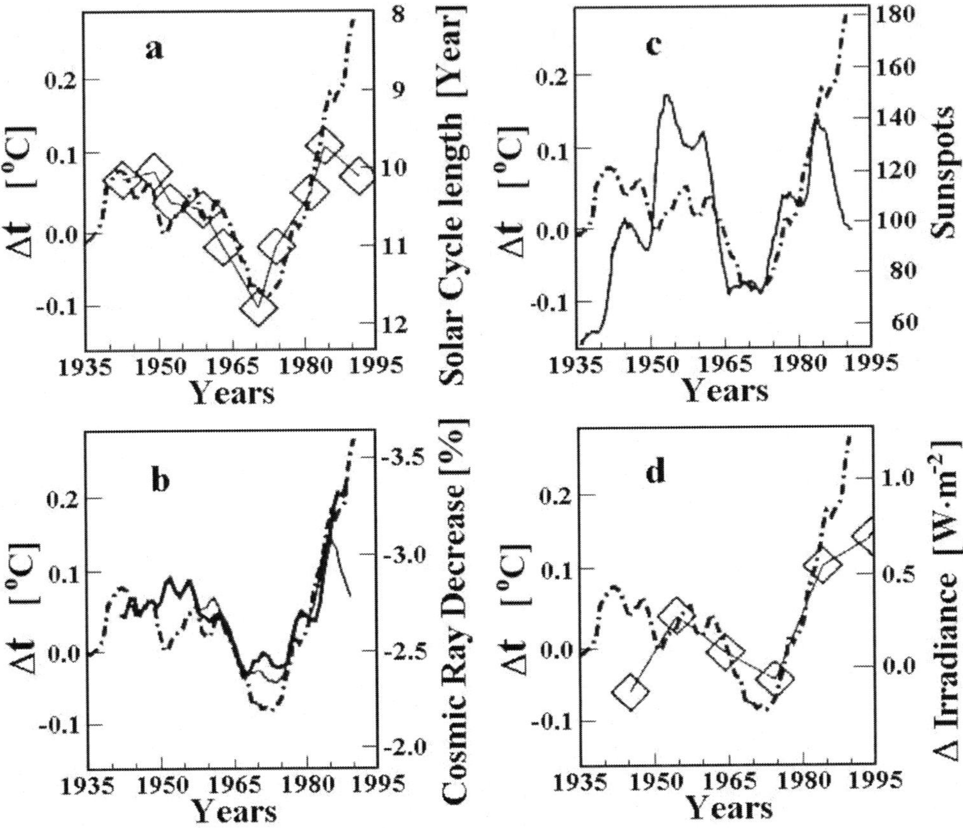

Figure 43. Eleven year average Northern hemisphere marine and land air temperature anomalies, Δt, (broken curve) compared with: (a) unfiltered solar cycle length; (b) Eleven year average CR intensity (thick solid curve – from Climax NM, normalized to ion chambers); (c) Eleven year average of sunspot numbers; and (d) decade variation in reconstructed solar irradiance from Lean et al., 1995 (zero level corresponds to 1367 W.m-2). From Swensmark (2000).

The solar irradiance is almost constant during the Maunder minimum and about 0.24% (or about 0.82 $W \cdot m^{-2}$) lower than the present value (see Panel **a** in Figure 44), but CR intensity and air surface temperature vary in a similar manner - see above sections; with increasing CR intensity there is a decrease in air surface temperature (see Panels **b** and **c** in Figure 44). The highest level of CR intensity was between 1690–1700, which corresponds to the minimum of air surface temperature (Mann et al., 1998) and also to the coldest decade (1690–1700).

11.7. The Connection between Ion Generation in the Atmosphere by CR and Total Surface of Clouds

The time variation of the integral rate of ion generation, q, (approximately proportional to CR intensity) in the middle latitude atmosphere at an altitude between 2–5 km was found by Stozhkov et al. (2001) for the period January 1984–August 1990 using regular CR balloon measurements. The relative change in q, $\Delta q/q$, have been compared with the relative changes of the total surface of clouds over the Atlantic Ocean, $\Delta S/S$, and are shown in Figure 45: the correlation coefficient is 0.91 ± 0.04. This result is in good agreement with results described above (see panel b in Figure 43 and panel c in Figure 44) and shows that there is a direct correlation between cloud cover and CR generated ions.

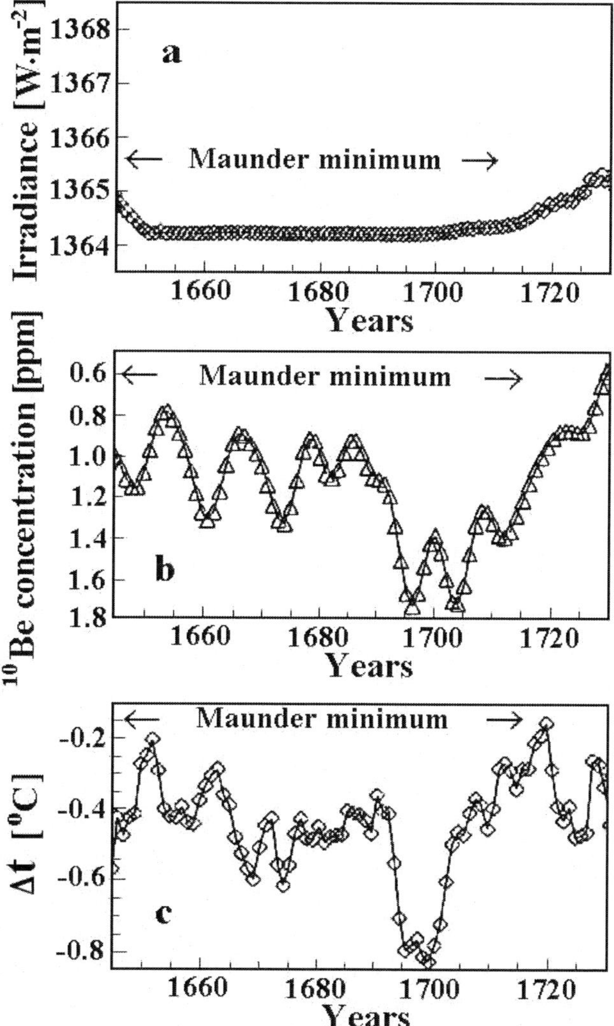

Figure 44. Situation in the Maunder minimum: a - reconstructed solar irradiance (Labitzke and Loon, 1993); b - cosmogenic ^{10}Be concentration (Beer et al., 1991); c - reconstructed relative change of air surface temperature, Δt, for the northern hemisphere (Jones et al., 1998). From Swensmark (2000).

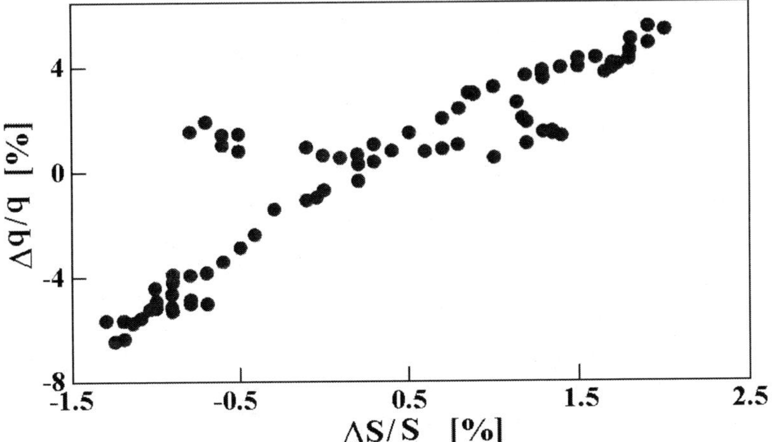

Figure 45. The positive relationship between the relative changes of total clouds covering the surface over Atlantic Ocean, $\Delta S/S$, in the period January 1984–August 1990 (Swensmark and Friis-Christiansen, 1997) and the relative changes of integral rate of ion generation $\Delta q/q$ in the middle latitude atmosphere in the altitude interval 2–5 km. From Stozhkov et al. (2001).

11.8. The Influence of Big CR Forbush Decreases and Solar CR Events on Rainfall

A decrease of atmospheric ionization leads to a decrease in the concentration of charge condensation centers. In these periods, a decrease of total cloudiness and atmosphere turbulence together with an increase in isobaric levels is observed (Veretenenko and Pudovkin, 1994). As a result, a decrease of rainfall is also expected. Stozhkov et al. (1995a, b, 1996) and Stozhkov (2002) analyzed 70 events of CR Forbush decreases (defined as a rapid decrease in observed galactic CR intensity, and caused by big geomagnetic storms) observed in 1956–1993 and compared these events with rainfall data over the former USSR. It was found that during the main phase of the Forbush decrease, the daily rainfall levels decreases by about 17%. Similarly, Todd and Kniveton (2001, 2004) investigating 32 Forbush decreases events over the period 1983–2000, found reduced cloud cover of 12-18%.

During big solar CR events, when CR intensity and ionization in the atmosphere significantly increases, an inverse situation is expected and the increase in cloudiness leads to an increase in rainfall. A study of Stozhkov et al. (1995a, b, 1996) and Stozhkov (2002), involving 53 events of solar CR enhancements, between 1942–1993, showed a positive increase of about 13 % in the total rainfall over the former USSR.

11.9. The Influence of Geomagnetic Disturbances and Solar Activity on the Climate through Energetic Particle Precipitation from the Inner Radiation Belt

The relationship between solar and geomagnetic activity and climate parameters (cloudiness, temperature, rainfall, etc) was considered above and is the subject of much ongoing research. The clearly pronounced relationship observed at high and middle latitudes is explained by the decrease of galactic CR intensity (energies in the range of MeV and GeV) with increasing solar and geomagnetic activity, and by the appearance of solar CR fluxes ionizing the atmosphere (Tinsley and Deen, 1991). This mechanism works efficiently at high latitudes, because CR particles with energy up to 1 GeV penetrate this region more easily to due to its very low cutoff rigidity. Near the equator, in the Brazilian Magnetic Anomaly (BMA) region, the majority of galactic and solar CR is shielded by the geomagnetic field. This field is at an altitude of 200–300 km and contains large fluxes of energetic protons and electrons trapped in the inner radiation belt. Significant magnetic disturbances can produce precipitation of these particles and subsequent ionization of the atmosphere. The influence of solar-terrestrial connections on climate in the BMA region was studied by Pugacheva et al. (1995).

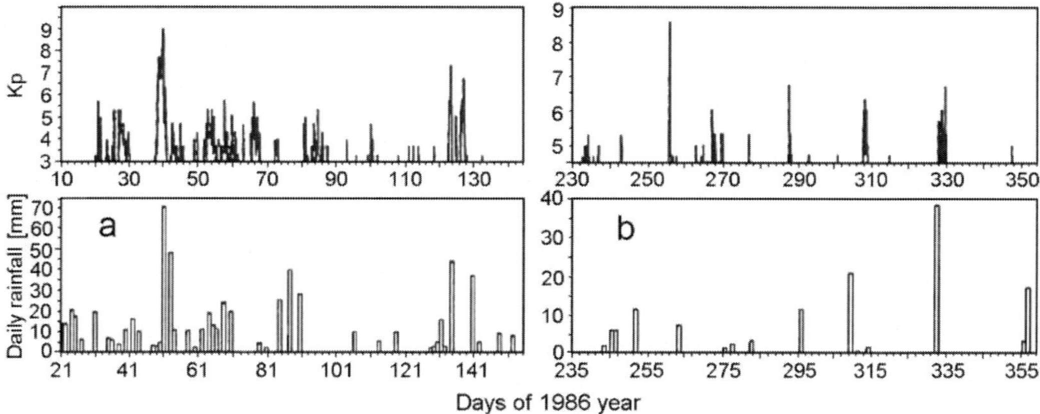

Figure 46. The Kp-index of geomagnetic activity (top panels) and rainfall level (bottom panels) in Campinas (left panels a) and in Ubajara (right panels b) in 1986. According to Pugacheva et al. (1995).

Two types of correlations were observed: 1) a significant short and long time scale correlation between the index of geomagnetic activity Kp and rainfall in Sao Paulo State; 2) the correlation-anti-correlation of rainfalls with the 11 and 22 year cycles of solar activity for 1860-1990 in Fortaleza. Figure 46 shows the time relationship between Kp-index and rain in Campinas (23° S, 47° W) and in Ubajara (3° S, 41° W), during 1986.

From Figure 46, it can be seen that, with a delay of 5–11 days, almost every significant (> 3.0) increase of the Kp-index is accompanied by an increase in rainfall. The effect is most noticeable at the time of the great geomagnetic storm of February 8 1986, when the electron fluxes of inner radiation belt reached the atmosphere between 18-21 February (Martin et al., 1995) and the greatest rainfall of the 1986 was recorded on 19 February. Again, after a series of solar flares, great magnetic disturbances were registered between 19–22 March, 1991. On the 22 March the Sao Paolo meteorological station showed the greatest rainfall of the year.

The relationship between long term variations of annual rainfall at Campinas, the Kp-index and sunspot numbers are shown in Figures 47 and 48.

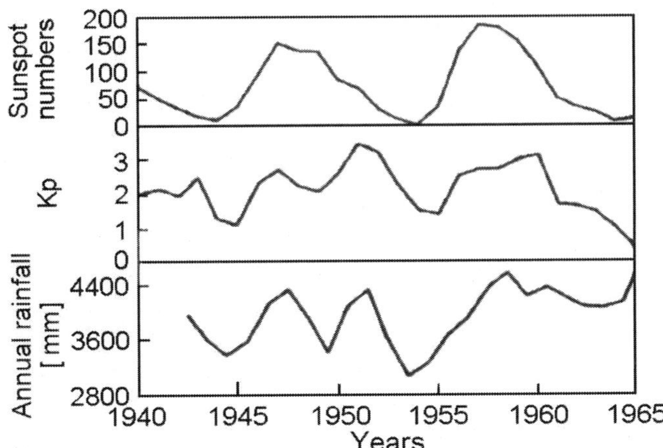

Figure 47. Long-term variations of rainfalls (Campinas, the bottom panel) in comparison with variations of solar and geomagnetic activity (the top and middle panels, respectively) for 1940–1965. From Pugacheva et al. (1995).

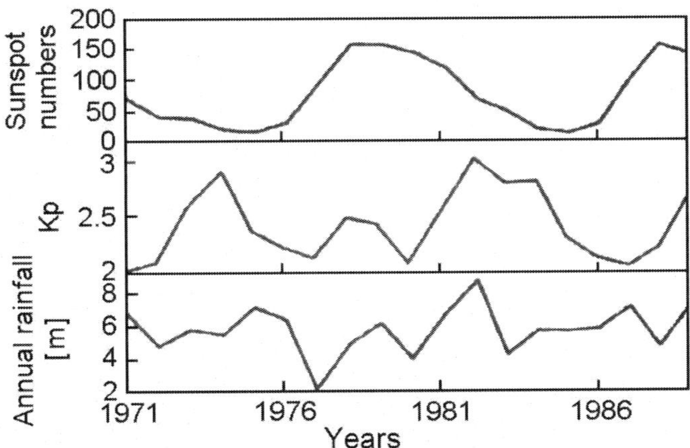

Figure 48. The same as in Figure 47, but for 1971-1990. From Pugacheva et al. (1995).

Figures 47 and 48 show the double peak structure of rainfall variation compared to the Kp-index. Only during the 20th solar cycle (1964–1975), the weakest of the 6 cycles shown, an anti-correlation between rainfalls and sunspot numbers is observed in most of Brazil. The Kp – rainfall correlation is more pronounced in the regions connected with magnetic lines occupied by trapped particles.

In Fortaleza (4° S, 39° W), located in an empty magnetic tube ($L = 1.054$), it is the other kind of correlation (see Figure 49).

From Figure 49 it can be seen that a correlation exists between sunspot numbers and rainfall between 1860–1900 (11th – 13th solar cycles) and 1933–1954 (17th and 18th cycles). The anti-correlation was observed during 1900–1933 (cycles 14th – 16th) and during 1954–1990 (cycles 19th – 21th). As far as sunspot numbers mainly anti-correlate with the galactic CR flux, an anti-correlation of sunspot numbers with rainfalls could be interpreted as a correlation of rainfalls with the CR. The positive and negative phases of the correlation interchange several times during the long time interval 1860–1990, that was observed earlier in North America (King, 1975). Some climate events have a 22 years periodicity similar to the 22 years solar magnetic cycle. The panel b in Figure 49 demonstrate 22 years periodicity of 11 years running averaged rainfalls in Fortaleza. The phenomenon is observed during 5 periods from 1860 to 1990. During the 11th – 16th solar cycles (from 1860 until 1930), the maxima of rainfalls correspond to the maxima of sunspot numbers of odd solar cycles 11th, 13th, 15th and minima of rainfalls correspond to maxima of even solar cycles 12th, 14th, 16th. During the 17th solar cycle the phase of the 22 years periodicity is changed to the opposite and the sunspot number maxima of odd cycles 19th and 21th correspond to the minima of rainfall. The effect is not pronounced (excluding years 1957–1977) in Sao Paolo.

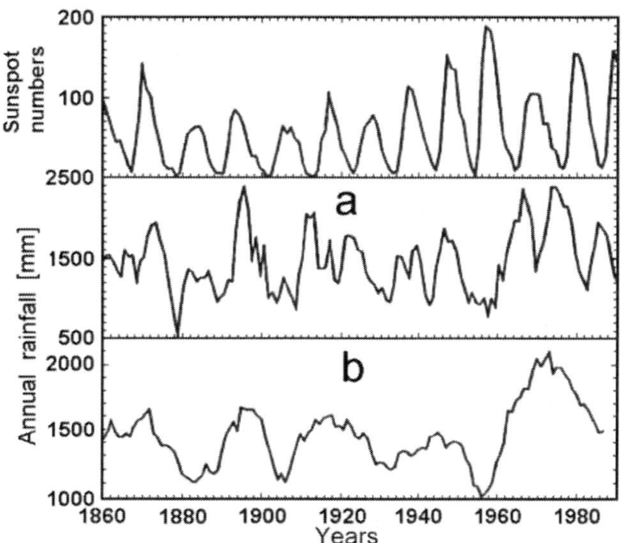

Figure 49. The comparison of yearly sunspot numbers long-term variation (the top panel) with 3 years and 11 years running averaged rainfalls (panels a and b, respectively) in Fortaleza (4° S, 39° W) during 1860–1990. From Pugacheva et al. (1995).

The difference in results obtained in Stozhkov (2002), Todd and Kniveton (2001, 2004) and in Pugacheva et al. (1995) can be easy understand if we take into account the large value of the cutoff rigidity in the BMA region. This is the reason why the variations in galactic and solar CR intensity in the BMA region, are not reflected in the ionization of the air and hence do not influence the climate. However in the BMA region other mechanisms of solar and magnetic activity can influence climatic parameters such as energetic particle precipitation coming from the inner radiation belt.

11.10. On the Possible Influence of Galactic CR on the Formation of Cirrus Hole and Global Warming

According to Ely and Huang (1987) and Ely et al. (1995), there are expected variations of upper tropospheric ionization caused by long-term variations of galactic CR intensity. These variations have resulted in the formation of the cirrus hole (a strong latitude dependent modulation of cirrus clouds). The upper tropospheric ionization is caused, largely, by particles with energy smaller than 1 GeV but bigger than about 500 MeV. In Figure 50 is shown the long term modulation of the difference between NM at Mt. Washington and Durham for protons with kinetic energy 650–850 MeV.

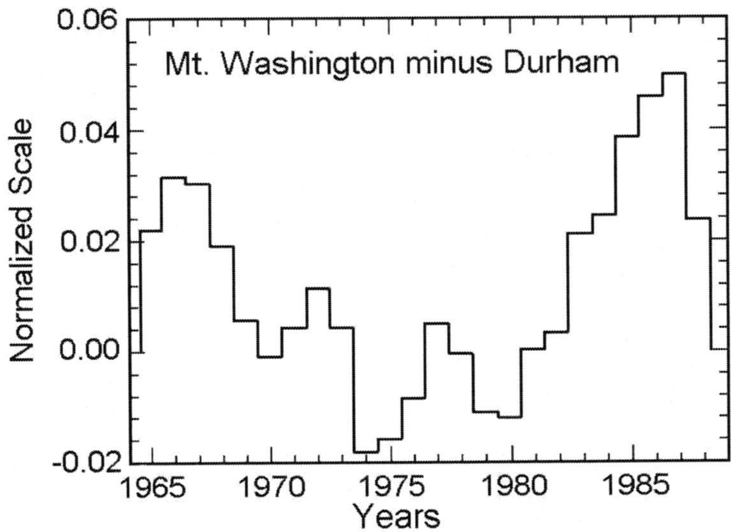

Figure 50. The observed 22 years modulation of galactic CR between 1.24 – 1.41 GV rigidity (i.e., protons with kinetic energy between 650–850 MeV, ionizing heavily in the layer 200–300 g/cm^2). From Ely et al. (1995).

Figure 50 clearly shows the 22 years modulation of galactic CR intensity in the range 650–850 MeV with amplitude of more than 3%. Variations of upper tropospheric ionization do have some influence on the cirrus covering and the "cirrus hole" is expected to correspond to a decrease in CR intensity.

According to Ely et al. (1995), the "cirrus hole" was observed in different latitude zones over the whole world between 1962 and 1971, centered at 1966 (see Figure 51).

Figure 51 gives the cirrus cloud cover data over a 25 a period, for the whole world, the equatorial zone (30°S–30°N) and the northern zone (30°N–90°N), showing fractional decreases in cirrus coverage of 7%, 4%, and 17%, respectively. The decrease of cirrus cover leads to an increase in heat loss to outer space (note, that only a 4 % change in total cloud cover is equivalent to twice the present greenhouse effect due to anthropogenic carbon dioxide). The influence of cirrus hole in the northern latitude zone (30°N–90°N), where the cirrus covering was reduced by 17 %, is expected to be great (this effect of the cirrus hole is reduced in summer by the increase of lower clouds resulting in enhanced insulation) The low temperatures produced from mid to high latitude significantly increase the pressure of the polar air mass and cause frequent 'polar break troughs' at various longitudes in which, for example, cold air from Canada may go all the way to Florida and freeze the grapefruit (Ely et al., 1995). However, when the cirrus hole is not present, the heat loss from mid to high latitudes is much less, and the switching of the circulation patterns (Rossby waves) is much less frequent.

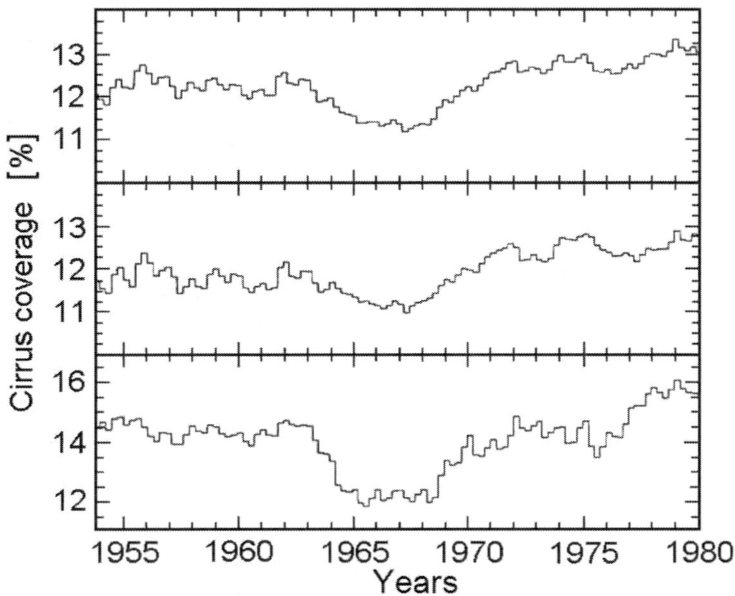

Figure 51. The 'cirrus hole' of the 1960's for: the whole world (the top panel); the equatorial zone (30°S–30°N; middle panel); the northern zone (bottom panel). From Ely et al. (1995).

11.11. Description of Long-term Galactic CR Variation by Both Convection-diffusion and Drift Mechanisms with the Possibility of Forecasting Some Part of Climate Change in the Near Future Caused by CR

It was shown in previous sections that CR might be considered sufficient links determining some part of space weather influence on climate change. From this point of view, it is important to understand mechanisms of galactic CR long-term variations and on this basis to forecast expected CR intensity in the near future. In Dorman (2005a, b, 2006) it was made on the basis of monthly sunspot numbers, taking into account time-lag between processes on the Sun and situation in the interplanetary space as well as the sign of general magnetic field (see Figure 52); in Belov et al. (2005) – mainly on the basis of monthly data of solar general magnetic field (see Figure 53).

From Figure 52 it follows that in the frame of convection-diffusion and drift models used in Dorman (2005a, b, 2006), it can be determined with very good accuracy expected galactic CR intensity in the past (when monthly sunspot numbers are known) as well as behavior of CR intensity in future if monthly sunspot numbers can be well forecasted. According to Belov et al. (2005), the same can be made with good accuracy on the basis of monthly data on the solar general magnetic field (see Figure 53). Let us note that described above results obtained in Dorman (2005a, b, 2006) and Belov et al. (2005) give the possibility to forecast some part of climate change connected with CR.

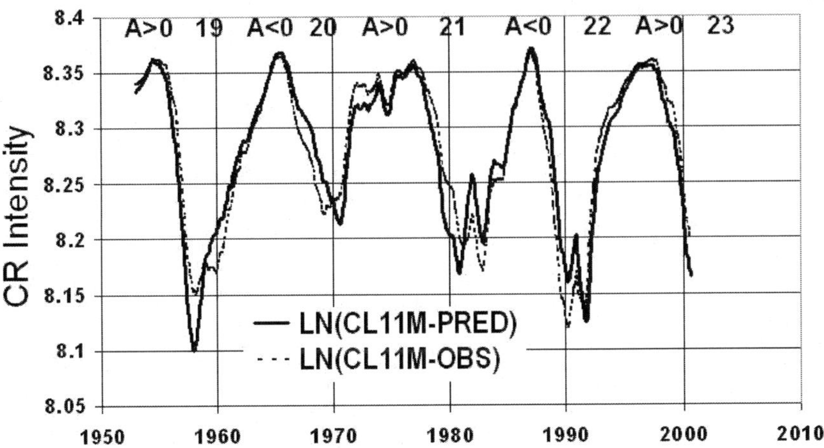

Figure 52. Comparison neutron monitor CR intensity of observed by Climax averaged with moving period of eleven month LN(CL11M-OBS) with predicted on the basis of monthly sunspot numbers from the model of convection-diffusion modulation, corrected on drift effects LN(CL11M-PRED). Correlation coefficient between both curves 0.97. From Dorman (2006).

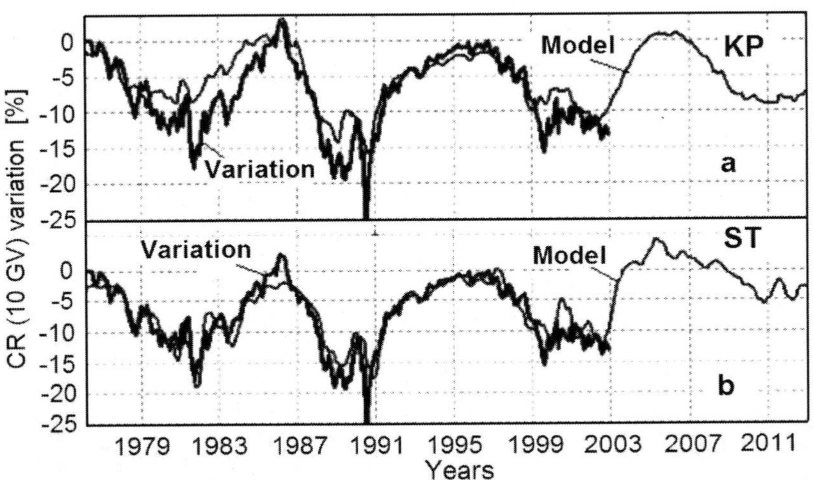

Figure 53. The forecast of galactic CR behavior based on the predicted values of the global characteristics of the solar magnetic field, thick line – data of CR intensity observations (Moscow NM), thin line – the predicted CR variation up to 2013 based on data of Kitt Peak Observatory (upper panel) and based on data of Stanford Observatory (bottom panel). From Belov et al. (2005).

11.12. Influence of Long-term Variation of Main Geomagnetic Field on Global Climate Change through CR Cutoff Rigidity Variation

The sufficient change of main geomagnetic field leads to change of planetary distribution of cutoff rigidities R_c and to the corresponding change of the i-th component of CR intensity $N_i(R_c, h_o)$ at some level h_o in the Earth's atmosphere $\Delta N_i(R_c, h_o)/N_{io} = -\Delta R_c W_i(R_c, h_o)$, where $W_i(R_c, h_o)$ is the coupling function (see details in Chapter 3 in Dorman, M2004). Variations of CR intensity caused by change of R_c are described in detail in Chapter 7 in Dorman (M2009), and here we will demonstrate results of Shea and Smart (2003) on R_c changing for the last 300 and 400 years (see Figure 54 and Table 5, correspondingly).

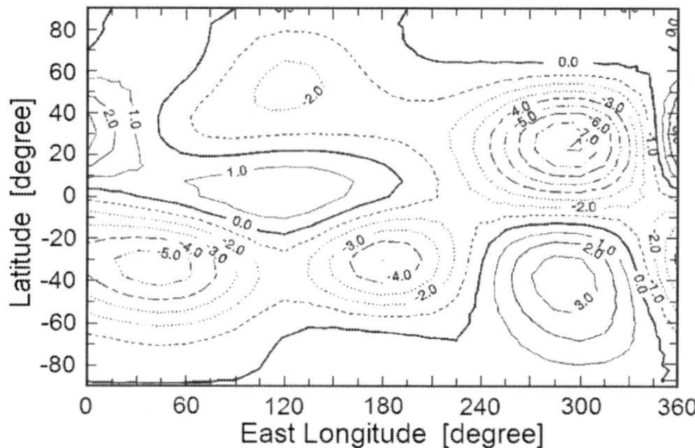

Figure 54. Contours of the change in vertical cutoff rigidity values (in GV) between 1600 and 1900. Full lines reflect positive trend (increasing of cutoff rigidity from 1600 to 1900); dotted lines reflect negative trend. According to Shea and Smart (2003).

Table 5. Vertical cutoff rigidities (in GV) for various epochs 1600, 1700, 1800, 1900, and 2000, as well as change from 1900 to 2000 owed to changes of geomagnetic field. According to Shea and Smart (2003)

Lat.	Long. (E)	Epoch 2000	Epoch 1900	Epoch 1800	Epoch 1700	Epoch 1600	Change 1900–2000	Region
55	30	2.30	2.84	2.31	1.49	1.31	−0.54	Europe
50	0	3.36	2.94	2.01	1.33	1.81	+0.42	Europe
50	15	3.52	3.83	2.85	1.69	1.76	−0.31	Europe
40	15	7.22	7.62	5.86	3.98	3.97	−0.40	Europe
45	285	1.45	1.20	1.52	2.36	4.1	+0.25	N. Amer.
40	255	2.55	3.18	4.08	4.88	5.89	−0.63	N. Amer.
20	255	8.67	12.02	14.11	15.05	16.85	−3.35	N. Amer.
20	300	10.01	7.36	9.24	12.31	15.41	+2.65	N. Amer.
50	105	4.25	4.65	5.08	5.79	8.60	−0.40	Asia
40	120	9.25	9.48	10.24	11.28	13.88	−0.23	Asia
35	135	11.79	11.68	12.40	13.13	14.39	+0.11	Japan
−25	150	8.56	9.75	10.41	11.54	11.35	−1.19	Australia
−35	15	4.40	5.93	8.41	11.29	12.19	−1.53	S. Africa
−35	300	8.94	12.07	13.09	10.84	8.10	−3.13	S. Amer.

Table 5 shows that the change of geomagnetic cutoffs, in the period 1600 to 1900, is not homogeneous: of the 14 selected regions, 5 showed increasing cutoffs with decreasing CR intensity, and 9 regions showed decreasing cutoffs with increasing CR intensity. From Table 5 it can be seen also that at the present time (from 1900 to 2000) there are sufficient changes in cutoff rigidities: decreasing (with corresponding increasing of CR intensity) in 10 regions, and increasing (with corresponding decreasing of CR intensity) in 3 regions. These changes give trend in CR intensity change which we need to take into account together with CR 11 and 22 years modulation by solar activity, considered in section 11.11.

11.13. Atmospheric Ionization by CR: The Altitude Dependence and Planetary Distribution

The main process in the link between CR and cloudiness is the air ionization which triggers chemical processes in the atmosphere. Figure 55 illustrates the total ionization of atmosphere by galactic CR (primary and secondary) as a function of altitude. The planetary distribution of ionization at the altitude of 3 km according to Usoskin et al. (2004), is shown in Figure 56 for the year 2000.

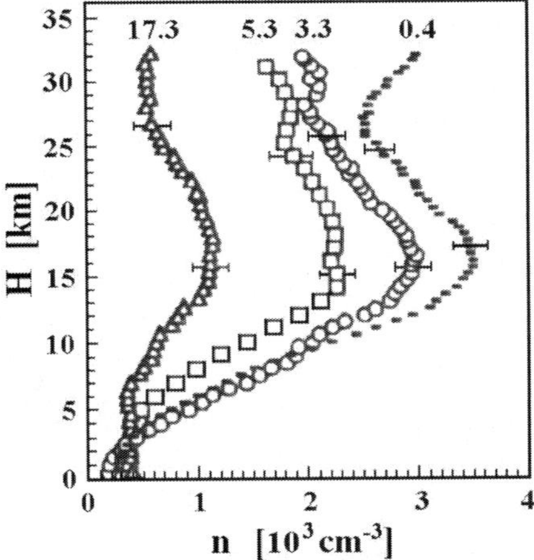

Figure 55. The ion concentration, n, profiles as a function of altitude, H, for different geomagnetic cutoff rigidities (numbers at the top are in units of GV). The horizontal bars indicate the standard deviations. From Ermakov et al. (1997).

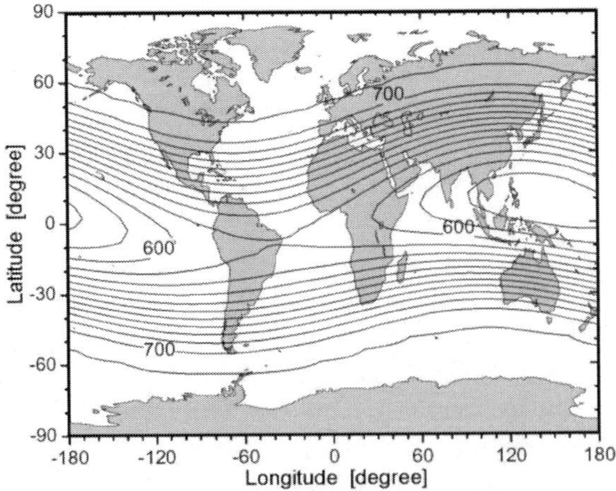

Figure 56. Planetary distribution of calculated equilibrium galactic CR induced ionization at the altitude of 3 km ($h = 725$ g/cm^2) for the year 2000. Contour lines are given the number of ion pairs per cm^3 in steps of 10 cm^{-3}. From Usoskin et al. (2004).

11.14. Project 'Cloud' As an Important Step in Understanding the Link between CR and Cloud Formation

The many unanswered questions in understanding the relationship between CR and cloud formation are being investigated by a special collaboration, within the framework of European Organization for Nuclear Research, involving 17 Institutes and Universities (Fastrup et al., 2000). The experiment, which is named 'CLOUD', is based on a cloud chamber (which is designed to duplicate the conditions prevailing in the atmosphere) and 'cosmic rays' from the CERN Proton Synchrotron. The Project will consider possible links between CR, variable Sun intensities and the Earth's climate change (see Figure 57).

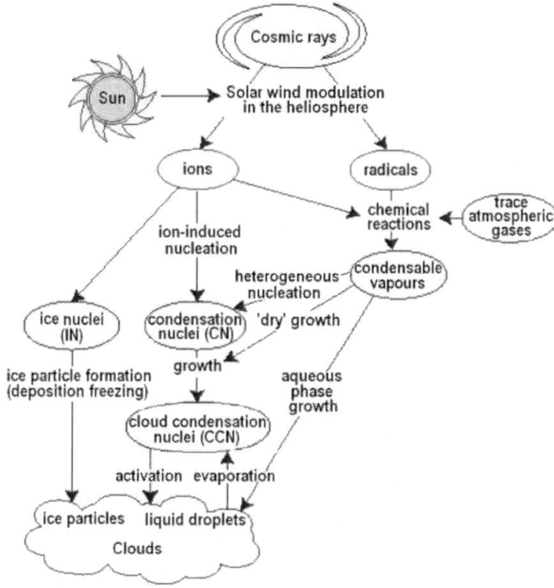

Figure 57. Possible paths of solar modulated CR influence on different processes in the atmosphere leading to the formation of clouds and their influence on climate. From Fastrup et al. (2000).

11.15. The Influence on the Earth's Climate of the Solar System Moving around the Galactic Centre and Crossing Galaxy Arms

The influence of space dust on the Earth's climate has been reviewed in Veizer et al. (2000). Figure 58 shows the changes of planetary surface temperature for the last 520 Ma according to Veizer et al. (2000). These data were obtained from the paleo-environmental records. During this period the solar system crossed Galaxy arms four times. In doing so, there were four alternating warming and cooling periods with planetary temperature changes of more than 5°C.

The amount of matter inside the galactic arms is more than the amount on the outside. The gravitational influence of this matter attracts the inflow of comets from Oort's cloud to the solar system (Fuhrer et al., 1999; Maseeva, 2004). It results in an increase in concentration of interplanetary dust in the zodiac cloud and a cooling of the Earth's climate (Hansen et al., 1999).

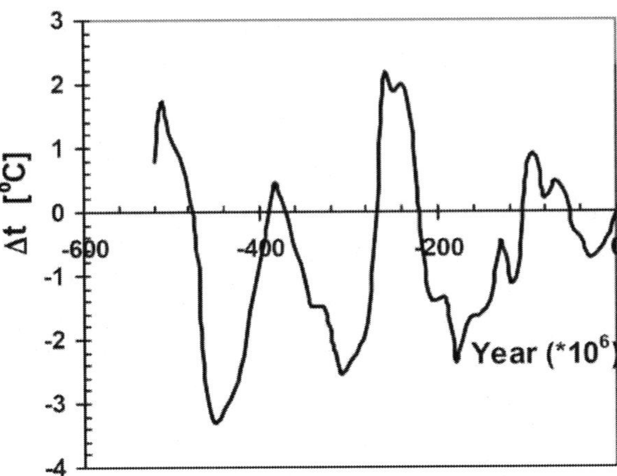

Figure 58. Changes of air temperature, Δt, near the Earth's surface for the last 520 million years according to the pale-environmental records (Veizer et al., 2000). From Ermakov et al. (2006).

11.16. The Influence of Molecular-Dust Galactic Clouds on the Earth's Climate

The solar system moves relative to interstellar matter with a velocity of about 30 km/s and sometimes passes through molecular-dust clouds. During these periods, we can expect a decrease in sea level air temperature. According to Dorman (2008b), the prediction of the interaction of a dust-molecular cloud with the solar system can be performed by measurements of changes in the galactic CR distribution function. From the past, we know that the dust between the Sun and the Earth has led to decreases of solar irradiation flux resulting in reduced global planetary temperatures (by 5-7°C). The plasma in a moving molecular dust cloud contains a frozen-in magnetic field; this moving field will modify the stationary galactic CR distribution outside the Heliosphere. The change in the distribution function can be significant, and it should be possible to identify these changes when the distance between the cloud and the Sun becomes comparable with the dimension of the cloud. The continuous observation of the time variation of CR distribution function for many years should make it possible to determine the direction, geometry and the speed of the dust-molecular cloud relative to the Sun. Therefore, it should, in the future, be possible to forecast climatic changes caused by this molecular-dust cloud.

Figure 59 shows the temperature changes at the Antarctic station Vostok (bottom curve), which took place over the last 420,000 years according to Petit et al. (1999). These data were obtained from isotopic analysis of O and H extracted from the ice cores at a depth 3,300 m. It is seen from Figure 59 that during this time the warming and cooling periods changed many times and that the temperature changes amounted up to 9°C. Data obtained from isotope analysis of ice cores in Greenland, which cover the last 100,000 years (Fuhrer et al., 1999), confirm the existence of large changes in climate.

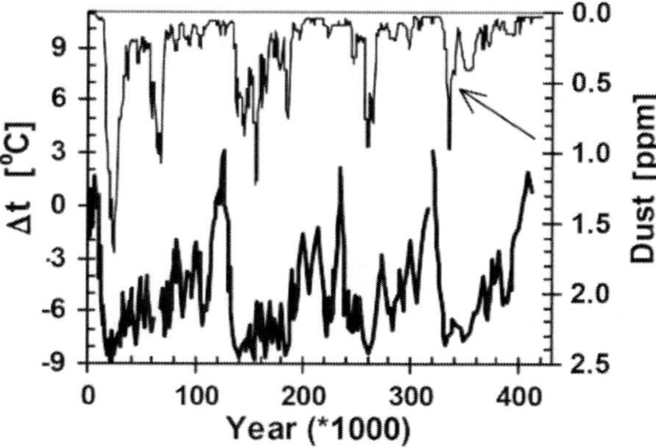

Figure 59. Changes of temperature, Δt, relative to the modern epoch (bottom thick curve) and dust concentration (upper thin curve) over the last 420 000 a (Petit et al., 1999). From Ermakov et al. (2006).

11.17. The Influence of Interplanetary Dust Sources on the Earth's Climate and Forecasting for the Next Half-century

According to Ermakov et al. (2006a, b), the dust of the zodiac cloud is a major contributory factor to climate changes in the past and at the present time. The proposed mechanism of cosmic dust influence is as follows: dust from interplanetary space enters the Earth's atmosphere during the yearly rotation of the Earth around the Sun. The space dust participates in the processes of cloud formation. The clouds reflect some part of solar irradiance back to space. In this way, the dust influences climate. The main sources of interplanetary dust are comets, asteroids, and meteor fluxes. The rate of dust production is continually changing. The effect of volcanic dust on the Earth's air temperature is illustrated in Figure 60 (Hansen et al., 1999).

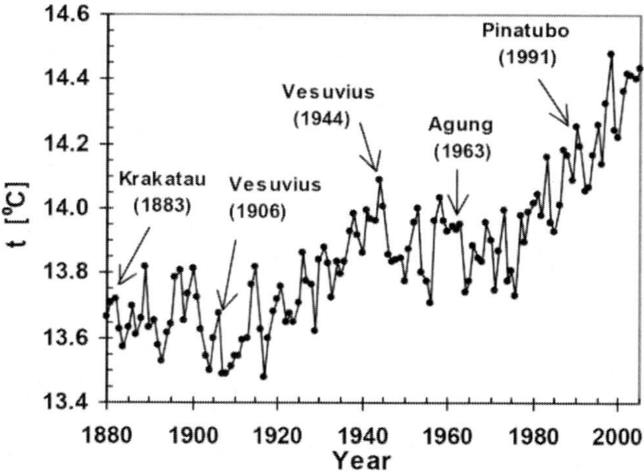

Figure 60. Yearly average values of the global air temperature, t, near the Earth's surface for the period from 1880 to 2005 according to Hansen et al. (1999). Arrows show the dates of the volcano eruptions with the dust emission to the stratosphere and short times cooling after eruptions. From Ermakov et al. (2006a).

According to Ermakov et al. (2006a), the spectral analysis of global surface temperature during 1880-2005 shows the presence of several spectral lines that can be identified with periods of meteor fluxes, comets, and asteroids. The results of analysis have been used in Ermakov et al. (2006a, b) to predict changes in climate over the next half-century: the interplanetary dust factor of cooling in the next few decades will be more important than the warming from the greenhouse effect (see Figure 61).

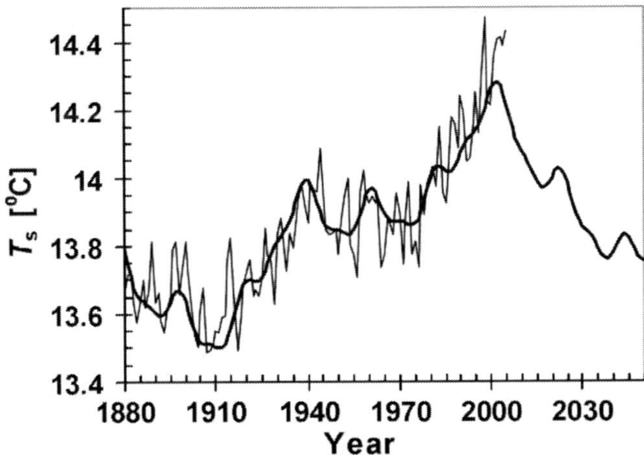

Figure 61. Forecasting of the planetary air-surface temperature for the next half-century. Thin line – measurements, thick line – calculated based on main harmonics. From Ermakov et al. (2006a, b).

11.18. Space Factors and Global Warming

It is now commonly thought of that the current trend of the global warming is causally related to the accelerating consumption of fossil fuels by the industrial nations. However, it has been suggested that this warming is a result of a gradual increase of solar and magnetic activity over the last 100 years. According to Pulkkinen et al. (2001), as shown in Figure 62, the solar and magnetic activity has been increasing since the year 1900 with decreases in 1970 and post 1980. This figure shows that the aa index of geomagnetic activity (a measure of the variability of the interplanetary magnetic field, IMF), varies, almost in parallel, with the global temperature anomaly.

It has been well established that the brightness of the Sun varies in proportion to solar activity. The brightness changes are very small and cannot explain all of the present global warming. However, the gradual increase of solar activity over the last hundred years has been accompanied by a gradual decrease of CR intensity in interplanetary space (Lockwood et al., 1999). The direct measurements of CR intensity on the ground by the global network of neutron monitors (NM) as well as regular CR intensity measurements from balloons in the troposphere and stratosphere over a period of more than 40a, show that there is a small negative trend of galactic CR intensity (Stozhkov et al., 2000) of about 0.08% per year. Extrapolating this trend to a hundred years, it gives a CR intensity decrease of 8%. From Figure 41 it can be seen that the decreasing of CR intensity by 8% will lead to a decrease of cloud coverage of about 2%. According to Dickinson (1975), decreasing cloud coverage by

2% corresponds to increasing the solar radiation falling on the Earth by about 0.5%. Using this information, Stozhkov et al. (2001) concluded that the observed increase of average planetary ground temperature of 0.4–0.8°C over the last 100 years, may be a result of this negative trend of CR intensity. Sakurai (2003) came to the same conclusion on the basis of analyzing data of solar activity and CR intensity.

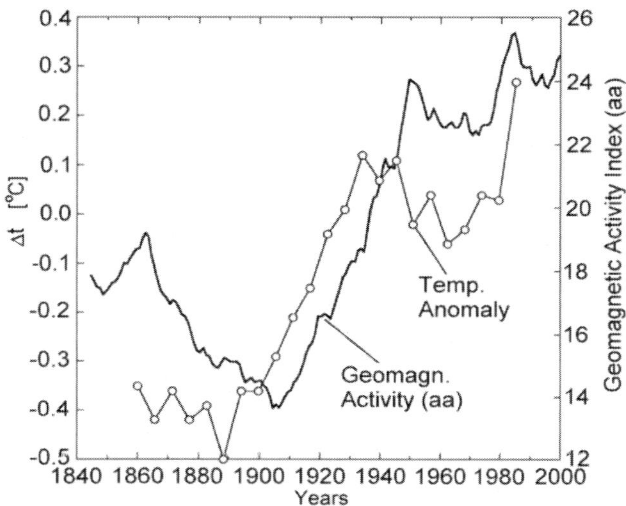

Figure 62. The geomagnetic activity (index aa) at the minimum of solar activity variation of the global temperature anomaly, Δt, from 1840 to 2000. From Pulkkinen et al. (2001).

11.19. Discussion on CR and Other Space Factors Effect on the Earth's Climate

Many factors from space and from anthropogenic activities influenced the Earth's climate. The initial response is that space factors are unlikely to be responsible for most of our present climate change. However, it is important that all possible space factors be considered, and from an analysis of past climate changes, we can identify our present phase and can predict future climate change. During the last several hundred million years, the Sun has rotated around the galactic centre several times with resultant climate changes. For example, considering the effects due to galactic molecular-dust concentrated in the galactic arms, as given in Figure 58, we can see that during the past 520 million years, there were four periods with surface temperatures lower than what we are presently experiencing and four periods with higher temperatures. On the other hand, during the past 420 thousand years (Figure 59) there were four decreases of temperature (the last one was about 20–40 thousand years ago: the so called big ice age), and five increases of temperature, the last of which happened few thousand years ago. At present, the Earth is in a slight cooling phase (of the order of one degree centigrade over several thousand years).

When considering CR variations as one of the possible causes of long-term climate change we need to take into account not only CR modulation by solar activity but also the change of geomagnetic cutoff rigidities (see Table 5). It is especially important when we consider climate change on a scale of between thousand and million years: paleo-magnetic investigations show that during the last 3.6 million years the magnetic field of the Earth changed sign nine times, and the Earth's magnetic moment changed - sometimes having a value of only one-fifth of its present value (Cox et al., 1967) - corresponding to increases of CR intensity and decreases of the surface temperature. The effects of space factors on our climate can be divided into two types:

- the 'gradual' type, related to changes on time scales ranging from 10^8 years to 11-22 years, producing effects which could be greater than that produced from anthropogenic factors, and
- the 'sudden' type, coming from Supernova explosions and asteroid impacts, for example, and which may indeed be catastrophic to our civilization. Volcanic and anthropogenic factors are also in a sense, 'sudden' factors in their effect on climate change.

It is necessary to investigate all of the possible 'sudden' factors and to develop methods of forecasting and also for protecting the biosphere and the Earth's civilization from big changes in climate and environment. We cannot completely exclude the possibility that a Supernova explosion, for example, took place twenty years ago at a distance of say thirty light years away. In this case its influence on our climate and environment will be felt in ten years time. According to Ellis and Schramm (1995), in this case, UV radiation would destroy the Earth's ozone layer over a period of about 300 years. The recent observations of Geminga, PSR J0437-4715, and SN 1987A strengthen the case for one or more supernova extinctions having taken place during the Phanerozoic era. In this case, a nearby supernova explosion would have depleted the ozone layer, exposing both marine and terrestrial organisms to potentially lethal solar ultraviolet radiation. In particular, photosynthesizing organisms including phytoplankton and reef communities would most likely have been badly affected.

As Quante (2004) noted, clouds play a key role in our climate system. They strongly modulate the energy budget of the Earth and are a vital factor in the global water cycle. Furthermore, clouds significantly affect the vertical transport in the atmosphere and determine, in a major way, the redistribution of trace gases and aerosols through precipitation. In our present-day climate, on average, clouds cool our planet; the net cloud radiative forcing at the top of the atmosphere is about $- 20 \ \mathrm{W \ m^{-2}}$. Any change in the amount of cloud or a shift in the vertical distribution of clouds, can lead to considerable changes in the global energy budget and thus affect climate.

Many of the 'gradual' types of space factors are linked to cloud formation. Quante (2004) noted that galactic CR (Swensmark and Friis-Christiansen, 1997; Swensmark, 1998; Marsh and Swensmark, 2000a, b) was an important link between solar activity and low cloud cover. However, new data after 1995 shows that the problem is more complicated and the correlation no longer holds (Kristjánsson et al., 2002). Kristjánsson et al. (2004) point out that still many details are missing for a complete analysis, but a cosmic ray modulation of the low cloud cover seems less likely to be the major factor in our present climate change, but its role in future climate changes must not be ruled out.

In this Section much emphasis has been given to the formation of clouds and the influence CR plays (through ionization and influence on chemical processes in the atmosphere) in their formation. This does not imply that CR is the only factor in their formation; dust, aerosols, precipitation of energetic particles from radiation belts and greenhouse gases, all play their part. However, the influence of CR is important and has been demonstrated here through:

- a direct correlation during one solar cycle and also for much longer periods,
- the correlation of CR intensity with the planetary surface temperature,
- by the direct relationship between cloud formation and CR air ionization,
- by the relationship between geomagnetic activity and rainfall through precipitation of energetic particles from radiation belts and
- by linking CR intensities with the cirrus holes.

The importance of CR cannot be stressed highly enough and it is important to develop methods for determining, with high accuracy, galactic CR intensity variations for the past, the present and for the near future.

In this Section several attempts have been made to explain the present climate change (the relatively rapid warming of the Earth discussed in Section 6 for 1937-1994), using space factors:

- through eleven year average CR intensities,
- by the increasing geomagnetic activity,
- by the decreasing CR intensity (of 8% over the past one hundred years) and
- by relating the spectral analysis of Ermakov et al. (2006a) to global temperature during 1880-2005.

Their results show the presence of several spectral lines that can be identified with periods of meteor fluxes, comets, and asteroids. On the basis of this work, Ermakov et al. (2006a, b) has predicted a cooling of the Earth's climate over the next half-century which they believe will be more important than warming from the greenhouse effect.

Finally, it appears that our present climate change (including a rapid warming of about 0.8 C over the past hundred years) is caused by a collective action of several space factors, volcano activities (with the dust emission rising to the stratosphere, resulting in the short term cooling after eruptions), as well as by anthropogenic factors with their own cooling and warming contributions. The relation between these contributions will determine the final outcome. At present the warming effect is stronger than the cooling effect. It is also very possible that the present dominant influence is anthropogenic in origin.

From Figure 59 it can be seen that now we are near the maximum global temperature reached over the last 400 thousand years, so an additional rapid increase of even a few °C could lead to an unprecedented and catastrophic situation. It is necessary that urgent and collective action be taken now by the main industrial countries and by the UN, to minimize the anthropogenic influence on our climate before it became too late. On the other hand, in the future, if the natural change of climate results in a cooling of the planet (see Figures 58 and 59), then special man-made factors, resulting in warming, may have to be used to compensate.

We considered above the change of planetary climate as caused mostly by two space factors: cosmic rays and space dust. This we advocate with the use of results obtained over a long time periods, from about ten thousand years to many millions of years. We also considered very short events of only one or a few days, such as Forbush effects and Ground Level Enhancements (GLE). In these cases, it is necessary to use a superposed method, summing over many events to sufficiently reduce the relative role of meteorological factors, active incident to the aforementioned short events (see Section 9).

Analysis of Kristjánsson and Kristiansen (2000) contradicts the simple relationship between cloud cover and radiation assumed in the CR-cloud-climate hypothesis, because this relationship really is much more complicated and is not the main climate-causal relationship. We agree with this result and with the result of Erlykin et al. (2009a, b) and Erlykin and Wolfendale (2011) that there is no simple causal connection between CR and low cloud coverage (LCC), that there is no correlation between CR and LCC for short-term variations, and that while there is a correlation between CR and LCC for long-term variations, that this connection can explain not more than about 20% of observed climate change. But the supposition of Erlykin et al. (2009a, b) that the observed long-term correlation between cosmic ray intensity and cloudiness may be caused by parallel separate correlations between CR, cloudiness and solar activity contradicts the existence of the hysteresis effect in CR caused by the big dimensions of the Heliosphere (Dorman and Dorman, 1967a, b; Dorman et al., 1997; Dorman, 2005a, b, 2006). This effect, which formed a time-lag of CR relative to solar activity of more than one year (different in consequent solar cycles and increasing inverse to particle energy), gives the possibility of distinguishing phenomena caused by CR from phenomena caused directly by solar activity (i.e. activity without time lag). The importance of cosmic ray influence on climate compared with the influence of solar irradiation can be seen clearly during the Maunder minimum (see Figure 44). Cosmic ray influence on climate over a very long timescale of many hundreds of years can be seen from Figure 40 (through variation of ^{14}C).

It is necessary to take into account that the main factors influencing climate are meteorological processes: cyclones and anticyclones; air mass moving in vertical and horizontal directions; precipitation of ice and snow (which changes the planetary radiation balance, see Waliser et al., 2011); and so on. Only after averaging for long periods (from one-ten years up to 100-1000 years and even million of years) did it become possible to determine much smaller factors that influence the climate, such as CR, dust, solar irradiation, and so on. For example, Zecca and Chiari (2009) show that the dust from comet 1P/Halley, according to data of about the last 2000 years, produces periodic variations in planetary surface temperature (an average cooling of about 0.08 °C) with a period 72 ± 5 years. Cosmic dust of interplanetary and interstellar origin, as well as galactic CR entering the Earth's atmosphere, have an impact on the Earth's climate (Ermakov et al., 2006, 2007, 2009; Kasatkina et al., 2007a, b; Dorman, 2009, 2012). Ermakov et al. (2006, 2009) hypothesized that the particles of extraterrestrial origin residing in the atmosphere may serve as condensation nuclei and, thereby, may affect the cloud cover. Kasatkina et al. (2007a, b) conjectured that interstellar dust particles may serve as atmospheric condensation nuclei, change atmospheric transparency and, as a consequence, affect the radiation balance. Ogurtsov and Raspopov (2011) show that the meteoric dust in the Earth's atmosphere is potentially one of the important climate forming agents in two ways: (i) particles of meteoric haze may serve as condensation nuclei in the troposphere and stratosphere; (ii) charged meteor particles residing in the mesosphere may markedly change (by a few percent) the total atmospheric resistance and thereby, affect the global current circuit. Changes in the global electric circuit, in turn, may influence cloud formation processes.

Let us underline that there is also one additional mechanism by which CR influence lower cloud formation, precipitation, and climate change: the nucleation by cosmic energetic particles of aerosol and dust, and through aerosol and dust – increasing of cloudiness. It was shown by Enghoff et al. (2011) in the frame of the CLOUD experiment at CERN that the irradiation by energetic particles (about 580 MeV) of air at normal conditions in the closed chamber led to aerosol nucleation (induced by high energy particles), and simultaneously to an increase in ionization (see also Kirkby et al., 2011).

Let us note that in this Section, we considered CR and dust aerosols separately, but acting in the same direction. Increasing CR intensity and increasing of aerosols and dust leads to increasing of cloudiness and a corresponding decrease of planetary surface temperature. Now, consistent with the experimental results of Enghoff et al. (2011) on aerosol nucleation in the frame of the CLOUD Project on the accelerator at CERN (see a short description of this Project in Dorman, M2004), it was found that with increasing intensity of energetic particles, the rate of formation of aerosol nucleation in the air at normal conditions increased sufficiently. This result can be considered as some additional physical evidence of the CR – cloud connection hypothesis.

12. DO CR AND OTHER SPACE WEATHER FACTORS INFLUENCE ON AGRICULTURAL MARKETS? NECESSARY CONDITIONS, POSSIBLE SCENARIOS, AND CASE STUDIES

12.1. The Matter of the Problem

The problem of the possible influence of solar activity on the Earth agriculture has a history of almost three centuries. One of the first records of this influence appears in the description of the Royal Society of Britain made by Jonathan Swift, a famous professor of theology and the father of the European satire, in his book devoted to the third travel of Gulliver to the island of Laputa (Swift, 1726). In this biting satire, Swift mentions the main fears of Laputians as follows:

a) The Earth can find itself in the burning hot comet tail and enter into a period of global warming threatening loss of all life.
b) The Sun's daily emission of its rays, without any nutriment to supply them, will at last be wholly consumed and annihilated, which must be accompanied by the Destruction of this Earth.

These fears reflected theories prevailing in Swift's contemporary scientific society, particularly to the Derham's theory that sunspots arise from solar volcanoes (Renaker, 1979). Interesting how enduring these fears are that can be met nowadays in science fiction novels and disaster films.

The next statement of the possible influence of the space weather on the agriculture was made by the father of the observational astronomy, William Herschel, famous for the discovery of Uranus. He compared the grain prices from the fundamental study of Adam Smith (1776) with the data on the sunspots number, and made a far-reaching conclusion about the coincidence of the five long periods of few sunspots with periods of the rise in wheat prices (Herschel, 1801). This work was presented to the Royal Society, and it was met extremely negatively by other members of the Society. According to a very sharp opinion of one of the Society's leaders (Lord Brougham), "since the publication of Gulliver's voyage to Laputa, nothing so ridiculous has ever been offered to the world" (Soon and Yaskell, 2003).

William Jevons, one of the founders of the mathematical school in economics, was the next victim of interest to this problem. He was one of the first economists that paid attention to cyclic processes in economic processes, and particularly, to cyclic recurrence of economic crises whose average length he estimated to be 10.2 years (Jevons, 1875, 1878, 1879). Because of the striking closeness of the crises cycle length to the 11-year length of the sunspots cycle, discovered shortly before the works of Jevons, he assumed that the solar activity modulates economic processes in some way. As a possible cause-effect relationship that could explain the coincidence of the both cycle's length, Jevons assumed that in years of "unfavorable" solar activity weather anomalies happen. They cause bad harvests, and consequently – a rise in food price that generates stock exchange crises. Moreover, Jevons extrapolated the series of coincidences between times of the previous five stock exchange crises and periods of the minimal solar activity, and also put attention on the rises in the stock exchange activity immediately after those minimums. He dared to forecast the next economic crisis in the period of the next minimum of solar activity expected in 1879, and this caused the natural reaction in the stock exchange and in the media. Because the promised crisis did not happen[3] either in 1879 or in later years, the Jevons theory was completely discredited, and the new term "sun-spot equilibria" appeared in economic science[4].

Further discovery of the exceptional constancy, within the range of 0.1%, of the level of solar radiation that reaches the Earth (received the name of the solar constant) deprived supporters of the solar activity influence on the Earth processes of their physical arguments for a long time. This "pessimistic" period continued until the discovery of new channels of solar activity influence on the Earth, united now by the term "space weather".

12.2. The Earth Weather Influence on Harvest

The fact of the harvest dependence on weather conditions is trivial and well-known. However, the threshold character of this dependence often is overlooked. For example, occasional light frosts in the period of blossoming or pouring rains at harvest can during a few

---

[3] In fact, the forecast of Jevons came true, in part. In the period of the minimal solar activity of 1876-1878 he mentioned, the monsoon transfer of the moist ocean air to the Southern India (the major part of the British Empire of that time) disappeared for three years. This caused a severe draught that brought to the humanitarian catastrophe known as Great Famine in India. As many as 6-10 million people died of starvation and another 60 million people turned into refugees.

[4] This term reflects high stock exchange sensitivity to any "significant" scientific, economic, or political information about threats to its stability. Such information forms pessimistic expectations of many participants of the stock exchange trade and can invite real panic in spite of unreliability of the initial forecast. The example of Jevons's forecast lays in the basis of the term "sunspot equilibria" in the modern economics. This is the state where speculators' expectations formed by a priori information, are taken into account. This information even being essentially irrelevant, can determine the speculators' behavior and influence the forecasted reality (Azariadis, 1981).

days reduce productivity almost to zero. At the same time, average annual or monthly weather indices (of temperature, humidity, precipitation) for this region can remain almost the same. For various crops, the critical agrotechnical and weather conditions can be different. Thus, for the perennial crop like grapes high enough temperatures in the period of ripening is very important, and for annual grain crops the air humidity can be critical.

To minimize losses of harvest because of weather anomalies, the practice of long-term selection and zoning of the most suitable crops is used, as well as agrotechnologies optimal for type of weather prevailing in the region. However, it is this long-term crops zoning for standard weather that under conditions of fast climate change can cause a shift of dominating crops to the state of risky agriculture very sensitive to small local weather anomalies[5]. The importance of localized efforts to increase local adaptive capacity, take advantage of climate changes that may lead to increased agriculture productivity where this is possible, and to buffer the situations where increased stresses are likely, is underlined in the study of Thornton et al. (2009) who simulated cropping for Eastern Africa. A similar conclusion that climate change takes an expression in local processes such as increased climatic variability, is made in the article of Head et al. (2011) using data of Australian wheat farming households.

12.3. Yield Deficit and Grain Market

In a free market, the deficit of first-necessity goods (agricultural goods belong to this category, of course) causes price spikes that reduces the demand and brings the market to the state determined by the equilibrium between supply and demand. Shortages (or surpluses) can cause the market reaction in a form of panic when prices spikes (or slumps) make them much higher (or lower) than the equilibrium level[6]. Such market panic reaction can be revealed in the case of food deficit caused by bad yield because of local weather anomalies. The food prices bursts stimulate import from regions not stricken with the bad weather what serves to dampen prices and to restore the market equilibrium. This mechanism, however, does not always work effectively. First, transport costs are included into the final price of agricultural products. Second, transfer of agricultural goods is restricted often by geographic, economic, and political barriers – relative market isolation, high customs, and low quotes protecting local producers[7]. Thus, in the markets isolated from the external supply and/or needed to pay expensive transport costs[8], there can be a panic reaction to the deficit of agricultural products, particularly, of grain, with burst price spikes.

---

[5] Shifts in the opposite direction are possible when climate changes lead to an increase in productivity and a decrease in its variability for crops traditional in the region. This is the conclusion of the study of Richter and Semenov (2005) where the influence of possible climate changes on winter wheat yield in England and Wales is modeled.

[6] Shiller (1990) studied stock market crashes and "hot" real estate markets in USA and Japan. Most of the respondents in this study picked up the answer "There is panic buying, and price becomes irrelevant" explaining the burst in housing prices.

[7] Another possible solution of the supply deficit and panic increase in vital goods prices (food, fuel) is the imposition of a rationing system. In this system ration, cards distributed by the authorities are used to balance between supply and demand, and not prices (widely used in USSR and many European countries during and after the World War II).

[8] Here is an example of influence of transport costs on the grain market: the difference between prices in Britain (importer) and in the USA (exporter) was reduced significantly in the end of the 19 century after beginning of wide use of inexpensive steamships for freight transport (Ejrnæs et al. 2008).

12.4. Three Necessary Conditions for Implementation of the Relationship between Space Weather and Earth Prices

Summarizing the above analysis, we formulate the following three necessary conditions for a possible implementation of the causal relationship between anomalies of the space weather and grain prices bursts caused by them:

1. High sensitivity of weather (local - in space and time) to the space weather factors, for example – to cosmic rays, sun ultraviolet radiation, and magnetosphere activity).
2. High sensitivity of the major grain crops yield in the region to weather anomalies (belonging to a risky agricultural region) in the studied period.
3. High sensitivity of the grain market to the supply deficit due to limited import capacity that can cause burst (panic) price leaps.

12.5. Causal Relationships between Space Weather and Grain Markets

The scheme in Figure 63 illustrates a possible chain of relationships that lead to the prices reaction to the unfavorable states of space weather in the regions where the above necessary conditions are fulfilled.

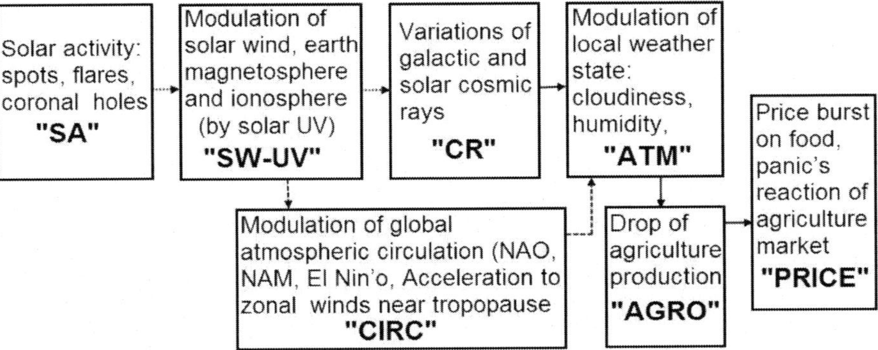

Figure 63. Possible causal chains between space weather and earth agriculture markets. Dash arrows – direct relations without threshold effects, solid arrows – non-linear relations and existence of threshold states.

The main detail of this scheme is the existence of non-linear sensitivity between selected elements of the scheme (imaged by bold arrows). As a result, the final connection of "solar activity – price level" in terms of theory of catastrophes (Arnold, M1992) must be presented not as a "hard dependence" like to $Y = \sum k_i X_i + Noise$, but as a "soft type dependence" $Y = \sum k_i(X_i, Y) * X_i(t - \tau) + Noise$ taking in account both feedback, possible phase delay and existence of external factors of influence caused by sources of another nature. Here X is the vector of input variables (solar activity space weather state, conditions in the Earth atmosphere, market state), Y – output response (prices, social reaction), k_i – coefficients (functions) of connection, τ - phase delay.

12.6. Four Possible Scenarios of Price Reaction on Space Weather

In the frame of the proposed causal scheme, four variants of the agriculture reaction to space weather/local weather are possible for different climatic zones and for different crops. The proposed set of scenarios is based on cosmic rays as a modulation factor that initiates cloudiness, deficit of radiation, lower temperature and excess of participation. We would like to note here that a possible space weather – earth weather relationship through the causal chain "cosmic ray" – "cloudiness" is not a unique channel of space weather influence. A full list of possible influence mechanisms is wider. In this paragraph, we limit our analysis by this only chain to illustrate possible scenarios of response that depend on local climate state and agriculture conditions. All these scenarios are presented in Figure 63 and include:

A. Zone I of risky agriculture is sensitive to deficit of solar irradiance, cold weather and redundant rains. The most probable candidates are northern Europe, particularly England. The most unfavorable state of space weather/solar activity is a minimum of sunspots with minimal solar wind and maximal cosmic ray flux that penetrate into the Earth's atmosphere and stimulate cloudiness formation in regions sensitive to this link (e.g., Northern Atlantic).

B. Zone II of risky agriculture is sensitive to dry and hot weather with droughts. The most probable candidates are southern Europe and the Mediterranean (Italy, Spain, North Africa). The most unfavorable state of space weather/solar activity is a maximum of sunspots with maximal solar wind and minimal cosmic ray flux that diminishes cloudiness formation above Atlantics.

C. Zone III includes specific cases of risky agriculture that is sensitive either to a deficit of precipitation (droughts) or to excess rains/cloudiness. This situation may take place in zones with the climate unfavorable for agriculture where it can succeed only in a very narrow range of weather condition (rains, temperature, humidity, solar irradiation). In this specific situation we can expect unfavorable conditions for agriculture production both in maximum and minimum states of solar activity. A possible candidate is Iceland in the18-th and 19-th centuries.

D. Zone IV where agriculture is not risky and its sensitivity to local weather fluctuations is relatively low. This situation leads to a neutral reaction of the agriculture market on the space weather and the phase of solar activity. The most probable candidate is Central Europe.

These four types of reactions describe possible scenarios of causal chains between the space weather and the Earth agriculture production for the suggested relationship "cosmic ray" – "cloudiness". Our analysis aims to reveal examples of the above scenarios in specific regions and historical periods in Europe and North America.

According to the above description, we may expect systematical price bursts in solar minimum states for cold and wet regions (zone I in Figure 64).

Similar price bursts are expected in hot and dry regions in states of solar maximum (zone III on Figure 64). For these two cases a typical time interval between price bursts is caused by a period of solar activity - 9-13 years. For specific regions, crops sensitive both to excess of precipitation and to its deficit (zone IV) can fail, and thus price bursts are possible both in minimal and maximal states of solar activity with typical time intervals between price bursts close to half of a solar cycle period - 4-6 years. The same typical time of price bursts can be caused by a formal mixing of data from different zones with opposite type of price reaction into one joined series. The example can be found in the work of Beveridge (1921) where the "Beveridge index of price" was calculated by averaging price indexes for approximately 50 markets from Western and Central Europe. In this work, Fourier response was revealed on periods 4-5 years.

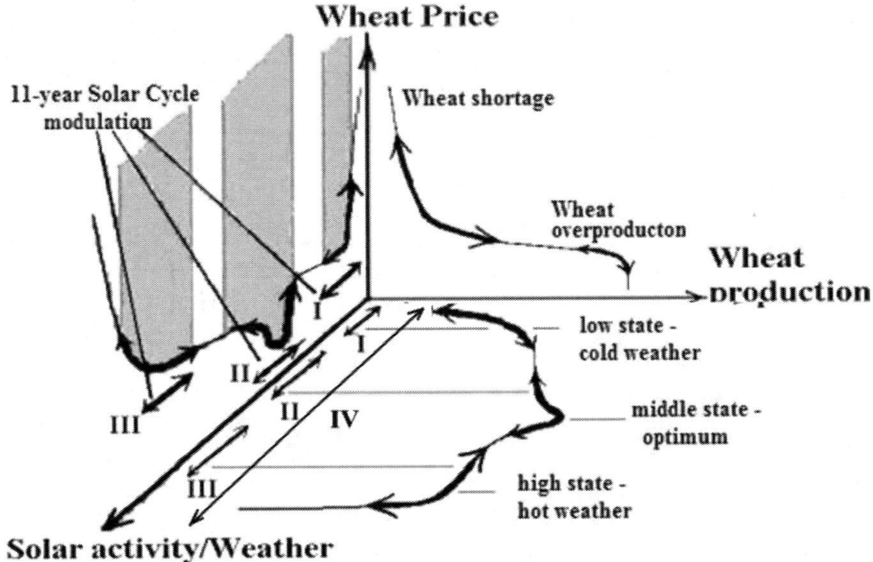

Figure 64. Four possible scenarios of responses of agricultural production and markets on possible modulation of local weather by solar activity/space weather.

12.7. Using Data

The main source of data on solar activity that is the driver of space weather is "sunspot number". In spite of starting observations in 1611 after Galileo discovered these numbers, the further minimum of Maunder interrupted regular sunspot observations. Regular and daily sunspot observations were renewed by several scientists only in the 18-th century after solar activity returned. It allowed constructing a catalog of solar activity (sunspots numbers) from 1700 up to present time[9]. Later, sunspot data were restored for the previous period (the 17-th century) using days when any manifestations of sunspots were observed on the Sun episodically[10].

Another approach used for estimation of solar activity in the period of Maunder Minimum and the previous period is an indirect method of restoration of the solar activity level based on the measurement of isotope Be^{10} in the Greenland ice (Beer et al, 1998). Although this method cannot give reliable quantitative estimation of sunspot numbers

---

[9] http://sidc.oma.be/DATA/yearssn.dat.
[10] Results of restoration are discussed in Ogurtsov et al. (2003), Nagovitsyn (2007) and presented in free FTP site of NOAA/NGDC: ftp://ftp.ngdc.noaa.gov/STP/SOLAR_DATA/ SUNSPOT_NUMBERS/ ANCIENT_DATA/ earlyssn.dat.

themselves, it enables identifying moments of maximal and minimal states of solar activity. We would like to note here that the data based on Be^{10} measurement are sensitive directly to cosmic ray flux penetrated into the atmosphere that is much more important than sunspot numbers. Cosmic rays are an agent of direct influence of space weather on atmospheric processes, controlled ionization and vapor condensation. On the opposite, the causal chain from sunspots to atmospheric process includes a lot of intermediate elements (solar wind, state of Earth magnetosphere and global atmospheric circulation). These intermediate elements in the causal chain are able to mask a possible link between sunspots and local atmosphere states.

The first data on wheat price used by William Hershel for his analysis was published in the famous work of Adam Smith (1776). The most full and reliable database on wheat prices in the Middle Age England was created one hundred years later as a result of selfless work of a great economist and statistician Prof. Rodgers (1887).

His database:

1. Cover the period from 1259 to the middle of the 18-th century.
2. Uses only price data for wheat purchased by monasteries and colleges that benefited from tax exemption in this period. This circumstance made unnecessary "tax optimization" considerations during registration of the transactions that increased reliability of this data source.

In addition to data on wheat prices, in our analysis we use the 700 year database (1264-1954) on consumable basket prices (Brown and Hopkins, 1956). An additional database that we used was an archive of wheat prices for 90 wheat markets of Middle Age Europe. We used the fullest part of the database from the period of 1590-1702 that covered the period of Maunder Minimum and the small Ice Age in Europe. The source of this database is the International Institute of Social History (2005).

To search for possible manifestations of space weather influence on agriculture markets in the New Time we used data of Agriculture Department of USA for the period of 1866-2002 of "durum" wheat prices (durum wheat is used for bread and bakery). We used also the comparative analyses of wheat production and wheat prices in Middle Age England and France (Appleby, 1979); the wheat prices analysis for Middle Age England from the article of Beveridge (1921); the analysis of the correlation between local weather and crops in the end of the 19-th century – first decades of the 20-th century from the work of Hooker (1907); analysis of wheat production in USA during first decades of the 20-th century from the article of Acrtowski (1910). To analyze a possible influence of forage a crop failure on drop of livestock in Iceland 18-19-th centuries that in turn caused famine and increased mortality toll we used the work of Vasey (2001).

12.8. Using Methods of Analysis

The main difficulties in searching for response of agriculture indices used as indicators of possible relationships to abnormal states of space weather/solar activity are caused by the following two factors:

1. Solar cycles of activity as a generator of variations in space weather are not stable both in frequencies of variations and in amplitudes of cycles. Time intervals between minimums of solar activity (a cycle period) change in a wide interval from 8 to 15 years. The amplitude of cycles is described as the Wolf number changes from tens to hundreds and often decreases to zero in periods of a few tens of years.

2. Solar activity includes different components (sunspots, flares, coronal holes) formed by different elements of the dynamo process or by their combinations (poloidal and azimuthal magnetic fields under photosphere, convection, differential rotation and meridional circulation). As a result, different components of solar activity have different phase patterns (change with phase during the cycle) and their behavior during a cycle of activity changes. For example, input of coronal holes (and recurrent fluxes in the solar wind accelerated from coronal holes) to space weather is maximal in minimum state of solar activity and is absent in maximal state. Solar flares of small amplitude have a very good correlation with sunspot numbers but for big solar flares with coronal mass ejections and acceleration of strong proton fluxes the situation is much more complicated and dynamic. The distribution of proton flares changes from cycle to cycle radically, and for some cycles their frequency is maximal not in the maximum of sunspot number but predominantly in the phase of rise or decay of sunspot activity (Shea and Smart, 1992). Additional difficulties are caused by instability of a relative input of different components of solar activity into space weather formation. It explains, for example, the observed drastic change during the last 100 years in the phase pattern of geomagnetic activity as a result of the change of relative input of coronal holes and sunspot during dynamo process in the Sun (Georgieva and Kirov, 2011; Kishcha, Dmitrieva and Obridko, 1999).

1. The impact of space weather factors on the Earth's atmosphere has a place on the background of complex and yet not explained global atmospheric circulation with effects of long-range action[11], phase instability and possible transitions like a strange attractor (Ruzmaikin, 2007).

2. Influence of space weather on the local weather and on agriculture markets, if it takes place, must happen on the background of another factors of influences that act simultaneously with space weather factors. These influences may have a comparable amplitude and may have both random and regular nature with periodical components on the same times as space weather (10-20 years). As examples, we may refer on climate variations, political or military events that lead to economic shocks, scientific and technological revolutions.

This situation makes use of classical statistical methods non-effective. As an example, we can mention methods aimed at the selection of a harmonic signal (Fourier analysis, periodogram analysis) with a selected period (for example, 11-year period), or at search for a direct linear relationship (regression or correlation analysis). It means that to search for a space weather influence we have to use another methods and statistics, much more robust to

---

[11] This effect may be illustrated by a situation when a local weather cataclysm like El-Nino near the west coast of South and Central America leads to a drastic weather response in the opposite side of the Earth in North Atlantics as a result of the atmospheric and oceanic mass and heat transfers. There are researchers that claim a relationship between solar activity and El Nino is possible (Ruzmaikin, 1999).

variability of period of solar cycle and amplitude of solar activity, to non-stability of atmospheric circulation and other external factors.

In the frame of our approach we may use the next robust statistics:

1. Comparison of interval statistics both between price bursts and between extreme states of solar activity (minimums, for example).
2. Search of a systematical phase asymmetry of wheat prices in "favorable" and "unfavorable" states of space weather/solar activity.
3. Regression analysis that uses dichotomous "dummy" variables "yes"/"no" related to the solar activity state (for example, "yes" or "no" for the minimum of solar activity).

In some regions, we must take into account a possibility of "long-range action" when regions located far from a region of space weather influence on the earth weather may be nevertheless sensitive to space weather action as a result of global circulation and cyclonic transfer of formed clouds over thousands kilometers (for example, from the North Atlantic to East Siberia).

Another effect that we have to take into account is a possible phase delay in price reaction on "unfavorable" states of space weather. These delays may be caused by reserves of the crop from the previous yield and by natural inertia of agricultural markets. Sensitivity of agriculture production to local weather conditions may be much higher in a case when a crop is concentrated in a compacted region with a certain type of weather conditions than in a case of a crop dispersed through thousands of kilometres with different climate conditions in different regions and, accordingly, with different (or even opposite) type of sensitivity to external factors. These are factors and parameters we take into account in the analysis of data for different regions in different historical periods that meet the above defined criteria.

12.9. Wheat Markets' Sensitivity to Space Weather in Medieval England

Medieval England is an ideal testing region for searching for manifestations of space weather influence on agricultural prices. In this period, this region met all three of the above mentioned conditions necessary for this relationship to be realized:

1. The weather in the region depends on the space weather factors in the zone of cloudiness formation in the North Atlantics that is sensitive to the cosmic rays variations during changes of solar activity.
2. The region belonged to the zone of risky agriculture, particularly, for wheat that is highly sensitive to weather anomalies in the vegetation period.
3. Relative isolation from European markets amplified price reaction to shortage in grain.

Another advantage of using this region in the research is availability of the reliable data collection of grain prices from 1259 to the 18th century due to the distinguished effort of Prof. Rogers (1887). The initial curve of changes of yearly wheat prices is presented in Figure 65.

Table 5. Statistical characteristics (average, median, standard deviation) for two samples of intervals between price bursts (wheat price and consumer basket price) and for the sample of intervals between minimal states of solar activity

Samples	Median	Average	Standard deviation
Min-Min intervals for sunspots	10.7	11.02	1.53
Intervals between wheat price bursts	11.00	11.14	1.44
Intervals between consumer basket price bursts	11.00	10.5	1.28

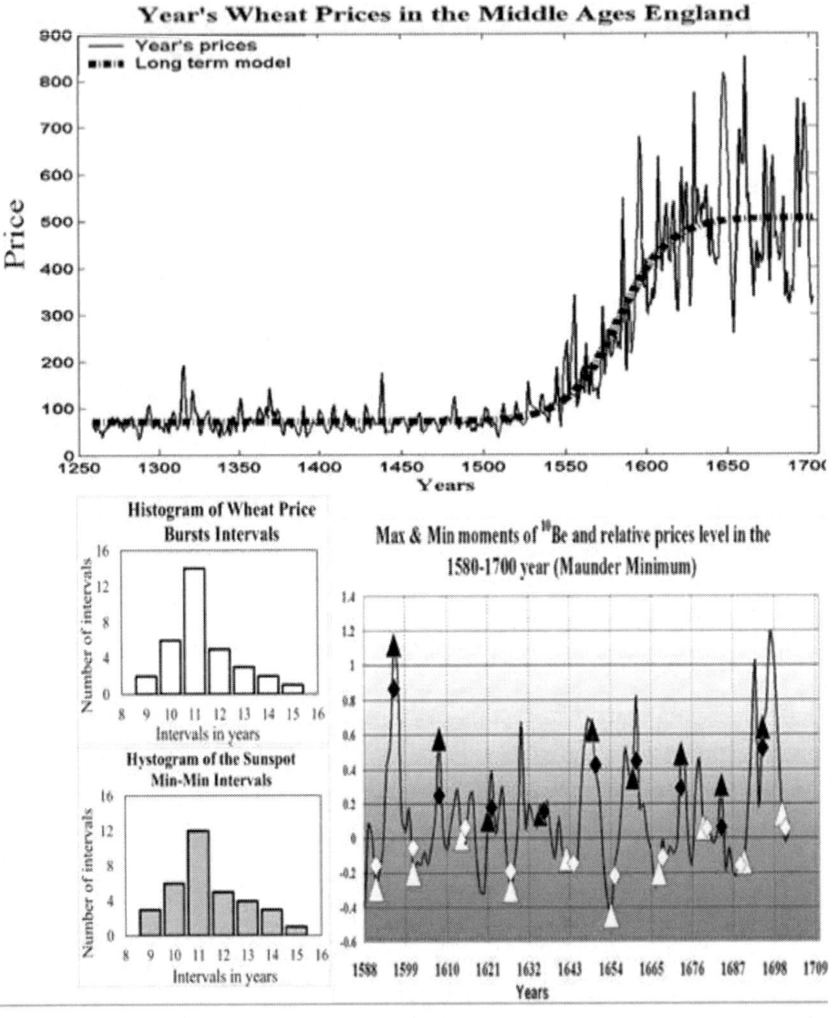

Figure 65. Panel above – wheat prices in Medieval England smoothed by a logistic function; below left – comparison of histograms of intervals between minimal states of solar activity and of intervals between wheat price bursts. The range of intervals is 8-16 years. For both histograms the maximum is achieved in the point of 11 years; below right – price asymmetry in states of minimum/maximum of solar activity identified by the method of Be10 in Maunder Minimum: white triangular are prices in states of maximum, white diamonds are prices average for three years that include the state of maximum, black triangular and diamonds are prices defined similarly for states of minimum of solar activity.

To search for space weather influences we used the above described methods of comparison between statistics of intervals and of search for price asymmetry. First, we compared statistics of intervals between wheat price bursts and of intervals between states of solar activity minimum (Pustil'nik et al., 2003; Pustil'nik and Yom Din, 2004a, b). The statistics were compared both for the distributions of the intervals by length (Figure 65b, c) and for statistical characteristics of these distributions (average length of the intervals, medians, standard deviations – Table 5).

It follows both from the results in Table 5 for statistical characteristics and from the histograms of these characteristics (Figure 65b) that the hypothesis that the distributions of intervals for wheat price and consumer basket price bursts and the distribution of solar activity minimums are not different one from another is significant (99%). Highly significant price asymmetry for minimum and maximum states of solar activity in the period of Maunder Minimum is another evidence of the space weather influence on wheat prices in Medieval England (Figure 65c). As it is seen in this Figure, wheat prices in years of solar activity minimum are higher (on average, twice as many) than prices in years of the nearest solar activity maximum for all nine cycles of solar activity. In this examination, the significance level is 99%, too.

These results show that realization of the above described causal chain between the space weather and grain markets is possible for the case study of Medieval England. The observed market reaction corresponds to case I (zone of risky agriculture sensitive to deficit of solar irradiance, cold weather and redundant rains) described in Section 12.8. The results are in good agreement with those expected for this climatic zone.

12.10. Wheat Markets Sensitivity to Space Weather in Medieval Europe

In the next stage of the analysis we tried to answer the following question. Is the discovered sensitivity of the grain markets in Medieval England a universal feature for all markets on earth (like dependence of the same prices on the seasonal "winter - summer" changes) or is this feature realized only in isolated zones and in certain historic periods when the three necessary conditions for the causal chain "space weather" – "earth prices" are satisfied? To answer the question we fulfilled regression analysis of data from 22 European grain markets that were presented by relatively full data from all markets in the database of the International Institute of Social History (2005).

We used the well-known method of regression analysis with dummy variables (first introduced by Suits (1957))[12]. In our study, we used this method for establishing relationships between states of grain markets with states of minimum or maximum of solar activity. We tested the hypothesis of existence of such relationship between solar activity and prices, and estimated its significance for samples for markets from different European zones. For this purpose, we introduced a dummy variable d_{min} that takes a value 1 in the years of solar activity minimum, and a value 0 otherwise. In a similar way, we introduced a dummy variable d_{max} for years of solar activity maximum. In a certain way, this method develops the method of the direct search for the price asymmetry of minimal and maximal phases of solar activity used above for the wheat market in Medieval England (Figure 5c). However, the method of

---

[12] These variables describe qualitative data, for example, of the kind "yes" – "no".

dummy variables is more accurate and enables clear estimating of the significance level of the hypothesis about the existence of such phase asymmetry.

For our analysis we used the period of 1590-1702 that includes the Maunder Minimum when the solar activity fell sharply. The following two reasons led to the choosing of this period:

1. Europe went through a little ice age in this period. A considerable part of areas moved to risky agriculture zones with an increased influence of weather anomalies on grain production.
2. Only for this period special measurements of the isotope Be^{10} were carried out to reflect directly contribution of cosmic rays that are a probable factor of influence on weather (Beer, 1998)[13].

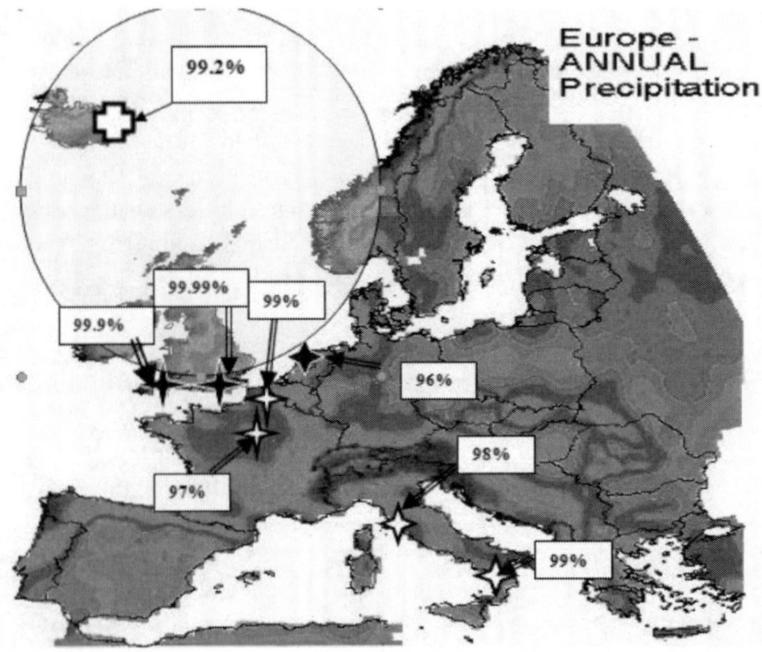

Figure 66. The Europe map of precipitation. Localization of wheat markets that show high sensitivity to space weather and its significance level, are shown. Black stars show markets sensitive to states of minimal solar activity (London, Exeter – England; Leiden – the Netherlands), white stars show markets sensitive to states of maximal solar activity (Napoli, Bassano – Italy). The stars size corresponds to the significance level.

We present significance level of the influence of minimal and maximal states of solar activity on wheat prices for a number of markets in Figure 66 where the Europe map of

---

[13] Additional analysis for other periods using regression with dummy variables showed a significant relationship between solar activity extremes and wheat prices in England up to 1840's (the results can be sent by the authors by request). Weak market sensitivity to the space weather after this period we explain by a sharp increase of the grain import – from 5%-8% during 1801-1840 up to 17%-18% beginning from 1841 (Ejrnaes et al. 2008). This growth of import broke the causal chain in the part "sensitivity of the isolated market to supply shortage".

precipitation is shown. As it is shown in this Figure, in the researched period all grain markets in England were highly sensitive (larger than 99%) to space weather states of both minimal (d_{min}) and maximal (d_{max}) solar activity. The wheat market in Leiden, the Netherlands, where climate is similar to England, was sensitive (95%) to solar activity minimums, too. At the same time, some of the markets in Belgium, France and especially in Italy (Napoli, Bassano) demonstrated high sensitivity to solar activity maximums.

Because modulation of cloudiness in the North Atlantic by a cosmic rays flux is one of the channels of the possible influence of the space weather on the Earth weather, one can expect excess precipitation and deficit of solar radiation (unfavorable for agriculture in cold and damp climate) in states of minimal solar activity. At the same time, the cosmic rays flux and the related cloudiness drop in states of maximal solar activity, and draughts are possible. The latter are unfavorable for agriculture in the South Europe influenced by hot and dry North Africa climate, especially in Italy and Spain.

These results show that the observed sensitivity of wheat markets to space weather is not a universal property invariable for all regions and periods. On the contrary, depending on satisfaction (full or partial) of the above formulated conditions a few cases are possible:

1. Any influence of the space weather/solar activity cannot be observed (for most markets of Central Europe).
2. Highly significant sensitivity to states of minimal solar activity is observed (England and neighboring markets of the continental Europe).
3. Significant influence of states of maximal solar activity in zones sensitive to draughts is observed (particularly, in Italy influenced by North Africa climate).

Thus either the observed distribution of zones sensitive to the space weather or the sign of this sensitivity are in good agreement with the above described scheme of the causal relationships between space weather and grain markets.

12.11. The USA Wheat Market Sensitivity to Space Weather in 20th Century

The previous analysis regarded grain markets in the Middle Ages or in the very beginning of the Modern Era. Our next question is whether the above described scenarios of space weather influence on Earth markets can be observed in the Modern Era? From the first glance, the all over the world implementation of modern agro techniques methods that increase the plant resistance to unfavorable weather, should cause breaking the second condition - belonging to risky agriculture - necessary for the above influence. Another factor of suppressing a possible sensitivity of grain markets to external anomalies is globalization of the world economics in the 20th century. This process made it easier in a drastic way to access external supply for markets with supply deficit. Nevertheless, we decided to analyze the US grain market in the New Time searching for possible manifestation of the space weather influence on prices. For this purpose, we used wheat prices from the period 1908-1993 based on data of USDA (2004). Wheat price variability for the researched period is shown in Figure 67.

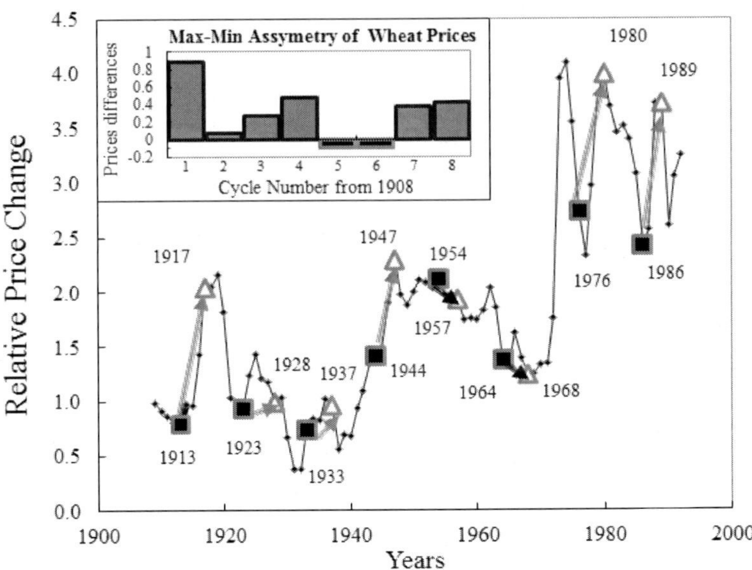

Figure 67. Panel a – changes in durum wheat prices in USA in the period 1908-1993. Black points and lines show price dynamics, white triangular - years of solar activity maximums, black squares – years of solar activity minimums, blue double arrows – price changes from solar activity minimum to the next maximum. *Left above* – a histogram of price asymmetry – a price difference between the maximum and the predecessor minimum of solar activity – depending on the solar activity cycle.

The small sample that includes only 8 solar activity cycles does not enable researching statistical properties of the intervals between price bursts as was made for Medieval England where 500 years of statistics is available. In this situation we can hope only that the phase asymmetry of prices can be revealed with significant differences between prices in states of maximal and minimal solar activity. As it is shown in Figure 67, for the researched sample, indeed, a significant price asymmetry for maximums and minimums of solar activity is observed: prices in states of maximal solar activity are systematically higher than prices in states of minimal solar activity. To estimate significance level of this asymmetry we used the Student criterion. For the average price asymmetry $\overline{\Delta\text{Price}} = 0.29$ (as it is defined in the legend of Figure 67b) and standard deviation $s(\overline{\Delta\text{Price}}) = 0.12$ the significance level for rejecting the zero hypothesis of asymmetry is 97%. Therefore, for the researched sample of wheat prices in USA one can infer that even in the New Times a significant influence of the space weather on agricultural markets is observed though the significance is a bit lower that that for observations during the Maunder Minimum. This unexpected, to some extent, result can be explained, maybe, by a very high concentration of the durum crop in a very small area in North Dakota on the border with Canada. This zone is influenced by the North Atlantic Oscillation (NAO) that, in turn, is sensitive to the space weather. Using the method of regression with dummy variables gave a similar significance level of 96% for the relationship "solar activity – prices".

12.12. Manifestation of Space Weather Influence on Famines and the Death Toll: A Case Study of Iceland in the 18th-19th Centuries

One of the most tragic manifestations of sharp rises of grain prices caused by poor harvest under adverse weather conditions is famine and mortality. The above described effect of space weather influence on the Earth weather that becomes apparent in the form of bad harvest and grain price rises, generally speaking should remain a trace in these sad statistics too. Obviously, such manifestation of space weather can take place only in those regions and periods for which all three necessary conditions of the relationship "space weather" – "agricultural prices" take place. We chose Iceland of the 18th-19th centuries as a possible region for searching for such manifestations[14]. The key source of data about bad feed harvest and the caused cases of decrease in population is the work of the US researcher Vasey (2001). As this is shown in this work, all periods of decrease in population of Iceland coincide with falls in livestock caused by bad feed harvest. The appropriate famine periods are marked by yellow triangular in Figure 68a. It is striking that the periods of famine years always coincide or close to extremes of solar activity (minima or maxima). To examine this hypothesis we marked out the periods of cycles around famine years. These periods were marked out from minimum to minimum for events that happened during the period of the cycle maximum, and on the opposite, from maximum to maximum for events that happened during the period of the cycle minimum. To combine data from various cycles that differ one from another in duration and amplitude, in one homogeneous sample, sunspots numbers were normalized in relative amplitudes and times were normalized in relative phases of the cycle. For this purpose, for events close to the cycle maximum the normalized amplitude y_{norm} was defined as deviation from the minimal value normalized by the range of changes during the cycle:

$$y_{norm} = (ssn_i - ssn_{min})/(ssn_{max} - ssn_{min}), \tag{132}$$

where ssn_i is a sun spots number in year i, ssn_{min} is a number of sun spots in the year of minimum, ssn_{max} is a number of sun spots in the year of maximum, and for events in the neighborhood of the cycle minimum the absolute value of deviation from the maximal value was used, also normalized by the range of changes:

---

[14] The reasons for choosing just this region and period are as follows:
- Location of Iceland in the region subject to weather fluctuations sensitive to the space weather factors (cosmic rays-cloudiness, North Atlantic Oscillation) corresponds to the first necessary condition from II.5.
- Iceland belonged to a risky agricultural region because of specific features of that time agriculture very sensitive to weather. In the researched period, animal husbandry fully dependent of pasture harvest was a major source of food, and not coastal fishery as it was in the later time. Iceland pastures are considered one of the best in Europe due to a long solar day in the short period of the cool summer. Thus, at those time the agriculture of the country was very sensitive to weather conditions (cloudiness, precipitations), and through them to the space weather. Let us emphasize that in the situation of Iceland the reason for bad feed harvest could be both in excess of precipitations and deficit of solar radiation, and in deficit of precipitation and draught (similar to case IV in Figure. 64). Thus for Iceland of that time we could expect a negative reaction to both extreme states of the solar activity (either minimum or maximum).
- Obvious isolation of Iceland in the researched period from major agricultural markets supply from whom could temper the impact of bad harvest.

$$y_{norm} = (ssn_{max} - ssn_i)/(ssn_{max} - ssn_{min}).$$ (133)

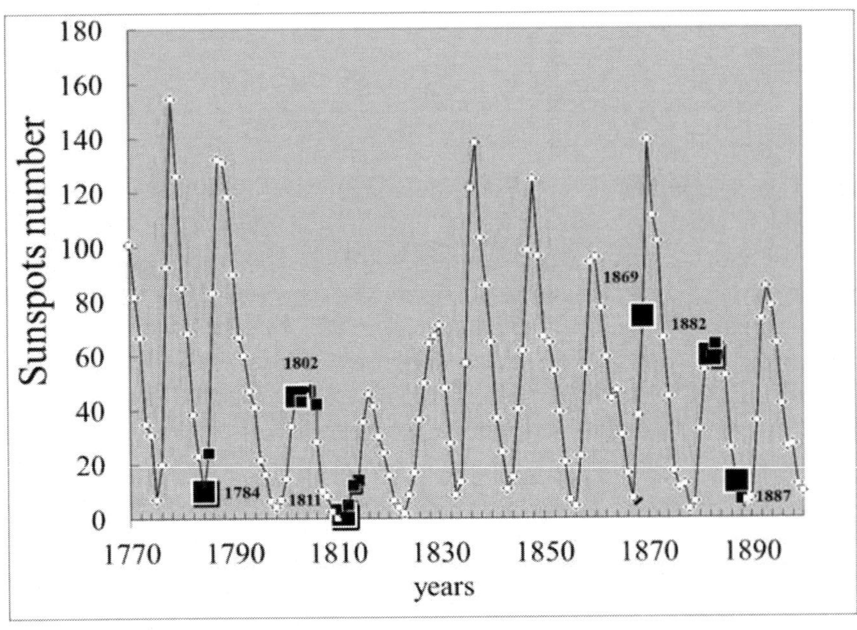

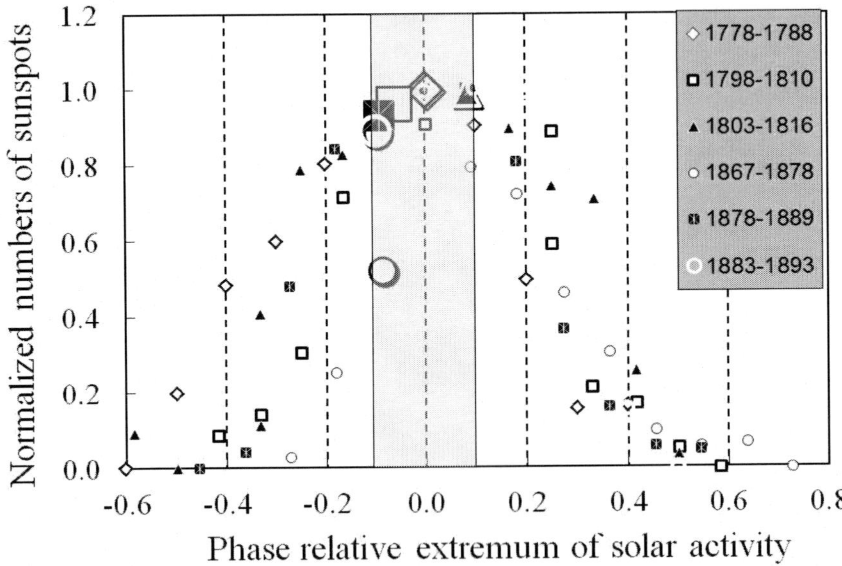

Figure 68. Panel above − a) relationship between solar activity and periods of famine/decrease in population in Iceland of 18-19 centuries, in the phase curve of changes in number of sun spots; below − b) a phase diagram of the result of superposition of 6 solar activity periods during which fall in population happened because of the famine caused by a decrease in livestock. In the horizontal axes phases of the cycle related to the cycle extremes (maximum or minimum) are marked, different small markers are sun spots numbers for different periods (cycles) and big size markers tracks years of famine beginning. The rectangle in the center of the Figure bounds the part of the phase space of cycles during which famines happened and population decreased.

The times t_i were recalculated as cycle phases Φ_i according to the following formula:

$$\Phi_i = |t_i - t_{ext}| / T_{cycle} \, , \tag{134}$$

where t_{ext} are times of minimum or maximum of the cycle where famine has had happened, T_{cycle} is the duration of the cycle. Normalized phase data for 6 cycles during which mass population decrease happened because of bad harvest and famine, are united in a sample presented in Figure 68b. As it is seen in this Figure, all 6 events of population decrease because of famine in Iceland in the period 1784-1900 are concentrated in a small part (20%) of the phase space of sun cycle around the cycle extremes (maximum or minimum). The probability of the random realization of such accurate phasing of two independent processes is estimated as $P = (1/5)^6 < 10^{-4}$. The revealed phase linkage of the periods of famine because of weather anomalies in Iceland, to the solar activity extremes enables to claim about the highly significant manifestation of the negative influence of the space weather/solar activity on the state of the agricultural market in the region at the researched period.

12.13. Summary of Research of Space Weather Influence on Agricultural Markets

In this study the model of a possible influence of space weather on the agricultural markets is presented based on the causal chain "space weather" – "earth weather" – "agricultural production' – "market prices". The main results are the following:

1. It is marked that realization of this causal chain demands simultaneous satisfaction of several necessary conditions for a specific zone and period. The influence of space weather on agricultural markets is not a universal phenomenon, but rather is revealed only in those zones and periods when and where all necessary conditions are satisfied (atmospheric, agro climatic, market).
2. A significant manifestation of this influence for the case of Medieval England is shown, particularly, for the period of the Maunder Minimum.
3. The dependence of the European markets' sensitivity on space weather (including the direction of this sensitivity) on localization in the corresponding climatic zone is shown.
4. It is shown that even in the Modern Era, the dependence on the space weather/solar activity is observed in the USA. This dependence is revealed for durum wheat produced in a small zone sensitive to the influence of the North Atlantic Oscillation.
5. The coincidence of the periods of decrease in population in Iceland of the 18th-19th centuries (caused by bad feed harvest and the following drop of livestock), on the one hand, and phases of solar activity extreme, on the other hand, are shown. We explain this coincidence as a manifestation of the earth markets dependence on the space weather we research in this study.
6. The presented results clearly indicate significant space weather influence on the Earth agricultural markets, and first of all, on grain markets. This is true for the zones

where the presented necessary conditions for the causal chains (Figure 64) are satisfied in the researched period.

Let us note that continuous progress in the development of agro technology using achievements in gene engineering, biotechnology, irrigation, agro chemistry and plant protection should result in augmentation of steadiness of crops toward external factors like weather anomalies. This process should disturb the causal chain "space weather" – "agricultural prices" breaking the key link "weather – yield". Accordingly, in the nearest future one could assume gradual weakening of grain and other crops prices' sensitivity to space weather. Unfortunately, the global and sharp climate change observed in the last years can bring the opposite results. The main feature of this change is the sharp growth of the amount and amplitude of weather anomalies especially shifts of the usual time bounds of seasonal weather states. Under these conditions, in many regions the long-term selection of crops most suitable for the zone can turn out to be not optimal. Sharp deviations of the local weather from the "standard" can move zones to the category of risky agriculture zones highly sensitive to weather anomalies. If such movement happens in the zone where the weather, in turn, is sensitive to the solar activity state and the external supply of agricultural production is limited for some reasons, the causal chain "space weather" – "agricultural prices" can become working even under the modern conditions of the technological progress and globalization. For a more specific identification of the regions where this negative phenomenon can take place in the nearest future, a progress in understanding various physical and technological processes is needed – global climate change and its consequences for local weather in various regions, zoned crops steadiness and mechanisms of solar-earth relationships regarding their influence on weather in the researched regions. To cope with this problem, joint efforts of specialists from various fields are needed - agrarian researchers and meteorologists, astrophysicists, and specialists in space weather.

ACKNOWLEDGMENTS

We would like cordially to thank H. S. Ahluwalia, J. Allen, A. V. Belov, E. B. Berezhko, J. W. Bieber, A. Chilingaryan, M. Duldig, E. A. Eroshenko, N. Iucci, K. Kudela, V. Kuznetsov, K. Munakata, Y. Muraki, M. Murat, M. Panasyuk, M. Parisi, N. G. Ptitsyna, M. A. Shea, D. F. Smart, M. Storini, J. F. Valdes-Galicia, P. Velinov, G. Villoresi, A. Wolfendale, V. G. Yanke, I. G. Zukerman for interesting discussions, Moshe Reuveni, Itzik ben Israel, Abraham Sternlieb, and Uri Dai for constant interest and support of Israel Cosmic Ray & Space Weather Center and Israel-Italian Emilio Segré Observatory. We are thankful to Israeli Ministry of Science for support in the frame of the project "National Centre of Knowledge in Cosmic Rays and Space Weather".

REFERENCES

Ahluwalia H. S., 1997. Three activity cycle periodicity in galactic cosmic rays, *Proc. 25th Intern. Cosmic Ray Conf.*, Durban, 2, 109-112.

Allen J. H. and Wilkinson D. C., 1993. Solar-terrestrial activity affecting systems in space and on Earth, in *Proc. Workshop Solar-Terrestrial Predictions-IV*, Ottawa, Canada, 1992, edited by Hruska, J., M. A. Shea, D. F., Smart, and G. Heckman, NOAA Environ. Res. Lab., Boulder, CO, 1, 75-107.

Appleby A. B. 1979, Grain Prices and Subsistence Crises in England and France, 1590-1740. *The Journal of Economic History*, 39(4): 865-887.

Arctowski H., 1910. Studies on climate and crops. *Bulletin of the American Geographical Society*, XLII(4): 270-282, XLII(7): 481-495.

Ardanuy P. E., Stowe L. L., Gruber A. and Weiss M., 1989. Shortwave, longwave and net cloud-radiative forcing as determined from NIMBUS-7 observations, *J. Geophys. Res.*, 96, 18537-18549.

Arnold, V. I., M1992. *Catastrophe Theory*. 3rd ed. Berlin: Springer-Verlag.

Artamonova I. and Veretenenko S. Galactic cosmic ray variation influence on baric system dynamics at middle latitudes, *J. Atmosph. and Solar-Terrestrial Physics*, 73, 366–370, 2011.

Azariadis C., 1981. Self-Fulfilling Prophecies", *Journal of Economic Theory*, 25, 380-396.

Baevsky, R. M. et al. 1996, *Proc. MEFA Int. Fair of Medical Technology & Pharmacy* (Brno, Chec Republic).

Baevsky R. M., Petrov V. M., Cornelissen G., Halberg F., Orth-Gomur K., Ekerstedt T., Otsuka K., Breus T., J. Siegelova, J. Dusek, and B. Fiser, 1997. Meta-analyzed heart rate variability, exposure to geomagnetic storms, and the risk of ischemic heart disease, *Scripta medica*, 70, No. 4-5, 201-206.

Barnard L., M. Lockwood, M. A. Hapgood, Owens M. J., Davis C. J., and Steinhilber F., 2011. Predicting space climate change, *Geophys. Res. Lett.*, 38, L16103, doi:10.1029/20 11GL048489.

Bavassano B., Iucci N., Lepping R. P., Signorini C., Smith E. J., and Villoresi G., 1994. Galactic cosmic ray modulation and interplanetary medium perturbations due to a long-living active region during October 1989, *J. Geophys. Res.*, 99, No. A3, 4227-4234.

Beer J., Raisbeck G. M., and Yiou F., 1991. Time variations of 10 Be and solar activity, in Sonett C. P., Giampapa M. S., and Matthews M. S. (eds.) *The Sun in time*, University of Arizona Press, 343-359.

Beer J., Tobias S. and Weiss N., 1998. An active Sun throughout the Maunder minimum, *Solar Phys.*, 181, 237-249.

Belov A. V., Dorman L. I., Eroshenko E. A., Iucci N., Villoresi G. and Yanke V. G., 1995a. Search for predictors of Forbush decreases. *Proc. 24-th Intern. Cosmic Ray Conf.*, Rome, 4, 888-891.

Belov A. V., Dorman L. I., Eroshenko E. A., Iucci N., Villoresi G. and Yanke V. G., 1995b. Anisotropy of cosmic rays and Forbush decreases in 1991. *Proc. 24-th Intern. Cosmic Ray Conf.*, Rome, 4, 912-915.

Belov A. V., Eroshenko E. A., and Yanke V. G., 1997. Modulation Effects in 1991-1992 Years, *Proc. 25th Intern. Cosmic Ray Conf.*, 1997, Durban, South Africa, 1, 437-440.

Belov A., Dorman L. I., Eroshenko E., Gromova L., Iucci N., Ivanus D., Kryakunova O., Levitin A., Parisi M., Ptitsyna N., Tyasto M., Vernova E., Villoresi G. and Yanke V., 2003. The relation between malfunctions of satellites at different orbits and cosmic ray variations, *Proc. 28th Intern. Cosmic Ray Conf.*, 2003, Tsukuba, Japan, 7, 4213-4216.

Belov A. V., Dorman L. I., Gushchina R. T., Obridko V. N., Shelting B. D. and Yanke V. G., 2005. Prediction of expected global climate change by forecasting of galactic cosmic ray intensity time variation in near future based on solar magnetic field data, *Adv. Space Res.*, 35, No. 3 , 491-495.

Berezhko E. G., Kozlov V. I., Kuzmin A. I., and Tugolukov N. N., 1987. Cosmic ray intensity micropulsations associated with disturbances of electromagnetic conditions in the heliosphere, *Proc. 20th Intern. Cosmic Ray Conf.*, 1987, Moscow, USSR, 4, 99-102.

Beveridge W. H. B., M1939. *Prices and Wages in England from the Twelfth to the Nineteenth Century.* Longmans, Green: London, New York.

Beveridge William H., 1921. Weather and Harvest Cycles, *Economic Journal*, XXXI, 429-453.

Bhattacharyya A. and Mitra B., 1997. Changes in cosmic ray cut-off rigidities due to secular variations of the geomagnetic field, *Ann. Geophys.*, 15, No. 6, 734-739.

Blokh Ya. L., Glokova E. S., Dorman L. I. and Inozemtseva O. I., 1959. Electromagnetic conditions in interplanetary space in the period from August 29 to September 10, 1957 determined by cosmic ray variation data. *Proc. 6-th Intern. Cosmic Ray Conf.*, Moscow, 4, 172-177.

Boberg F., Lundstedt H., 2002. Solar Wind Variations Related to Fluctuations of the North Atlantic Oscillation, *Geophysical Research Letters*, 29(15): 13-1.

Brázdil R., Dobrovolný P., Luterbacher J., Moberg A., Pfister C., Wheeler D., Zorita E., 2010. European climate of the past 500 years: new challenges for historical climatology. *Climatic Change*, 101, 7-40.

Breus, T., Siegelova, J., Dusek, J., and Fiser, B. 1997, *Scripta Medica* 70, 1999.

Brown E. H. P. and Hopkins S. V., 1956. Seven centuries of the prices of consumables, compared with builders' wage-rates. *Economica*, XXIII (89-92): 296-314.

Cox A., Dalrymple G. B., and Doedl R. R., 1967. Reversals of the Earth's magnetic field, *Sci. American*, 216, No. 2, 44-54.

Dickinson R. E., 1975. Solar variability and the lower atmosphere, *Bull. Am. Met. Soc.*, 56, 1240-1248.

Dobrica V., C. Demetrescu, C. Boroneant, G. Maris, 2009. Solar and geomagnetic activity effects on climate at regional and global scales: Case study-Romania, *J. Atmospheric and Solar-Terrestrial Physics*, 71, 1727– 1735.

Dorman L. I., 1959. On the energetic spectrum and lengthy of cosmic ray intensity increase on the Earth caused by shock wave and albedo from magnetized front of corpuscular stream. *Proc. 6-th Intern. Cosmic Ray Conf.*, Moscow, 4, 132-139.

Dorman L. I., 1960. On the energetic spectrum and prolongation of the cosmic ray intensity increase on the Earth caused by shock wave of corpuscular stream. *Proc.of Yakutsk Filial of Academy of Sciences*, Ser. Phys., Academy of Sciences of USSR Press, Moscow, No. 3, 145-147.

Dorman L. I., M1963a. *Cosmic Ray Variations and Space Explorations.* NAUKA, Moscow, pp 1023.

Dorman L. I., M1963b. *Geophysical and Astrophysical Aspects of Cosmic Rays.* North-Holland Publ. Co., Amsterdam (In series Progress in Physics of Cosmic Ray and Elementary Particles, ed. by J. G.Wilson and S. A.Wouthuysen, Vol. 7), pp 320.

Dorman L. I., M1972. *Acceleration Processes in Space.* VINITI, Moscow (in series Summary of Science, Astronomy, Vol. 7).

Dorman L. I., M1974. *Cosmic Rays: Variations and Space Exploration.* North-Holland Publ. Co., Amsterdam. pp 675.

Dorman L. I., M1978. *Cosmic Rays of Solar Origin.* VINITI, Moscow (in series Summary of Science, Space Investigations, Vol.12).

Dorman L. I., 2002. Solar Energetic Particle Events and Geomagnetic Storms Influence on People's Health and Technology; Principles of Monitoring and Forecasting of Space Dangerous Phenomena by Using Online Cosmic Ray Data, in *Proc. 22nd ISTC Japan Workshop on Space Weather Forecast in Russia/CIS*, Nagoya, Japan, June 2002, edited by Muraki Y., Nagoya University, 2, 133-151.

Dorman L. I., M2004. *Cosmic Rays in the Earth's Atmosphere and Underground*, Kluwer Academic Publishers, Dordrecht/Boston/London.

Dorman L. I., 2005a. Prediction of galactic cosmic ray intensity variation for a few (up to 10–12) years ahead on the basis of convection-diffusion and drift model, *Annales Geophysicae*, 23, No. 9, 3003-3007.

Dorman L. I., 2005b. Estimation of long-term cosmic ray intensity variation in near future and prediction of their contribution in expected global climate change, *Adv. Space Res.*, 35, 496-503.

Dorman L. I., 2006. Long-term cosmic ray intensity variation and part of global climate change, controlled by solar activity through cosmic rays, *Adv. Space Res.*, 37, 1621-1628.

Dorman L. I., M2006. *Cosmic Ray Interactions, Propagation, and Acceleration in Space Plasmas*, Springer, Netherlands.

Dorman L. I., 2008a. Forecasting of radiation hazard and the inverse problem for SEP propagation and generation in the frame of anisotropic diffusion and in kinetic approach, *Proc. 30-th Intern. Cosmic Ray Conf.*, Merida, Mexico, 1, 175-178.

Dorman L. I., 2008b. Natural hazards for the Earth's civilization from space, 1. Cosmic ray influence on atmospheric processes, *Adv. Geosci.*, 14, 281-286.

Dorman L. I., 2009. Chapter 3. The Role of Space Weather and Cosmic Ray Effects in Climate Change, in book *Climate Change: Observed Impacts on Planet Earth*, Edited by Trevor M. Letcher, Elsevier.

Dorman L. I., M2009. *Cosmic Rays in Magnetospheres of the Earth and other Planets*, Springer, Netherlands.

Dorman L. I., M2010. *Solar Neutrons and Related Phenomena*, Springer, Netherlands.

Dorman L. I., 2012. Cosmic rays and space weather: effects on global climate change, *Ann. Geophys.*, 30, No. 1, 9–19.

Dorman I. V. and Dorman L. I., 1967a. Solar wind properties obtained from the study of the 11-year cosmic ray cycle. 1, *J. Geophys. Res.*, 72, No. 5, 1513-1520.

Dorman I. V. and Dorman L. I., 1967b. Propagation of energetic particles through interplanetary space according to the data of 11-year cosmic ray variations, *J. Atmosph. and Terr. Phys.*, 29, No. 4, 429-449.

Dorman L. I. and Dorman I. V., 2005. Possible influence of cosmic rays on climate through thunderstorm clouds, *Adv. Space Res.*, 35, 476-483.

Dorman L.I. and Freidman G.I., 1959. On the possibility of charged particle acceleration by shock waves in the magnetized plasma. *Proc. on All-Union Conf. on Magnetohydrodinamics and Plasma Physics*, Latvia SSR Academy of Sciences Press, Riga, pp. 77-81.

Dorman L. I. and Miroshnichenko L. I., M1968. *Solar Cosmic Rays.* FIZMATGIZ, Moscow, pp 508. English translation published for the National Aeronautics and Space

Administration and National Science Foundation in 1976 (TT 70-57262/NASA TT F-624), Washington, D.C.

Dorman L. I. and Pustil'nik L. A., 1999. Statistical characteristics of SEP events and their connection with acceleration, escaping and propagation mechanisms, *Proc. 26-th Intern. Cosmic Ray Conference*, Salt Lake City, 6, 407-410.

Dorman L. I., I. Ya. Libin, M. M. Mikalayunas, and Yudakhin K. F., 1987. On the connection between cosmophysical and geophysical parameters in 19-20 cycles of solar activity, *Geomagnetism and Aeronomy*, 27, No. 2, 303-305.

Dorman L. I., Ya I. Libin, and Mikalajunas M. M., 1988a. About the possibility of the influence of cosmic factors on weather, spectral analysis: cosmic factors and intensity of storms, *The Regional Hidrometeorology (Vilnius)*, 12, 119-134.

Dorman L. I., Ya I. Libin, and Mikalajunas M. M., 1988b. About the possible influence of the cosmic factors on the weather. Solar activity and sea storms: instantaneous power spectra, *The Regional Hidrometeorology (Vilnius)*, 12, 135-143.

Dorman L. I, Iucci N., Villoresi G., 1993a. The use of cosmic rays for continuos monitoring and prediction of some dangerous phenomena for the Earth's civilization. *Astrophysics and Space Science*, 208, 55-68.

Dorman L. I., Iucci N. and Villoresi G., 1993b. Possible monitoring of space processes by cosmic rays. *Proc. 23-th Intern. Cosmic Ray Conf.*, Calgary, 4, 695-698.

Dorman L. I., Iucci N. and Villoresi G., 1993c. Space dangerous phenomena and their possible prediction by cosmic rays. *Proc. 23-th Intern. Cosmic Ray Conf.*, Calgary, 4, 699-702.

Dorman L. I., Iucci N., Villoresi G., 1995a. The nature of cosmic ray Forbush decrease and precursory effects. *Proc. 24-th Intern. Cosmic Ray Conf.*, Rome, 4, 892-895.

Dorman L. I., Villoresi G., Belov A. V., Eroshenko E. A., Iucci N., Yanke V. G., Yudakhin K. F., Bavassano B., Ptitsyna N. G., Tyasto M. I., 1995b. Cosmic-ray forecasting features for big Forbush decreases. *Nuclear Physics B,.* 49A, 136-144.

Dorman L. I., Iucci N. and Villoresi G., 1997a. Auto-model solution for nonstationary problem described the cosmic ray preincrease effect and Forbush decrease. *Proc. of 25- th Intern. Cosmic Ray Conference,* Durban (South Africa), 1, 413-416.

Dorman L. I., Villoresi G., Dorman I. V., Iucci N. and Parisi M., 1997b. On the expected CR intensity global modulation in the Heliosphere in the last several hundred years. *Proc. 25-th Intern. Cosmic Ray Conference*, Durban (South Africa), 7, 345-348.

Dorman L. I., Iucci N., Ptitsyna N. G., Villoresi G., 1999. Cosmic ray Forbush decrease as indicators of space dangerous phenomenon and possible use of cosmic ray data for their prediction, *Proc. 26-th Intern. Cosmic Ray Conference*, Salt Lake City, 6, 476-479.

Dorman L. I., Shakhov B. A., and Stehlik M., 2003a. The second order pitch-angle approximation for the cosmic ray Fokker-Planck kinetic equations, *Proc. 28th Intern. Cosmic Ray Conf.*, Tsukuba (Japan), 6, 3535-3538.

Dorman L. I., Dorman I. V., Iucci N., Parisi M., Ne'eman Y., Pustil'nik L. A., Signoretti F., Sternlieb A., Villoresi G., and Zukerman I. G., 2003b. Thunderstorms Atmospheric Electric Field Effect in the Intensity of Cosmic Ray Muons and in Neutron Monitor Data, *J. Geophys. Res.*, 108, No. A5, 1181, SSH 2_1 – 8.

Eddy J. A., 1976. The Maunder Minimum, *Science*, 192, 1189-1202.

Ejrnaes M., Persson K. G., and Rich S., 2008. Feeding the British: convergence and market efficiency in the nineteenth-century grain trade. *The Economic History Review*, 61(s1): 140-171.

Ely J. T. A. and Huang T. C., 1987. Geomagnetic activity and local modulations of cosmic ray circa 1 GeV, *Geophys. Res. Lett.*, 14, No. 1, 72-75.

Ely J. T. A., Lord J. J., and Lind F. D., 1995. GCR, the Cirrus hole, global warming and record floods, *Proc. 24th Intern. Cosmic Ray Conf.*, Rome, 4, 1137-1140.

Enghoff M. B., Pedersen J. O. P., Uggerhøj U. I., Paling S. M., and Svensmark H., 2011. Aerosol nucleation induced by a high energy particle beam, *Geophys. Res. Lett.*, 38, L09805, doi:10.1029/2011GL047036.

Erlykin A. D. and Wolfendale A. W., 2011. Cosmic ray effects on cloud cover and the irrelevance to climate change, *J. Atm. and Solar-Terr. Physics*, 73, 1681–1686.

Erlykin A. D., Gyalai G., Kudela K., Sloan T. and Wolfendale A. W., 2009a. Some aspects of ionization and the cloud cover, cosmic ray correlation problem, *J. Atmospheric and Solar-Terrestrial Physics*, 71, 823-829.

Erlykin A. D., Gyalai G., Kudela K., Sloan T., and Wolfendale A. W., 2009b. On the correlation between cosmic ray intensity and cloud cover, *J. Atm. and Solar-Terr. Physics*, 71, 1794–1806.

Ermakov V. I., Bazilevskaya G. A., Pokrevsky P. E., and Yu. I. Stozhkov, 1997. Cosmic rays and ion production in the atmosphere, *Proc. 25th Intern. Cosmic Ray Conf.*, Durbin, 7, 317-320.

Ermakov V, Okhlopkov V., and Stozhkov Yu., 2006a. Influence of Zodiac Dust on the Earth's Climate, *Proc. European Cosmic Ray Symposium*, Lesboa, Paper 1-72.

Ermakov V. I., Okhlopkov V. P., and Stozhkov Y. I., 2006b. Influence of cosmic dust on the Earth's climate, *Bulletin of Lebedev Physics Institute*, Russian Ac. of Sci., Moscow, No. 3, 41-51.

Ermakov V. I., Okhlopkov V. P., and Stozhkov Yu. I., 2006c. Effect of cosmic dust on terrestrial climate, *Kratk. Soob'sh. Fiz. FIAN*, No. 3, 41–51. In Russian.

Ermakov V. I., Okhlopkov V. P., and Stozhkov Yu. I., 2007. Effect of cosmic dust on cloudiness, albedo, and terrestrial climate, *Vestn. Mosk. Univ.*, Ser. 3: Fiz. Astron., No. 5, 41–45. In Russian.

Ermakov V. I., Okhlopkov V. P., and Stozhkov Yu. I., 2009. Cosmic rays and dust in the Earth's atmosphere, *Izv. Russ. Akad. Nauk*, Ser. Fiz., 73, No. 3, 434–436. In Russian.

Fastrup B., Pedersen E., Lillestol E. et al. (Collaboration CLOUD), 2000. A study of the link between cosmic rays and clouds with a cloud chamber at the CERN PS, *Proposal CLOUD*, CERN/SPSC 2000-021.

Fedorov Yu., Stehlik M., Kudela K., and Kassovicova J., 2002. Non-diffusive particle pulse transport: Application to an anisotropic solar GLE, *Solar Physics*, 208, No. 2, 325-334.

Fenton, A. G. et al. 1959, *Can. J. Phys.* 37, 970.

Ferraro R. R., Weng F., Grody N. C., and Basist A., 1996. An eight-year (1987-1994) time series of rainfall, clouds, water vapor, snow cover, and sea ice derived from SSM/I measurements, *Bull. Am. Met. Soc.*, 77, 891-905.

Friis-Christiansen E. and Lassen K., 1991. Length of the solar cycle: an indicator of solar activity closely associated with climate, *Science*, 254, 698-700.

Fuhrer K., Wolff E. W., Johnsen S. J., 1999. Timescales for dust variability in the Greenland Ice Core Project (GRIP) in the last 100,000 years, *J. Geophys. Res.*, 104, No. D24, 31043-31052.

Georgieva K. and Kirov B., 2011. Solar dynamo and geomagnetic activity, *Journal of Atmospheric and Solar-Terrestrial Physics*, 73, Issue 2-3, 207-222.

Haigh J. D., 1996. The impact of solar variability on climate, *Science*, 272, 981-984.

Haigh J. D., Winning A. R., Toumi R. and Harder J. W., 2010, An influence of solar spectral variations on radiative forcing of climate, *Nature*, 467, 7 Oct 2010, 696-699.

Hansen J., Ruedy R., Glascoe J., and Sato M., 1999. GISS analysis of surface temperature change, *J. Geophys. Res.*, 104, No. D24, 30997-31022.

Harrison G. and Stephenson D., 2006. Empirical evidence for a non-linear effect of galactic cosmic rays on clouds, *Proceedings of the Royal Society A*, A 462, 2068, 1221-1233 doi: 10.1098/rspa.2005.1628.

Hartmann D. L., 1993. Radiative effects of clouds on the Earth's climate in aerosol-cloud-climate interactions, in P. V. Hobbs (ed) *Aerosol-Cloud-Climate Interactions*, Academic Press, 151.

Head L., Atchison J., Gates A., Muir P., 2011. A Fine-Grained Study of the Experience of Drought, Risk and Climate Change Among Australian Wheat Farming Households. *Annals of the Association of American Geographers*, 101(5): 1089-1108.

Herschel W., 1801. Observations Tending to Investigate the Nature of the Sun, in Order to Find the Causes and Symptoms of its Variable Emission of Light and Heat, *Philosophical Transactions of the Royal Society of London*, 91, 261-331.

Hong P. K., Miyahara H., Yokoyama Y., Takahashi Y., and Sato M., 2011. Implications for the low latitude cloud formations from solar activity and the quasi-biennial oscillation, *J. Atmospheric and Solar-Terrestrial Physics*, 73, 587–591.

Hooker P. H., 1907. Correlations of the weather and crops, *Journ. Roy. Statistical Society*, 70(1).

Hruska J. and Shea M. A., 1993, *Adv. Space Res.* 13, 451.

International Institute of Social History, 2005. Available online http://www.iisg.nl/index.php.

Iucci N., Dorman L. I., Levitin A. E., Belov A. V., Eroshenko E. A., Ptitsyna N. G., G. Villoresi, Chizhenkov G. V., Gromova L. I., Parisi M., Tyasto M. I., and Yanke V. G., 2006. Spacecraft operational anomalies and space weather impact hazards, *Adv. Space Res.*, 37, No. 1, 184-190.

Jevons W. S., 1878. Commercial Crises and Sun-Spots 1, *Nature*, 19, No. 472, 33-37.

Jevons W. S., 1879. Sun-Spots and Commercial Crises, *Nature*, 19, No. 495, 588-590.

Jevons W. S., 1882. The solar commercial cycle, *Nature*, 26, No. 662, 226-228.

Jones P. D., Briffa K. R., Barnett T. P., and Tett S. F. B., 1998. High resolution palaeo-climatic records for the last millennium: interpretation, integration and comparison with general circulation model control run temperatures, *The Holocene*, 8, 455-471.

Kappenman J. G. and Albertson V. D., 1990. Bracing for the geomagnetic storms, *IEEE Spectrum*, 27, No. 3, 27-33.

Kasatkina E. A., Shumilov O. I., and Krapiec M., 2007a. On periodicities in long term climatic variations near 68° N, 30° E, *Adv. Geosci.*, 13, 25–29.

Kasatkina E. A., Shumilov O. I., Lukina N. V., Krapiec M., and Jacoby G., 2007b. Stardust component in tree rings, *Dendrochronologia*, 24, 131–135.

King J. W., 1975. Sun-weather relationships, *Astronautics and Aeronautics*, 13, No. 4, 10-19.

Kirkby J. and CLOUD collaboration, 2011, Role of sulphuric acid, ammonia and galactic cosmic rays in atmospheric aerosol formation, *Nature*, 476, No. 10343, 429-433.

Kishcha P. V., Dmitrieva I. V., and Obridko V. N., 1999, Long-term variations of the solar-geomagnetic correlation, total solar irradiance, and northern hemispheric temperature (1868–1997), *Journal of Atmospheric and Solar-Terrestrial Physics*, 61, 799-808.

Koenig, H., & Ankermueller, F. 1982, *Naturwissenscahften, 17, 47.*

Kristjánsson J. E., Staple A., and Kristiansen J., 2002. A new look at possible connections between solar activity, clouds and climate, *Geophys. Res. Lett.*, 23, 2107-2110.

Kristjánsson J. E., Kristiansen J. and Kaas E., 2004. Solar activity, cosmic rays, clouds and climate - an update, *Adv. Space Res.*, 34, No. 2, 407-415.

Kudela K. and Bobik P., 2004. Long-term variations of geomagnetic rigidity cutoffs, *Solar Physics*, 224, 423–431.

Labitzke K. and van Loon H., 1993. Some recent studies of probable connections between solar and atmospheric variability, *Ann. Geophys.*, 11, 1084-1094.

Laken B. A. and Kniveton D. R., 2011. Forbush decreases and Antarctic cloud anomalies in the upper troposphere, *J. Atmospheric and Solar-Terrestrial Physics*, 73, 371–376.

Lassen K. and Friis-Christiansen E., 1995. Variability of the solar cycle length during the past five centuries and the apparent association with terrestrial climate, *J. Atmos. Solar-Terr. Phys.*, 57, 835-845.

Lean J., Skumanich A., and White O., 1992. Estimating the Sun's radiative output during the Maunder minimum, *Geophys. Res. Lett.*, 19, 1591-1594.

Lean J., Beer J., and Breadley R., 1995. Reconstruction of solar irradiance since 1610: implications for climate change, *Geophys. Res. Lett.*, 22, 3195-3198.

Lockwood M., R. Stamper, and Wild M. N., 1999. A doubling of the Sun's coronal magnetic field during the past 100 years, *Nature*, 399, No. 6735, 437-439.

Lukianova R. and Alekseev G., 2004. Long-Term Correlation between the NAO and Solar Activity. *Solar Physics,* 224(1-2), 445-454.

Mann M. E., Bradley R. S., and Hughes M. K., 1998. Global-scale temperature patterns and climate forcing over the past six centuries, *Nature*, 392, 779-787.

Mansilla G. A., 2011. Response of the lower atmosphere to intense geomagnetic storms, *Advances in Space Research*, 48, 806–810.

Markson R., 1978. Solar modulation of atmospheric electrification and possible implications for the Sun-weather relationship, *Nature*, 273, 103-109.

Marsh N. D. and Swensmark H., 2000a. Low cloud properties influenced by cosmic rays, *Phys. Rev. Lett.*, 85, 5004-5007.

Marsh N. and Swensmark H., 2000b. Cosmic rays, clouds, and climate, *Space Sci. Rev.*, 94, No. 1-2, 215-230.

Martin I. M., Gusev A. A., Pugacheva G. I., Turtelli A., and Mineev Y. V., 1995. About the origin of high-energy electrons in the inner radiation belt, *J. Atmos. and Terr. Phys.*, 57, No. 2, 201-204.

Maseeva O. A., 2004. Role of huge molecular clouds in the evolution of Oort's cloud, *Astronomichesky vestnik*, 38, No. 4, 372-382.

McCracken K. G. and Parsons N. R., 1958. Unusual Cosmic-Ray Intensity Fluctuations Observed at Southern Stations during October 21-24, 1957, *Phys. Rev.,Ser. 2,* 112, No. 5, 1798-1801.

Mendoza B. and Pazos M., 2009. A 22 yr hurricane cycle and its relation with geomagnetic activity, *J. Atmospheric and Solar-Terrestrial Physics*, 71, 2047–2054.

Munakata K., Bieber J. W., Yasue S.-I., Kato C., Koyama M., Akahane S., Fujimoto K., Fujii Z., Humble J. E., and Duldig M. L., 2000. Precursors of geomagnetic storms observed by the muon detector network, *J. Geophys. Res.*, 105, No. A12, 27,457-27,468.

Nagashima K., Sakakibara S., Fujimoto K., Tatsuoka R., and Morishita I., 1990. Localized pits and peaks in Forbush decrease, associated with stratified structure of disturbed and undisturbed magnetic fields, *IL Nuovo Cimento C,* 13C, Ser. 1, No. 3, 551-587.

Nagovitsyn Yu. A., 2007. Solar cycles during the Maunder minimum, *Astronomy Letters*, 33, 5, 385-391.Ney E. R., 1959. Cosmic radiation and weather, *Nature*, 183, 451-452.

Ogurtsov M. G. and Raspopov O. M., 2011. Possible impact of interplanetary and interstellar dust fluxes on the Earth's climate, *Geomagnetism and Aeronomy*, 51, No. 2, 275–283.

Ogurtsov M. G., Kocharov G. E., and Nagovitsyn Yu. A., 2003, Solar Cyclicity during the Maunder Minimum, *Astron. Reports.*, 47, Issue 6, 517-524.

Ogurtsov M. G., Raspopov O. M., Oinonen M., Jungner H., and Lindholm M., 2010. Possible manifestation of non-linear effects when solar activity affects climate changes, *Geomagnetism and Aeronomy*, 50, No. 1, 15-20.

Ohring G. and Clapp P. F., 1980. The effect of changes in cloud amount on the net radiation on the top of the atmosphere, *J. Atmospheric Sci.*, 37, 447-454.

Parker E. N., M1963. *Interplanetary Dynamical Processes*, John Wiley and Suns, New York-London.

Parker E. N., M1965. *Interplanetary Dynamical Processes*, Inostrannaja Literatura, Moscow (In Russian, Translation by L. I. Miroshnichenko, Editor L. I. Dorman).

Petit J. R., Jouzel J., Raunaud D. et al., 1999. Climate and atmospheric history of the past 420,000 years from Vostok ice core, Antarctica, *Nature*, 399, 429-436.

Price C., 2000. Evidence for a link between global lightning activity and upper tropospheric water vapour, *Nature*, 406, 290-293.

Ptitsyna N. G., Villoresi G., Dorman L. I., Iucci N. and Tyasto M. I. 1998. Natural and man-made low-frequency magnetic fields as a potential health hazard, *UFN (Physics Uspekhi)*, 168, No 7, 767-791.

Pudovkin M. and Veretenenko S., 1995. Cloudiness decreases associated with Forbush decreases of galactic cosmic rays, *J. Atmos. Solar-Terr. Phys.*, 57, 1349-1355.

Pudovkin M. and Veretenenko S., 1996. Variations of the cosmic rays as one of the possible links between the solar activity and the lower atmosphere. *Adv. Space Res.*, 17, No. 11, 161-164.

Pudovkin M. I. and Raspopov O. M., 1992. The mechanism of action of solar activity on the state of the lower atmosphere and meteorological parameters (a review), *Geomagn. and Aeronomy*, 32, 593-608.

Pudovkin M. I. and Raspopov O. M., 1993. Physical mechanism of the action of solar activity and other geophysical factors on the state of the lower atmosphere, meteorological parameters, and climate, *Phys. Usp.*, 36, No.7, 644–647.

Pugacheva G. I., Gusev A. A., Martin I. M. et al., 1995. The influence of geomagnetic disturbances on the meteorological parameters in the BMA region, *Proc. 24th Intern. Cosmic Ray Conf.*, Rome, 4, 1110-1113.

Pulkkinen T. I., Nevanlinna H., Pulkkinen P. J., and Lockwood M., 2001. The Sun-Earth connection in time scales from years to decades and centuries, *Space Sci. Rev.*, 95, 625-637.

Pustil'nik L., Yom Din G., and Dorman L., 2003. Manifestations of Influence of Solar Activity and Cosmic Ray Intensity on the Wheat Price in the Medieval England (1259–1703 Years), *Proc. 28th Intern. Cosmic Ray Conf.*, Tsukuba, 7, 4131-4134.

Pustil'nik L. and Yom Din G., 2004a. Influence of solar activity on the state of the wheat market in medieval England, *Solar Physics*, 223, 335–356.

Pustil'nik L. and Yom Din G., 2004b. Space climate manifestation in earth prices – from medieval England up to modern U. S. A., *Solar Physics*, 224, 473–481.

Quante M., 2004. The role of clouds in the climate system, *J. Phys. IV France*, 121, 61-86.

Ramanathan V., Cess R. D., Harrison E. F., Minnis P., Barkstrom B. R., E. Ahmad, and D. Hartmann, 1989. Cloud-radiative forcing and climate: results from the Earth Radiation Budget Experiment, *Science*, 243, No. 4887, 57-63.

Reiter R., 1954. The importance of the atmospheric long radiation disturbances for the statistical biometeorology, *Archiv fur Physikalische Therapie*, 6, No. 3, 210-216.

Reiter R., 1955. Bio-meteorologie auf physikalisher Basis, *Phys. Blaetter*, 11, 453-464.

Renaker D., 1979. Swift's Laputians as a Caricature of the Cartesians. *PMLA*, 94(5): 936-944. Published by Modern Language Association.

Richter G. M. and Semenov M. A., 2005, Modeling impacts of climate change on wheat yields in England and Wales—assessing drought risks, *Agric. Syst.*, 84 (1), 77–97.

Rogers J. E. T., M1887. *Agriculture and Prices in England*, Vol. 1-8, Oxford, Clarendon Press; Reprinted by Kraus Reprint Ltd, Vaduz in 1963.

Rossow W. B. and Cairns B., 1995. Monitoring changes of clouds, *J. Climate*, 31, 305-347.

Rossow W. B. and Shiffer R., 1991. ISCCP cloud data products, *Bull. Am. Met. Soc.*, 72, 2-20.

Ruzmaikin A., 1999, Can El Nino amplify the solar forcing of climate, *Geophysical Research Letters*, 26, 15, 2255-2258.

Ruzmaikin A., 2007, Effect of solar variability on the Earth's climate pattern, *Advances in Space Research*, 40, 1146–1151.

Sakurai K., 2003. The Long-Term Variation of Galactic Cosmic Ray Flux and Its Possible Connection with the Current Trend of the Global Warming, *Proc. 28th Intern. Cosmic Ray Conf.*, Tsukuba, 7, 4209-4212.

Schlegel K., Diendorfer G., Them S., and Schmidt M., 2001. Thunderstorms, lightning and solar activity – Middle Europe, *J. Atmos. Solar-Terr. Phys.*, 63, 1705-1713.

Shapiro A. I., Schmutz W., Rozanov E., Schoell M., Haberreiter M., Shapiro A. V., and Nyeki S., 2011. A new approach to the long-term reconstruction of the solar irradiance leads to large historical solar forcing, *Astronomy & Astrophysics*, 529, A67 (1-8), DOI: 10.1051/0004-6361/201016173.

Shaviv N. J., 2005, On Climate Response to Changes in the Cosmic Ray Flux and Radiative Budget, *J. Geophys. Res. Space Phys.*, 110 (A8): A08105.

Shea M. A. and Smart D. F., 1992, Recent and Historical Solar Proton Events, *Radiocarbon*, 34, No. 2,. 255-262.

Shea M. A. and Smart D. F., 2003. Preliminary Study of the 400-Year Geomagnetic Cutoff Rigidity Changes, Cosmic Rays and Possible Climate Changes, *Proc. 28th Intern. Cosmic Ray Conf.*, Tsukuba, 7, 4205-4208.

Shiller R. J., 1990. Speculative Prices and Popular Models. *The Journal of Economic Perspectives*, 4 (2), 55-65.

Shindell D., Rind D., Balabhandran N., Lean J., and Lonengran P., 1999. Solar cycle variability, ozone, and climate, *Science*, 284, 305-308.

Smith A., M1776. *An Inquiry into the Nature and Causes of the Wealth of Nations*. London, W. Strahan & T. Cadell.

Smith G. L., Priestley K. J., Loeb N. G., Wielicki B. A., Charlock T. P., Minnis P., Doelling D. R., and Rutan D. A., 2011. Clouds and Earth Radiant Energy System (CERES), a review: Past, present and future, *Advances in Space Research*, 48, 254–263.

Soon W. and Yaskell S., M2003. *The Maunder Minimum and the Variable Sun-Earth Connection*, World Scientific Publishing Company.

Stowe L. L., Wellemayer C. G., Eck T. F., Yeh H. Y. M., and the Nimbus-7 Team, 1988. Nimbus-7 global cloud climatology, Part 1: Algorithms and validation, *J. Climate*, 1, 445-470.

Stozhkov Yu. I., 2002. The role of cosmic rays in atmospheric processes, *J. Phys. G: Nucl. Part. Phys.*, 28, 1-11.

Stozhkov Yu. I., Zullo J., Martin I. M. et al., 1995a. Rainfalls during great Forbush decreases, *Nuovo Cimento*, C18, 335-341.

Stozhkov Yu. I., Pokrevsky P. E., Martin I. M. et al., 1995b. Cosmic ray fluxes and precipitations, *Proc. 24th Intern. Cosmic Ray Conf.*, Rome, 4, 1122-1125.

Stozhkov Yu. I., Pokrevsky P. E., Zullo J. Jr. et al., 1996. Influence of charged particle fluxes on precipitation, *Geomagn. and Aeronomy*, 36, 211-216.

Stozhkov Yu. I., Pokrevsky P. E., and Okhlopkov V. P., 2000. Long-term negative trend in cosmic ray flux, *J. Geophys. Res.*, 105, No. A1, 9-17.

Stozhkov Yu. I., Ermakov V. I., and Pokrevsky P. E., 2001. Cosmic rays and atmospheric processes, *Izvestia Russian Academy of Sci., Ser. Phys.*, 65, No. 3, 406-410.

Suits D. B., 1957. Use of Dummy Variables in Regression Equations. *Journal of the American Statistical Association*, 52(280), 548-551.

Swensmark H., 1998. Influence of cosmic rays on the Earth's climate, *Phys. Rev. Lett.*, 81, 5027-5030.

Svensmark H., 2000. Cosmic rays and Earth's climate, *Space Sci. Rev.*, 93, 175-185.

Swensmark H. and Friis-Christiansen E., 1997. Variation of cosmic ray flux and global cloud coverage – a missing link in solar-climate relationships, *J. Atmos. Solar-Terr. Phys.*, 59, 1225-1232.

Swift J., M1726. *Gulliver's Travels, Part III. A Voyage to Laputa, Balnibarbi, Luggnagg, Glubbdubdrib, and Japan.*

Thomas W. B., 1995. Orbital anomalies in Goddard Spacecraft for Calendar Year 1994, *NASA Technical Paper 9636*, pp. 1–53.

Thornton P. K., Jones P. G., Alagarswamy G., and Andresen J., 2009. The temporal dynamics of crop yield responses to climate change in East Africa. *Global Environmental Change*, 19, 54–65.

Tinsley B. A., 1996. Solar wind modulation of the global electric circuit and the apparent effects of cloud microphysics, latent heat release, and tropospheric dynamics, *J. Geomagn. Geoelectr.*, 48, 165-175.

Tinsley B. A., 2000. Influence of solar wind on the global electric circuit, and inferred effects on cloud microphysics, temperature, and dynamics in the troposphere, *Space Sci. Rev.*, 94, No. 1-2, 231-258.

Tinsley B. A. and Deen G. W., 1991. Apparent tropospheric response to MeV-GeV particle flux variations: a connection via electrofreezing of supercooled water in high-level clouds?, *J. Geophys. Res.*, 96, No. D12, 22283-22296.

Todd M. C. and Kniveton D. R., 2001. Changes in cloud cover associated with Forbush decreases of galactic cosmic rays, *J. Geophys. Res.*, 106, No. D23, 32031–32042.

Todd M. C. and Kniveton D. R., 2004. Short-term variability in satellite-derived cloud cover and galactic cosmic rays: an update *J. Atmosph. and Solar-Terrestrial Physics*, 66, 1205–1211.

USDA, 2004. Prices Received by Farmers: Historic Prices & Indexes 1908-1992 (92152). *National Agricultural Statistics Service.* http://usda.mannlib.cornell.edu/, accessed 23 July 2004.

Usoskin I. G., Gladysheva O. G., and Kovaltsov G. A., 2004. Cosmic ray induced ionization in the atmosphere: spatial and temporal changes, *J. Atmosph. Solar-Terr. Phys.*, 66, No. 18, 1791-1796.

Vasey D. A., 2001. A quantative assessment of buffer among temperature variations, livestock, and Human Population of Iceland, 1784 to 1900, *Climatic Change*, 48, 243-263.

Veizer J., Godderis Y., and Francois I. M., 2000. Evidence for decoupling of atmospheric CO_2 and global climate during the Phanerozoiceon, *Nature*, 408, 698-701.

Velinov P., Nestorov G., and Dorman L., M1974. *Cosmic Ray Influence on Ionosphere and Radio Wave Propagation*. Bulgaria Academy of Sciences Press, Sofia.

Veretenenko S. V. and Pudovkin M. I., 1994. Effects of Forbush decreases in cloudiness variations, *Geomagnetism and Aeronomy*, 34, 38-44.

Villoresi G., Breus T. K., Dorman L. I., Iucci N., Rapoport S. I., 1994. The influence of geophysical and social effects on the incidences of clinically important pathologies (Moscow 1979-1981). *Physica Medica*, 10, No 3, 79-91.

Villoresi G., Dorman L. I., Ptitsyna N. G., Iucci N., Tyasto M. I., 1995. Forbush decreases as indicators of health-hazardous geomagnetic storms. *Proc. 24-th Intern. Cosmic Ray Conf.*, Rome, 4, 1106-1109.

Waliser D. E., Li J.-L. F., L'Ecuyer T. S., and Chen W.-T., 2011. The impact of precipitating ice and snow on the radiation balance in global climate models, *Geophys. Res. Lett.*, 38, L06802, doi:10.1029/2010GL046478.

Weng F. and Grody N. C., 1994. Retrieval of cloud liquid water using the special sensor microwave imager (SSM/I), *J. Geophys. Res.*, 99, 25535-25551.

Zecca A. and Chiari L., 2009. Comets and climate, *J. Atm. and Solar-Terr. Physics*, 71, 1766–1770.

In: Homage to the Discovery of Cosmic Rays …
Editor: Jorge A. Perez-Peraza

ISBN: 978-1-63483-084-3
© 2015 Nova Science Publishers, Inc.

Chapter 14

SOLAR ACTIVITY, COSMIC RAYS, AND GLOBAL CLIMATE CHANGES

Yuri Stozhkov[1] and Victor Okhlopkov[2]
[1]P. N. Lebedev Physical Institute of the Russian Academy of Sciences, Moscow, Russia
[2]D. V. Skobeltsyn Research Institute of Nuclear Physics of M. V. Lomonosov
Moscow State University, Moscow, Russia

ABSTRACT

In 2012 the scientific community celebrated memorable date − 100 years from the cosmic ray discovery by Austrian scientist Victor Hess. Cosmic rays occupy important place in astrophysical investigations, in nuclear physics research, solar-terrestrial relationships, and atmosphere physics. In this article the questions of solar activity, cosmic ray modulation and global climate changes are discussed. All these questions are tightly related with each other. The analysis of the solar activity where we take the sunspot number R_z (international values) as solar activity parameter in the past and present shows that new long-term solar activity minimum has started like Dalton's one in the past. It is well known that cosmic ray fluxes observed at the Earth's orbit depend on solar activity level which is described by sunspot number R_z. In the current 24th solar activity cycle with its low value of R_z the cosmic ray fluxes recorded near the Earth are higher in comparison with the fluxes observed during the previous solar cycles. As the solar activity is expected to be low in the subsequent (2–3) solar cycles, the cosmic ray fluxes will exceed ones observed earlier in the previous solar cycles.

The separate part of this article is devoted to the discussion of the physical mechanism responsible for the global warming process. It is shown that solar activity and cosmic ray flux changes cannot explain the observed increase of the global temperature of air near the Earth's surface. The cosmic dust entering into the Earth's atmosphere from the zodiacal dust cloud in space could be responsible for the climate changes. The changes of dust concentration in the zodiacal dust cloud could essentially influence the cloud coverage over the globe and the transparency of the atmosphere. It in its turn influences on climate changes. Analysis of the changes in the global temperature shows that from the beginning of the 21st century the global temperature decreases, thus the process of global cooling has started. In the article the global cooling is shown to be taking place up to ~2050 when the temperature conditions on the Earth will be the almost same as in the middle of the 20th century.

INTRODUCTION

Below the following questions will be discussed: (1) solar activity in the past, present, and nearest future; (2) cosmic rays which depend on solar activity level; (3) mechanism responsible for the global climate changes. Sunspot number of R_z and sunspot group number of η were chosen as solar activity parameters. There are long-term sets of these parameters. It allows to analyze their time changes at the long-term base in order to compare the current solar cycle with the cycles in the past and to predict the solar activity behavior for the next 50 years. The 11-year solar cycle is being observed for more than 2.5 thousand years. However, sometimes during rather long-term time intervals solar activity becomes very low (the values of R_z are small). These intervals were called grand-minima. The solar activity analysis of the past cycles and the current 24th solar cycle leads us to the conclusion that from the beginning of the 21st century we have a new prolonged solar activity minimum. It is predicted that this new minimum will persist during the next (2–3) solar cycles up to ~2050.

The second question under discussion is galactic cosmic ray fluxes and their temporal changes. In the solar system these fluxes are controlled by solar activity level. Because in the next ten years solar activity level is expected to be low the galactic cosmic ray fluxes (in the following cosmic ray fluxes) will be rather high in comparison with the ones observed from the middle of 1950's when the regular cosmic ray monitoring was started. In the analysis given below the data on cosmic ray fluxes measured in the Earth's atmosphere of the northern and southern (Antarctica) polar latitudes and at the northern middle latitude have been used.

The regular measurements have been performed by the coworkers of the Lebedev Physical Institute of the Russian Academy of Sciences beginning from the middle of 1957 to the present. In 2009 the highest cosmic ray fluxes were recorded for the whole ~55-year history of the regular monitoring of cosmic ray fluxes. In 2009 the solar activity level was very low ($R_z < 20$), and at the distance of 1 astronomical unit from the Sun (in the following 1 a.u.) the strength of interplanetary magnetic field responsible for the cosmic ray modulation declined to 3.5 nT, instead of ~5 nT observed usually in the preceding solar activity minima.

The third discussed subject is the global warming problem and physical mechanism responsible for this phenomenon. The time dependence of $\Delta T(t)$ is analyzed where $\Delta T(t)$ is the difference between the average global temperature of air at the surface layer obtained for the period of 1901–2000 and the current value of global average temperature. Based on the results of the analysis of solar activity and cosmic ray data the correlations between $\Delta T(t)$ values, solar activity and cosmic ray fluxes is shown to be low. It means that neither solar activity nor cosmic rays can essentially influence on the Earth's climate changes. The following mechanism responsible for global changes of climate is suggested. Between the Sun and Mar's orbit there is the zodiacal dust cloud. The Earth rotates around the Sun inside this cloud. The cosmic dust falls on the top of the Earth's atmosphere. The dust concentration is changing in time. Accordingly the amount of cosmic dust falling on the atmosphere is changing also. Cosmic dust in the Earth's atmosphere influences on transparency of air and cloud formation process. In their turn, the changes of air transparence and rate of the cloud formation modify the amount of solar energy falling on the Earth's surface. The analysis of the temperature data $\Delta T(t)$ made for the period of ~130 years shows that from the beginning of the 21st century the global cooling is observed. Towards the ~2050 the average global temperature of air surface layer will be near the value observed in the middle of the last century.

SOLAR ACTIVITY IN THE PAST, PRESENT, AND FUTURE

Solar Activity in the Past

One of the main characteristics of solar activity is its quasi-periodical changes. The sunspot number R_z is changed with the average period of about 11 years. This quasi-periodicity is marked on the time scale of ~2650 years [1–3]. But sometimes the long-term periods of very low solar activity, so called grand minima, take place. One of them is well known Maunder's minimum with the duration of about 100 years (1620–1720), another minimum is Dalton's one (1790–1835) [4, 5]. The periods of grand minima occurred during the last 1000 years are given in Table 1 and Figure 1. Their starts and completions are defined approximately. The causes of the appearance of such long-term solar activity minima are unknown. There are several suggestions to explain these phenomena. For example, the stochastic changes of solar dynamo parameters could account for their appearance [7].

Table 1. The grand minima of solar activity for the last 1000 years [5]

The minimum name	Oort	Wolf	Spörer	Maunder	Dalton
Period	1010 - 1050	1280 - 1350	1460 - 1540	1620 - 1720	1790 - 1835

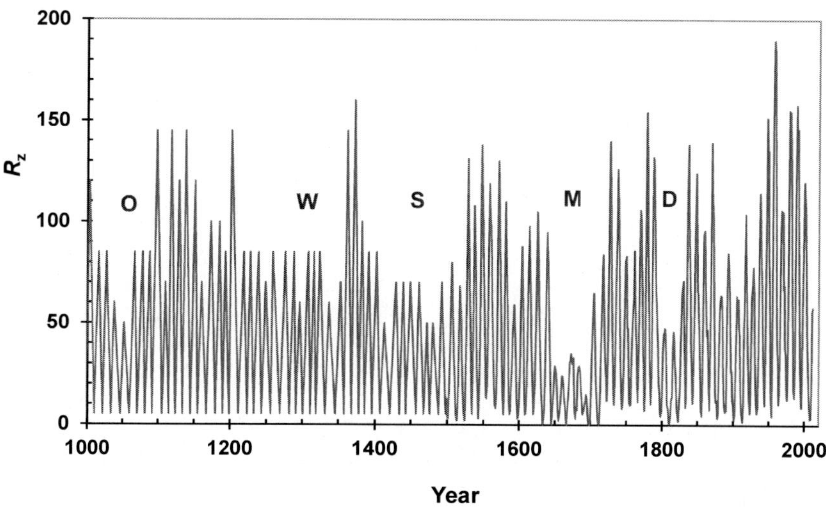

Figure 1. The values of sunspot number R_z averaged per year as a function of time [1, 4-6]. The blue letters denote grand minima of solar activity: O – Oort's minimum, W – Wolf's minimum, S – Spörer's minimum. M – Maunder's minimum, and Dalton's minimum.

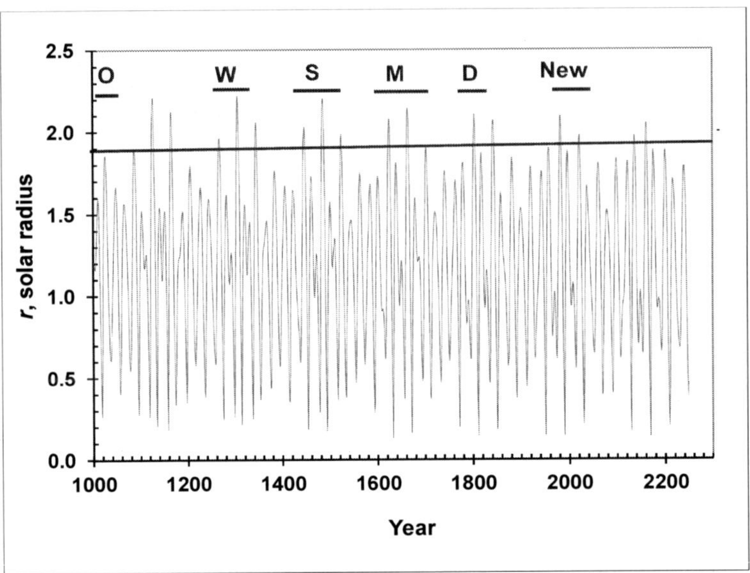

Figure 2. The calculated distances r between the center of mass of the solar system and the center of the Sun as a function of time. Horizontal bars in the upper part of this Figure show the duration of the long-term grand minima of solar activity: O – Oort's minimum, W – Wolf's minimum, S – Spörer's minimum, M – Maunder's minimum, Dalton's minimum, and N – modern new minimum. Horizontal blue line was drawn by hand and separates time intervals with the maximum values of r.

Relationship of the Grand Minima Appearance with the Distance between the Center of Mass of the Solar System and the Center of the Sun

In Figure 2 the calculated distances r between the center of mass of the Solar system and the center of the Sun as a function of time are shown. In the upper part of Figure 2 the grand minima of solar activity are denoted by horizontal bars (see Table 1). It is seen that the long-term solar activity minima were observed when the maximum values of distances r had been reached and had maximum fluctuations. There are 5 grand minimum events and the only Oort's minimum brakes this relationship. But this minimum was ~1000 years ago and our knowledge on solar activity related to this period is very sparse. From the data given in Figure 2 one can suggest that at present we have a new grand minimum of solar activity and it will continue during next several decades. This conclusion about long-term low solar activity was made several years ago and the subsequent experimental data has confirmed its validity [8, 9]. Next solar activity grand minimum is expected in the second half of the 22nd century.

Current 24th Solar Cycle and Its Future Development

The current 24th solar cycle is unusual one in comparison with the previous cycles. Its start (increase of R_z) was like the start of the 14th solar cycle (1903–1913). However, in the current cycle the low solar activity ($R_z \leq 20$) was observed for more than 5 years. During this cycle very small values of the solar polar magnetic field strength and interplanetary magnetic field strength at the Earth's orbit have been recorded [10, 11]. As a consequence of it the high fluxes of galactic cosmic rays were recorded at 1 a.u. In 2009 the highest cosmic ray flux was measured for whole history (~60 years) of the regular cosmic ray observations [12].

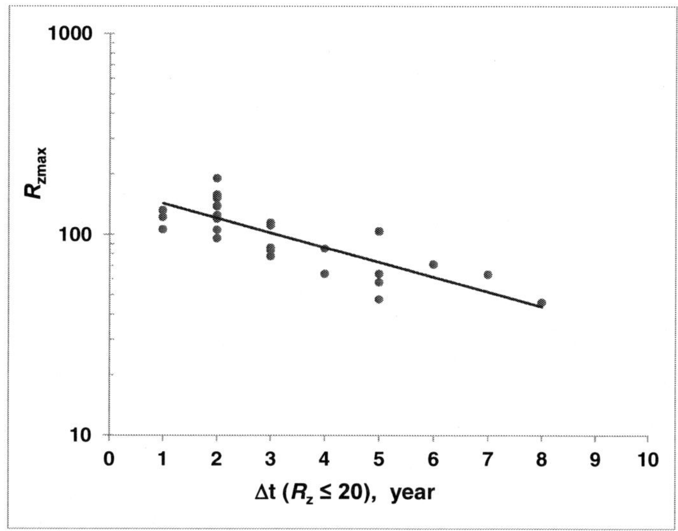

Figure 3. The dependence of values R_{zmax} from the previous solar minimum duration Δt ($R_z \leq 20$). The yearly averages have been used. Straight line is the calculation made by the least square method: $R_{zmax} = 168.6 \cdot \exp(-0.16\Delta t)$ with the correlation coefficient $k = -(0.80\pm0.07)$.

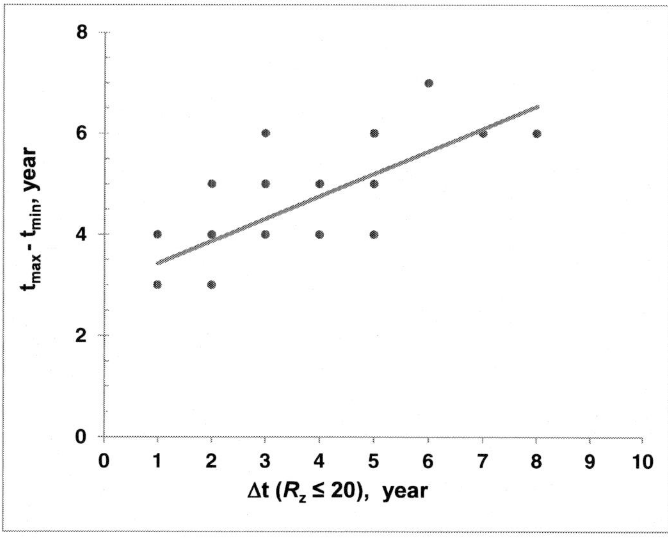

Figure 4. The dependence of the time interval between minimum and next maximum of solar activity ($t_{max} - t_{min}$) and the duration of solar activity minimum Δt ($R_z \leq 20$). The yearly averages have been used. The straight line calculated with the least square method gives the following dependence: ($t_{max} - t_{min}$) = $0.470 \cdot \Delta t$ ($R_z \leq 20$) + 2.943 with the correlation coefficient $k = 0.72\pm0.10$.

Table 2. Maximum values of sunspot group number η_{max} and average heliolatitudes ϕ of η_{max} observed in the 19-24 solar cycles [14, 15]. Monthly averages of η_{max} and ϕ were used. In each cycle two maximum values η_{max} are observed (two lines in Table 2). The values of η_{max} and ϕ were smoothed with 5 points

Solar cycle number	Time of η_{max}	η_{max}	$\phi°$
19	11.1957	16.3	18.8
	02.1959	13.9	15.2
20	10.1967	9.8	19.0
	03.1970	10.0	13.0
21	10.1979	15.6	18.2
	11.1981	11.8	13.7
22	08.1989	14.2	19.5
	06.1991	14.1	15.4
23	05.2000	11.1	17.2
	11.2001	10.7	13.7
24	11.2011	6.9	17.5

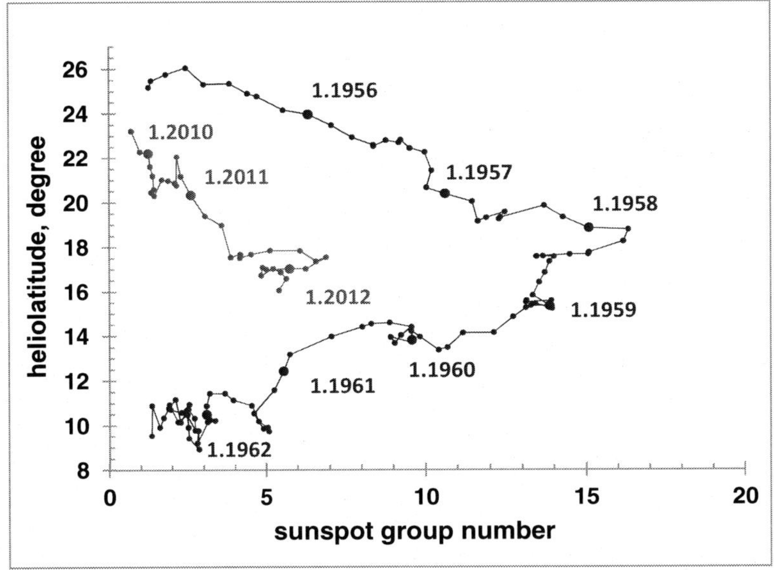

Figure 5. The relationship between sunspot group number η and their average heliolatitude ϕ in the 19th and 24th solar cycles (blue and red curves correspondingly) [16]. The monthly averages of η and ϕ (5-point smoothed) were used. The large points correspond to the beginning of each year.

The development of solar activity in the nearest future was discussed in many papers [e.g., 13, 14]. In the earlier papers rather high solar activity maximum was predicted with $R_z \geq 120$–150. During the last year this value was decreased to $R_z \approx (60$–$80)$. Our prediction on the future solar activity development differs from the most others. We predict the new long-term solar activity minimum like Dalton's one (1790–1835) with the low values of maximum sunspot number, $R_{zmax} \approx (50$–$70)$ [8]. The new grand minimum will have the duration of several decades (2 or more of 11-year solar cycles). To evaluate the value $R_{z\max}$ in the current solar cycle, let us consider the relationship between the duration of solar minimum Δt and subsequent maximum of solar activity $R_{z\max}$ [15]. The duration of solar minimum Δt is defined by the number of years when the sunspot number R_z averaged per year was $R_z \leq 20$. In Figure 3 the relationship of $R_{z\max}$ and Δt derived from data of the previous solar cycles is depicted. The Figure 4 gives the relationship of the time interval between solar minimum and next solar maximum $(t_{\max} - t_{\min})$ with the solar minimum duration Δt $(R_z \leq 20)$. There is a linear relation between these values. From Figures 3 and 4 it follows that the current solar activity maximum will have $R_{z\max} = (50–70)$ and is expected in 2012–2014. Also the time of the beginning of solar activity maximum can be evaluated from the relationship of sunspot group number η with their averaged heliolatitude ϕ. It is known that new solar cycle begins from the appearance of sunspots or their groups at the high heliolatitudes $\phi \approx (25–35°)$. With the solar activity growth and according to Spörer's rule the value η increases and the average heliolatitude ϕ decreases.

At the end of 11-year solar cycle the sunspots are at the heliolatitudes $\phi \approx (8–12°)$. Analysis of sunspot group number η and their heliolatitude distribution performed for the 19–24 solar cycles showed that maximum values η were observed twice in each solar cycle when their average heliolatitude was about $\phi \approx (17–19°)$ and then $\phi \approx (13–15.5°)$. In Table 2 the values $\eta_{\max}$ and ϕ for 19–24 solar cycles are given. In Figures 5 and 6 the examples of the dependences of η and ϕ in the 19th and 24th and in the 21st and 24th solar cycles are shown. One can see that in the 24th solar cycle the first maximum of η occurred in the November 2011 and the second (if it occurs) is expected in 2013–2014. In the solar cycles from the 19th to the 24th the same patterns in the dependences between η и ϕ were observed. Using the results presented above the prognosis on the development of solar activity during the next decades was made [8, 9].

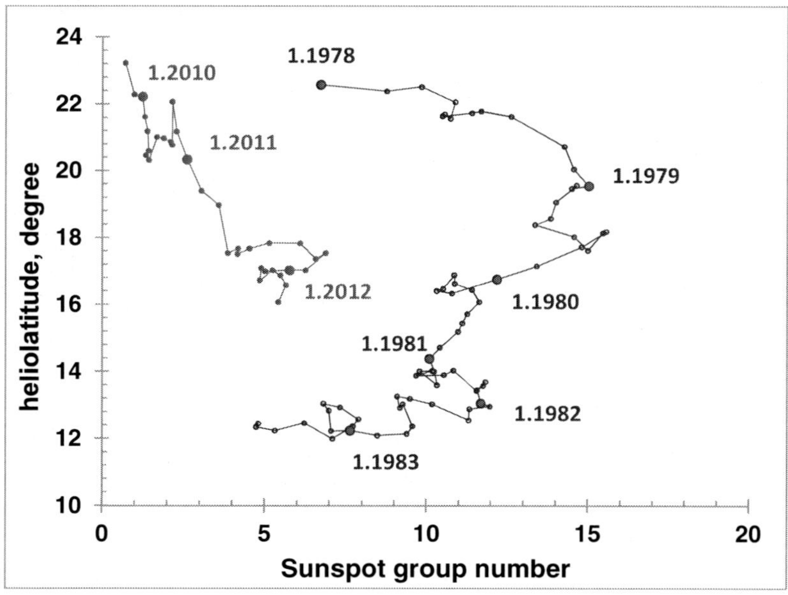

Figure 6. The same as in Figure 5 but for the 21st and 24th solar cycles.

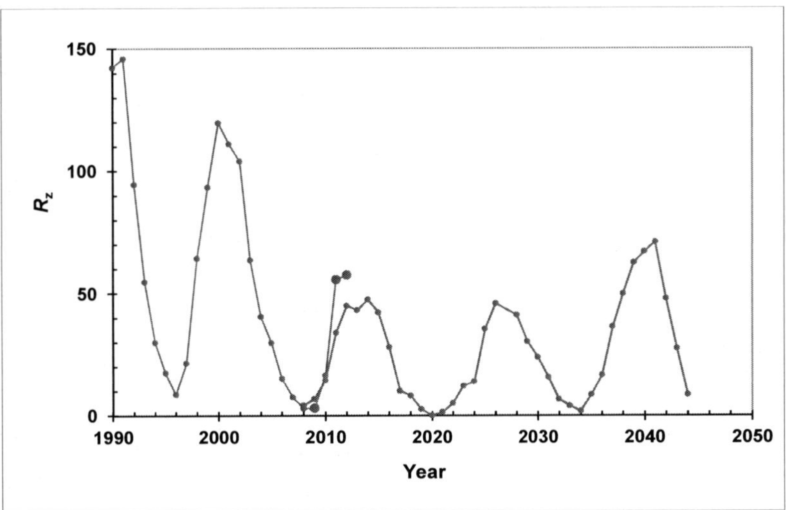

Figure 7. Time dependence of observed sunspot number R_z (monthly averages) from 1990 up to now (red points and curve) and predicted values R_z for the current 24th solar cycle (blue points and curve). The prognosis was made 4 years ago [8, 9]. The values of R_z are 5-point smoothed.

The results given in Figures 2, 5 and 6 were the basis of this prognosis. It was suggested that a new long-term solar activity minimum was started in ~2005. This new grand minimum will be like Dalton's one and will have the duration of about several solar cycles. It means that in the following (2 or 3) solar cycles the values R_{zmax} will be low as it was observed in the current 24th solar cycle. In Figure 7 the prediction of solar activity development for the period from now up to 2045 is presented [8, 9]. Several years have elapsed from the first publication of this prognosis and experimental data on R_z obtained for these past years are in agreement with the prediction.

COSMIC RAY FLUXES IN THE PAST AND PRESENT

Measurements of Cosmic Ray Fluxes

In the heliosphere the cosmic ray fluxes depend on solar activity level, in our case on sunspot number R_z. There is a tight relation between cosmic ray flux and R_z with the high correlation coefficient $k \approx 0.9$. The delay time of cosmic ray flux relative to solar activity is ~ (0.5 – 1) year. Below we will use the data on cosmic ray fluxes in the atmosphere obtained in Lebedev Physical institute of the Russian Academy of Sciences. The regular measurements of cosmic ray fluxes in the Earth's atmosphere were started in the middle of 1957 and they are continued up to now. These measurements are performed in the atmosphere at the northern polar latitudes (Murmansk region, geomagnetic cutoff rigidity $R_c = 0.6$ GV), the southern polar latitude (observatory Mirny in the Antarctica, $R_c = 0.04$ GV), and the middle latitude (Moscow region, $R_c = 2.4$ GV). The data obtained in these measurements span the altitudes from the ground level up to (30–35) km. Up to now ~85000 radiozonds for measuring cosmic rays were launched [17, 18].

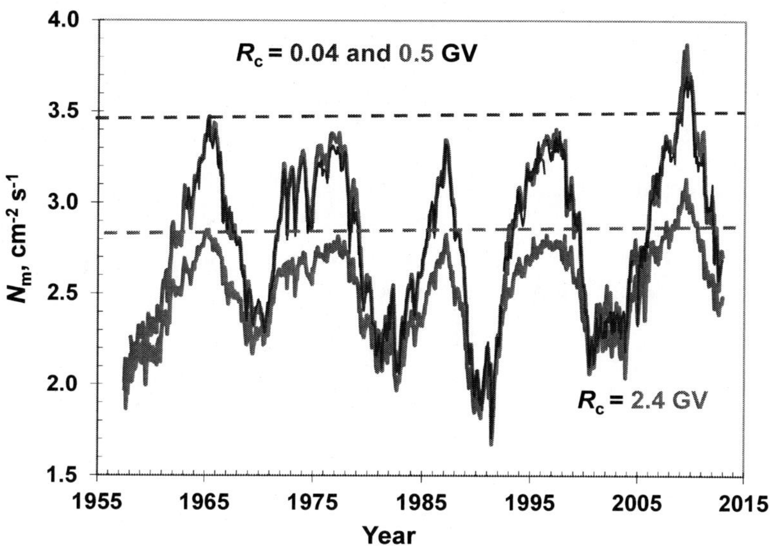

Figure 8. Time dependence of the monthly averaged cosmic ray fluxes in the maximum of their absorption in air of the northern polar latitude ($R_c = 0.6$ GV, green curve), southern polar latitude ($R_c = 0.04$ GV, Antarctica, blue curve), and middle northern latitude ($R_c = 2.4$ GV, Moscow region, red curve). The dotted horizontal lines mark the maximum of cosmic ray fluxes measured in 1965.

We have the unique long-term set of homogeneous data about time and spatial characteristics of cosmic ray fluxes at the different altitudes in the atmosphere for the period more than 55 years. In Figure 8 the time dependences of monthly averaged cosmic ray fluxes in the atmosphere of the polar and middle latitudes are shown. The maxima of cosmic ray fluxes of the absorption curve in the atmosphere N_m (so-called Pfotzer's maxima) are shown. These data were obtained at the altitudes about (20–24) km in the polar atmosphere and about (18–20) km in the atmosphere of the middle latitude.From the data presented in Figure 8 it follows that the time dependences of $N_m(t)$ show alternating sequences of peak and flat parts of $N_m(t)$.

Such behavior reflects 22-year solar magnetic cycle in cosmic ray time dependence. The solar magnetic cycle includes negative and positive phases: in the negative phases solar magnetic field lines come out from the solar southern polar cap and come into the solar northern polar cap; in the positive phases the polarity is turned vice versa. The change of phase of the solar polar magnetic field takes place in or near period with solar activity maximum (maximum of sunspot number R_z). Now we are in the period of the solar polar magnetic field inversion.

During the negative phases of the 22-year solar magnetic cycles (in our case these periods are 1959–1968, 1982–1988, and 2002 – present time) the peaked dependences of $N_m(t)$ are observed. In the positive phases (1972–1980, 1992–1999) these dependences have more or less flat forms. While going from one phase of solar magnetic cycle to another one the directions of magnetic field lines in the northern and southern parts of the heliosphere are inversed. The different forms of time dependences of cosmic ray fluxes in negative and positive phases are defined by different directions of the drift of cosmic ray fluxes in the heliosphere.

Anomalous High Cosmic Ray Fluxes in 2008 - 2009

At the end of 2008 and in 2009 in the atmosphere of polar and middle latitudes the highest cosmic ray fluxes were recorded during the whole history of their monitoring (about 60 years) [12]. In this period a very low solar activity (sunspot number R_z was near zero) and very weak interplanetary magnetic field strength B (see below Figure 9) were observed. If in the previous solar cycle minima (1976, 1986, 1996) the value B was ~5 nT, from the end of 2008 till the beginning of 2010 B was equal to ~3.5 nT. The solar wind speed was the same as in the previous solar activity minima (see Figure 10).

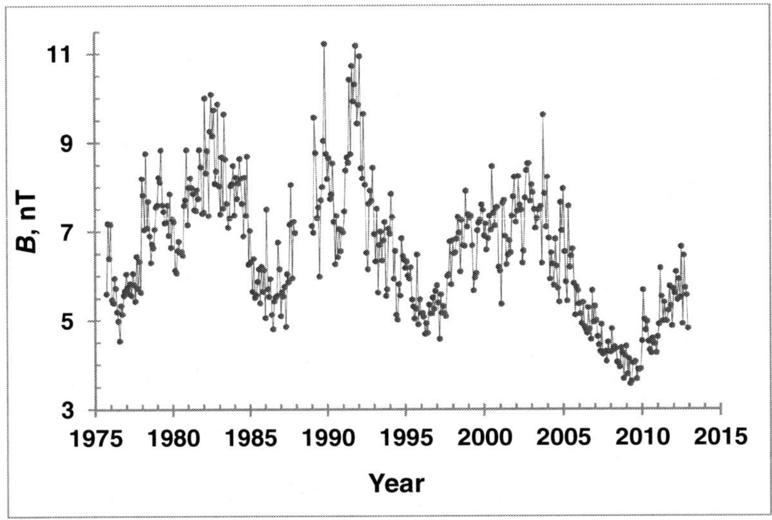

Figure 9. Interplanetary magnetic field strength (monthly averages) at 1 a.u. as a function of time [16].

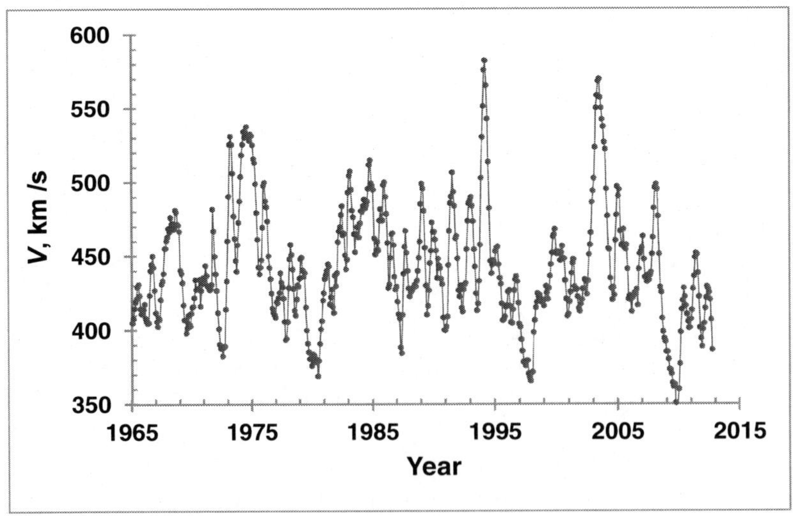

Figure 10. Monthly averages of solar wind speed at 1 a.u. as a function of time [6].

Thus, the low solar activity and the weak interplanetary magnetic field strength are discovered to be the main factors responsible for the appearance of high cosmic ray fluxes in 2008–2010. In 2009 in the polar atmosphere the values N_m increased by ~15% in comparison with N_m observed in 1965. At the ground level the neutron monitors gave the increase of ~3%. In 2012 cosmic ray flux decreased because the solar activity increased and in October – November the maximum solar activity (possible the first maximum) occurred. However, the value R_{zmax} was by 1.5–2 times lower than in the previous 19–23 solar cycles. As a result, now the cosmic ray fluxes are higher than the corresponding ones observed in the maxima of the previous solar cycles.

WHAT IS THE MAIN CAUSE (CAUSES) OF THE GLOBAL WARMING PROCESS?

Global Temperature of Air near the Earth's Surface and Antropogenic Influence on Climate Changes

The climate changes are very important subject of investigations because climate determines the conditions of life on the Earth. During the last ~100 years process of global warming was observed. One of its main characteristics is the increase of global air temperature at the surface. Usually the value $\Delta T(t) = T(t) - T_{av}$ is considered where T_{av} is the surface air temperature averaged over globe for some period. In our case this period is 1901–2000. The values $T(t)$ are the current global averaged temperature. The experimental data on $\Delta T(t)$ are shown in Figure 11 [19]. During the last ~130 years the value ΔT increased by ~0.8 °C, which indicates the global climate warming. There are two points of view explaining the cause of this effect.

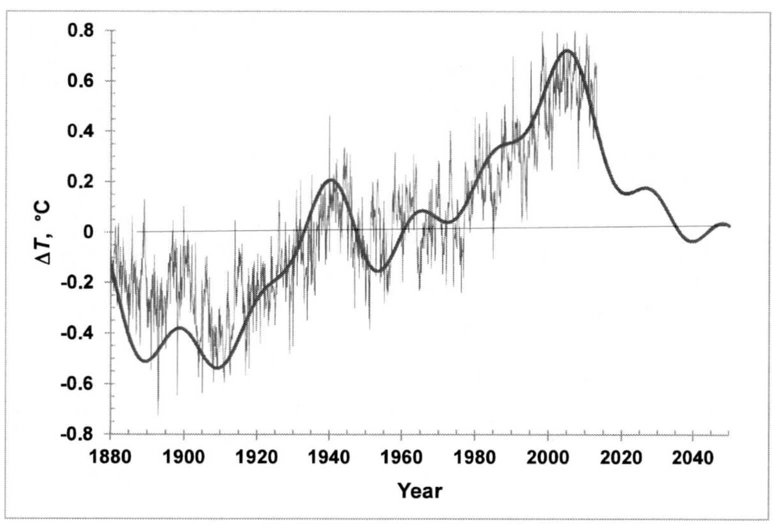

Figure 11. Time dependence of monthly averages of the global changes of surface air temperature ΔT (red curve) [19], calculated values ΔT and their prediction up to 2050 (blue curve, see text below) [20].

The first one is based on the antropogenic factors, and the second point of view is based on the natural ones. The human activity increases the concentration of greenhouse gases in the atmosphere, such as CO_2, CH_4, water vapor and others. These gases absorb the surface infrared radiation and it increases the temperature of air near the Earth's surface. The numerous calculations performed in the frame of various models of climatic system predict the future growth of the global temperature up to several degrees to the end of 21st century [21]. However, in the experimental data on ΔT presented in Figure 11 some irregularities are seen. In the past rather long-term periods of growth and decrease of temperature were observed (see, e.g., 1945–1980 and 2005 – present time). These temperature irregularities are difficult to explain by human impact on the climate system.

Solar Activity and Cosmic Ray Fluxes Influences on the Values ΔT

Another point of view to understand the causes of the global temperature growth includes the influence of natural factors. Let us consider some of them widely discussed in scientific papers. For more than 35 years the high precision measurements of bolometric solar irradiation F have been performed using satellites [22]. It was proved that in the 11-year solar cycles the value F is changed in the limits less than 0.1%. These small changes of the solar irradiation could produce the very small changes of ΔT, less than 0.05 °C.

Could the changes of solar activity be responsible for the increase of global temperature? Here the sunspot number R_z was chosen as solar activity index. In Figure 12 the time dependences of ΔT and R_z are given. From the comparison of these time dependences one can do the conclusion about the absence of the correlation between the changes of global temperature ΔT and solar activity R_z.

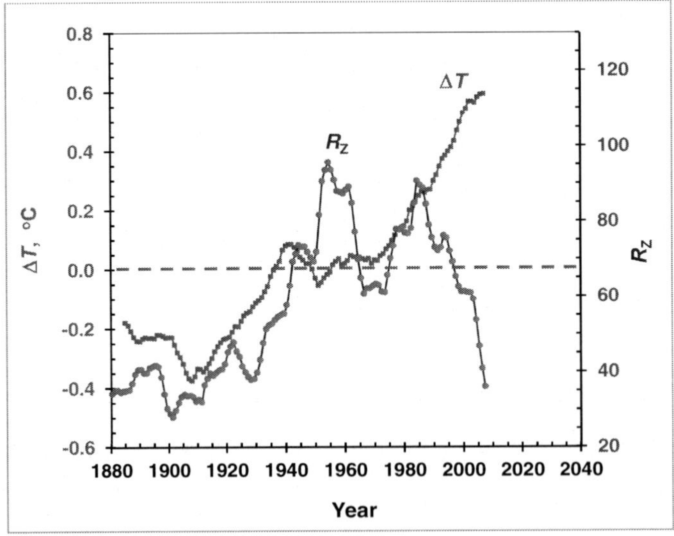

Figure 12. Time changes of the annual averages ΔT and R_z (red and blue curves accordingly). The vertical axes for ΔT and R_z are given on the left and right vertical axes accordingly. The values ΔT and R_z were smoothed with the 11-year period to remove 11-year solar cycle in cosmic rays.

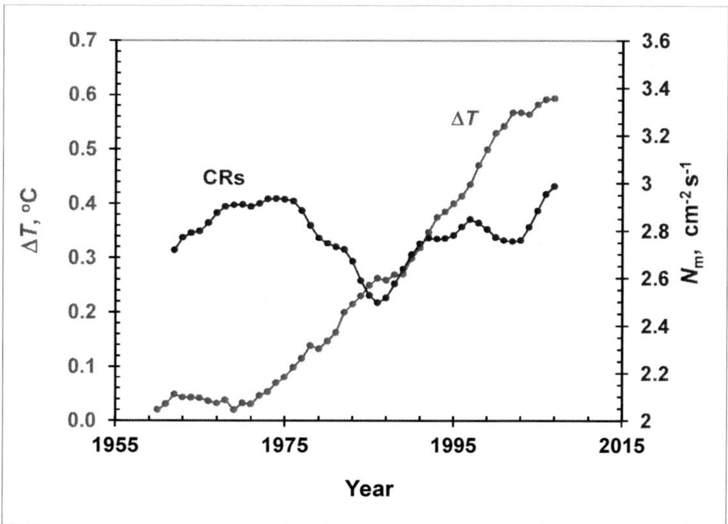

Figure 13. Time dependences of ΔT (red points and curve, vertical axis is at the left side) and N_m (blue dots and curve, vertical axis is at the right side). The values ΔT and N_m were smoothed with the 11-year period to remove 11-year solar cycle in cosmic rays. The cosmic ray flux N_m was measured at the northern polar latitude in the maximum of absorption curve of cosmic particles in the atmosphere.

In recent years many papers were published where the possible impact of cosmic ray fluxes on cloud coverage and climate were considered [23]. Cosmic ray fluxes play the decisive role in the processes associated with atmospheric electricity. Cosmic rays are the main sources of electric charges in the atmosphere via ion production. They provide electric conductivity of air from the ground level up to ~40 km, they play important role in the cloud formation processes. Also cosmic rays are responsible for the lightning production [24]. Therefore, the study of cosmic ray influence on atmospheric processes is very important. But the analysis of the time dependences of $\Delta T(t)$ and $N_m(t)$ leads to the conclusion that relationship between these values is absent (see Figure 13). Thus from the data shown in Figures 12 and 13 it follows that neither solar activity nor cosmic ray flux changes can explain the observed growth of global air surface temperature.

Cosmic Dust and Its Role in the Climate Changes

Several years ago it was suggested that cosmic dust being in zodiacal cloud could be responsible for the climate changes on Earth [25]. Our Earth's orbit around the Sun lies inside of the zodiacal dust cloud. This cloud extends from Sun to Mars's orbit. The zodiacal dust is mainly concentrated in the galactic plane. The comets coming to the solar system produce this dust. The dust comes to the Earth's atmosphere forming condensation centers of water vapor on which water droplets are grown. Every day a huge amount of cosmic dust from 4×10^2 to 10^4 tons precipitates at the Earth's atmosphere. The changes of dust concentration in the zodiacal cloud will have an impact on the total mass of cosmic dust in the atmosphere, and in its turn it will change global cloud coverage. About 60% of the globe are covered by clouds which reflect ~30% solar irradiance coming to the Earth.

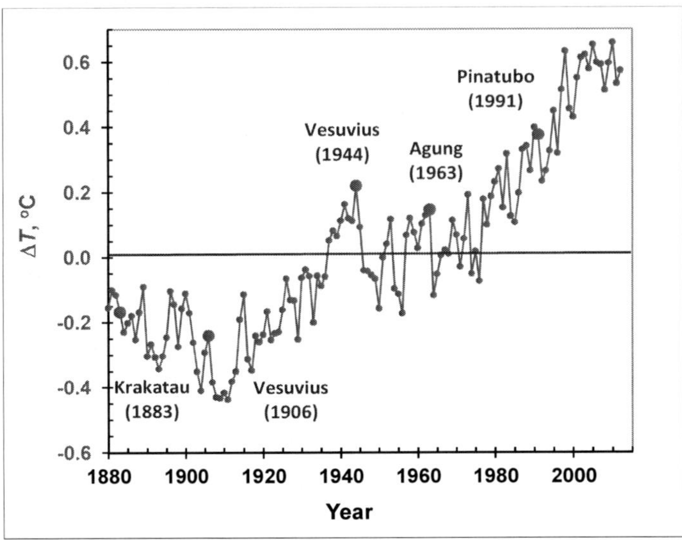

Figure 14. Global temperature change ΔT (annual averages) and the dates of volcano powerful eruptions (large red points). After such eruptions in (1–2) years the decreases of ΔT occurred.

Table 3. Periodicities found in the monthly average data of ΔT and the periods of conjunctions of several planets during their motion around the Sun

Amplitude of ΔT, deg.	Period for ΔT, year^{-1}	Planet conjunctions	Period, year^{-1}
0.355	197.4	Venus-Earth-Jupiter-Saturn	195.0
0.198	66.6	Mercury-Venus-Earth-Jupiter	68.0
0.082	34.0	Mercury-Venus-Earth-Saturn	32.2
0.084	21.8	Venus-Earth-Jupiter	22.1

Finally, the changes of zodiacal dust concentration will change the Earth's climate. The effect of dust on the global surface air temperature is seen from Figure 14 where the events of powerful strato-volcano eruptions for the last ~130 years are shown. In these events huge masses of volcanic dust were ejected into the stratospheric altitudes (~20 km and higher). This dust extended over the whole globe. As a result in (1–2) years the values ΔT decreased. As it was mentioned above, the comets are the main sources of cosmic dust in the solar system. The motion of comets is controlled by planets. There are periodicities in the motion of planets and combination of planets. These periodicities have to manifest in variations of cosmic dust concentration in zodiacal cloud and in the climate changes. The spectral analysis of the temperature changes ΔT presented in Figure 11 showed the presence of several significant spectral lines with large amplitudes. In Table 3 these periodicities and the periods of conjunctions of several planets during their motion around the Sun are listed. The comparison of periodicities obtained in the temperature changes ΔT and the periods of conjunctions of several planets during their motion around the Sun shows the coincidence of the periods (the discrepancies are less than 8%).

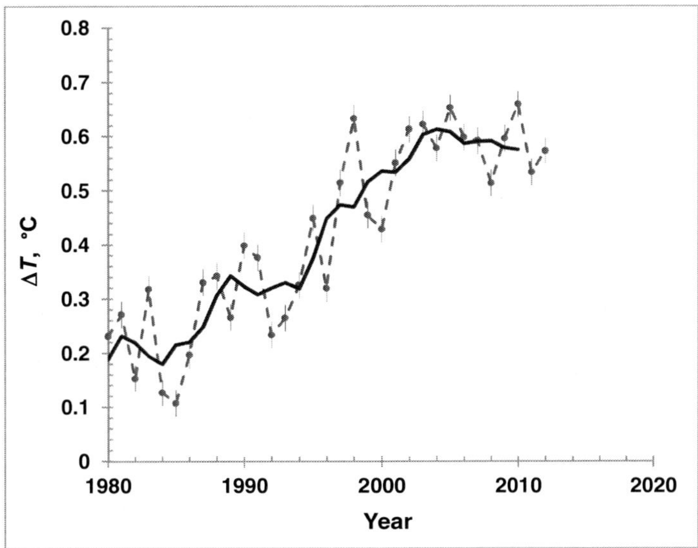

Figure 15. Time dependence of the annual averages of the global temperature changes ΔT (dotted red curve with standard deviations) and 5 point smoothed values ΔT (blue curve).

We have used these periodicities (the second column of Table 3) to describe the time dependence of the temperature changes ΔT and to make the prediction of their behavior up to 2050 (see Figure 11 where red curve is experimental data and blue curve is calculation). The blue curve describes the experimental data for the period 1880–2012 with a good accuracy and predicts the decrease of values ΔT to zero to 2050. Figure 15 demonstrates the time transition from the increase of ΔT to their decrease. The values ΔT began to decrease from the beginning of 21st century. The blue curve in Figure 15 shows the tendency to the decrease of values ΔT. It means that the global warming process was finished and the period of global cooling was started [20, 25].

CONCLUSION

In this paper the solar activity in the current 24th solar cycle is discussed and the prognosis of its development in the next ~50 years is given. It is shown that the new long-term minimum of solar activity has started like Dalton's one in the past. It is expected that maximum value of annual averages of sunspot number R_z in the current solar cycle will be in the limits of (50–70). The solar activity of the subsequent solar cycles will be low also.

In 2009 the highest cosmic ray flux was measured for the ~60-year interval of cosmic ray monitoring. The low solar activity and the weak interplanetary magnetic field strength at 1 a.u. were responsible for the appearance of high cosmic ray fluxes at the Earth's orbit. In the context of the prognosis of low solar activity in the next decades the high galactic cosmic ray fluxes are expected to be in the heliosphere.

The consideration of the physical mechanisms that could cause the global warming process leads to the following conclusions. The 11-year changes of the total solar irradiation (TSI), solar activity (changes of R_z), and cosmic ray fluxes cannot explain global increase of air surface temperature observed in the 19th – 20th centuries. It is difficult to explain that human activity could cause the global warming process because the time dependence of ΔT has several rather long-term time intervals with the increases and decreases of temperature.

Also starting from ~2005 the gradual decrease of ΔT value has been marked. To explain the observed climate changes the mechanism of cosmic dust influence on atmospheric processes was suggested. The dust comes to the atmosphere from the zodiacal cloud in which our planet is moving. The incoming dust influences on the atmosphere transparency and the cloud formation process. The decrease of dust in the atmosphere increases air transparency and decreases cloud coverage. It gives the growth of temperature on the Earth's surface. And conversely, the increase of dust gives the global temperature decrease. The spectral analysis of the temperature data reveals the presence of periodicities in the time dependence of ΔT. The prediction based on the spectral analysis results gives the decrease of ΔT with time. It means that the global warming has to be changing with the global cooling. This cooling process has already started and by the year ~2050 the Earth's climate will be the same as it was in 40's years of the last century.

ACKNOWLEDGMENTS

This work was partially supported by the Russian Foundation of Basic Research (projects 10-02-00326a, 11-02-00095a, 12-02-10007k) and the Program of the Presidium of the Russian Academy of Sciences "Fundamental properties of matter and astrophysics".

REFERENCES

[1]   Shove D.J. *J. Geophys. Res.,* 1995, 60, No. 2, 127.

[2]   Ermakov V.I., Okhlopkov V.P., Stozhkov Y.I. Spectral analysis of data on the solar sunspot formation over the last 2650 years. *Bulletin of the Lebedev Physics Institute of Russian Academy of Sciences.* M.: LPI, 2002, № 10, pp. 12-17 (in Russian).

[3]   Ermakov V.I., Okhlopkov V.P., Stozhkov Y.I. Stability of the 11-year solar activity averaged period over the last 2650. International Conference "Active processes on the Sun and stars". *Proceedings of the conference of the Commonwealth of Independent States and the Baltic Countries,* Saint-Petersburg, 1-6 July, 2002, СПб.: NUUPF St. Petersburg, 2002, pp. 44-47.

[4]   Eddy John A. *Scientific American,* 1977, 236, No. 5, 80.

[5]   Soon, W. W.-H., Yaskell, S. H. *Maunder Minimum and the Variable Sun-Earth Connections.* World Scientific Publishing Co., 2003, p. 336.

[6]   http://www.swpc.noaa.gov/ftpdir/weekly/RecentIndices.txt.

[7]   Usoskin I.G., Sokoloff D., Moos D. *Solar Physics,* 254, 345, DOI 10.1007/s11207-008-9293-6 (2009).

[8]   Stozhkov Y. Some questions of cosmic ray modulation and cosmic ray time dependence in the 24-th solar cycle. *Proceedings of the International Symposium FORGES 2008,* Yerevan, Armenia, 12 (2009), 12-18.

[9]   Stozhkov Y.I, Ermakov V.I., Okhlopkov V.P. Workshop «Solar and stellar activity cycles», SAI MSU, 18-19 December 2009. *Proceedings of workshop*, Saint-Petersburg, 2009, 263.

[10]   http://wso.stanford.edu/gifs/Polar.gif.

[11]   http://omniweb.gsfc.nasa.gov/form/dx1.html.

[12]   Stozhkov Y.I., Svirzhevsky N.S., Bazilevskaya G.A., Krainev M.B., Svirzhevskaya A.K., Makhmutov V.S., Logachev V.I., Vashenyuk E.V. *Astrophysics and Space Sciences Transactions*, 2011, 7, 379.

[13]   Bhat., Jain R., Aggarwal M. *Solar Physics*, 2009, 24. DOI 10.1007/s11207-009-9439-1.

[14]   http://solarscience.msfc.nasa.gov/predict.shtml.

[15]   http://solarscience.msfc.nasa.gov/greenwch.shtm.

[16]   *Solar data.* 1948 – 1991, Nauka, Leningrad, USSR (in Russian).

[17]   Charakhchyan A. N. *Soviet Physics Uspekhi*, 1964, 7(3), 358 (in Russian).

[18]   Stozhkov Y. I., Svirzhevsky N.S., Bazilevskaya G.A., Kvashnin A.N., Makhmutov V.S., Svirzhevskaya A.K. Long-term (50 years) measurements of cosmic ray fluxes in the atmosphere. *Advances in Space Research*, 2009, 44, No. 10, p. 1124-1137.

[19]   ftp://ftp.ncdc.noaa.gov/pub/data/anomalies/annual.land_ocean.90S.90N.df_1901-2000mean.dat.

[20]   Ermakov V.I., Okhlopkov V.P., Stozhkov Y.I. Influence of cosmic dust on cloud coverage, albedo, and the Earth's climate. *Vestnik of Moscow State University*, serie of physics and astronomy 3, 2007, № 5, pp. 41-45; http://www.phys.msu.su/rus/research/vmuphys/archive/.

[21]   D.S. Chapman and M. G. Davis. Climate Change: Past, Present, and Future. *Eos*, 2010, 91, No. 37, 325.

[22]   ftp://ftp.ngdc.noaa.gov/STP/SOLAR_DATA/SOLAR_IRRADIANCE/composite_d25_07_0310a.dat.

[23]   H. Svensmark and E. Friis Christensen. Variation of cosmic ray flux and global coverage – a missing link in solar-climate relationship. *J. Atm. and Solar-Terrest. Physics*, 1997, 59, 1225.

[24]   Ermakov V.I., Stozhkov Y.I. Thundercloud physics. *Preprint LPI*, 2004, № 2. M.: LPI, 38 p.; http://ellphi.lebedev/ru/6.

[25]   Ermakov V.I., Okhlopkov V.P., Stozhkov Y.I. Influence of cosmic dust on the Earth's climate. *Bulletin of the Lebedev Physics Institute of Russian Academy of Sciences*. M.: LPI, 2006, № 3, pp. 41-51 (in Russian).

In: Homage to the Discovery of Cosmic Rays ...
Editor: Jorge A. Perez-Peraza

ISBN: 978-1-63483-084-3
© 2015 Nova Science Publishers, Inc.

Chapter 15

THE (SUPER)-SECULAR PERIODICITY OF THE COSMIC RAYS DURING THE HOLOCENE

Victor Manuel Velasco Herrera

Instituto de Geofísica, Universidad Nacional Autónoma de México,
Ciudad Universitaria, Coyocán, D. F. Mexico

ABSTRACT

The analysis of the cosmic rays is a tool to study the solar magnetic field. The galactic cosmic rays flux is modulated by the change in interplanetary magnetic field and the barycentre motion of the solar system. Since there is no direct observational data to study the cosmic rays variability over a long-time scale, proxy data such as cosmogenic radionuclides has to be relied on. The cosmogenic isotopes are produced mainly by galactic cosmic ray flux. The analysis of cosmogenic isotopes in natural archives provides a means to extend the knowledge of solar variability, cosmic rays and barycentre motion over much longer periods. Analysis of the cosmogenic isotope records is more difficult than the analysis of sunspot numbers. This is due to the fact that the cosmogenic radionuclide reflects the production rate, which is modulated not only by the solar activity, but also by atmospheric transport and deposition processes.

INTRODUCTION

The Sun has been suspected to influence decade-to-century weather and climate changes. One of the principal difficulties to quantify the variability of the Sun in weather and climate change has been the absence of long-term measurements of both the weather, climatic and solar activity phenomena. In order to study the long-term behaviour of the Sun and its interaction with the weather and near-Earth environment, many indices may be used to represent different aspects of solar activity. Since there is no direct observational data to study the solar variability over a long-time scale, proxy data such as cosmogenic radionuclides (Beer et al., 2003) has to be relied on. Analysis of cosmogenic nuclides stored in natural archives provides a means to extend the knowledge of solar variability over much longer periods of time. The influence of solar activity changes on weather and climatic phenomena is currently debated, as well as the (super)-secular periodicities.

DATA AND ANALYSIS

The solar magnetic field and barycenter motion in the heliosphere modulates the cosmic rays. The parameter that quantifies the galactic cosmic rays deceleration produced by the solar activity is the solar modulation parameter; Φ is obtained from cosmogenic isotopes time series corrected about geomagnetic field variations and derived of the Greenland Ice Core Project (GRIP, 72°35'N 37°38'W, Vonmoos et al., 2006) and to ^{14}C production rates. Here we use the Φ reconstructions of Usoskin and Kromer (2005) and Steinhilber et al. (2012). Also we used the annual cosmic rays time series between 1964-2012 (black line in Fig. 1) of the cosmic ray station of the University of Oulu (http://cosmicrays.oulu.fi).

While the records of cosmogenic nuclides contain information of the solar activity, these cosmogenic nuclides are produced mainly by high-energy neutrons interacting with nitrogen, oxygen and argon within the atmosphere. Rather than counting the number of nuclear reactions, the amount of cosmogenic nuclides produced is determined. This is possible because these nuclides are eventually removed from the atmosphere and some of them are stored in natural archives such as ice sheets where they can be found and counted thousands of years later (Beer, J., 2003). The simplest technique to investigate periodicities in solar activity and cosmic ray flux is the Fourier Transformation. Although useful for stationary time series, this method is not appropriate for time series that do not fulfil this condition such as of the time series of the registrations Φ. In order to find the time evolution of the main frequencies of the time series, the wavelet method is applied using the Morlet wavelet analysis to analyse localized variations of power in a time series at many different frequencies. The Morlet wavelet consists of a complex exponential modulated by a Gaussian equation: $\psi_0(\eta) = \pi^{-\frac{1}{4}} e^{i\omega_o \eta} e^{-\eta^2/2}$, where t is the time, s is the wavelet scale and ω_0 is the dimensionless frequency. The wavelet power spectrum is defined (Torrence and Compo, 1998) as: $S(s) = \sum_n^N |W_n(s)|^2$, where $|W_n(s)|^2$ is the wavelet spectrum power with $W_n(s)$ as the wavelet transformation of the X series. For the wavelet spectrum, the significance level for each scale was estimated using only values inside the cone of influence (COI). The COI is the region of the wavelet spectrum in which edge effects become important and COI is defined here as the e-folding time for the autocorrelation of wavelet power of each scale. This e-folding time is chosen so that the wavelet power for a discontinuity at the edge drops by a factor e^{-2} and ensures that the edge effects are negligible beyond this point (Torrence and Compo, 1998). To determine the levels of significance of the global wavelet power spectrum, it is necessary to choose an appropriate background spectrum. Then different realizations of the process are assumed to be randomly distributed about this mean or expected background, and the actual spectrum may be compared against this random distribution.

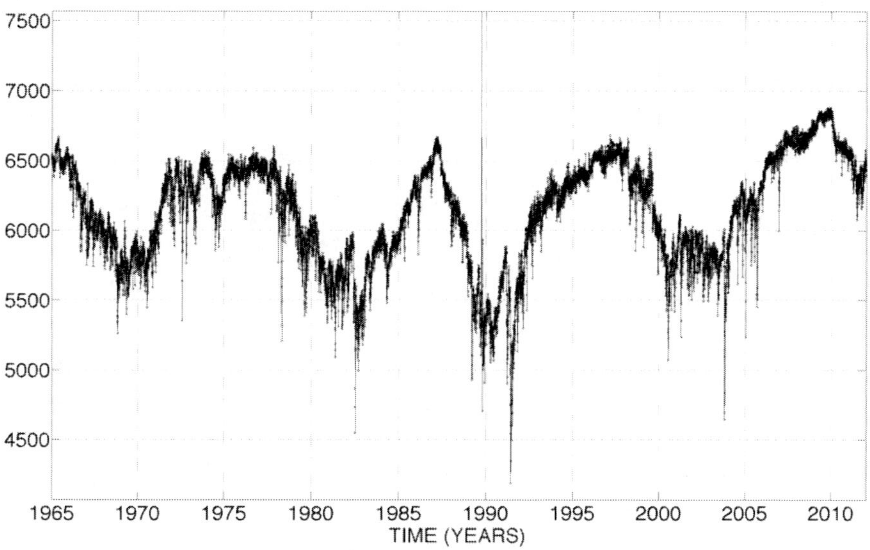

Figure 1. Cosmic Rays time series between 1964 and 2012. The time series correspond to annual mean data.

For many phenomena, an appropriate background spectrum is either white noise (with a flat Fourier spectrum) or red noise (increasing power with decreasing frequency). A simple model for red noise is the autoregressive linear Markov process (Torrence and Compo, 1998). The Morlet wavelet spectrum for each of the series appears in the mid-panel of every figure. All the global wavelet coherence spectra appear in the part furthest left of the all figures. The COI are the curved thick lines crossing the spectral wavelet figures. The dashed curves in the global wavelet spectra indicate the power of the red noise level. The uncertainties of the peaks appearing in the global wavelet spectra are obtained from the peak full width at half maximum. The colour bar for the significance level of the Morlet wavelet spectrum is shown, for all figures, at the bottom of Figure 1.

RESULTS

Figure 2a shows the Probability Density Function (PDF), which presents a binomial distribution with two local maxima; el first maximum is less that the second one. The first (second) PDF maximum is the average value of the minimum (maximum) value of the cosmic rays. The Figure 2b shows the Cumulative Distribution Function (CDF), the 40% of the value corresponds to the first maximum of the PDF and the remaining 60% is the second maximum of the PDF.

Figure 3 shows the wavelet spectral analysis of the cosmic rays time series the global wavelet spectrum shows three periodicities with more of 95% confidence level: 11, 60, 120 and 240 yrs. Different studies suggest that solar mid-term; i.e. periodicities between 1 and 2 years, originate from chaotic quasi-periodic process and not from stochastic or intermittent processes (e.g. Bouwer, 1992; Mendoza and Velasco Herrera 2011).

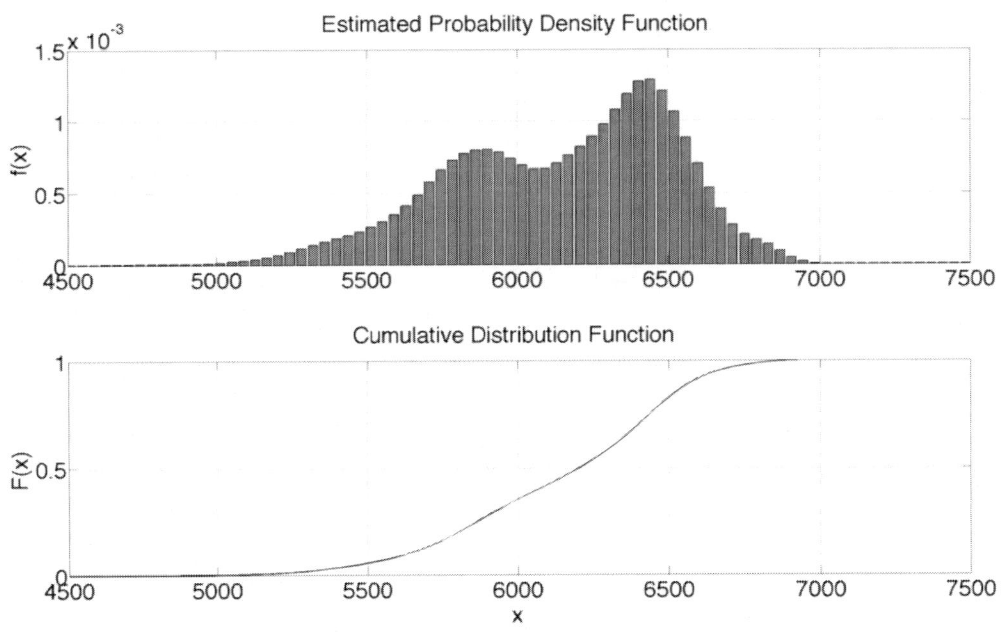

Figure 2. (a) Probability Density Function. (b) Cumulative Distribution Function.

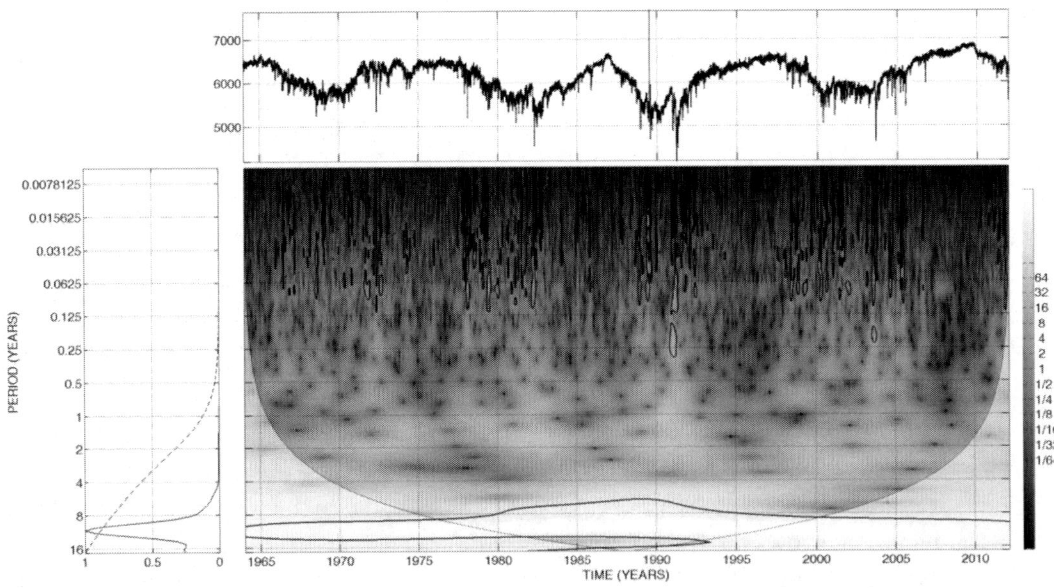

Figure 3. (a) Cosmic Rays time serie (b) Global Wavelet Spectrum (c) Local Wavelet Spectrum of (a) The colour bar for the significance level of the Morlet wavelet spectrum is shown, for all figures.

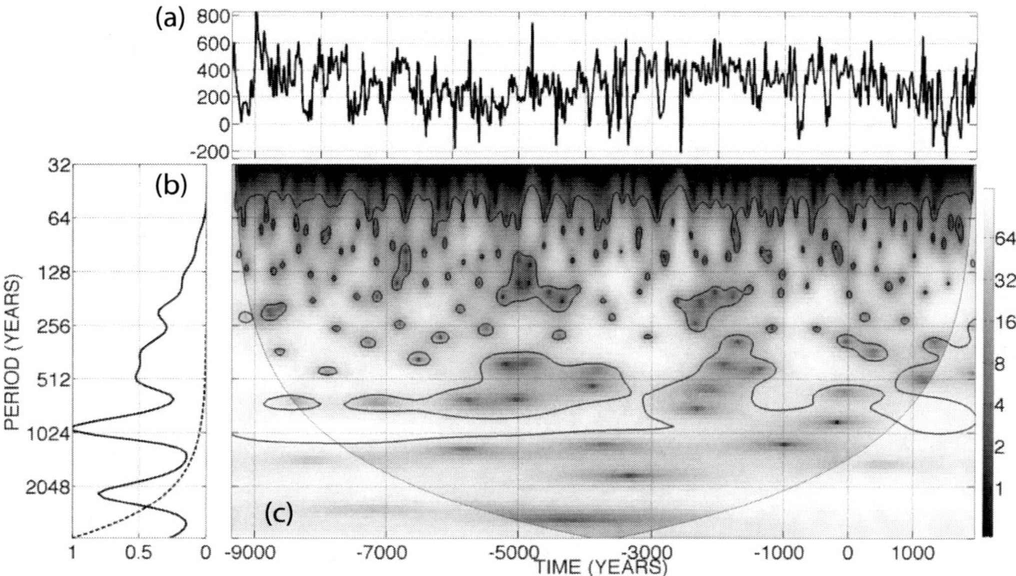

Figure 4. (a) Φ reconstruction of Steinhilber et al. (2012). (b) Global Wavelet Spectrum (c) Local Wavelet Spectrum of (a).

In Fig. 3we show the wavelet analysis of the Cosmic Rays time series from 1964 to 2012. The global wavelet presents periodicities of 0.04, 0.076, 0.24, 0.45, 1.29, 1.7 and 11 years. The most prominent periodicity is the 11 years. Studies about the cyclic behavior of the solar activity suggest the existence of periodicities between 1 and 2 years, called the mid-term periodicities (MTPs). They were found in galactic cosmic ray intensity, coronal hole area, long duration X-ray solar emission, solar wind velocity and the *aa* geomagnetic index (e.g. Mendoza and Velasco 2011).

Moreover, recently Abreu et al. (2012) studying solar activity for the past 9400 years using the modulation potential found the periodicities of 88; 104; 150 and 506 years using a Fourier Transform, which is not appropriate for this type of time series. It is more correct to use the global wavelet. We apply the wavelet analysis to Φ reconstruction of Steinhilber et al. (2012). In Fig. 4 we show the wavelet analisys of the Φ reconstruction of Steinhilber et al. (2012). The global wavelet (Fig. 4b) presents periodicities of 60, 80, 128, 240, 480, 500, 1000 and 2100 years. The periodicities of 60 and 80 years (Yoshimura-Gleissberg cycle) have been reported using the cosmogenic isotopes ^{14}C, ^{10}Be ans Sunspot Number. This periodicity could be associated with the barycenter motion.

The 120 and 240 (de Vries or Suess cycle) years periodicities have been reported using the cosmogenic isotopes ^{14}C and ^{10}Be (Velasco and Mendoza, 2008; Stuiver and Braziunas, 1993). The periodicity of 128 years can be associated with solar magnetic activity and the negative (positive) 128-years phase coincides with the gran minima (máxima) of the solar cycle (Velasco et al., 2013). The periodicity of 240 years can be associated with the barycentre motion (Jose, 1965).

The 480-500 years (unnamed), 1000-years (Eddy cycle) and 2100-years (Hallstattzeit cycle) periodicities have been reported using different solar activity proxy (Soon et al., 2013).

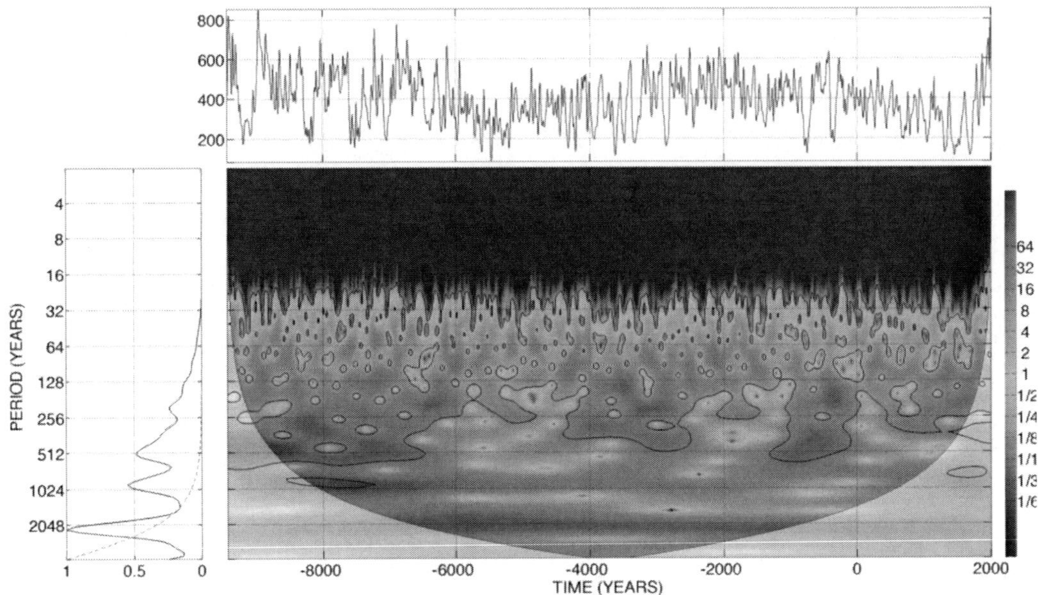

Figure 5. (a) Φ reconstructions of Usoskin and Kromer (2005). (b) Global Wavelet Spectrum (c) Local Wavelet Spectrum of (a).

In Fig. 5 we show the wavelet analisys of the Φ reconstruction of Usoskin and Kromer (2005). The global wavelet (Fig. 5b) presents periodicities of 60, 128, 240, 500, 1000 and 2100 years. The periodicities of 60 years (Yoshimura-Gleissberg cycle) could be associated with the barycenter motion. The periodicity of 128 years can be associated with solar magnetic activity. The 240 years periodicities (de Vries or Suess cycle) can be associated with the barycentre motion (Jose, 1965). The 500 years (unnamed), 1000-years (Eddy cycle) and 2100-years (Hallstattzeit cycle) periodicities have been reported using different solar activity proxy (Soon et al., 2013).

All of these periodicities are potentially physically relevant for linking climate to intrinsic variations of the Sun's magnetic activity (Soon et al., 2013) and the Barycenter motion.

CONCLUSION

The results obtained from the wavelet analysis; using the Φ reconstructions of Usoskin and Kromer (2005) and Steinhilber et al. (2012), show the existence of (super)-secular periodicities in the Holocene.

We found the 60-years, 128-years, 240-years, 500-years, 1000-years and 2100-years cycles could be associated to intrinsic variations of the Sun's magnetic activity and the Barycenter motion.

The 120-128-years periodicity has not been widely reported before. The solar activity grand minima periodicity is of 120 years, this periodicity could possibly be one of the principal periodicities of the magnetic solar activity.

REFERENCES

Beer, J., Vonmoss, M.V., McCracken, K.G.and Mende, W. Heliospheric Modulations over the past 10,000 years as derived from Cosmogenic Nuclides.

Bond, G. and et al. A Pervasive Millennial-Scale Cycle in North Atlantic

Bond, G. and et al. Persistent Solar Influence on North Atlantic Climate During the Holocene. *Science,* (2001), 294:2130-2135

Braun, H. and et al. Possible Solar origin of the 1470 year glacial climate cycle demonstrated in a couple model. *Nature,* (2005) 438:208-2111.

Dansgarrd, W., and et al. Evidence for general instability of past climate from a 250-kyr ice-core record. *Nature,* (1993) 364: 218-220.

Finkel, R.C., and Nishiizumi, K. Beryllium 10 concentrations in the Greenland Ice Sheet Project 2 ice core from 3-40 ka. *Journal of Geophysical,* (1997) Research 102:26699-26706.

Grootes, P.M., and et al. Comparison of oxygen isotope records from the GISP2 and GRIP Greenland ice cores. *Nature,* (1993) 366:552-554.

Holocene and Glacial Climates. *Science,* (1997), 278:12571264.

International Cosmic Ray Conference, (2003), p. 4147-4150.

Johnsen, S.J., and et al. Irregular glacial interstadials recorded in a new Greenland ice core. *Nature,* (1992) 359:311-331.

Mendoza, B. and Velasco, V.M. On Mid-Term Periodicities in Sunspot Groups and Flare Index. *Solar Phys.* 271 (2011) 169-182.

Sakai, K., and Peltier, W.R. Dansgaard-Oeschenger Oscillations in a Coupled Atmosphere-Ocean climate model. *J. Climate,* (1997), 10:949-970.

Steig, E.J., Brook, E,J,, White, J.W.C., Sucher, C.M., Bender, M.L., Lehman, S.J., Morse, D.L., Waddington, E.D., andClow, G.D. Synchronous climate changes in Antarctica and the North Atlantic. *Science,* (1998) 282 (5386): 92-95 (1998).

Soon, W., et al., S. Solar and Climatic Variations on Centennial to Millennial Timescales. JQS (2013).

Steinhilber, F., et al. (2012), 9400 years of cosmic radiation and solar activity from ice cores and tree rings, *Proc. Natl. Acad. Sci.* U. S. A., 109(16), 5967-5971

Van Geel, B., and et al. The role of solar forcing upon climate change. *Quat. Sci. Rev* (1999) 18:331-338.

Torrence, C. and Compo, G.P. A practical guide to wavelets. *Bull. Am. Metereol. Soc.,* (1998) 79. 61-78.

Uriarte, C.A. Earths Climate History Servicio Central de Publicaciones del Gobierno Vasco, (2003) p.203.

Van Geel, B., and et al. The role of solar forcing upon climate change. *Quat. Sci. Rev* (1999) 18:331-338.

Velasco, V.M. and Mendoza, B. Assessing the relationship between solar activity and some large scale climatic phenomena. *Advances in Space Research* 42 (2008) 866878.

Velasco, V.M., Mendoza, B., and Valdes-Galicia, J.F. The 120 yrs solar cycle of the Cosmogenic Isotopes Proceedings of the 30th International Cosmic Ray Conference Vol. 1 (SH), pages 553556.

Velasco, V.M., Mendoza, B., and Velasco, G. Reconstruction and Prediction of the Total Solar Irradiance: From Maunder Minimum to 21st Century. *New Astronomy* (2013).

In: Homage to the Discovery of Cosmic Rays …   ISBN: 978-1-63483-084-3
Editor: Jorge A. Perez-Peraza   © 2015 Nova Science Publishers, Inc.

Chapter 16

COSMIC RAYS IN SUPERNOVA REMNANTS: A STORY OF THE EMPEROR'S NEW CLOTHES?

Chih-Yueh Wang[*]

Department of Physics, Chung-Yuan Christian University, Taiwan
University Interdiscipline Department, Honor College,
National Taiwan University of Science and Technology, Taiwan

ABSTRACT

Strong shock waves of supernova remnants are believed to be efficient accelerators and primary sources of Galactic cosmic ray particles, mostly protons, with energies up to the knee of the cosmic ray spectrum, 10^{15} eV.

The non-thermal X-ray emission from the shells of Galactic SNRs and a few newly discovered SNRs, believed to be the synchrotron emission from electrons accelerated to 100 TeV, represents evidence that cosmic-ray particles are produced in SNR shocks. However, spectral evidence of ion/proton acceleration in SNRs remains circumstantial. Global and full-time hydrodynamic studies of SNRs demonstrate that the expected suprathermal cooling from cosmic-ray acceleration would significantly compress the intershock region of a SNR toward the contact discontinuity, but the degree of convective mixing is not enhanced by the contraction of the shell, even for very efficient acceleration of cosmic rays at the shocks.

The observed dynamics in the young Galactic SNR Tycho's remnant (SN1572) are consistent with the exponential ejecta density model under adiabatic conditions, suggesting that the injection of cosmic rays is too weak to affect the flow properties. Such global and fundamental dynamic evidences challenge the thought that Supernova shock waves are efficient cosmic-ray accelerators.

Keywords: Cosmic rays, shock waves, hydrodynamics, instabilities, supernovae: general, supernovae: individual (SN 1572), supernova remnants

---

[*] Email: cw5b@msn.com

ACCELERATION OF COSMIC-RAY PARTICLES
IN SUPERNOVA REMNANTS

Cosmic rays (CRs) are electrons and atomic nuclei, mostly protons, which move at nearly the speed of light. Their energies can exceed 10^{15} eV. Fermi (1949) proposed a statistical, diffuse acceleration mechanism for supernova remnants (SNRs), in which collisions between charged particles and interstellar clouds through magnetic mirroring create a suprathermal, power-law population of cosmic ray particles. Because the diffusive acceleration has canceling effect for approaching and receding collisions to first order, the energy gain in the ratio of cloud to particle speed is second order, with the resulting spectrum depending on a free parameter, the CR escape timescale. Later it was realized that in a strong shock the scattered particles would experience only approaching collisions in the reference frame, so that the particles acquire first-order energy increases and much more rapid acceleration, and the spectrum depends on the shock compression ratio. This diffusive shock acceleration (DSA), or first-order Fermi acceleration, was quickly accepted as the explanation for the radio-emitting electrons in SNRs of 10^{15} eV. Galactic cosmic-ray particles with energies up to the knee of the CR spectrum, 10^{15} eV. were also suggested to arise in strong shock waves of supernova remnants (Axford 1981).

The energies of CR observed on Earth and the strength of the interstellar magnetic fields deduced from radio measurements support the idea that CRs emit synchrotron radiation in X-ray wavelengths. SN 1006 was the first demonstration of the presence of high-energy particles in the putative sources of Galactic cosmic rays. Its X-ray spectrum shows completely featureless spectra well fitted by a power-law in the bright limbs and a typical thermal spectrum of a young supernova remnant in the interior. To explain the featureless X-ray spectrum of SN 1006, Reynolds and Chevalier (1981) first proposed that synchrotron emission, rather than other thermal and non-thermal processes, was responsible for its high-energy photons. Their model assumed that radiative losses take effect only after the particles have been accelerated to high energies. Observations of X-ray synchrotron radiation from electrons ever since have shown that SNRs can create GeV emissions. However, such high-energy CRs are not linked with the synchrotron radiation in radio and optical wavelengths that are related to the interstellar magnetic fields in SNRs (for a review see Reynolds 2008). Reynolds and Keohane (1999) calculated the maximum energies of shock-accelerated electrons in 14 young shell-type SNRs, assuming a constant, moderate magnetic field strength of 10G; they concluded that young remnants could not be responsible for the highest-energy Galactic cosmic rays. The expected energies for DSA in SNRs are at most 10^{15} eV.

The spectral signature of supernova shocks being efficient accelerators and primary sources of Galactic cosmic-ray particles would be the decay of pions, generated while the shocks collide with the atom and molecules in the surrounding ISM. Electrons could also be accelerated in supernova remnants via the scattering of low-energy photons, mostly the 2.7K cosmic background photons, to energetic γ-rays through inverse Compton scattering (Koyama *et al.* 1995; Mastichiadis and de Jager 1996, and Tanimori *et al.* 1998 for observations of SN 1006). Several synchrotron-dominated shell SNRs were discovered in the late 1990s: RX J 1713.7-3946 (G347.3-05) and Vela Jr. (G266.2-1.2). Synchrotron-emitting thin filaments at the blast wave edges of the historical SNRs Tycho, Kepler, and Cas A were also inferred as evidences of cosmic rays. Cascade showers of optical photons in RX J

1713.7-3946, which appear similar to SN 1006 and presumably arise from pion decay, even reveal the presence of TeV γ-rays (Enmoto *et al.* 2002). Even though growing evidence relates cosmic-ray ions to the thermal properties of the shock-heated X-ray-emitting gas (Decourchelle, Ellison, and Ballet 2000; Ellison, Slane, and Gaensler 2001; Ellison *et al.* 2010), spectral evidence of ion/proton acceleration in SNRs remains circumstantial. Of the ~250 observed Galactic remnants, most appear to be dominated by thermal emission from their shells. Theoretical calculations also disagree whether the γ-rays emissions in the direction of RX J 1713.7-3946 and others come from ions (Berezkho and Volk 2006) or electrons (Katz and Waxman 2008).

Identifying high-energy synchrotron emission from cosmic rays protons is difficult. The synchrotron flux scales as $1/m$, where m is the rest mass of the particle, whereas the critical frequency of the emission scales as $1/m^3$. Since a TeV proton produces a 1000 times smaller flux than a TeV electron and many more electrons than protons would contribute to the emission, even though a TeV proton produces a 1000 times lower critical frequency, identifying the synchrotron emission in the radio wavelength is virtually impossible.

The best and only direct evidence of cosmic rays in SNRs relies on the γ-rays emission that is not associated with inverse Compton emission from electrons. The indirect evidence may involve measurements of abnormally low electron temperatures and curvature in the SNR spectrum (Glenn Allen 2011, private communication; Allen, Houck, and Sturner 2008). However, even if a spectral signature of proton/ion acceleration was found in SNRs, the evidence in support of isolated SNRs as the main accelerators of galactic CRs may still be insufficient (Butt 2009).

INTERSHOCK RAYLEIGH-TAYLOR INSTABILITY AND ASSOCIATED SHOCK COMPRESSION RATIO OF SUPERNOVA REMNANTS

During the young, ejecta-dominated phase of SNR expansion, the shocked gas are susceptible to the Rayleigh-Taylor (R-T) convective instability, causing the reverse-shocked ejecta of the inner shell to protrude into, and mixes with, the forward-shocked ambient medium in the outer shell. The R-T convection has been considered effective for modifying the spherical morphology of an SNR and spreading the ejecta material out and beyond the nominal radius of the forward shock. Extreme examples of SNRs with irregular morphology include core-collapse SN 1986 J (Bartel *et al.* 1991), SN 1993 J (Bietenholz, Bartel, and Rupen 2001), and Cas A, the type II SN prototype, of which optical (Fesen and Gunderson 1996) and X-ray (Hughes *et al.* 2000) observations reveal numerous bumps and protrusions on its roughly-spherical outline. X-ray (Hughes, 1997; Hwang, Hughes, and Petre 1998) and radio (Reynoso *et al.* 1997) observations of Tycho's remnant similarly show heavy ejecta material and dense knots that non-deceleratingly reach close to, and protrude from, the southeastern periphery of the remnant. Radio observations of Tycho also revealed a band of regularly spaced knots immediately beneath the sharply marked forward shock front at the northeaster edge (Velázquez *et al.* 1998).

The R-T convection thrives when the density contrast between the two shocked shells and the deceleration of the gas are sufficiently high. The initial density velocity structure of the supernova progenitor, or ejecta, is important to determine the evolution of the supernova remnant. For core-collapse explosions, the freely-expanding ejecta form a relatively uniform density core surrounded by a steep power-law declining envelope described by $\rho_{SN} = v^{-n}t^{-3}$ with $n\sim10$-12, where $v=r/t$ is the flow velocity of the free-expanding ejecta (Matzner and McKee, 1999). For thermonuclear explosions (white dwarf, Type Ia SNe), early models depict a power-law density distribution of index 7 for the outer ejecta (Colgate and McKee 1969). Recent comparisons with many successful thermonuclear explosion models show that the overall density distribution of the free ejecta follows an exponential decline,

$$\rho_{SN} = A \cdot \exp(\frac{v}{v_e}) \cdot t^{-3}$$

where the constants $A = 6^{3/2} M^{5/2} / 8\pi E^{3/2}$ and the velocity scale height $v_e = \sqrt{E/M}$ are determined by integrating the mass density and the kinetic energy density over space (Dwarkadas and Chevalier 1998). The exponential density decline for Type Ia supernovae (SNe) reflects the fact that the explosion energy is steadily released behind a subsonic burning front, as opposed to pure shock acceleration in core-collapse SNe. In the model a higher velocity scale gives a flatter density distribution at a given flow velocity, and the density gradient $n = d\ln\rho_{SN} / d\ln r = v/v_e = r/v_e t$ increases with radius while evolves toward a flatter power index.

Notably, most hydrodynamic studies of the R-T instability in SNRs adopt a power-law density distribution and neglect the inner flat component of supernova ejecta, giving a timeless self-similar solution only applicable to early phases of core-collapse SNRs, when the inner flat ejecta core has not expanded far enough to change the intershock structure. In contrast, the exponential model predicts compatible shock radius and deceleration parameters for SN 1006 and SN 1572 (Dwarkadas and Chevalier 1998). A characteristic of the exponential model is that the convection decays over time as the density flattens. Therefore, given the age, the metal-enriched mixture near Tycho's periphery cannot arise from the R-T convection (Wang & Chevalier, 2001). The extensively mixed structures in many SNRs, including the high velocities of Ni56 and the large filling factor of Fe (Woosley 1988; Li *et al.* 1993), are most likely to form during an asymmetric explosion (Kifonidis *et al.* 2000, 2003; Hungerford, Fryer, and Warren 2003) and the subsequent radioactive process due to the nickel bubble effect (Basko 1994), in which the radioactive decay of the Ni56-Co56-Fe56 sequence inflates a nickel bubble and compresses the diffuse ejecta into dense ejecta clumps. Radiation-hydrodynamic simulations of the nickel bubble process found the speed and interaction age of the inferred ejecta clumps well agree with those of the observed X-ray knots/filament in Tycho's periphery (Wang 2005, 2008).

The synchrotron-emitting heavy-element gas near the blast wave has been inferred as evidences of efficient cosmic-ray particle acceleration as the supernova shock interacts with the ambient medium. The acceleration process of higher-energy cosmic rays is estimated to transform at least 10% of the total ejecta kinetic energy into relativistic particles, and it cannot occur too soon (~100 yr) after the supernova explosion, as in that case cosmic rays would lose most of their energy (Blandford and Eichler 1987; Drury, Markiewicz, and Vo"lk 1989; Dorfi

2000). When efficient particle acceleration occurs, relativistic, superthermal particles are created, accompanied by decreased postshock temperature. The region between the forward and reverse shocks of an SNR shrinks and becomes denser. This effect increases the compression ratio of the shock from $\sigma = 4$, where

$$\sigma = \frac{\gamma + 1}{\gamma - 1}$$

for strong shocks, to $\sigma > 4$. Thus, the effective adiabatic index of a cosmic-ray SNR is expected to be less than 5/3. The increased compression of the gas is expected to produce a more compact intershock structure and enhance the amplitude for the R-T instability.

Results from many hydrodynamic models provide estimates for the effective adiabatic index and compression ratio. Chevalier (1983) considered two-fluid, self-similar solutions, where the ejecta was a thermal gas with $\gamma = 5/3$ and the ambient medium a relativistic gas with $\gamma = 4/3$. In the case in which relativistic particles at the shock front contribute 100% of the total pressure, the effective adiabatic index is 1.333. The shock compression ratio is similar to that in the late radiative phase of SNR expansion, $\sigma > 7$. Blondin and Ellison (2001) showed that the compression in the power-law model with $\gamma = 1.1$ ($\sigma = 21$) is similar to the case of radiative cooling considered by Chevalier and Blondin (1995), where the ejecta behaved as an ideal gas with $\gamma = 1$ and the ambient medium was modeled with $\gamma = 5/3$. Although theoretically arbitrarily large compression ratios are allowed, a maximum compression that corresponds to $\gamma = 1.1$ seems suitable for a young remnant. However, in kinetic simulations of diffusive shock acceleration (DSA), the propagation and dissipation of Alfven waves upstream reduces the CR acceleration and the precursor compression, such that the cosmic-ray-modified shocks depend only weakly on the Mach number M_0 of the shock, and the total compression ratio is less than 10 even for $M_0 \sim 100$ (Kang, Ryu, and Jones, 2010). If the injection of cosmic-ray particles into the shock is weak, $\sigma \sim 4$ can still hold for a high Mach number shock.

GLOBAL EVOLUTION OF SUPERNOVA REMNANTS

The idealized models with an adiabatic index of $\gamma = 5/3$, such as that of Chevalier et al. (1992), have revealed rather limited R-T mixing of the Rayleigh-Taylor instability in SNRs. Jun and Norman (1996a and 1996b) reached a similar conclusion after considering rapid cooling in the shocked ejecta (Chevalier and Blondin 1995). Jun and Jones (1999) used accelerated cosmic-ray electrons as test particles to calculate the radio synchrotron emission but emphasized the reverse shock. Since rapid cooling and efficient cosmic-ray particle acceleration (Ellison, Berezkho, and Baring 2000) both enhance the shock compression of the gas, Blondin and Ellison (2001) adopted a low adiabatic index $\gamma < 5/3$ to examine the change of gas flow expected from particle acceleration. In this case, the mixing extends far enough to reach the forward shock front and considerably perturb the remnant outline. Recently Fraschetti et al. (2010) employed $\gamma = 4/3$ to represent the relativistic cosmic-ray particles, and Ferrand et al. (2010) applied a kinetic model of nonlinear diffusive particle acceleration to the same hydrodynamic model as in Fraschetti et al; they found the development of instability not to be strongly enhanced, independently of whether the cosmic rays dominate the reverse shock, or the shocks receive a back-reaction of cosmic rays that evolves with time.

Since the extent of R-T mixing is determined by the deceleration and the density profile of the intershock region, we employ low adiabatic indices to the exponential density model to examine the overall dynamics of the gas in a typical Type Ia remnant. In the simulations the instability was seeded early enough to ensure that the flow evolves into the nonlinear regime and becomes independent of its initial conditions before Tycho's present epoch. One-dimensional hydrodynamic simulations of the cases γ = 5/3, 4/3, 1.2, and 1.1 were performed and presented in Wang (2011). Figure 1 in Wang (2011) presents the one-dimensional density profile in a normalized age t'=1.75 that is approximately the age of Tycho's remnant, 438 years, using the standard explosion parameters. If the explosion has a higher energy like a W7-like condition, then t'=2.02. Movies 1, 2, 3, and 4 (Wang, 2011) illustrate the evolutions in density. One major difference from the powerlaw, self-similar solution is that, because the sound speeds of the hot, shocked ejecta and hot, shocked ISM are still high and permit no pressure difference at their interface, the pressure-equilibrated contact discontinuity remains fixed in space and the intershock layers contract toward the contact discontinuity. Consequently, at Tycho's present age, the ratios of forward shock radius to reverse shock radius are reduced to 1.51, 1.33, 1.22, and 1.15, respectively, with γ = 5/3, 4/3, 1.2 and 1.1. At γ = 1.1 the tremendous energy losses both in the forward and reverse shocks increase the reverse to forward shock radius ratio to an extremely high value of 0.87. The motion of the reverse shock and its subsequent impact with the explosion center are significantly delayed. Notably, these observed position and deceleration of the shocks are consistent with the result obtained using the exponential model with γ = 5/3, but differ greatly from those obtained with γ = 1.1 (Warren et al., 2005).

Contraction of the intershock layer and increase in the density contrast across the contact discontinuity generate narrower, denser and sharper R-T fingers. During the nonlinear instability developing stage, an increasingly larger instability mode gradually builds up. The reverse shock front becomes corrugated and moves slightly inward; however, the forward shock is not much perturbed. Only in the initial stage of linear growth the R-T fingers exert sufficient pressure that slightly disrupts the blast wave outline. In later stages, the R-T instability vanishes and the Kelvin-Helmholtz instability takes over, creating vortex rings in the less dense regions left by the original mushroom caps,

IMPLICATIONS

Contrary to previous thought, the compression of the intershock shell does not affect the degree of convective mixing because strong perturbations rapidly stall. The extension of the R-T fingers is greatest during the stage of linear instability growth. In the saturated nonlinear stage, the R-T finger reaches only ~87% of the blast wave radius in Tycho's present age, regardless of the value of adiabatic index. The extent of mixing is virtually identical to that in the non-cooling case considered by Wang and Chevalier (2001). This result is also consistent with Blondin and Ellison's (2001) that the saturated instability growth rates in the power-law model do not detectably vary with γ. Although in early nonlinear phases with γ = 1.2 one R-T finger near the polar axis stretches to ~ 93% of the blast wave radius and even slightly perturb the shock, this mixture is much less dense than the gas immediately beneath the forward shock front, and most fingers remains below 87 per cent of the blast wave radius. The changes in the extension of the convective flow for low values of γ are insignificant when $\gamma \gtrsim 1.2$. Such an outcome argues against the thin X-ray filaments being evidence for cosmic ray acceleration.

The extreme case of $\gamma = 1.1$ involves significant, and probably not physically reasonable, losses of energy from the shock to cosmic rays. In Ferrand et al. (2010), the kinetic model of Blasi (2004) that solves the particle distribution and the fluid velocity as functions of the particle momentum is used to calculate the time-evolving shock properties and particle spectrum; a compact shock structure is revealed, but the R-T mixing remains not much affected. The escape of cosmic rays from shocks (by Alfven heating, etc.) is not well understood, and detailed models of cosmic-ray-modified shock remain theoretical. For future modeling of cosmic-ray modified SNRs, a magneto-hydrodynamic treatment of cosmic rays given in Skilling (1975a; 1975b), which includes the particles as a second fluid component in a plasma system (Lo, Ko, and Wang 2011), can be utilized.

While the exponential model applies to thermonuclear supernovae, the above result also holds for core-collapse SNRs after t' ≥ 0.7, corresponding to ~173 yr for $M_{1.4} = 1$ and M_{51} =1, or ~890 yr for a ten-fold explosion mass, when the reverse shock begins to enter the innermost plateau of the ejecta and terminates the self-similar expansion phase (Wang and Chevalier, 2002). In Blondin and Ellison (2001), the 3-D simulations reveal more deformation of the forward shock than the 2-D simulations; the overall compression ratio drops and the width of the intershock shell increases, but the convection remaining unaffected by the dimensionality of the simulation.

External factors that may adversely influence the flow include a strong magnetic field in the ISM, in which case the pressure from the shocked toroidal magnetic field will distend the forward-shocked layer, preventing the forward shocked layer from shrinking, and further impeding the R-T mixing. However, radio polarization measurements of several supernova remnants show the dominance of the radial magnetic field (Dickel, ven Breugel, and Milne 1991 for Tycho's SNR; Dickel, Strom, and Milne 2001 for G315.4-23). These observations also reveal the presence of out-moving Mach cones that resembles the features in the Vela SNR (Strom et al. 1995; Aschenbach, Egger, and Tru"umper 1995), and therefore suggests an ejecta clumps origin (Wang and Chevalier 2002), a different scenario for how ejecta gas could approach the blast wave. On the other hand, if the forward shock encounters a clumpy interstellar medium, then the vortices generated behind the shock front may channel the R-T instability to enhance the mixing and create protrusions in the supernova remnant shell (Jun, Jones, and Norman 1996).

Fujita et al. (2009, 2010) suggested that particles can be accelerated close to molecular clouds even after the SNR becomes radiative. However, this result is based on Monte-Carlo simulations; like most theoretical studies, a sophisticate, full-time dynamical demonstration of the mechanism is desirable. In the shock-cloud colliding scenario, because the extension of the R-T fingers to the forward shock is transient, the shock-cloud interaction should not overcome the global decay of instability in a remnant like Tycho's and in a typical, turbulent ISM.

The possibility that the mixture of heavy-element ejecta near the supernova blast wave is produced by R-T convection during the ejecta-dominated stage of an SNR is excluded, and it appears that the injection of cosmic-ray particles at Tycho's forward shock is too weak to alter the shock properties, including the compression ratio and convective instabilities. Notably, the case of $\gamma = 1.2$ yields a 12% lower speed of the blast wave than does $\gamma = 5/3$, equivalent to an 23% dissipation of the supernova kinetic energy; the case of $\gamma = 1.1$ yields a 28% lower shock speed, amounting to unrealistic losses of energy. We conclude that even if the injection and acceleration of cosmic rays at the shock are as strong as in the extreme case of $\gamma = 1.1$, perturbations on the remnant outlines are not to be substantially enhanced relative to the adiabatic case.

The dependence of the compression ratio on the sonic and Alfven Mach number implies that the compression is most pronounced in SNRs with large shock speeds but relatively weak magnetic fields (Berezhko and Ellison, 1999). This result contradicts the amplified magnetic field (AMF) scheme that assumes that shocks in SNe and remnants could greatly amplify pre-existing magnetic fields (Chevalier 1977) and so facilitates DSA. The strongest argument supporting AMF is the synchrotron X-ray emission residing in very thin filaments parallel to the remnant edge.

Bell and Lucek (2001) illustrated the streaming instability of upstream cosmic rays in a parallel shock for a specific mechanism of magnetic-field amplification. However, as demonstrated in Wang (2011), the thin filaments in SNRscannot be formed by enhanced R-T mixtures, either due to compression of the remnant's intershock shell induced by suprathermal cooling, or interaction with the interstellar clouds. Much of the underlying process of AMF remains to be explored.

While our full-time hydrodynamic simulations of SNRs contradict the interpretation of Warren et al. (2005) based on X-ray data, the simulations simply reflect the basic dynamic properties of flow that X-ray power spectra may not adequately reveal. Our hydrodynamic simulations further argue against the X-ray filaments being evidence for cosmic ray acceleration, regardless of the amount of losses in the SN kinetic energy. Recent *PAMELA* measurements show that the proton and helium spectra between 1 GeV and 1.2 TeV are different and unfit for a single power law, which also challenge the *cosmic-ray* acceleration paradigm (Adriani et al., 2011).

On the other hand, metallicity measurements of cosmic rays by *FERMI* indicate that OB associations and their super bubbles are likely sources of a substantial fraction of cosmic rays (Ackermann et al., 2011; Binns, 2011). In summary, the observed convective mixing in SNRs is best explained by the expansion of clumpy ejecta that are produced by the nickel bubble effect into the intershock region of the remnant. In the nickel bubble scenario, the radioactive heating of 56Ni inflates the central ejecta, drives a strong thin shock, and compresses the outer heavy elements into dense ejecta clumps. The ejecta clumps is estimated to have a size less than 1% of the blast wave radius and a high density contrast of ~100. Since the heavy elements were shocked before they underwent reverse shock, this process is likely to facilitate the creation of more energetic particles, as well as denser clumping, in a supernova remnant.

ACKNOWLEDGMENTS

I would like to thank Roger Chevalier, John Blondin, and Vikram Dwarkadas for correspondence and discussions. Special thanks are also due to Federico Fraschetti and Glenn Allen for useful correspondence on hydrodynamic simulations and observations of cosmic rays in SNRs. This work is supported by the National Center for High-Performance Computing and the National Science Council of Taiwan.

REFERENCES

Ackermann, M., 2011, *Science,* 334, 1103.

Allen, G. E., Houck, J. C., and Sturner, S. J., 2008, *ApJ*, 683, 773.

Adriani, O. et al., 2011, *Science*, 332, 69.

Aschenbach, B., Egger, R., and Tru"umper, J., 1995, *Nature*, 373, 587.

Axford, W. I., 1981, Proc. 17th Int. Cosmic-Ray Conf. (Paris), 12, 155.

Bartel, N., Rupen, M. P., Shapiro, I. I., Preston, R. A., and Rius, A., 1991, *Nature,* 350, 212.

Basko M., 1994, *ApJ*, 425, 264.

Bell, A. R. and Lucek, S. G., 2001, *MNRAS,* 321, 433.

Berezhko, E. G. and Volk, H., 2006, *A&A,* 451, 981.

Berezhko, E. G., and Ellison, D. C., 1999, *ApJ*, 526, 385.

Bietenholz, M. F., Bartel, N., and Rupen, M. P., 2001, *ApJ*, 557, 770.

Binns, W. R., 2011, *Science,* 334, 1071.

Blandford, R. D., and Eichler, D., 1987, *Phys. Rev.,* 154, 1.

Blasi, P., 2004, *Astropart. Phys.,* 21, 45.

Blondin, J. M., and Ellison, D. C., 2001, *ApJ*, 560, 244.

Butt, Y., 2009, *Nature*, 460, 701

Chevalier, R. A., 1977, *Nature,* 266,701.

Chevalier, R. A., 1983, *ApJ*, 272, 765.

Chevalier, R. A. and Blondin, J. M., 1995, *ApJ*, 444, 312.

Chevalier, R. A., Blondin, J. M., and Emmering, R. T., 1992, *ApJ*, 392, 118.

Colgate, S. A., and McKee, C., 1960, *ApJ*, 157, 623.

Decourchelle, A., Ellison, D. C., and Ballet, J., 2000, *ApJ*, 543, L57.

Dickel, J. R., Strom, R. G., and Milne, D. K., 2001, *ApJ*, 546, 447.

Dickel, J. R., van Breugel, and Strom, R. G., 1991, *ApJ*, 101, 2151.

Dorfi, E. A., 2000, *Ap&SS*, 272, 227.

Drury, L. O'C., Markiewicz, W. J., and Vo\"lk, H. J., 1989, *A&A*, 225, 179.

Dwarkadas, V. V., 2000, *ApJ*, 541, 418.

Dwarkadas, V. V. and Chevalier, R. A., 1998, *ApJ*, 497, 807.

Ellison, D. C., Berezhko, E. G., and Baring, M. G., 2000, *ApJ*, 540, 292.

Ellison, D. C., Slane, P., and Gaensler, B. M., 2001, *ApJ*, 712, 287.

Ellison, D. C., Patnaude, D. J., Slane, P., and Raymond, J., 2010, *ApJ*, 712, 287.

Enmoto, R. et al., 2002, *Nature,* 418,499.

Fermi, E., 1949, *Phys. Rev.,* 75, 1169.

Ferrand, G., Decourchelle1, A., Ballet, J., Teyssier, R., and F. Fraschetti, F., 2010, *A&A*, 509, L10.

Fraschetti, F., Teyssier, T., Ballet, J., and Decourchelle, A., 2010, *A&A*, 515, A104.

Fesen, R. A., and Gunderson, K. S., 1996, *ApJ*, 470, 967.

Fujita, Y., Ohira, Y., Tanaka, S. J., and Takahara, F., 2009, *ApJ*, 707, L179.

Fujita, Y., Ohira, Y., and Takahara, F., 2010, *ApJ*, 712, L153.

Hughes, J. P., 1997, in X-Ray Imaging and Spectroscopy of Cosmic Hot Plasmas, ed. F. Makino and K. Mitsuda (Tokyo: Universal Academy Press), 359.

Hughes, J. P., Rakowski, C. E., Burrows, D. N., and Slane, P. O., 2000, *ApJ*, 528, L109.

Hungerford, A. L., Fryer, C. L., and Warren, M. S., 2003, *ApJ*, 594, 390.

Hwang, U., Hughes, J. P., and Petre, R., 1998, *ApJ*, 497, 833.

Kang, H., Ryu, D., Jones, T. W., 2009, *ApJ*, 695, 2009.

Jun, B.-I., and Jones, T. W., 1999, *ApJ*, 511, 774.

Jun, B.-I., Jones, T. W., and Norman, M. L., 1996, *ApJ*, 468, L59.

Jun, B.-I., Norman, M. L., 1996a, *ApJ*, 465, 800.

Jun, B.-I., Norman, M. L., 1996b, *ApJ*, 472, 245.

Katz, B. and Waxman, E. *J. Cosmol. Astropart. Phys.*, 2008, 1, 18.

Kifonidis, K., Plewa, T., Janka, H.-Th., and M"uller, E., 2003, *A&A*, 408, 621.

Kifonidis, K., Plewa, T., Janka, H.-Th, and Muller, E., 2000, *ApJ*, 531, L123.

Koyama, K., Petre, R., Gotthelf, E. V., Hwang, U., Matsura, M., Ozaki, M., and Holt, S. S., 1995, *Nature,* 378, 255.

Li H., McCray R., Sunyaev R. A., 1993, *ApJ*, 419, 824.

Lo, Y.-Y, Ko, C.-M., and C.-Y. Wang, 2011, *Computer Physics Communications,* 182, 179.

Mastichiadis, A., and de Jager, O. C., 1996, *A&A*, 311, L5.

Matzner, C. D., McKee, C. F., 1999, *ApJ*, 510, 379.

Reynolds, S. P., 2008, ARAA, 46, 89.

Reynolds, S. P. and Chevalier, R. A., 1981, *ApJ*. 245, 912.

Reynolds, S. P., and Keohane, J. W., 1999, *ApJ*, 525, 368.

Reynoso *et al.*, 1997, *ApJ*, 491, 816.

J. Skilling, *MNRAS,* 1975a, 172, 577.

J. Skilling, *MNRAS,* 1975b, 173, 255.

Strom, R. G., Johnston, H. M., Verbunt, F., and Aschenbach, B., 1995, *Nature,* 373, 590.

Tanimori, T. et al., 1998, *ApJ*, 497, L25.

Vel'aquez, P. F., Go'mez, D. O., Dubner, G. M., Gime'nez de Castro, and Costa, A., 1998, *A&A*, 334, 1060.

Wang, C.-Y., Chevalier, R. A., 2001, *ApJ*, 549, 1119.

Wang, C.-Y., Chevalier, R. A., 2002, *ApJ*, 574, 155.

Wang, C.-Y., 2005, *ApJ*, 626, 183.

Wang, C.-Y., 2008, *ApJ*, 686, 337.

Wang, C.-Y., 2011, *MNRAS,* 415, 83.

Warren, J. S., Hughes, J. P., Badenes, C., Ghavamian, P., McKee, C., Moffett, D., Plucinsky, P., Rakowski, C., Reynoso, E., Slane, P., 2005, *ApJ*, 634, 376.

Woosley S. E., 1998, *ApJ*, 330, 218.

INDEX

F

J

K

L

M

Q

R

S

T